Fog and Edge Computing

Wiley Series On Parallel and Distributed Computing

Series Editor: Albert Y. Zomaya

A complete list of titles in this series appears at the end of this volume.

Fog and Edge Computing

Principles and Paradigms

Edited by Rajkumar Buyya and Satish Narayana Srirama

WILEY

This edition first published 2019

Registered Office
John Wiley & Sons, Inc., 111 River Street, Hoboken, NJ 07030, USA

Editorial Office
111 River Street, Hoboken, NJ 07030, USA

For details of our global editorial offices, customer services, and more information about Wiley products visit us at www.wiley.com.

Wiley also publishes its books in a variety of electronic formats and by print-on-demand. Some content that appears in standard print versions of this book may not be available in other formats.

Library of Congress Cataloging-in-Publication Data

Names: Buyya, Rajkumar, 1970- editor. | Srirama, Satish Narayana, 1978- editor.
Title: Fog and edge computing : principles and paradigms / edited by Rajkumar Buyya, Satish Narayana Srirama.
Description: Hoboken, NJ, USA : John Wiley & Sons, Inc., 2019. | Series: Wiley series on parallel and distributed computing | Includes bibliographical references and index. |
Identifiers: LCCN 2018054742 (print) | LCCN 2018057015 (ebook) | ISBN 9781119525011 (Adobe PDF) | ISBN 9781119525066 (ePub) | ISBN 9781119524984 (hardcover)
Subjects: LCSH: Cloud computing. | Electronic data processing–Distributed processing.
Classification: LCC QA76.585 (ebook) | LCC QA76.585 .F63 2019 (print) | DDC 004.67/82–dc23
LC record available at https://lccn.loc.gov/2018054742

Cover design by Wiley
Cover image: © Shaxiaozi/iStock.com

Set in 10/12pt WarnockPro by SPi Global, Chennai, India

10 9 8 7 6 5 4 3 2 1

Contents

List of Contributors *xix*
Preface *xxiii*
Acknowledgments *xxvii*

Part I Foundations *1*

1 Internet of Things (IoT) and New Computing Paradigms *3*
Chii Chang, Satish Narayana Srirama, and Rajkumar Buyya
1.1 Introduction *3*
1.2 Relevant Technologies *6*
1.3 Fog and Edge Computing Completing the Cloud *8*
1.3.1 Advantages of FEC: SCALE *8*
1.3.1.1 Security *8*
1.3.1.2 Cognition *8*
1.3.1.3 Agility *9*
1.3.1.4 Latency *9*
1.3.1.5 Efficiency *9*
1.3.2 How FEC Achieves These Advantages: SCANC *9*
1.3.2.1 Storage *9*
1.3.2.2 Compute *10*
1.3.2.3 Acceleration *10*
1.3.2.4 Networking *11*
1.3.2.5 Control *12*
1.4 Hierarchy of Fog and Edge Computing *13*
1.4.1 Inner-Edge *13*
1.4.2 Middle-Edge *14*
1.4.2.1 Local Area Network *14*
1.4.2.2 Cellular Network *14*

1.4.3 Outer-Edge *14*
1.4.3.1 Constraint Devices *14*
1.4.3.2 Integrated Devices *15*
1.4.3.3 IP Gateway Devices *15*
1.5 Business Models *16*
1.5.1 X as a Service *16*
1.5.2 Support Service *16*
1.5.3 Application Service *17*
1.6 Opportunities and Challenges *17*
1.6.1 Out-of-Box Experience *17*
1.6.1.1 OOBE-Based Equipment *18*
1.6.1.2 OOBE-Based Software *18*
1.6.2 Open Platforms *18*
1.6.2.1 OpenStack++ *18*
1.6.2.2 WSO2–IoT Server *18*
1.6.2.3 Apache Edgent *19*
1.6.3 System Management *19*
1.6.3.1 Design *19*
1.6.3.2 Implementation *19*
1.6.3.3 Adjustment *20*
1.7 Conclusions *20*
References *21*

2 Addressing the Challenges in Federating Edge Resources *25*
Ahmet Cihat Baktir, Cagatay Sonmez, Cem Ersoy, Atay Ozgovde, and Blesson Varghese
2.1 Introduction *25*
2.2 The Networking Challenge *27*
2.2.1 Networking Challenges in a Federated Edge Environment *28*
2.2.1.1 A Service-Centric Model *29*
2.2.1.2 Reliability and Service Mobility *29*
2.2.1.3 Multiple Administrative Domains *29*
2.2.2 Addressing the Networking Challenge *30*
2.2.3 Future Research Directions *33*
2.3 The Management Challenge *34*
2.3.1 Management Challenges in a Federated Edge Environment *35*
2.3.1.1 Discovering Edge Resources *35*
2.3.1.2 Deploying Services and Applications *35*
2.3.1.3 Migrating Services across the Edge *36*
2.3.1.4 Load Balancing *36*
2.3.2 Current Research *36*
2.3.3 Addressing the Management Challenges *37*

2.3.3.1 Edge-as-a-Service (EaaS) Platform *37*
2.3.3.2 Edge Node Resource Management (ENORM) Framework *38*
2.3.4 Future Research Directions *39*
2.4 Miscellaneous Challenges *40*
2.4.1 The Research Challenge *40*
2.4.1.1 Defined Edge Nodes *41*
2.4.1.2 Unified Architectures to Account for Heterogeneity *41*
2.4.1.3 Public Usability of Edge Nodes *41*
2.4.1.4 Interoperability with Communication Networks *42*
2.4.1.5 Network Slices for Edge Systems *42*
2.4.2 The Modeling Challenge *43*
2.4.2.1 Computational Resource Modeling *43*
2.4.2.2 Demand Modeling *44*
2.4.2.3 Mobility Modeling *44*
2.4.2.4 Network Modeling *44*
2.4.2.5 Simulator Efficiency *44*
2.5 Conclusions *45*
References *45*

3 Integrating IoT + Fog + Cloud Infrastructures: System Modeling and Research Challenges *51*
Guto Leoni Santos, Matheus Ferreira, Leylane Ferreira, Judith Kelner, Djamel Sadok, Edison Albuquerque, Theo Lynn, and Patricia Takako Endo
3.1 Introduction *51*
3.2 Methodology *52*
3.3 Integrated C2F2T Literature by Modeling Technique *55*
3.3.1 Analytical Models *58*
3.3.2 Petri Net Models *61*
3.3.3 Integer Linear Programming *63*
3.3.4 Other Approaches *64*
3.4 Integrated C2F2T Literature by Use-Case Scenarios *65*
3.5 Integrated C2F2T Literature by Metrics *68*
3.5.1 Energy Consumption *68*
3.5.2 Performance *70*
3.5.3 Resource Consumption *70*
3.5.4 Cost *71*
3.5.5 Quality of Service *71*
3.5.6 Security *72*
3.6 Future Research Directions *72*
3.7 Conclusions *73*
Acknowledgments *74*
References *75*

4 **Management and Orchestration of Network Slices in 5G, Fog, Edge, and Clouds** *79*
Adel Nadjaran Toosi, Redowan Mahmud, Qinghua Chi, and Rajkumar Buyya
4.1 Introduction *79*
4.2 Background *80*
4.2.1 5G *80*
4.2.2 Cloud Computing *82*
4.2.3 Mobile Edge Computing (MEC) *82*
4.2.4 Edge and Fog Computing *82*
4.3 Network Slicing in 5G *83*
4.3.1 Infrastructure Layer *84*
4.3.2 Network Function and Virtualization Layer *85*
4.3.3 Service and Application Layer *85*
4.3.4 Slicing Management and Orchestration (MANO) *86*
4.4 Network Slicing in Software-Defined Clouds *87*
4.4.1 Network-Aware Virtual Machines Management *88*
4.4.2 Network-Aware Virtual Machine Migration Planning *88*
4.4.3 Virtual Network Functions Management *89*
4.5 Network Slicing Management in Edge and Fog *91*
4.6 Future Research Directions *93*
4.6.1 Software-Defined Clouds *93*
4.6.2 Edge and Fog Computing *95*
4.7 Conclusions *96*
Acknowledgments *96*
References *96*

5 **Optimization Problems in Fog and Edge Computing** *103*
Zoltán Ádám Mann
5.1 Introduction *103*
5.2 Background / Related Work *104*
5.3 Preliminaries *105*
5.4 The Case for Optimization in Fog Computing *107*
5.5 Formal Modeling Framework for Fog Computing *108*
5.6 Metrics *109*
5.6.1 Performance *109*
5.6.2 Resource Usage *110*
5.6.3 Energy Consumption *111*
5.6.4 Financial Costs *111*
5.6.5 Further Quality Attributes *112*
5.7 Optimization Opportunities along the Fog Architecture *113*
5.8 Optimization Opportunities along the Service Life Cycle *114*
5.9 Toward a Taxonomy of Optimization Problems in Fog Computing *115*

5.10 Optimization Techniques *117*
5.11 Future Research Directions *118*
5.12 Conclusions *119*
Acknowledgments *119*
References *119*

Part II Middlewares *123*

6 Middleware for Fog and Edge Computing: Design Issues *125*
Madhurima Pore, Vinaya Chakati, Ayan Banerjee, and Sandeep K. S. Gupta
6.1 Introduction *125*
6.2 Need for Fog and Edge Computing Middleware *126*
6.3 Design Goals *126*
6.3.1 Ad-Hoc Device Discovery *127*
6.3.2 Run-Time Execution Environment *127*
6.3.3 Minimal Task Disruption *127*
6.3.4 Overhead of Operational Parameters *127*
6.3.5 Context-Aware Adaptive Design *128*
6.3.6 Quality of Service *128*
6.4 State-of-the-Art Middleware Infrastructures *128*
6.5 System Model *129*
6.5.1 Embedded Sensors or Actuators *130*
6.5.2 Personal Devices *130*
6.5.3 Fog Servers *131*
6.5.4 Cloudlets *131*
6.5.5 Cloud Servers *131*
6.6 Proposed Architecture *131*
6.6.1 API Code *132*
6.6.2 Security *132*
6.6.2.1 Authentication *132*
6.6.2.2 Privacy *133*
6.6.2.3 Encryption *133*
6.6.3 Device Discovery *133*
6.6.4 Middleware *133*
6.6.4.1 Context Monitoring and Prediction *133*
6.6.4.2 Selection of Participating Devices *134*
6.6.4.3 Data Analytics *134*
6.6.4.4 Scheduling and Resource Management *135*
6.6.4.5 Network Management *135*
6.6.4.6 Execution Management *135*
6.6.4.7 Mobility Management *135*

6.6.5 Sensor/Actuators *136*
6.7 Case Study Example *136*
6.8 Future Research Directions *137*
6.8.1 Human Involvement and Context Awareness *137*
6.8.2 Mobility *137*
6.8.3 Secure and Reliable Execution *137*
6.8.4 Management and Scheduling of Tasks *138*
6.8.5 Modularity for Distributed Execution *138*
6.8.6 Billing and Service-Level Agreement (SLA) *138*
6.8.7 Scalability *138*
6.9 Conclusions *139*
References *139*

7 A Lightweight Container Middleware for Edge Cloud Architectures *145*
David von Leon, Lorenzo Miori, Julian Sanin, Nabil El Ioini, Sven Helmer, and Claus Pahl
7.1 Introduction *145*
7.2 Background/Related Work *146*
7.2.1 Edge Cloud Architectures *146*
7.2.2 A Use Case *148*
7.2.3 Related Work *149*
7.3 Clusters for Lightweight Edge Clouds *149*
7.3.1 Lightweight Software – Containerization *149*
7.3.2 Lightweight Hardware – Raspberry Pi Clusters *151*
7.4 Architecture Management – Storage and Orchestration *152*
7.4.1 Own–Build Cluster Storage and Orchestration *152*
7.4.1.1 Own–Build Cluster Storage and Orchestration Architecture *152*
7.4.1.2 Use Case and Experimentation *153*
7.4.2 OpenStack Storage *153*
7.4.2.1 Storage Management Architecture *153*
7.4.2.2 Use Case and Experimentation *154*
7.4.3 Docker Orchestration *154*
7.4.3.1 Docker Orchestration Architecture *155*
7.4.3.2 Docker Evaluation – Installation, Performance, Power *157*
7.5 IoT Integration *159*
7.6 Security Management for Edge Cloud Architectures *159*
7.6.1 Security Requirements and Blockchain Principles *160*
7.6.2 A Blockchain-Based Security Architecture *161*
7.6.3 Integrated Blockchain-Based Orchestration *163*
7.7 Future Research Directions *165*
7.8 Conclusions *166*
References *167*

8 **Data Management in Fog Computing** *171*
Tina Samizadeh Nikoui, Amir Masoud Rahmani, and Hooman Tabarsaied
8.1 Introduction *171*
8.2 Background *172*
8.3 Fog Data Management *174*
8.3.1 Fog Data Life Cycle *175*
8.3.1.1 Data Acquisition *175*
8.3.1.2 Lightweight Processing *175*
8.3.1.3 Processing and Analysis *175*
8.3.1.4 Sending Feedback *175*
8.3.1.5 Command Execution *177*
8.3.2 Data Characteristics *177*
8.3.3 Data Pre-Processing and Analytics *178*
8.3.3.1 Data Cleaning *178*
8.3.3.2 Data Fusion *178*
8.3.3.3 Edge Mining *179*
8.3.4 Data Privacy *179*
8.3.5 Data Storage and Data Placement *180*
8.3.6 e-Health Case Study *180*
8.3.7 Proposed Architecture *181*
8.3.7.1 Device Layer *184*
8.3.7.2 Fog Layer *184*
8.3.7.3 Cloud Layer *185*
8.4 Future Research and Direction *186*
8.4.1 Security *186*
8.4.2 Defining the Level of Data Computation and Storage *186*
8.5 Conclusions *186*
References *188*

9 **Predictive Analysis to Support Fog Application Deployment** *191*
Antonio Brogi, Stefano Forti, and Ahmad Ibrahim
9.1 Introduction *191*
9.2 Motivating Example: Smart Building *193*
9.3 Predictive Analysis with FogTorchΠ *197*
9.3.1 Modeling Applications and Infrastructures *197*
9.3.2 Searching for Eligible Deployments *199*
9.3.3 Estimating Resource Consumption and Cost *201*
9.3.4 Estimating QoS-Assurance *204*
9.4 Motivating Example (continued) *206*
9.5 Related Work *207*
9.5.1 Cloud Application Deployment Support *207*
9.5.2 Fog Application Deployment Support *210*

9.5.3 Cost Models *211*
9.5.4 Comparing iFogSim and FogTorchΠ *212*
9.6 Future Research Directions *214*
9.7 Conclusions *216*
References *217*

10 Using Machine Learning for Protecting the Security and Privacy of Internet of Things (IoT) Systems *223*
Melody Moh and Robinson Raju
10.1 Introduction *223*
10.1.1 Examples of Security and Privacy Issues in IoT *224*
10.1.2 Security Concerns at Different Layers in IoT *224*
10.1.2.1 Sensing Layer *225*
10.1.2.2 Network Layer *225*
10.1.2.3 Service Layer *226*
10.1.2.4 Interface Layer *226*
10.1.3 Privacy Concerns in IoT Devices *226*
10.1.3.1 Information Privacy *228*
10.1.3.2 Categorization of IoT Privacy Issues *229*
10.1.4 IoT Security Breach Deep-Dive: Distributed Denial of Service (DDoS) Attacks on IoT Devices *230*
10.1.4.1 Introduction to DDoS *230*
10.1.4.2 Timeline of Notable DoS Events [25] *231*
10.1.4.3 Reason for the Recent Success of the DDoS Attacks *232*
10.1.4.4 Directions for Prevention of Specific Attacks on IoT Devices *232*
10.1.4.5 Steps to Prevent Attacks on IoT Devices *233*
10.2 Background *234*
10.2.1 Brief Overview of Machine Learning *234*
10.2.2 Frequently Used Machine-Learning Algorithms *235*
10.2.2.1 Classification *235*
10.2.2.2 Regression *235*
10.2.2.3 Clustering *235*
10.2.2.4 Dimensionality Reduction *236*
10.2.2.5 Combining Models (Ensemble ML) *236*
10.2.2.6 Artificial Neural Networks *237*
10.2.3 Examples of Machine-Learning Algorithms in IoT *237*
10.2.3.1 Overview *237*
10.2.3.2 Examples *237*
10.2.4 Machine-Learning Algorithms by IoT Domains *238*
10.2.4.1 Healthcare *238*
10.2.4.2 Utilities – Energy/Water/Gas *239*
10.2.4.3 Manufacturing *239*
10.2.4.4 Insurance *239*

10.2.4.5 Traffic *240*
10.2.4.6 Smart City – Citizens and Public Places *240*
10.2.4.7 Smart Homes *241*
10.2.4.8 Agriculture *241*
10.3 Survey of ML Techniques for Defending IoT Devices *242*
10.3.1 Systematic Categorization of ML Solutions for IoT Security *242*
10.3.2 Examples of ML Algorithms for IoT Security *243*
10.3.2.1 Malware Detection Using SVM *243*
10.3.2.2 Malware Detection Using a Random Forest *243*
10.3.2.3 Intrusion Detection Using PCA, Naïve Bayes, and KNN *244*
10.3.2.4 Anomaly Detection Using Classification *244*
10.3.3 Use of Artificial Neural Networks (ANN) to Forecast and Secure IoT Systems *244*
10.3.4 New Flavors of Attacks on IoT Devices *245*
10.3.4.1 Mirai *245*
10.3.4.2 Brickerbot *245*
10.3.4.3 FLocker *246*
10.3.4.4 Summary *246*
10.3.5 Proposal for Effective ML Techniques to Achieve IoT Security *246*
10.3.5.1 Insights from the Research *246*
10.3.5.2 Proposals *247*
10.4 Machine Learning in Fog Computing *248*
10.4.1 Introduction *248*
10.4.2 Machine Learning for Fog Computing and Security *249*
10.4.3 Examples of Machine Learning in Fog Computing *249*
10.4.3.1 ML in Fog Computing in Industry *249*
10.4.3.2 ML in Fog Computing in Retail *250*
10.4.3.3 Fog Computing for Self-Driving Cars *250*
10.4.4 Machine Learning in Fog Computing Security *250*
10.4.5 Other Machine-Learning Algorithms for Fog Computing *252*
10.5 Future Research Directions *252*
10.6 Conclusions *252*
References *253*

Part III Applications and Issues *259*

11 Fog Computing Realization for Big Data Analytics *261*
Farhad Mehdipour, Bahman Javadi, Aniket Mahanti, and Guillermo Ramirez-Prado
11.1 Introduction *261*
11.2 Big Data Analytics *262*
11.2.1 Benefits *263*

11.2.2 A Typical Big Data Analytics Infrastructure *263*
11.2.2.1 Big Data Platform *263*
11.2.2.2 Data Management *264*
11.2.2.3 Storage *264*
11.2.2.4 Analytics Core and Functions *264*
11.2.2.5 Adaptors *264*
11.2.2.6 Presentation *265*
11.2.3 Technologies *265*
11.2.4 Big Data Analytics in the Cloud *265*
11.2.5 In-Memory Analytics *265*
11.2.6 Big Data Analytics Flow *266*
11.3 Data Analytics in the Fog *267*
11.3.1 Fog Analytics *268*
11.3.2 Fog-Engines *269*
11.3.3 Data Analytics Using Fog-Engines *270*
11.4 Prototypes and Evaluation *272*
11.4.1 Architecture *272*
11.4.2 Configurations *274*
11.4.2.1 Fog-Engine as a Broker *274*
11.4.2.2 Fog-Engine as a Data Analytics Engine *274*
11.4.2.3 Fog-Engine as a Server *274*
11.4.2.4 Communication with Fog-Engine versus the Cloud *274*
11.5 Case Studies *277*
11.5.1 Smart Home *277*
11.5.1.1 Fog-Engine as a Broker *277*
11.5.1.2 Fog-Engine as a Data Analytic Engine *278*
11.5.1.3 Fog-Engine as a Server *279*
11.5.2 Smart Nutrition Monitoring System *279*
11.6 Related Work *282*
11.7 Future Research Directions *287*
11.8 Conclusions *287*
References *288*

12 Exploiting Fog Computing in Health Monitoring *291*
Tuan Nguyen Gia and Mingzhe Jiang
12.1 Introduction *291*
12.2 An Architecture of a Health Monitoring IoT-Based System with Fog Computing *293*
12.2.1 Device (Sensor) Layer *294*
12.2.2 Smart Gateways with Fog Computing *295*
12.2.3 Cloud Servers and End-User Terminals *296*
12.3 Fog Computing Services in Smart E-Health Gateways *297*
12.3.1 Local Database (Storage) *297*

12.3.2 Push Notification *298*
12.3.3 Categorization *299*
12.3.4 Local Host with User Interface *299*
12.3.5 Interoperability *299*
12.3.6 Security *300*
12.3.7 Human Fall Detection *301*
12.3.8 Fault Detection *303*
12.3.9 Data Analysis *303*
12.4 System Implementation *304*
12.4.1 Sensor Node Implementation *304*
12.4.2 Smart Gateways with Fog Implementation *305*
12.4.3 Cloud Servers and Terminals *307*
12.5 Case Studies, Experimental Results, and Evaluation *308*
12.5.1 A Case Study of Human Fall Detection *308*
12.5.2 A Case Study of Heart Rate Variability *309*
12.6 Discussion of Connected Components *313*
12.7 Related Applications in Fog Computing *313*
12.8 Future Research Directions *314*
12.9 Conclusions *314*
References *315*

13 Smart Surveillance Video Stream Processing at the Edge for Real-Time Human Objects Tracking *319*
Seyed Yahya Nikouei, Ronghua Xu, and Yu Chen
13.1 Introduction *319*
13.2 Human Object Detection *320*
13.2.1 Haar Cascaded-Feature Extraction *321*
13.2.2 HOG+SVM *322*
13.2.3 Convolutional Neural Networks (CNNs) *324*
13.3 Object Tracking *327*
13.3.1 Feature Representation *327*
13.3.2 Categories of Object Tracking Technologies *328*
13.3.3 Point-Based Tracking *329*
13.3.3.1 Deterministic Methods *329*
13.3.3.2 Kalman Filters *330*
13.3.3.3 Particle Filters *330*
13.3.3.4 Multiple Hypothesis Tracking (MHT) *331*
13.3.4 Kernel-Based Tracking *331*
13.3.5 Silhouette-Based Tracking *332*
13.3.6 Kernelized Correlation Filters (KCF) *332*
13.4 Lightweight Human Detection *335*
13.5 Case Study *337*
13.5.1 Human Object Detection *337*

13.5.2 Object Tracking 339
13.5.2.1 Multi-Object Tracking 339
13.5.2.2 Object Tracking Phase In and Out 341
13.5.2.3 Tracking Object Lost 341
13.6 Future Research Directions 342
13.7 Conclusions 343
References 343

14 Fog Computing Model for Evolving Smart Transportation Applications *347*
M. Muzakkir Hussain, Mohammad Saad Alam, and M.M. Sufyan Beg
14.1 Introduction 347
14.2 Data-Driven Intelligent Transportation Systems 348
14.3 Mission-Critical Computing Requirements of Smart Transportation Applications 351
14.3.1 Modularity 351
14.3.2 Scalability 352
14.3.3 Context-Awareness and Abstraction Support 352
14.3.4 Decentralization 353
14.3.5 Energy Consumption of Cloud Data Centers 353
14.4 Fog Computing for Smart Transportation Applications 354
14.4.1 Cognition 355
14.4.2 Efficiency 355
14.4.3 Agility 356
14.4.4 Latency 356
14.5 Case Study: Intelligent Traffic Lights Management (ITLM) System 359
14.6 Fog Orchestration Challenges and Future Directions 362
14.6.1 Fog Orchestration Challenges for Intelligent Transportation Applications in IoT Space 363
14.6.1.1 Scalability 363
14.6.1.2 Privacy and Security 363
14.6.1.3 Dynamic Workflows 364
14.6.1.4 Tolerance 364
14.7 Future Research Directions 364
14.7.1 Opportunities in the Deployment Phase 365
14.7.1.1 Optimal Node Selection and Routing 365
14.7.1.2 Parallelization Approaches to Manage Scale and Complexity 366
14.7.1.3 Heuristics and Late Calibration 366
14.7.2 Opportunities in Runtime Phase 366
14.7.2.1 Dynamic Orchestration of Fog Resources 367
14.7.2.2 Incremental Computation Strategies 367
14.7.2.3 QoS-Aware Control and Monitoring Protocols 367

14.7.2.4 Proactive Decision-Making *367*
14.7.3 Opportunities in Evaluation Phase: Big-Data-Driven Analytics (BD^2A) and Optimization *368*
14.8 Conclusions *369*
References *370*

15 Testing Perspectives of Fog-Based IoT Applications *373*
Priyanka Chawla and Rohit Chawla
15.1 Introduction *373*
15.2 Background *374*
15.3 Testing Perspectives *376*
15.3.1 Smart Homes *376*
15.3.2 Smart Health *378*
15.3.3 Smart Transport *390*
15.4 Future Research Directions *393*
15.4.1 Smart Homes *393*
15.4.2 Smart Health *398*
15.4.3 Smart Transport *402*
15.5 Conclusions *405*
References *406*

16 Legal Aspects of Operating IoT Applications in the Fog *411*
G. Gultekin Varkonyi, Sz. Varadi, and Attila Kertesz
16.1 Introduction *411*
16.2 Related Work *412*
16.3 Classification of Fog/Edge/IoT Applications *413*
16.4 Restrictions of the GDPR Affecting Cloud, Fog, and IoT Applications *414*
16.4.1 Definitions and Terms in the GDPR *415*
16.4.1.1 Personal Data *415*
16.4.1.2 Data Subject *415*
16.4.1.3 Controller *415*
16.4.1.4 Processor *415*
16.4.1.5 Pseudonymization *416*
16.4.1.6 Limitation *416*
16.4.1.7 Consent *416*
16.4.1.8 Right to Be Forgotten *417*
16.4.1.9 Data Portability *417*
16.4.2 Obligations Defined by the GDPR *418*
16.4.2.1 Obligations of the Controller *418*
16.4.2.2 Obligations of the Processor *420*
16.4.3 Data Transfers Outside the EU *422*
16.4.3.1 Data Transfers to Third Countries *422*

16.4.3.2 Remedies, Liabilities, and Sanctions *424*
16.4.4 Summary *424*
16.5 Data Protection by Design Principles *425*
16.5.1 Reasons for Adopting Data Protection Principles *426*
16.5.2 Privacy Protection in the GDPR *427*
16.5.3 Data Protection by Default *428*
16.6 Future Research Directions *430*
16.7 Conclusions *430*
Acknowledgment *431*
References *431*

17 Modeling and Simulation of Fog and Edge Computing Environments Using iFogSim Toolkit *433*
Redowan Mahmud and Rajkumar Buyya
17.1 Introduction *433*
17.2 iFogSim Simulator and Its Components *435*
17.2.1 Physical Components *435*
17.2.2 Logical Components *436*
17.2.3 Management Components *436*
17.3 Installation of iFogSim *436*
17.4 Building Simulation with iFogSim *437*
17.5 Example Scenarios *438*
17.5.1 Create Fog Nodes with Heterogeneous Configurations *438*
17.5.2 Create Different Application Models *439*
17.5.2.1 Master–Worker Application Models *440*
17.5.2.2 Sequential Unidirectional Dataflow Application Model *441*
17.5.3 Application Modules with Different Configuration *443*
17.5.4 Sensors with Different Tuple Emission Rate *444*
17.5.5 Send Specific Number of Tuples from a Sensor *444*
17.5.6 Mobility of a Fog Device *445*
17.5.7 Connect Lower-Level Fog Devices with Nearby Gateways *447*
17.5.8 Make Cluster of Fog Devices *449*
17.6 Simulation of a Placement Policy *450*
17.6.1 Structure of Physical Environment *450*
17.6.2 Assumptions for Logical Components *450*
17.6.3 Management (Application Placement) Policy *451*
17.7 A Case Study in Smart Healthcare *461*
17.8 Conclusions *463*
References *464*

Index *467*

List of Contributors

Zoltán Ádám Mann
University of Duisburg-Essen
Germany
e-mail: zoltan.mann@gmail.com

Edison Albuquerque
Universidade de Pernambuco
Brazil
e-mail: edison@ecomp.poli.br

Mohammad Saad Alam
Aligarh Muslim University
India
e-mail: saad.alam@zhcet.ac.in

Ahmet Cihat Baktir
Bogazici University
Turkey
e-mail: cihatbaktir@gmail.com

Ayan Banerjee
Arizona State University
USA
e-mail: abanerj3@asu.edu

M. M. Sufyan Beg
Aligarh Muslim University
India
e-mail: mmsbeg@cs.berkely.edu

Antonio Brogi
University of Pisa
Italy
e-mail: brogi@di.unipi.it

Rajkumar Buyya
University of Melbourne
Australia
e-mail: rbuyya@unimelb.edu.au

Vinaya Chakati
Arizona State University
USA
e-mail: vchakati@asu.edu

Chii Chang
University of Tartu
Estonia
e-mail: chang@ut.ee

Priyanka Chawla
Lovely Professional University
India
e-mail: priyankamatrix@gmail.com

Rohit Chawla
Apeejay College
India
e-mail: rc.j2ee@gmail.com

Yu Chen
Binghamton University
USA
e-mail: ychen@binghamton.edu

Qinghua Chi
University of Melbourne
Australia
e-mail: chiqinghua@huawei.com

Nabil El Ioini
Free University of Bozen-Bolzano
Italy
e-mail: nelioini@unibz.it

Patricia Takako Endo
Universidade de Pernambuco
Dublin City University
Ireland
e-mail: patricia.endo@upe.br

Cem Ersoy
Bogazici University
Turkey
e-mail: ersoy@boun.edu.tr

Leylane Ferreira
Universidade Federal de Pernambuco
Brazil
e-mail: leylane.silva@gprt.ufpe.br

Matheus Ferreira
Universidade de Pernambuco
Brazil
e-mail:
matheus0906.mhci@gmail.com

Stefano Forti
University of Pisa
Italy
e-mail: stefano.forti@di.unipi.it

Tuan Nguyen Gia
University of Turku
Finland
e-mail: tuan.nguyengia@utu.fi

Sandeep Kumar S. Gupta
Arizona State University
USA
e-mail: sandeep.gupta@asu.edu

Sven Helmer
Free University of Bozen-Bolzano
Italy
e-mail: shelmer@inf.unibz.it

M. Muzakkir Hussain
Aligarh Muslim University
India
e-mail:
md.muzakkirhussain@zhcet.ac.in

Ahmad Ibrahim
University of Pisa
Italy
e-mail: ahmad@di.unipi.it

Bahman Javadi
Western Sydney University
Australia
e-mail:
b.javadi@westernsydney.edu.au

Mingzhe Jiang
University of Turku
Finland
e-mail: mizhji@utu.fi

Judith Kelner
Universidade Federal de Pernambuco
Brazil
e-mail: jk@gprt.ufpe.br

Attila Kertesz
University of Szeged
Hungary
e-mail: keratt@inf.u-szeged.hu

Theo Lynn
Dublin City University
Ireland
e-mail: theo.lynn@dcu.ie

Aniket Mahanti
University of Auckland
New Zealand
e-mail: a.mahanti@auckland.ac.nz

Redowan Mahmud
University of Melbourne
Australia
e-mail:
mahmudm@student.unimelb.edu.au

Farhad Mehdipour
New Zealand School of Education
and STEM Fern Ltd.
Auckland
New Zealand
e-mail: farhadm@nzseg.com

Lorenzo Miori
Free University of Bozen-Bolzano
Italy
e-mail: memorys60@gmail.com

Melody Moh
San Jose State University
USA
e-mail: melody.moh@sjsu.edu

Seyed Yahya Nikouei
Binghamton University
USA
e-mail: snikoue1@binghamton.edu

Tina Samizadeh Nikoui
Islamic Azad University
Iran
e-mail: tina.samizadeh@srbiau.ac.ir

Atay Ozgovde
Galatasaray University
Turkey
e-mail: atay.ozgovde@gmail.com

Claus Pahl
Free University of Bozen-Bolzano
Italy
e-mail: cpahl@unibz.it

Madhurima Pore
Arizona State University
USA
e-mail: mpore@asu.edu

Amir Masoud Rahmani
Islamic Azad University & University
of Human Development
Iran
e-mail: rahmani@srbiau.ac.ir

Robinson Raju
San Jose State University
USA
e-mail: robinson.raju@sjsu.edu

Guillermo Ramirez-Prado
Unitec Institute of Technology
Auckland
New Zealand
e-mail: gprado@unitec.ac.nz

Djamel Sadok
Universidade Federal de Pernambuco
Brazil
e-mail: jamel@gprt.ufpe.br

Julian Sanin
Free University of Bozen-Bolzano
Italy
e-mail: Julian.Sanin@stud-inf.unibz.it

Guto Leoni Santos
Universidade Federal de Pernambuco
Brazil
e-mail: guto.leoni@gprt.ufpe.br

Cagatay Sonmez
Bogazici University
Turkey
e-mail: cagataysonmez@hotmail.com

Satish Narayana Srirama
University of Tartu
Estonia
e-mail: srirama@ut.ee

Hooman Tabarsaied
Islamic Azad University
Iran
e-mail: h.tabarsaied@yahoo.com

Adel Nadjaran Toosi
University of Melbourne
Australia
e-mail:
adel.nadjaran@unimelb.edu.au

Sz. Varadi
University of Szeged
Hungary
e-mail:
varadiszilvia@juris.u-szeged.hu

Blesson Varghese
Queen's University Belfast
UK
e-mail: B.Varghese@qub.ac.uk

G. Gultekin Varkonyi
University of Szeged
Hungary
e-mail: gizemgv@juris.u-szeged.hu

David von Leon
Free University of Bozen-Bolzano
Italy
e-mail: david@davole.com

Ronghua Xu
Binghamton University
USA
e-mail: rxu22@binghamton.edu

Preface

The Internet of Things (IoT) paradigm promises to make "things" such as physical objects with sensing capabilities and/or attached with tags, mobile objects such as smart phones and vehicles, consumer electronic devices, and home appliances such as refrigerators, televisions, and healthcare devices as part of the Internet environment. In cloud-centric IoT (CIoT) applications, the sensor data from these "things" is extracted, accumulated, and processed at the public/private clouds, leading to significant latencies. Fog computing addresses this issue in developing real-time IoT applications, by mainly utilizing proximity-based computational resources across the IoT layers such as gateways, cloudlets, and network switches/routers. A similar approach of utilizing proximity resources in the telecommunication domain is mobile edge computing.

To realize the full potential of fog and edge computing and similar paradigms, researchers and practitioners need to address several challenges and develop suitable conceptual and technological solutions for tackling them. These include development of scalable architectures, moving from closed systems to open systems, dealing with privacy and ethical issues involved in data sensing, storage, processing, and actions, designing interaction protocols, and autonomic management.

The primary purpose of this book is to capture the state-of-the-art in fog and edge computing, their applications, architectures, and technologies. The book also aims to identify potential research directions and technologies that will facilitate insight generation in various domains from smart home, smart cities, science, industry, business, and consumer applications. We expect the book to serve as a reference for larger audiences such as system architects, practitioners, developers, new researchers, and graduate-level students. This book also comes with an associated website (hosted at http://cloudbus.org/fog/book/) containing pointers to advanced on-line resources.

Organization of the Book

This book contains chapters authored by several leading experts in the fields of IoT, cloud, and fog computing. The book is presented in a coordinated and integrated manner, starting with the fundamentals and followed by the middleware and technological solutions to implement fog and edge-related applications.

The contents of the book are organized into three parts:

I. Foundations
II. Middlewares
III. Applications and Issues

Part I focuses on Foundations and is made up of five chapters. The first chapter, "**Internet of Things (IoT) and New Computing Paradigms,**" discusses the IoT paradigm along with CIoT limitations. The relevant technologies and new computing paradigms that address these limitations such as fog computing, edge computing and mist computing, are discussed along with their main advantages and basic mechanisms. The hierarchy of fog and edge computing environments is discussed, and the opportunities and challenges offered by fog and edge computing are discussed thoroughly. The challenges along with their future research directions are further structured into networking, management, and resource and modeling challenges, in Chapter 2, "**Addressing the Challenges in Federating Edge Resources.**" The use of modelling techniques and the relevant literature to represent and evaluate an integrated cloud-to-things system comprising cloud computing, fog computing, and the IoT is reviewed in Chapter 3, "**Integrating IoT + Fog + Cloud Infrastructures: System Modeling and Research Challenges.**" The state-of-the-art literature on network slicing in 5G, edge/fog, and cloud computing is reviewed in Chapter 4, "**Management and Orchestration of Network Slices in 5G, Fog, Edge, and Clouds.**" Part I concludes with a discussion of generic conceptual framework for optimization problems in fog computing, based on consistent, well-defined, and formalized notation for constraints and optimization objectives, in Chapter 5, "**Optimization Problems in Fog and Edge Computing.**"

Part II focuses on Middlewares and is made up of five chapters. Chapter 6, "**Middleware for Fog and Edge Computing: Design Issues,**" discusses different aspects of the design of middleware for Fog and Edge computing along with a proposed architecture. Chapter 7, "**A Lightweight Container Middleware for Edge Cloud Architectures,**" discusses the core principles of an edge cloud reference architecture that is based on containers as the packaging and distribution mechanism. The chapter also provides experimental results with Raspberry Pi clusters to validate the proposed architectural solution. Chapter 8, "**Data Management in Fog Computing,**" proposes the conceptual architecture for the data management in fog computing environments. The chapter also provides a review of the fog data management, along with future research directions. Chapter 9, "**Predictive Analysis to Support Fog Application Deployment,**" discusses FogTorchΠ prototype that supports application deployment in the fog. The prototype permits expression of processing capabilities, predicts QoS attributes, and estimates operational costs of a fog infrastructure, along with processing and QoS requirements of an application. Chapter 10, "**Using Machine Learning for Protecting the Security and Privacy of Internet of Things (IoT) Systems,**" reviews the machine learning (ML) techniques for defending IoT devices, along with a discussion on scope of ML in fog computing.

Part III focuses on Applications and relevant issues and is made up of seven chapters. Chapter 11, "**Fog Computing Realization for Big Data Analytics,**" discusses a fog-engine prototype that can be deployed in the traditional centralized data analytics platform to realize the data analytics in the fog environment. Smart home and smart nutrition monitoring system case studies are provided, which conceptually utilize the fog-engine. Chapter 12, "**Exploiting Fog Computing in Health Monitoring,**" discussed fog computing services in smart e-health gateways. The proposed system is implemented and evaluated with a remote ECG (electrocardiogram) monitoring case study. Chapter 13 discussed "**Smart Surveillance Video Stream Processing at the Edge for Real-Time Human Objects Tracking.**" The computations and algorithms used at the fog and edge levels to create such automated surveillance system are discussed and compared. Chapter 14, "**Fog Computing Model for Evolving Smart Transportation Applications,**" identified the computing needs of the data-driven transportation architecture and devised a fog-assisted cloud-based computational platform for smart transportation applications, in the context of intelligent traffic management system (ITSM) use case. Chapter 15 discussed and reviewed "**Testing Perspectives of Fog-Based IoT Applications,**" in the smart home, smart health, and smart transport domains. Chapter 16, "**Legal Aspects of Operating IoT Applications in the Fog,**" classified fog/edge/IoT applications, analyzed the latest restrictions introduced by the General Data Protection Regulation (GDPR), and discussed how these legal constraints affect the design and operation of IoT applications in fog and cloud environments. Another critical issue related to fog application development is that it is very costly due to the fact that the fog computing environment incorporates IoT devices, fog nodes, and cloud datacenters, along with a huge amount of IoT data. To address this, Chapter 17 discussed "**Modeling and Simulation of Fog and Edge Computing Environments Using iFogSim Toolkit.**" iFogSim simulator components are discussed and installation details are provided, along with detailed guidelines to model the fog environment.

Acknowledgments

First and foremost, we are grateful to all the contributing authors for their time, effort, and understanding during the preparation of the book.

We thank Albert Zomaya, editor of Wiley book series on parallel and distributed computing, for his enthusiastic support, enabling us to easily navigate through Wiley's publication process.

Raj would like to thank his family members, especially Smrithi, Soumya, and Radha Buyya, for their love, understanding, and support during the preparation of the book. Satish would like to thank his wife, Gayatri, and parents (S. Lakshminarayana and Lolakshi) for their love and support and his new born daughter, Meghana, for the pleasantness she brought into the family.

Finally, we would like to thank the staff at Wiley, particularly Brett Kurzman (senior editor) and Victoria Bradshaw (editorial assistant). They were wonderful to work with!

Rajkumar Buyya
The University of Melbourne and Manjrasoft Pty Ltd, Australia

Satish Narayana Srirama
The University of Tartu, Estonia

Part I

Foundations

1

Internet of Things (IoT) and New Computing Paradigms

Chii Chang, Satish Narayana Srirama, and Rajkumar Buyya

1.1 Introduction

The Internet of Things (IoT) [1] represents a comprehensive environment that interconnects a large number of heterogeneous physical objects or things such as appliances, facilities, animals, vehicles, farms, factories etc. to the Internet, in order to enhance the efficiency of the applications such as logistics, manufacturing, agriculture, urban computing, home automation, ambient assisted living, and various real-time ubiquitous computing applications.

Commonly, an IoT system follows the architecture of the Cloud-centric Internet of Things (CIoT) in which the physical objects are represented in the form of Web resources that are managed by the servers in the global Internet [2]. Fundamentally, in order to interconnect the physical entities to the Internet, the system will utilize various front-end devices such as wired or wireless sensors, actuators, and readers to interact with them. Further, the front-end devices have the Internet connectivity via the mediate gateway nodes such as Internet modems, routers, switches, cellular base stations, and so on. In general, the common IoT system involves three major technologies: embedded systems, middleware, and cloud services, where the embedded systems provide intelligence to the front-end devices, middleware interconnects the heterogeneous embedded systems of front-end devices to the cloud and finally, the cloud provides comprehensive storage, processing, and management mechanisms.

Although the CIoT model is a common approach to implement IoT systems, it is facing the growing challenges in IoT. Specifically, CIoT faces challenges in BLURS—bandwidth, latency, uninterrupted, resource-constraint, and security [3].

- **Bandwidth**. The increasingly large and high-frequent rate data produced by objects in IoT will exceed the bandwidth availability. For example, a connected car can generate tens of megabytes' data per second for the information of its route, speeds, car-operating condition, driver's condition,

Fog and Edge Computing: Principles and Paradigms, First Edition.
Edited by Rajkumar Buyya and Satish Narayana Srirama.

surrounding environment, weather etc. Further, a self-driving vehicle can generate gigabytes of data per second due to the need for real-time video streaming. Therefore, fully relying on the distant Cloud to manage the things becomes impractical.

- **Latency**. Cloud faces the challenges of achieving the requirement of controlling the end-to-end latency within tens of milliseconds. Specifically, industrial smart grids systems, self-driving vehicular networks, virtual and augmented reality applications, real-time financial trading applications, healthcare, and eldercare applications cannot afford the causes derived from the latency of CIoT.
- **Uninterrupted**. The long distance between cloud and the front-end IoT devices can face issues derived from the unstable and intermittent network connectivity. For example, a CIoT-based connected vehicle will be unable to function properly due to the disconnection occurred at the intermediate node between the vehicle and the distant cloud.
- **Resource-constrained**. Commonly, many front-end devices are resource-constrained in which they are unable to perform complex computational tasks and hence, CIoT systems usually require front-end devices to continuously stream their data to the cloud. However, such a design is impractical in many devices that operate with battery power because the end-to-end data transmission via the Internet can still consume a lot of energy.
- **Security**. A large number of constraint front-end devices may not have sufficient resources to protect themselves from the attacks. Specifically, outdoor-based front-end devices, which rely on the distant cloud to keep them updated with the security software, can be attackers' targets, in which the attackers are capable of performing a malicious activity at the edge network where the front-end devices are located and the cloud does not have full control on it. Furthermore, the attacker may also damage or control the front-end device and send false data to the cloud.

The growing challenges of CIoT raised a question—*what can be done to overcome the limitation of current cloud-centric architecture?*

In the last decade, several approaches have tried to extend the centralized cloud computing to a more geo-distributed manner in which the computational, networking, and storage resources can be distributed to the locations that are much closer to the data sources or end-user applications. For example, the geo-distributed cloud-computing model [4] tends to partition the portions of processes to the data centers near the edge network. Further, the mobile cloud computing model [5] introduced the physical proximity-based cloud computing resources provisioned by the local wireless Internet access point providers. Moreover, academic research projects [6] have experimented with the feasibility of the mobile ad hoc network (MANET)-based cloud using the advanced RISC machine (ARM)-powered devices. Among the various approaches, the industry-led fog computing architecture, which was first introduced by Cisco research [7], has gained the most attention.

Fog computing architecture [8] covers a broad range of equipment and networks. In general, it is a conceptual model that address all the possibilities to extend the cloud to the edge network of CIoT, from the geo-distributed data center, intermediate network nodes to the extreme edge where the front-end IoT devices are located. Figure 1.1 illustrates different network computing paradigms supporting IoT-enabled smart systems and applications. To enumerate, the general CIoT paradigm (mark 1) manages the smart systems entirely at the distant central cloud datacenter in which the IoT devices act as simple sensory data collectors or actuators and leave the processes and decision-making to the cloud. The generic edge computing paradigm (mark 2) distributes certain tasks to the IoT devices or the co-located computers within the same subnet of the IoT devices. Such tasks can be data classification, filtering, or signal converting, for example. Fog computing paradigm (marks 3 and 4) utilizes a hierarchical-based distributed computing model that supports horizontal scalability of the computational resources.

For example, a fog-enabled IoT system can distribute the simple data classification tasks to the IoT devices and assign the more complicated context reasoning tasks at the edge gateway devices. Further, for the analytics tasks that involve terabytes of data, which requires higher processing power, the system can further move the processes to the resources at the core network

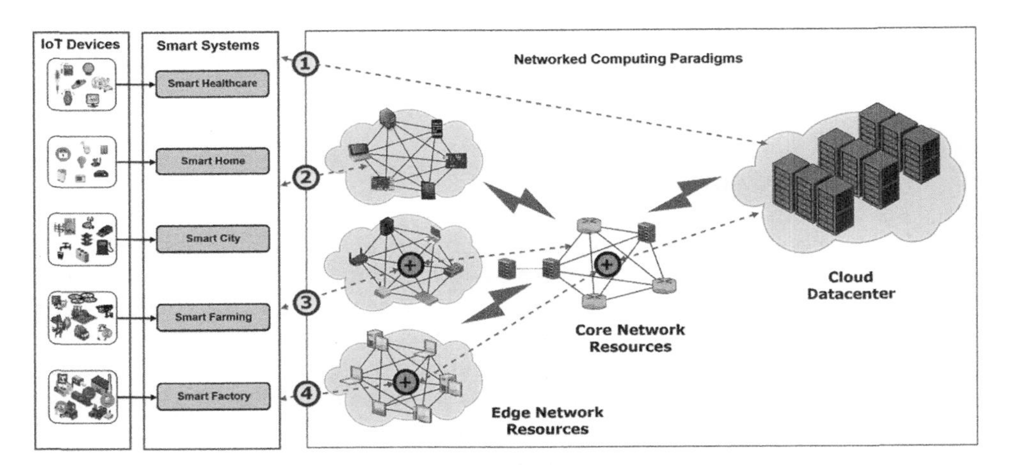

Smart Systems enabled by different computing paradigms:
(1) Cloud Computing, (2) Edge Computing, and (3) & (4) Fog Computing
(+) - Indicates capability further enhanced by the next level resources in the network.

Figure 1.1 IoT applications and environments with supporting computing paradigms.

such as the data centers of wide area network (WAN) service providers or it can utilize the cloud. Certainly, the decision of where the system should assign the tasks among the resources across different tiers depends on efficiency and adaptability. For example, smart systems may need to assign certain decision-making tasks to the edge devices in order to provide timely notification about the situation, such as the patient's condition in the smart healthcare, the security state of the smart home, the traffic condition of the smart city, the water supply condition of smart farming, or the production line operation condition of a smart factory.

The industry has seen fog as the main trend for the practical IoT systems, and the leading OpenFog consortium has established collaboration with major industrial standard parties such as European Telecommunications Standards Institute (ETSI) multi-access edge computing (MEC) and IEEE Standard for fog computing and networking [9] to hasten the fog. Furthermore, the fog market research report [10] stated that the market value of fog will grow from $3.7 billion by 2019 up to $18.2 billion by 2022 across different fields, where the top five utilization domains of fog will be energy/utilities, transportation, healthcare, industry, and agriculture.

In this chapter, we discuss foundations of computing paradigms for realizing emerging IoT applications, especially fog and edge computing, their background, characteristics, architectures and open challenges. Section 1.2 presents related technologies to fog and edge computing. Section 1.3 describes how fog and edge can improve CIoT. Section 1.4 explains the hierarchy of fog and edge computing environments. Section 1.5 illustrates the business models of fog and edge computing. Section 1.6 provides the information regarding to the opportunities and challenges in fog and edge computing. Finally, Section 1.7 summarizes the content of the chapter.

1.2 Relevant Technologies

The notion of having computational resources near the data sources may seem not new. Particularly, the term—edge computing appeared in 2004 to illustrate a system that distributes program methods and the corresponding data to the network edge towards enhancing performance and efficiency [11]. Similarly, the notion of having virtualization technology-based computing resources within the Wi-Fi subnet was introduced in 2009 [5]. However, the real industrial interest in extending computational resources to the edge network only started after the introduction of fog computing for IoT. Prior to that, applying utility cloud at the edge network was more or less a research topic in academia without explicit definition or architecture and with minor industrial involvement. In contrast, the industry has invested in fog computing architecture by establishing OpenFog consortium founded by ARM Holdings,

Cisco, Dell, Intel, Microsoft, Princeton University, and over 60 members from major industrial and academic partners in the world. Further, in collaboration with international standard organizations such as ETSI and IEEE, fog has become a major trend in general information and communication technology (ICT) today.

In last several years, researchers have been using different terminologies to illustrate the similar architectures with fog. For example, the author of virtual machine (VM)-based cloudlet [5] tended to use edge computing to describe the notion of the cloud at the edge. Moreover, the author's later work indicated that fog is a part of edge computing [12]. On the other hand, OpenFog consortium has specifically differentiated the two terminologies. Explicitly, the initial objective of cloudlet was to provide the mobile application a substitution from the distant cloud, in which the mobile applications can offload computing-intensive tasks to the nearby cloudlet VM machines co-located within the same Wi-Fi subnet. By contrast, the initial introduction of fog computing aimed to complete the cloud by extending the cloud to the network gateways themselves. In essence, cloudlet can be seen as one of the practical approaches for fog computing when the co-located physical server machines are available.

Certain other works have been describing multi-access edge computing (MEC; formerly mobile edge computing) as an exchangeable term with fog. Essentially, ETSI introduced MEC as a standard from the perspective of telecommunication, in which ETSI specifies the application programming interface (API) standards about how telecommunication companies can provide computing virtualization-based service to their clients based on extending the existing infrastructure used in network function virtualization (NFV), which has been already implemented in existing equipment such as cellular base transceiver stations (BTSs). Although it is inaccurate to describe MEC as an exchangeable term with fog, according to the recent collaboration between OpenFog and ETSI, MEC will become a practical approach to hasten the realization of fog computing [13].

Mist computing was an alternative term to fog in the earlier stages. However, recent works have described mist as a subset of fog. Accordingly, mist elaborates the need of distributing computing mechanism to the extreme edge of IoT, where the IoT devices are located, in order to minimize the communication latency between IoT devices in milliseconds [14–16]. Essentially, the motivation of mist computing is to grant the IoT devices with the capability of self-awareness in terms of self-organizing, self-managing, and several self-* mechanisms. Therefore, the IoT devices will be able to continuously operate even when the Internet connection is unstable.

In general, mist devices may sound similar to the embedded services or mobile Web services [17] in which the application services are hosted in heterogeneous resource-constrained devices such as sensors, actuators, and mobile phones. However, mist emphasizes the capability of self-awareness and situation-awareness in which it allows dynamic and remotely (re)deploying software program code to the devices based on the situation and context changes [14]. Such a feature shares similarity with fog in providing a platform that allows flexible software deployment and reconfigurations.

Realizing that, the fog requires the support of all the related edge computing technologies. In other words, one is unable to deploy and manage fog without integrating edge computing technologies. Therefore, in the rest of this chapter, we use the term fog and edge computing (FEC) to describe the whole domain.

1.3 Fog and Edge Computing Completing the Cloud

FEC provides a complement to the cloud in IoT by filling the gap between cloud and things toward providing service continuum [3]. This section describes the advantages of FEC and addresses the question of how it achieves these advantages.

1.3.1 Advantages of FEC: SCALE

In particular, FEC offers five main advantages, which can be exemplified by SCALE—security, cognition, agility, latency, and efficiency [8].

1.3.1.1 Security

FEC supports additional security to IoT devices to ensure safety and trustworthiness in transactions. For example, today's wireless sensors deployed in outdoor environments often require a remote wireless source code update in order to resolve the security-related issues. However, due to various dynamic environmental factors such as unstable signal strength, interruptions, constraint bandwidth etc., the distant central backend server may face challenges to perform the update swiftly and, hence, increases the chance of cybersecurity attack. On the other hand, if the FEC infrastructure is available, the backend can configure the best routing path among the entire network via various FEC nodes in order to rapidly perform the software security update to the wireless sensors.

1.3.1.2 Cognition

FEC enables the awareness of the objectives of their clients toward supporting autonomous decision-making in terms of where and when to deploy computing, storage, and control functions. Essentially, the awareness of FEC, which involves a number of mechanisms in terms of self-adaptation, self-organization, self-healing, self-expression, and so forth [16], shifts the role of IoT devices from passive to active smart devices that can continuously operate and react to customer requirements without relying on the decision from the distant Cloud.

1.3.1.3 Agility

FEC enhances the agility of the large scope IoT system deployment. In contrast to the existing utility Cloud service business model, which relies on the large business holder to establish, deploy, and manage the fundamental infrastructure, FEC brings the opportunity to individual and small businesses to participate in providing FEC services using the common open software interfaces or open Software Development Kits (SDKs). For examples, the MEC standard of ETSI and the Indie Fog business model [18] will hasten the deployment of large-scope IoT infrastructures.

1.3.1.4 Latency

The common understanding of FEC is to provide rapid responses for the applications that require ultra-low latency. Specifically, in many ubiquitous applications and industrial automation, the system needs to collect and process the sensory data continuously in the form of the data stream in order to identify any event and to perform timely actions. Explicitly, by applying FEC, these systems are capable of supporting time-sensitive functions. Moreover, the softwarization feature of FEC, in which the behavior of physical devices can be fully configured by the distant central server using software abstraction, provides a highly flexible platform for rapid re-configuration of the IoT devices.

1.3.1.5 Efficiency

FEC enhances the efficiency of CIoT in terms of improving performance and reducing the unnecessary costs. For example, by applying FEC, the ubiquitous healthcare or eldercare system can distribute a number of tasks to the Internet gateway devices of the healthcare sensors and utilize the gateway devices to perform the sensory data analytics tasks. Ideally, since the process happens near the data source, the system can generate the result much faster. Further, since the system utilizes gateway devices to perform most of the tasks, it highly reduces the unnecessary cost of outgoing communication bandwidth.

1.3.2 How FEC Achieves These Advantages: SCANC

The high-level description of the advantages provided by FEC leads to a question: *How does FEC provide these advantages?* To answer the question, here, we describe the five basic mechanisms supported by FEC-enabled devices (FEC node; see Figure 1.2). Specifically, the mechanisms can be termed as SCANC, which corresponds to storage, compute, acceleration, networking, and control.

1.3.2.1 Storage

The mechanism of storage in FEC corresponds to the temporary data storing and caching at the FEC nodes in order to improve the performance of

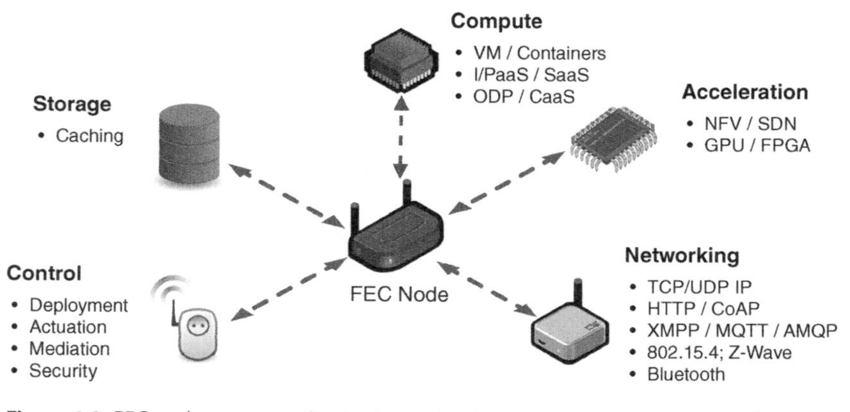

Figure 1.2 FEC nodes supports five basic mechanisms—storage, compute, acceleration, networking, and control.

information or content delivery. For example, content service providers can perform multimedia content caching at the FEC nodes that are most close to their customers in order to improve the quality of experience [19]. Further, in connected vehicle scenarios, the connected vehicles can utilize the roadside FEC nodes to fetch and to share the information collected by the vehicles continuously.

1.3.2.2 Compute

FEC nodes provide the computing mechanisms mainly in two models—infrastructure or platform as a service (I/PaaS) and software as a service (SaaS). In general, FEC providers offer I/PaaS based on two approaches—hypervisor virtual machines (VMs) or containers engines (CEs), which enable flexible platforms for FEC clients to deploy the customized software they need in a sandbox environment hosted in FEC nodes. Besides the I/PaaS, the SaaS is also promising in FEC service provision [3]. To enumerate, SaaS providers can offer two types of services—on-demand data processing (ODP) and context as a service (CaaS). Specifically, an ODP-based service has pre-installed methods that can process the data sent from the client in the request/response manner. Whereas, the CaaS-based service provides a customized data provision method in which the FEC nodes can collect and process the data to generate meaningful information for their clients.

1.3.2.3 Acceleration

FEC provides acceleration with a key concept—programmable. Fundamentally, FEC nodes support acceleration in two aspects—networking acceleration and computing acceleration.

- **Networking acceleration**. Initially, most network operators have their own configuration for message routing paths and their clients are unable to request their own customized routing tables. For example, an Internet service provider (ISP) in East Europe may have two routing paths with different latency to reach a Web server located in Central Europe, and the path a client will be on is based on the ISP's load balancing setting, which in many cases is not the optimal option for the client. On the other hand, FEC supports a network acceleration mechanism based on network virtualization technology, which enables FEC nodes to operate multiple routing tables in parallel and to realize a software-defined network (SDN). Therefore, the clients of the FEC nodes can configure customized routing path for their applications in order to achieve optimal network transmission speed.

- **Computing acceleration**. Researchers in fog computing have envisioned that the FEC nodes will provide computing acceleration by utilizing advanced embedded processing units such as graphics processing units (GPUs) or field programmable gate arrays (FPGA) units [8]. Specifically, utilizing GPUs to enhance the process of complex algorithms has become a common approach in general cloud computing. Therefore, it is foreseeable that FEC providers may also provide the equipment that contains middle- or high-performance independent GPUs. Further, FPGA units allow users to redeploy program codes on them in order to improve or update the functions of the host devices. Particularly, researchers in sensor technologies [20] have been utilizing FPGA for runtime reconfiguration of sensors for quite some time. Further, in comparison with GPUs, FPGA has the potential to be a more energy-efficient approach for providing the needed acceleration based on allowing clients to configure their customized code on the FEC nodes.

1.3.2.4 Networking

Networking of FEC involves vertical and horizontal connectivities. Vertical networking interconnects things and cloud with the IP networks; whereas, horizontal networking can be heterogeneous in network signals and protocols, depending on the supported hardware specification of the FEC nodes.

- **Vertical networking**. FEC nodes enable the vertical network using IP network-based standard protocols such as the request/response-based TCP/UDP sockets, HTTP, Internet Engineering Task Force (IETF) – Constraint Application Protocol (CoAP) or publish-subscribe-based Extensible Messaging and Presence Protocol (XMPP), OASIS – Advanced Message Queuing Protocol (AMQP; ISO/IEC 19464), Message Queue Telemetry Transport (MQTT; ISO/IEC PRF 20922), and so forth. Specifically, the IoT devices can operate server-side functions (e.g. CoAP server) that allow FEC nodes, which act as the proxy of cloud, to collect data from them and then forward the data to the cloud. Further, FEC nodes can also operate as the message broker of publish-subscribe-based protocol that allows the IoT devices to publish data streams to the FEC nodes and enable the cloud backend to subscribe the data streams from the FEC nodes.

- **Horizontal networking.** Based on various optimization requirements such as energy efficiency or the network transmission efficiency, IoT systems are often using heterogeneous cost-efficient networking approaches. In particular, smart home, smart factories, and connected vehicles are commonly utilizing Bluetooth, ZigBee (based on IEEE 802.15.4), and Z-Wave on the IoT devices and connecting them to an IP network gateway toward enabling the connectivity between the devices and the backend cloud. In general, the IP network gateway devices are the ideal entities to host FEC servers since they have the connectivity with the IoT devices in various signals. For example, the cloud can request that an FEC server hosted in a connected car communicate with the roadside IoT equipment using ZigBee in order to collect the environmental information needed for analyzing the real-time traffic situation.

1.3.2.5 Control

The control mechanism supported by FEC consists of four basic types – deployment, actuation, mediation, and security:

1. **Deployment control** allows clients to perform customizable software program deployment dynamically. Further, clients can configure FEC nodes to control which program the FEC node should execute and when it should execute it. Further, FEC providers can also provide a complete FEC network topology as a service that allows clients to move their program from one FEC node to another. Moreover, the clients may also control multiple FEC nodes to achieve the optimal performance for their applications.
2. **Actuation control** represents the mechanism supported by the hardware specification and the connectivities between the FEC nodes and the connected devices. Specifically, instead of performing direct interaction between the cloud and the devices, the cloud can delegate certain decisions to FEC nodes to directly control the behavior of IoT devices.
3. **Mediation control** corresponds to the capability of FEC in terms of interacting with external entities owned by different parties. In particular, the connected vehicles supported by different service providers can communicate with one another, though they may not have a common protocol initially. With the softwarization feature of FEC node, the vehicles can have on-demand software update toward enhancing their interoperability.
4. **Security control** is the basic requirement of FEC nodes that allows clients to control the authentication, authorization, identity, and protection of the virtualized runtime environment operated on the FEC nodes.

1.4 Hierarchy of Fog and Edge Computing

In general, from the perspective of central cloud in the core network, CIoT systems can deploy FEC servers at three edge layers – inner-edge, middle-edge, and outer-edge (see Figure 1.3). Here, we summarize the characteristics of each layer.

1.4.1 Inner-Edge

Inner-edge (also known as near-the-edge [4]) corresponds to countrywide, statewide, and regional WAN of enterprises, ISPs, the data center of evolved packet core (EPC) and metropolitan area network (MAN). Initially, service providers at inner-edge only offer the fundamental infrastructures for connecting local networks to the global Internet. However, the recent needs in improving the quality of experience (QoE) of Web services have motivated the geo-distributed caching and processing mechanism at the network data centers of WAN. For example, in the commercial service aspect, Google Edge Network (peering.google.com) collaborates with ISPs to distribute data servers at the ISPs' data centers in order to improve the response speed of Google's cloud services. Further, many ISPs (e.g. AT&T, Telstra, Vodafone, Deutsche Telekom etc.) are aware that many local businesses require low latency cloud and hence, they have offered local cloud within the country. Based on the reference architecture of fog computing [8], the WAN-based cloud data centers can be considered as the fog of inner-edge.

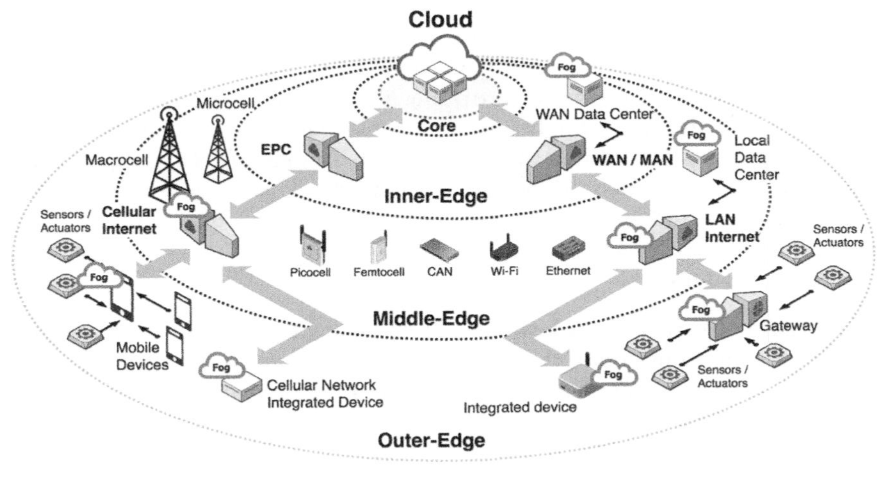

Figure 1.3 Hierarchy of fog and edge computing.

1.4.2 Middle-Edge

Middle-edge corresponds to the environment of the most common understanding of FEC, which consists of two types of networks—local area network (LANs) and cellular network. To summarize, LANs include ethernet, wireless LANs (WLANs) and campus area network (CANs). The cellular network consists of the macrocell, microcell, picocell, and femtocell. Explicitly, middle-edge covers a broad range of equipment to host FEC servers.

1.4.2.1 Local Area Network

The emerging Fog computing architecture introduced by Cisco's research [7] was utilizing Internet gateway devices (e.g. Cisco IR829 Industrial Integrated Router) to provide the similar model as utility Cloud services in which the gateway devices provide virtualization technologies that allow the gateway devices to support FEC mechanisms mentioned previously. Further, it is also an ideal solution to utilize the virtualization technology-enabled server computers located within the same subnet of LAN or CAN (i.e. within the one-hop range between the IoT device and the computer) with the FEC nodes. Ordinarily, such an approach is also known as local cloud, local data center or cloudlet.

1.4.2.2 Cellular Network

The idea of providing FEC mechanisms derived from the existing network virtualization technologies that have been used in various cellular networks. In general, most developed cities have wide coverage of cellular networks provided by numerous types of BTSs, which are the ideal facilities to serve as roadside FEC hosts for various mobile IoT use cases such as connected vehicles, mobile healthcare, and virtual or augmented reality, which require rapid process and response on the real-time data stream. Therefore, major telecommunication infrastructure and equipment providers such as Nokia, ADLink, or Huawei have started providing MEC-enabled hardware and infrastructure solutions. Accordingly, it is foreseeable that in the near future, cellular network-based FEC will be available in a broad range of related equipment, from macrocell and microcell BTSs to the indoor cellular extension equipment such as picocell and femtocell [21] base stations.

1.4.3 Outer-Edge

Outer-edge, which is also known as extreme-edge, far-edge, or mist [14–16], represents the front-end of the IoT network, which consists of three types of devices—constraint devices, integrated devices, and IP gateway devices.

1.4.3.1 Constraint Devices

Constraint devices such as sensors or actuators are usually operated by microcontrollers that have the very limited processing power and memory.

For example, Atmel ATmega328 single-chip microcontroller, which is the CPU of Arduino Uno Rev3, has only 20 MHz processing power and 32kB flash memory. Commonly, IoT administrators would not expect to deploy complex tasks to this type of devices. However, due to the field-programmable ability of today's wireless sensors and actuators, the IoT system can always update or reconfigure the program code of the devices dynamically and remotely. Explicitly, such a mechanism grants the constraint IoT devices with self-awareness feature and motivated the mist-computing discipline [14], which emphasizes the abilities of IoT devices in self-management of the interaction and collaboration among IoT devices themselves toward achieving a highly autonomous Machine-to-Machine (M2M) environment without relying on the distant Cloud for all their activities.

1.4.3.2 Integrated Devices

These devices are operated by processors that have decent processing power. Further, integrated devices have many embedded capabilities in networking (e.g. Wi-Fi and Bluetooth connectivities), embedded sensors (e.g. gyroscope, accelerator), and decent storage memory. Typically, Acorn RISC Machine (ARM), CPU-based smartphones, and tablets (e.g. Android OS, iOS devices) are the most cost-efficient commercial products of integrated devices. They can perform sensing tasks and can also interact with the cloud via the middle-edge facilities. Although the integrated devices may have constraints in the OS environment that reduce the flexibility of deploying a virtualization platform on them, considering how swiftly ARM CPUs and embedded sensors in the integrated devices are evolving, it is foreseeable that in the near future, virtualization-based FEC will be available on integrated devices. Overall, at this stage, a few platforms such as Apache Edgent (edgent.apache.org) or Termux (termux.com) are promising approaches toward realizing FEC on the integrated devices.

1.4.3.3 IP Gateway Devices

Hubs, or IP gateway devices, act as the mediator between the constrained devices and the middle-edge devices. Commonly, because of the need for energy-efficient wireless communication, many constraint devices do not operate in IP network, which usually requires energy-intensive Wi-Fi (e.g. IEEE 802.11g/n/ac). Instead, the constraint devices are communicated using the protocols that consume less energy, such as Bluetooth Low Energy, IEEE 802.15.4 (e.g. ZigBee), or Z-Wave. Furthermore, since the low-energy communication protocols do not directly connect with the IP network, the system would use IP gateway devices to relay the communication messages between the constraint devices and the Internet gateway (e.g. routers). Hence, the backend cloud is capable of interacting with the frontend constraint devices. In general, the Linux OS-based IP gateway devices such as Prota's hub (prota.info), Raspberry Pi, or ASUS Tinker Board can easily host virtualization environment such as Docker Containers Engine. Hence, it is common to see that research projects [22–24] have been utilizing IP gateway devices as FEC nodes.

1.5 Business Models

While most discussions of FEC focus on the advantages and applications, a fundamental question remains to be explored regarding what the business models of FEC will look like. Here, we discuss the three basic business models derived from the recent works [3, 10, 18].

1.5.1 X as a Service

Here, the *X* of the X as a service (XaaS) corresponds to infrastructure, platform, software, networking, cache or storage, and many other types of resources mentioned in general cloud services. Specifically, XaaS providers of FEC allow their clients to pay to use the hardware equipment that supports SCANC mechanisms described in the previous section. Further, XaaS model is not limited to major business providers such as ISPs or the large cloud providers. Ideally, individuals and small businesses can also provide XaaS in the form of IndieFog [18] that is based on the popular consumer as provider (CaP) service provisioning model in multiple domains.

For example, the MQL5 Cloud Network distributed computing project (cloud.mql5.com) utilizes customer-premises equipment (CPE) to perform various distributed computing tasks. Further, Fon (fon.com) utilizes CPE to establish a global Wi-Fi network. These examples indicate that many individuals are willing to let application service providers pay to use their equipment for offering services.

1.5.2 Support Service

The support service of FEC is similar to the software management support services in general information systems in which the clients who own the hardware equipment can pay the support service provider to provide them with the corresponding software installation, configuration, and updates on the clients' equipment based on their requirements. Further, the clients may also pay the provider for monthly or annual support services for maintenance and technical support. In general, support service providers offer their clients highly customized solutions to achieve the optimal operation of their FEC-integrated systems.

A typical example of the support service provider is how Cisco provides the fog computing solution, in which the clients purchase Cisco's IOX-enabled equipment and then pay an additional service fee to gain access to the software update and technical support for configuring their FEC environments. It is foreseeable that in the near future, such a model will not be constrained to the single provider's hardware and software. The support service provider will be decoupled from the hardware equipment vendors, just as today's enterprise information systems support service providers such as RedHat, IBM, or Microsoft.

1.5.3 Application Service

Application service providers provide application solutions to help their clients in processing the data within or outside of the client's operation environments. For example, the recent Digital Twinning technologies create a real-time virtualized twin that clones the real-world behavior of a broad range of physical entities, from industrial facilities, equipment to the entire factory plane and the involved production lane and supply chains. Explicitly, such technology can provide insight into the efficiency and performance toward optimizing and improving the industrial activities. Accordingly, an FEC application service provider can provide the Digital Twinning solution configured across all the involved entities at the edge networks in order to provide the analysis in an ultra-low latency manner (less than tens of milliseconds) toward helping the industrial system with rapid reactions. Similarly, the FEC application service providers can also assist local government in real-time traffic control systems that facilitate the self-driving, connected vehicles. Further, IndieFog providers can also offer various application services such as analytics useful to ambient assisted living (AAL) service providers. For example, an IndieFog provider who has installed Apache Edgent can offer the built-in stream data classification function as an application service for the mobile AAL clients in the close proximity.

1.6 Opportunities and Challenges

Additional opportunities and challenges concern the out-of-box experience, open platforms, and system management. This section discusses these issues.

1.6.1 Out-of-Box Experience

Industrial marketing research forecasts that the market value of FEC hardware components will reach $7,659 million by the year 2022 [10], which indicates that more FEC-ready equipment such as routers, switches, IP gateway, or hubs will be available in the market. Further, it is foreseeable that many of these products will feature the out-of-box experience (OOBE) in two forms—OOBE-based equipment and OOBE-based software.

1.6.1.1 OOBE-Based Equipment

OOBE-based equipment represents that the product vendors have integrated the FEC runtime platform with their products such as routers, switches, or other gateway devices in which the consumers who purposed the equipment can easily configure and deploy FEC applications on the equipment via certain user interfaces, which is similar to the commercial router products that have graphical user interfaces for users to configure customized settings.

1.6.1.2 OOBE-Based Software

This is similar to the experience of Microsoft Windows in which the users who own FEC-compatible devices can purchase and install OOBE-based FEC software on their devices toward enabling FEC runtime environment and the SCANC mechanisms without any extra low-level configuration.

The OOBE-based FEC faces challenges in defining standardization for software and hardware. First, OOBE-based equipment raises a question for the vendors as to what FEC platform and related software packages should be included in their products. Second, OOBE-based software raises a question for the vendors regarding compatibility. Specifically, users may have devices in heterogeneous specification and processing units (e.g. x86, ARM etc.) in which the vendor may need to provide a version for each type of hardware. Moreover, developing and maintaining such an OOBE-based software can be extremely costly unless a corresponding common specification or standard for hardware exists.

1.6.2 Open Platforms

At this stage, besides the commercial platforms such as Cisco IOX for fog computing, there are a few open platforms for supporting FEC. However, most of the platforms are in the early stage and they have limited support in deployment. Below, we summarize the characteristics of each platform.

1.6.2.1 OpenStack++

OpenStack++ [25] is a framework developed by Carnegie Mellon University Pittsburgh for providing VM-based cloudlet platform on regular x86 computers for mobile application offloading. Explicitly, since the recent trend intends to apply lightweight virtualization technology-based FEC, OpenStack++ is less applicable to most use cases such as hosting FEC servers on routers or hubs. Further, it indicates that the virtualization technology used in FEC is focused more on containerization such as the Docker Containers Engine.

1.6.2.2 WSO2–IoT Server

Available at (wso2.com/iot, the WSO2–IoT server is an extension of the popular open-source enterprise service-oriented integration platform WSO2 server that consists of certain IoT-related mechanisms, such as connecting a broad range of common IoT devices (e.g. Arduino Uno, Raspberry Pi, Android OS devices, iOS devices, Windows 10 IoT Core devices etc.) with the cloud using standard protocols such as MQTT and XMPP. Further, WSO2–IoT server includes the embedded Siddhi 3.0 component that allows the system to deploy real-time streaming processes in embedded devices. In other words, WSO2–IoT server provides the FEC computing capability to outer-edge devices.

1.6.2.3 Apache Edgent

See edgent.apache.org. Formerly known as Quarks, Apache Edgent is an open-source runtime platform contributed by IBM. Generally, the platform provides distributed stream data processing between cloud and edge devices. Specifically, the cloud-side supports most major open platforms in the stream data processing field such as Apache Spark, Apache Storm, Apache Flink, and so forth. Further, at the outer-edge, Edgent supports common open operating systems such as Linux and Android OS. In summary, by utilizing Edgent, a system can dynamically migrate the stream data processing between the cloud and edge, which ideally fulfils the need in most use cases that involve edge analytics.

Current open platforms lack capability in deploying and managing FEC across all the hierarchy layers of edge networks. However, it is likely due to the inflexibility of existing commercial devices in supporting the need for FEC mechanisms configuration. On the other hand, it also indicates an opportunity for product vendors to provide the enhanced devices that support FEC.

1.6.3 System Management

Management of FEC involves the three basic life cycle phases: design, implementation, and adjustment.

1.6.3.1 Design

The system administration team needs to identify the ideal location among the three edge tiers (i.e. inner-edge, middle-edge, outer-edge) for placing FEC servers [3]. Further, the administration team needs to develop or apply an ideal abstract modeling approach that can describe what types of resources the FEC servers are required with and how the FEC servers can interact with the system.

1.6.3.2 Implementation

The administration team needs to consider the heterogeneity of FEC environments, especially at the middle-edge and outer-edge where the nodes may have various hardware specification, communication protocols, and operating systems. Specifically, existing FEC equipment vendors (e.g. Cisco or Dell) may provide the isolated platform, which leads the implementation complex since the developers need to implement their FEC for each platform. Although there are a number of ongoing industrial-led open platform projects for FEC, the dependency requirement of each platform can still lead to a significant cost of time in the implementation.

1.6.3.3 Adjustment

The FEC system needs to support runtime adjustment in which the system can schedule where and when to activate FEC functions in order to optimize the overall processes. For example, the system should have capabilities to dynamically deploy/terminate the runtime environment (e.g. VM or containers) and application methods on a feasible FEC node. Further, the system should be able to dynamically move the runtime environment or application methods from one FEC node to another based on the runtime context factors. Commonly, the required capabilities of adjustment phase raise challenges in how to support the reliability of software migration among FEC nodes and how to minimize the latency caused by such activities. In particular, dynamic code deployment and reconfiguration at the outdoor-based far-edge is highly challenging in terms of latency and reliability because of the dynamic nature of wireless and mobile communication in which the signal interruption can cause the failure of code deployment [16].

1.7 Conclusions

Fog and edge computing (FEC) enhance the cloud-centric Internet of Things (CIoT) by extending the cloud computing model to the edge networks of IoT where the network intermediate nodes such as routers, switches, hubs, and IoT devices are participating with the information-processing and decision-making toward improving security, cognition, agility, latency, and efficiency (abbreviated by SCALE).

This chapter provides an introductory overview of the state-of-the-art in FEC in terms of technical background, characteristics, deployment environment hierarchy, business models, opportunities, and open challenges. Specifically, we have described the five fundamental advantages of FEC—SCALE, which is realized by the five mechanisms of FEC nodes—storage, compute, acceleration, networking, and control (SCANC). Further, to clarify the resource availability and their capabilities, this chapter has explained the three layers of FEC environment from the perspective of the central cloud in the core network. To enumerate, it consists of inner-edge with WAN providers, middle-edge with LAN and the frontline cellular networks, and outer-edge where the hubs and IoT devices are located.

The capabilities of FEC will enable three types of business models known as X as a Service (XaaS), support service, and application service. In summary, XaaS corresponds to the model that provides IaaS, PaaS, SaaS and S/CaaS (storage or caching as a service), which are similar to existing cloud service models; support service corresponds to FEC software installation, configuration, and maintenance service that helps clients to set up their FEC on their own equipment; application service denotes the service providers that cater the complete solution that serves FEC mechanism to clients without them needing to configure their own FEC system.

FEC brings new opportunities and also raises new challenges in development and operation. Specifically, development faces challenges in complexity and standardization, which potentially leads to the difficulty in system integration across different FEC providers and IoT end-points. Further, the operation challenge derives from the management cycle of FEC in terms of design, implementation, and adjustment. Explicitly, the heterogeneous network and entities involved in FEC led to more complicated challenges than the core network Internet-based cloud. On the other hand, the industry is aware of the challenges and has started a number of open platforms such as WSO2–IoT, Apache Edgent. Further, the recent Linux Foundation Project—EdgeX Foundry (edgexfoundry.org), which aims to provide a complete Software Development Kit for FEC, has shown that industrial interest in IoT is no longer satisfied with the connectivity between devices and cloud. Instead, the trend has moved from connected things to cognitive things in which the processes and decisions are performed as close to the physical objects as possible, even to the IoT devices themselves.

References

1 J. Gubbi, R. Buyya, S. Marusic and M. Palaniswami. Internet of Things (IoT): A vision, architectural elements, and future directions. *Future Generation Computer Systems*, 29(7): 1645–1660, 2013.

2 C. Chang, S.N. Srirama, and R. Buyya. Mobile cloud business process management system for the Internet of Things: A survey. *ACM Computing Surveys*, 49(4): 70:1–70:42, December 2016.

3 M. Chiang and T. Zhang, Fog and IoT: An overview of research opportunities. *IEEE Internet of Things Journal*, 3(6): 854–864, 2016.

4 H.P. Sajjad, K. Danniswara, A. Al-Shishtawy and V. Vlassov. SpanEdge: Towards unifying stream processing over central and near-the-edge data centers. In *Proceedings of the IEEE/ACM Symposium on Edge Computing (SEC)*, pp. 168–178, IEEE, 2016.

5 M. Satyanarayanan, P. Bahl, R. Caceres and N. Davies. The Case for VM-Based Cloudlets in Mobile Computing, *IEEE Pervasive Computing*, 8(4): 14–23, 2009.

6 S.W. Loke, K. Napier, A. Alali, N. Fernando and W. Rahayu. Mobile computations with surrounding devices: Proximity sensing and multilayered work stealing. *ACM Transactions on Embedded Computing Systems (TECS),* 14(2): 22:1–22:25, February 2015.

7 F. Bonomi, R. Milito, J. Zhu, and S. Addepalli. Fog computing and its role in the Internet of Things. In *Proceedings of the First Edition of the MCC Workshop on Mobile Cloud Computing*, pp. 13–16, ACM, August 2012.

8 OpenFog Consortium. OpenFog Reference Architecture for Fog Computing. *Technical Report,* February 2017.

9 IEEE Standard Association. FOG – Fog Computing and Networking Architecture Framework, [Online] http://standards.ieee.org/develop/wg/FOG.html. Accessed: 2 April 2018.

10 451 Research. Size and impact of fog computing market. The 451 Group, USA, October 2017. [Online] https://www.openfogconsortium.org/wp-content/uploads/451-Research-report-on-5-year-Market-Sizing-of-Fog-Oct-2017.pdf. Accessed: 2 April 2018.

11 H. Pang and K.L. Tan. Authenticating query results in edge computing. In *Proceedings of the 20th International Conference on Data Engineering,* pp. 560–571, IEEE, March 2004.

12 M. Satyanarayanan. The Emergence of Edge Computing *Computer,* 50(1): 30–39, 2017.

13 OpenFog News. New IEEE working group is formed to create fog computing and networking standards [Online]. https://www.openfogconsortium.org/news/new-ieee-working-group-is-formed-to-create-fog-computing-and-networking-standards/. Accessed: 2 April 2018.

14 J.S. Preden, K. Tammemae, A. Jantsch, M. Leier, A. Riid, and E. Calis. The benefits of self-awareness and attention in fog and mist computing. *Computer,* 48(7): 37–45, 2015.

15 M. Liyanage, C. Chang, and S. N. Srirama. mePaaS: Mobile-embedded platform as a service for distributing fog computing to edge nodes. In *Proceedings of the 17th International Conference on Parallel and Distributed Computing, Applications and Technologies (PDCAT-16),* pp. 73–80, Guangzhou, China, December 16–18, 2016.

16 K. Tammemäe, A. Jantsch, A. Kuusik, J.-S. Preden, and E. Õunapuu. Self-aware fog computing in private and secure spheres. *Fog Computing in the Internet of Things,* pp. 71–99, Springer International Publishing, 2018.

17 S. N. Srirama, M. Jarke, and W. Prinz, Mobile web service provisioning. In *Proceedings of the Advanced International Conference on Telecommunications and International Conference on Internet and Web Applications and Services (AICT-ICIW'06),* pp. 120–120. IEEE, 2006.

18 C. Chang, S.N. Srirama, and R. Buyya. Indie fog: An efficient fog-computing infrastructure for the Internet of Things. *IEEE Computer,* 50(9): 92–98, September 2017.

19 A.S. Gomes, B. Sousa, D. Palma, V. Fonseca, Z. Zhao, E. Monteiro, T. Braun, P. Simoes, and L. Cordeiro. Edge caching with mobility prediction in virtualized LTE mobile networks. *Future Generation Computer Systems,* 70: 148–162, May 2017.

20 Y.E. Krasteva, J. Portilla, E. de la Torre, and T. Riesgo. Embedded runtime reconfigurable nodes for wireless sensor networks applications. *IEEE Sensors Journal,* 11(9): 1800–1810, 2011.

21 D. Lopez-Perez, I. Guvenc, G. de la Roche, M. Kountouris, T.Q. Quek, and J. Zhang. Enhanced intercell interference coordination challenges in heterogeneous networks. *IEEE Wireless Communications,* 18(3): 22–30, 2011.

22 W. Hajji and F.P. Tso. Understanding the performance of low power Raspberry Pi cloud for big data. *MDPI Electronics,* 5(2): 29:1–29:14, 2016.

23 A. Van Kempen, T. Crivat, B. Trubert, D. Roy, and G. Pierre. MEC-ConPaaS: An experimental single-board based mobile edge cloud. In *Proceedings of the 5th IEEE International Conference on Mobile Cloud Computing, Services, and Engineering (MobileCloud),* pp. 17–24, 2017.

24 R. Morabito. Virtualization on Internet of Things edge devices with container technologies: a performance evaluation. *IEEE Access,* 5(0): 8835–8850, 2017.

25 K. Ha and M. Satyanarayanan. Openstack++ for Cloudlet Deployment. *Technical Report CMU-CS-15-123,* School of Computer Science, Carnegie Mellon University, Pittsburgh, USA, 2015.

2

Addressing the Challenges in Federating Edge Resources

Ahmet Cihat Baktir, Cagatay Sonmez, Cem Ersoy, Atay Ozgovde, and Blesson Varghese

2.1 Introduction

Edge computing is rapidly evolving to alleviate latency, bandwidth, and quality-of-service (QoS) concerns of cloud-based applications as billions of 'things' are integrated to the Internet [1]. Current research has primarily focused on decentralizing resources away from centralized cloud data centers to the edge of the network and making use of them for improving application performance. Typically, edge resources are configured in an ad hoc manner and an application or a collection of applications may privately make use of them. These resources are not publicly available, for example, like cloud resources. Additionally, edge resources are not evenly distributed but are sporadic in their geographic distribution.

However, ad hoc, private, and sporadic edge deployments are less useful in transforming the global Internet. The benefits of using the edge should be equally accessible to both the developing and developed world for ensuring computational fairness and for connecting billions of devices to the Internet. However, there is minimal discourse on how edge deployments can be brought to bear in a global context – federating them across multiple geographic regions to create a global edge-based fabric that decentralizes data center computation. This, of course, is currently impractical, not only because of technical challenges but also because it is shrouded by social, legal, and geopolitical issues. In this chapter, we discuss two key challenges – networking and management in federating edge deployments, as shown in Figure 2.1. Additionally, we consider resource and modeling challenges that will need to be addressed for a federated edge.

The key question we will be asking for addressing the networking challenge is, "How can we create a dynamic enough networking environment that is compatible with the foreseen edge computing scenarios in a federated setting?" [2]. This is already a difficult issue for standalone and/or small-scale

Fog and Edge Computing: Principles and Paradigms, First Edition.
Edited by Rajkumar Buyya and Satish Narayana Srirama.

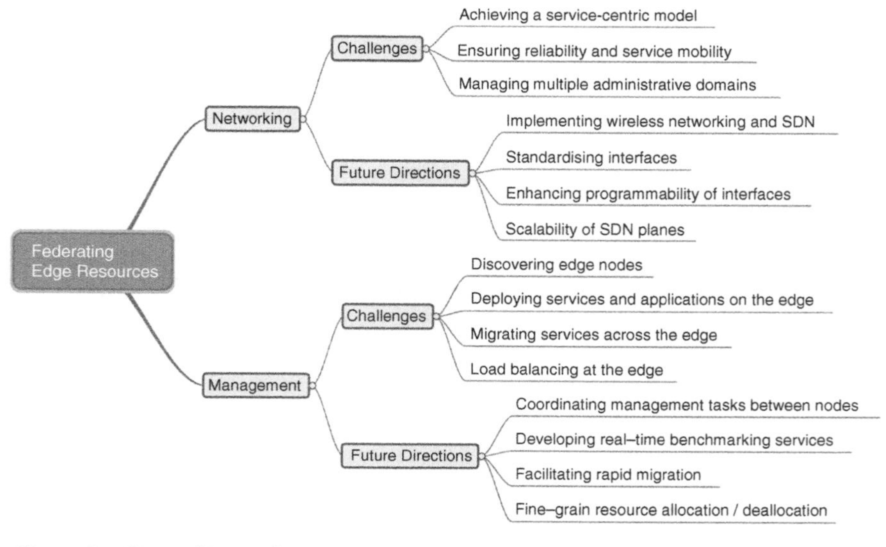

Figure 2.1 Networking and management challenges in federating edge resources.

edge deployments and requires further consideration in a federated setting. The dynamicity required is provided by the programmability of networking resources available through software-defined networking (SDN) in today's context [3, 4]. SDN with its northbound programming interface is an ideal candidate for the orchestration of edge computing resources [5]. In the federated edge context, however, a global coordination within SDN administrative domains would be needed. A harmony between local edge deployments and the federated infrastructure is critical since both views of the system will be based on the same networking resources, possibly from competing perspectives. This will likely require a complete rethinking of the networking model for the edge and further effort on the east–west interface of SDN. This chapter will discuss potential avenues for resolving the networking challenges.

As in any large-scale computing infrastructure, addressing management challenges becomes pivotal in offering seamless services. Currently, edge-based deployments assume that services running on an edge node either can be cloned or will be available on an alternate edge node [6]. While this is a reasonable assumption for developing research in its infancy, it becomes a key challenge when federating edge resources. In this context, future Internet architectures will need to consider how services can be rapidly migrated from one node to another based on demand [7]. Current technologies have limited scope in realizing this because of the high overhead and the lack of suitability for resource-constrained environments, such as the edge. We will provide a discussion on management issues – benchmarking, provisioning, discovery, scaling, and migration – and provide research directions to address these [8, 9].

Additionally, we present resource and modeling challenges in this area. The resource challenge is related to both the hardware and software levels of resources that are employed at the edge [10–13]. Although there are a number of reference architectures available for edge-based systems, we have yet to see a practical implementation of these systems. Often, the hardware solutions are bespoke to specific applications and bring about heterogeneity at a significant scale that is hard to bridge. On the other hand, the software solutions that offer abstraction for making edge resources publicly available have large overheads since they were designed for cloud data centers.

The final consideration is the modeling challenge. Integrating the edge into the cloud-computing ecosystem brings about a radical change in the Internet architecture. It would be practically impossible to investigate and know the implications of large-scale edge deployments, both from a technological and socioeconomic perspective. A number of these can be modeled in simulators that offer the advantage of repeatability of experiments, minimizing hardware costs on experimental testbeds, and testing in controlled environments [14]. However, our current understanding of interactions among users, edge nodes, and the cloud is limited.

Throughout this chapter, we use the general term *edge* to refer to a collection of technologies that aim toward decentralizing data center resources for bringing computational resources closer to the end user. Mobile cloud computing (MCC), cloudlets, fog computing, and multi-access edge computing (MEC) can all be considered as instances of edge computing [9]. Therefore, generally speaking, the principles discussed in this chapter can be broadly applied to the aforementioned technologies.

The remainder of this chapter is organized in line with the above discussion and is as follows. Section 2.2 considers the networking challenge. Section 2.3 considers the management challenge. Section 2.4 presents resource and modeling challenges. Section 2.5 concludes this chapter.

2.2 The Networking Challenge

The networking environment in which edge servers will facilitate distributed computing is likely to be dynamic. This is because of the constantly varying demands at the end-user level. The network infrastructure will need to ensure that the QoS of deployed applications and services are not affected [15]. For this, the quality of user experience cannot be compromised and the coordination of activities to facilitate edge computing must be seamless and hidden from the end user [16].

Table 2.1 Network challenges, their causes and potential solutions in federating edge resources.

Networking challenge	Why does it occur?	What is required?
User mobility	Keeping track of different mobility patterns	Mechanisms for application layer handover
QoS in a dynamic environment	Latency-intolerant services, dynamic state of the network	Reactive behavior of the network
Achieving a service-centric model	Enormous number of services with replications	Network mechanisms focusing on "what" instead of "where"
Ensuring reliability and service mobility	Devices and nodes joining the network (or leaving)	Frequent topology update, monitoring the servers and services
Managing multiple administrative domains	Heterogeneity, separate internal operations and characteristics, different service providers	Logically centralized, physically distributed control plane, vendor independency, global synchronization

Table 2.1 summarizes the network challenges we consider in this chapter. The general networking challenge is in coping with the highly dynamic environment that the edge is anticipated to be. This directly affects, for example, user mobility. As computational resources are placed closer to the source of the traffic, services become contextual. This results in the need for handling application layer handovers from one edge node to another [17]. Depending on where the users are located and how request patterns are formed, the location of a service may change at any time. Another challenge is related to maintaining QoS in a dynamically changing environment.

2.2.1 Networking Challenges in a Federated Edge Environment

Federating edge resources brings about a larger number of networking challenges related to scalability. For example, global synchronization between different administrative domains will need to be maintained in a federated edge. Individual edge deployments will have different characteristics, such as the number of services hosted and end users in its coverage. In a federated context, different service offloads from multiple domains will need to be possible and will require synchronization across the federated deployments. In this section, we consider three challenges that will need to be addressed. We assume that edge computing applications will be shaped by novel traffic characteristics that will leverage edge resources, possibly from different service providers.

2.2.1.1 A Service-Centric Model

The first challenge is in *achieving a service-centric model* on the edge. The traditional host-centric model follows the 'server in a given geographic location' model, which is restrictive in a number of ways. For example, simply transferring a virtual machine (VM) image from one location to another can be difficult. However, in global edge deployments, the focus needs to be on 'what' rather than 'where' so that services can be requested without prior knowledge of its geographic location [18]. In this model, services may have a unique identifier, may be replicated in multiple regions, and may be coordinated. However, this is not a trivial task, given the current design of the Internet and protocol stacks, which do not facilitate global coordination of services.

2.2.1.2 Reliability and Service Mobility

The second challenge is in *ensuring reliability and service mobility*. User devices and edge nodes may connect and disconnect from the Internet instantly. This could potentially result in an unreliable environment. A casual end-user device will be expecting seamless service perhaps via a plug-and-play functionality to obtain services from the edge, but an unreliable network could result in latencies. The challenge here will be to mitigate this and create a reliable environment that supports the edge. One mechanism to implement reliability is by either replicating services or by facilitating migration (considered in the management challenge) of services from one node to another. The key challenge here is to keep the overheads to a minimum so that the QoS of an application is not affected in any way.

2.2.1.3 Multiple Administrative Domains

The third challenge is in *managing multiple administrative domains*. The network infrastructure will need to be able to keep track of recent status of the network, edge servers and services deployed over them. When a collection of end-user devices requires a service at the edge, first the potential edge host will need to be determined. The most feasible edge node will then be chosen as the resource for the execution.

There are two alternate scenarios that need to be considered for this operation: (i) the server is nearest to the end-users; or (ii) the potential server resides in another geographic region. Independent of the scenario, the network should forward the request to the server and return the response to the end user. During this progress, the data packets may travel across several distinct domains with multiple transport technologies. The challenge here is, given this heterogeneity, user experience must not be compromised and the technical details may need to be concealed from the user device.

Addressing the above challenges requires a solution that inherits the characteristics of both a centralized and distributed system. In order to achieve a global view of the network and maintain synchronization across separate administrative domains, the network orchestrator will need to follow a centralized structure. However, the control operations for coordinating the internal operations of a private domain will need to be distributed. In other words, the control of the network should be distributed over the network but should be placed within a logically centralized context.

2.2.2 Addressing the Networking Challenge

We propose SDN as a solution for addressing the networking challenges, as it naturally lends itself to handling them [5]. The key concept of SDN is to separate the control plane from the data plane and concentrate the core logic on a software-based controller [19]. The controller maintains the general view of the underlying network resources through its logically centralized structure [20]. This simplifies the management of the network, enhances the capabilities of the resources, and lowers the complexity barriers by utilizing resources more efficiently [21, 22]. Most importantly, SDN facilitates instantaneous decision-making in a dynamic environment by monitoring the status of the network at any time.

The control plane communicates with the underlying network nodes through the OpenFlow protocol [23], which is considered as de-facto standard for the southbound interface of the SDN. On the other hand, the applications that define the behavior of the network communicate with the controller through the northbound interface, although it still remains to be standardized [24, 25].

The programmable control plane can be either centralized or distributed physically. The initial proposal of SDN and OpenFlow considers a campus environment, and the design criteria were based on a single controller assuming that the control channel can handle the typical area of coverage. However, the novel edge computing scenarios demand more than this. In order to make edge deployments publicly accessible and to construct a global pool of computing resources at the edge, the control plane should be distributed to enable the orchestration with multiple control instances. A typical SDN-orchestrated edge computing environment is depicted in Figure 2.2, where the network devices are aligned with the SDN controller and related northbound applications.

The logically centralized scheme of the control plane is a key feature in managing user mobility by simplifying the management of the connected devices and resources [26]. When a new device is connected to the network or authenticated to another network due to mobility, the network should react as soon as possible and provide the plug-and-play functionality. This capability is granted to the SDN controller with a functionality of topology discovery through the OpenFlow Discovery (OFDP) protocol [27]. As soon as the state of an end user changes, the controller immediately updates the corresponding flow rules. Through a module implemented as a northbound application,

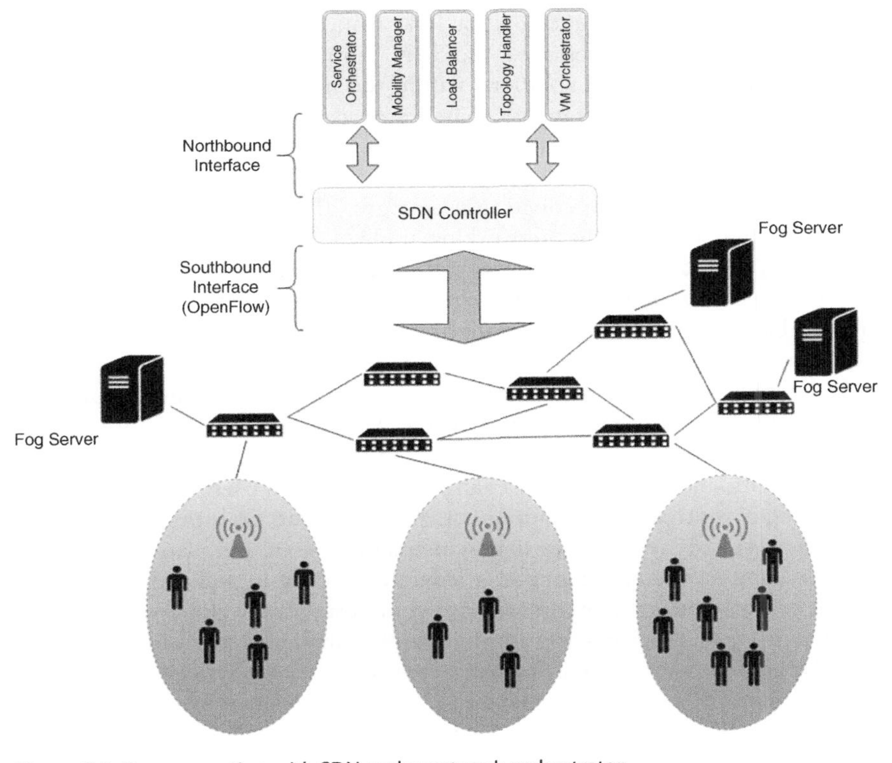

Figure 2.2 Fog computing with SDN as the network orchestrator.

the topology is checked frequently and any newly added, disconnected, or modified node can be updated on the topology view. The opportunity here is that the node can be an end-user device, a computational resource, or a switch. The controller can handle the integration of each type of component while updating the topology view. This approach also enables the application layer handover, which is triggered by the mobility during service offloading.

The utilization of OpenFlow-based switches, such as OpenvSwitch [28], and SDN controller as the umbrella of the whole system enhances the effectiveness of the control through a better resolution of management [29]. Considering the northbound applications that describe the behavior of the network through user-defined policies, the network could be reactive or proactive. For instance, in a university campus environment, students and university staff are always in motion. The traffic flow increases during the daytime and decreases after working hours. In this single administrative domain, where mobility is high, the reactive operations come into prominence. Through the ability of gathering statistical information, such as the traffic load forwarded by a certain node or link, from the data plane elements by exchanging OpenFlow messages, the SDN controller may dictate flow rules that lead to near optimal solutions within the network. In the case of edge computing, where multi-tenants share resources and application instances enforce strict QoS criteria in terms of latency, the SDN controller can modify the flow rules at the edge of the network if an edge server becomes highly loaded or a probability of congestion emerges. The SDN controller not only monitors the status of the networking nodes and links but it can also be integrated with a server monitor functionality through a northbound application. Therefore, one can define a customized policy that is able to provide a load-balancing algorithm considering both computing and networking resources.

Federating edge resources create a globally accessible infrastructure, but also makes the environment more dynamic. Managing mobility within a single domain, handling the application layer handover among the servers in the same vicinity, and reacting to the changes due to a set of users can be leveraged by a single control plane component. However, the realistic and practical approaches of edge computing deployments necessitate a network behavior to flexibly support the operations of a federated setting in a global context. As might be expected, a single control plane cannot satisfy the global management of various types of devices and administrative domains. The evolution of SDN and OpenFlow allows for a logically centralized but physically distributed control plane. The data traffic may be forwarded through at least two different domains that belong to separate service providers. Therefore, there is a need for abstraction and control over the disjoint domains with multiple controllers.

A controller can be deployed for handling the operations within a single domain. However, there will be the need for interdomain or inter-controller communication to maintain reliability in forwarding traffic to the gateways. This communication is provided by the east–west interface. The control plane can be organized as either hierarchical or flat. In the hierarchical structure, a master plane provides the synchronization among the domains. The lower-level controllers are responsible for their own domains. If an event occurs within a domain, the corresponding controller can update the other controllers by informing the master controller. When a flat structure is utilized, the controllers communicate directly with each other to achieve synchronization through their east–west interfaces.

In a federated edge setting, the distributed control plane will play an important role in addressing scalability and consistency issues. Considering a service-centric environment, multiple controllers should simultaneously handle coordinating the service replications and tracking their locations. Without the flexibility provided by SDN and programmable networks [30], extra effort is required to implement the service-centric design. Since SDN can intrinsically retrieve a recent view of the underlying network, controllers can keep track of the locations of the services. A northbound application that maps the service identifiers to the list of locations paves the road for embedding the service-centric model into the global edge setting. Whenever a user offloads a service by specifying its identifier, the controller responsible for that domain can inspect this information and retrieve the list of possible destinations. The list of possible destinations is frequently updated by communicating with the remaining controllers that are coordinating other administrative domains. With the help of an adjacent load balancing northbound application, the network can determine the most feasible server and forward the request by modifying the header fields of the packet. If the destination is deployed in another region, the forwarding operation requires the packets to be routed over multiple domains, and it is handled by the cooperative work of the distributed control plane. If a new service is deployed within a region or a replication of a service is created, the responsible controller initially updates its database and creates an event to inform the other controllers to keep them synchronized. The OpenFlow messages exchanged through the east–west interfaces provide the global synchronization in case of an event.

Service mobility needs to be addressed in the context of a federated edge. Creating, migrating, and replicating services must be accomplished at the edge to deal with varying traffic patterns and load balancing [31]. SDN again is a candidate solution since it determines possible destination nodes, and the path that can be taken for migrating a service such that the performance is least affected (congestion is prevented) [32]. The operation in SDN can be carried out using flow rules.

In research and experiments, SDN is known for managing and handling heterogeneity well [33, 34]. From a network perspective, federated edge environments will typically be heterogeneous in that they will comprise different network types in addition to the varying traffic flow patterns. In this case, the control plane could provide an interoperable-networked environment comprising multiple domains that belong to different providers for both edge servers and end-user devices. Additionally, vendor dependency and compatibility issues between different networking devices are eliminated [35].

2.2.3 Future Research Directions

The integration of fog computing and SDN has immense potential for accelerating practical deployments and federating resources at the edge. However, there are avenues that still need to be explored for bridging the gap between fog computing and SDN. In this section, we consider four such avenues as directions for future research:

1. **The implementation of wireless networking and SDN**. Existing research and practical implementations achieve network virtualization via SDN.

However, the focus is usually on the virtualization and management of SDN controllers in a wired network [36]. We believe the benefits of SDN and current standards, such as OpenFlow, must be fully harnessed by wireless networks for federating edge nodes, which will be serving mostly a mobile community in the future.

2. **Standardization of interfaces for interoperability**. OpenFlow is currently the de facto standard for the southbound interface; however, there are no recognized standards for northbound communication (although the Open Networking Foundation (ONF) for northbound standards [37] organizes a working group). Lack of standardization prevents interoperability among northbound applications that run on top of the same controller. We believe that developing standards for the northbound applications is another important avenue for future research. Additionally, existing SDN-based scenarios do not depend on the east–west interface, and there is very little research in this area. Communication between the adjacent controllers needs to become more reliable and efficient in order not to burden the control channels. We believe that focusing efforts toward this area will provide opportunities for federating a pool of computational resources at the edge.
3. **Enhancing programmability of existing standards and interfaces**. Our experience in programming using OpenFlow leads us to recommend the implementation of additional functionalities. The recent version of OpenFlow (v1.5.1) provides only partial programmability within the network. In order to enable federation of edge resources for generic fog deployments, we believe additional research is required for enhancing the programmability of standards and interfaces.
4. **Scalability of SDN planes for reaching wider geographic areas**. It is anticipated that edge nodes will be distributed over a wide geographic area and there will be distinct administrative domains in fog computing systems. Here, a distributed form of the SDN control plane is required to communicate with the adjacent controllers. Therefore, we recommend further investigation of the scalability of SDN planes [38]. This is challenging because there are no standards in place for east–west interfaces, which will need to be utilized for the inter-controller communication.

2.3 The Management Challenge

Adding a single layer of edge nodes in between the cloud and user devices introduces significant management overhead. This becomes even more challenging when clusters of edge nodes need to be federated from different geographic locations to create a global architecture.

Table 2.2 Management challenges, the need for addressing them, and potential solutions in federating edge resources.

Management challenge	Why should it be addressed?	What is required?
Discovery of edge nodes	To select resources when they are geographically spread and loosely coupled	Lightweight protocols and handshaking
Deployment of service and applications	Provide isolation for multiple services and applications	Monitoring and benchmarking mechanisms in real-time
Migrating services	User mobility, workload balancing	Low overhead virtualization
Load balancing	To avoid heavy subscription on individual nodes	Auto-scaling mechanisms

2.3.1 Management Challenges in a Federated Edge Environment

In this section, we consider four management challenges that will need to be addressed. These are presented in Table 2.2.

2.3.1.1 Discovering Edge Resources

The first management challenge is related to *discovering edge resources* both at individual and collective levels. At the individual level, potential edge nodes that can provide computing will need to be visible in the network, both to applications running on user devices and their respective cloud servers. At the collective levels, a collection of edge nodes in a given geographical location (or at any other granularity) will need to be visible to another collection of edge nodes.

In addition to the system challenges, assuming that an edge node has near-similar capabilities of a network device and general purpose computational device, the challenge here is to determine the best practice for discovery – whether discovery of edge nodes is (i) self-initiated and results in a loosely coupled collection; (ii) initiated by an external monitor that results in a tightly coupled collection; or (iii) a combination of the former.

2.3.1.2 Deploying Services and Applications

The second management challenge is related to *deploying services and applications on the edge*. Typically, a service that can furnish requests from user devices will need to be offloaded onto one or a collection of edge nodes. However, this will not be possible without knowing the capabilities of the target edge and matching them against the requirements of services or applications (such as the expected load and amount of resources required), given that there may be multiple clusters of edge nodes available in the same geographic location.

Benchmarking multiple edge nodes (or multiple collections) simultaneously will be essential here to meet the service objectives. This is challenging and will need to be performed in real-time.

2.3.1.3 Migrating Services across the Edge

The third management challenge is related to *migrating services across the edge.* Existing technology allows for deploying applications and services using virtual machines (VMs), containers, and unikernel technologies. These technologies have proven to be useful in the cloud context to deploy an application and migrate them across data centers. Given the availability of significant resources in a cloud data center, it is not challenging to maintain a large repository of images that can be used to start up or replicate services in the event of failures or load balancing. This, however, is challenging on the edge given the real-time and resource constraints. Additionally, the shortest path in the network for migrating services from an edge node to another will need to be considered.

2.3.1.4 Load Balancing

The fourth management challenge is related to the *load balancing at the edge.* If there is significant subscription of services at the edge, then the resource allocation for individual services on a single edge node or in the collection will need to be managed. For example, if there is one service that is heavily subscribed when compared to other services that are dormant on the edge, then the resources allocated to the heavily subscribed service will need to be scaled. While this is just one scenario, it becomes more complex when more services require resources from the same collection of edge nodes. This will require significant monitoring of resources at the edge, but traditional methods cannot be employed given the resource constraints on edge nodes. Similarly, mechanisms will need to be put in place for scaling the resources for one service (which may be heavily subscribed) while de-allocating resources from dormant services. Both the monitoring and scaling mechanisms will need to ensure integrity so that the workload is fairly balanced.

2.3.2 Current Research

Existing techniques for discovery of edge nodes can be classified based on whether they operate in a multi-tenant environment (i.e. more than one service can be hosted on the edge node). For example, FocusStack discovers edge nodes in a single tenant environment [39], whereas ParaDrop [40] and edge-as-a-service (EaaS) [41] operate in multi-tenant edge environments. However, there are additional challenges that will need to be addressed to enable discovery when multiple collections of edge nodes are federated.

Current research on deploying services focuses on pre-deployment resource provisioning (matching requirements of an application against available resources before the application is deployed) [42]. Post-deployment becomes even more important both in the context of individual edge nodes and federated edge resources due to variability (more applications need to be hosted on a collection of edge nodes) of workloads that are anticipated on the edge. Additionally, workload deployment services that operate on distributed clusters focus on large jobs, such as Hadoop or MapReduce [43, 44]. However, post-deployment techniques suitable for more fine-grained workloads will be required for federated edge resources.

Migration of services via VMs across clusters is possible, but in reality has a significant time overhead [45, 46]. Additionally, live migration across geographically distributed cloud data centers is more challenging and time consuming. Similar strategies have been adopted in the context of edge resources for live migration of VMs [47, 48]. While this is possible (although migration takes a few minutes), it is still challenging to use existing strategies for real-time use. Additionally, VMs may not be the de facto standard for hosting services on the edge [11, 49]. Alternate lightweight technologies, such as containers, and how they may be used to migrate workloads at the edge will need to be investigated and the strategies underpinning these will need to be incorporated within container technologies.

Monitoring of edge resources will be a key requirement for achieving load balancing. For example, performance metrics will need to be monitored for implementing auto-scaling methods to balance workloads on the edge. Existing monitoring systems for distributed systems either do not scale or are resource consuming. These are not suitable for large-scale resource-constrained edge deployments. Current mechanisms for auto scaling resources are limited to single-edge nodes and employ lightweight monitoring [11]. However, scaling these mechanisms is challenging.

2.3.3 Addressing the Management Challenges

Three of the above four research challenges – namely, discovery, deployment, and load balancing – were addressed in the context of individual edge nodes at Belfast on the EaaS platform and the ENORM framework.

2.3.3.1 Edge-as-a-Service (EaaS) Platform

The EaaS [41] platform targets the discovery challenge and implements a lightweight discovery protocol for a collection of homogeneous edge resources (Raspberry Pis). The EaaS platform operates in a three-tier environment – top tier is the cloud, bottom tier comprises user devices, and the middle tier contains edge nodes. The platform requires a master node, which may either be a compute available network device or a dedicated node, and executes a manager process that communicates with the edge nodes. The master node manager communicates with potential edge nodes and installs a manager on the edge node to execute commands. Administrative control panels are available on the master node to monitor individual edge nodes. Once the EaaS platform discovers an edge node, then Docker or LXD containers can be deployed. The platform was tested in the context of an online game, similar to the popular Pokémon GO, to improve the overall performance of the application.

The benefit of this platform is that the discovery protocol that is implemented is lightweight and the overhead is a few seconds for launching, starting, stopping, or terminating containers. Up to 50 containers with the online game workload were launched on an individual edge node. However, this has been carried out in the context of a single collection of edge nodes. Further research will be required to employ such a model in a federated edge environment.

The major drawback of the EaaS platform is that it assumes a centralized master node that can communicate with all potential edge nodes. The research also assumes that the edge nodes can be queried and can, via owners, be made available in a common marketplace. Additionally, the security-related implications of the master node installing a manager on the edge node and executing commands on the edge node has not been considered.

2.3.3.2 Edge Node Resource Management (ENORM) Framework

The ENORM framework [11] primarily addresses the deployment and load balancing challenges on individual edge nodes. Similar to the EaaS platform, ENORM operates in a three-tier environment, but a master controller does not control the edge nodes. Instead, it is assumed that they are visible to cloud servers that may want to make use of the edge. The framework allows for partitioning a cloud server and offloading it to edge nodes for improving the overall QoS of the application.

The framework is underpinned by a provisioning mechanism for deploying workloads from a cloud server onto an edge server. The cloud and an edge server establish a connection via handshaking to ensure that there are sufficient resources available to fulfill the request of the server that will be offloaded onto the edge. The provisioning mechanism caters to the entire life cycle of an application server from offloading it onto the edge via a container until it is terminated and the cloud server is notified.

Load balancing on a single edge node is accomplished by implementing an auto-scaling algorithm. It is assumed that an edge node could be a traffic routing node, such as a router or mobile base station, and therefore an offloaded service should not compromise the QoS of the basic service (traffic routing) that is executed on the node. Each application server executing on the edge node has a priority. Each edge server is monitored (in terms of both network and system performance) and it is estimated whether the QoS can be met. If an edge server with a higher priority cannot meet its QoS, then the resources for the application is scaled. If the resource requirements of an application cannot be met on the edge, then it is moved back to the cloud server that offloaded it. This occurs iteratively in periodic intervals to ensure that the QoS is achieved and the node is stable.

The ENORM framework is also validated on the online game use-case as for the EaaS platform. It is noted that the application latency can be reduced between 20% and 80% and the overall data transferred to the cloud for this use-case is reduced by up to 95%.

2.3.4 Future Research Directions

Both the EaaS platform and the ENORM framework have limitations in that they do not assume federated edge resources. In this section, the following four research directions for addressing management challenges when federating edge resources are considered:

1. **Coordinating management tasks between heterogeneous nodes of multiple edge collections**. Federating edge resources inevitably requires the bringing together heterogeneous edge nodes (routers, base stations, switches, and dedicated low-power compute devices). While managing homogeneous resources in itself can be challenging, it will be more complex to coordinate multiple collections of heterogeneous resources. The challenge here is enabling the required coordination via a standard protocol to facilitate management between devices that are geographically apart, that have varying CPU architectures, and may inherently be used for network traffic routing.
2. **Developing real-time benchmarking services for federated edge resources**. Given the varying computational capabilities and workloads on (traffic through) edge nodes, cloud servers will need to reliably benchmark a portfolio of edge nodes. This portfolio may be from different or the same geographic location, so that via benchmarking the application server can identify edge nodes that may meet service-level objectives (SLOs) if a partitioned workload needs to be deployed on the edge. Mechanisms facilitating this in real time will need to be developed.
3. **Facilitating rapid migration between federated edge resources**. Current migration techniques typically have overheads in the order of a few minutes in the best cases when attempting to migrate from one node to another. This overhead will obviously increase with geographic distance. Current mechanisms for migration take a snapshot of the VM or the container on an edge node and then transport this across to another node. To facilitate fast migration, perhaps alternate virtualization technologies may need to be developed that allow for migration of more abstract entities (such as functions or programs). This technology may also be employed in upcoming serverless computing platforms for developing interoperable platforms across federated edge resources.

4. **Investigating fine-grain resource allocation/deallocation for load balancing using auto-scaling.** Current auto-scaling methods add or remove discrete predefined units of resources on the edge for auto-scaling. However, this is limiting in resource-constrained environments in that resources may be over-provisioned. Alternate mechanisms will need to be investigated that can derive the amount of resources that need to be allocated/deallocated based on specific application requirements to meet SLOs without compromising the stability of the edge environment.

2.4 Miscellaneous Challenges

The previous two sections have considered the networking and management challenges in federating edge resources that are geographically distributed. However, there are additional challenges that need to be considered. For example, the challenge of developing pricing models to make use of edge resources. This will rely on a solution space that cannot be fully foreseen today given that the technology for supporting public edge computing is still in its infancy. In this section, we consider two further challenges, namely the resource and modeling challenges as shown in Figure 2.3, which are dependent on both networking and management.

2.4.1 The Research Challenge

The prospect of including an edge layer between cloud data centers and user devices for emerging applications is appealing because latency and transmission of data to the core of the network can be minimized to improve the overall QoS of an application. Although there are reference edge

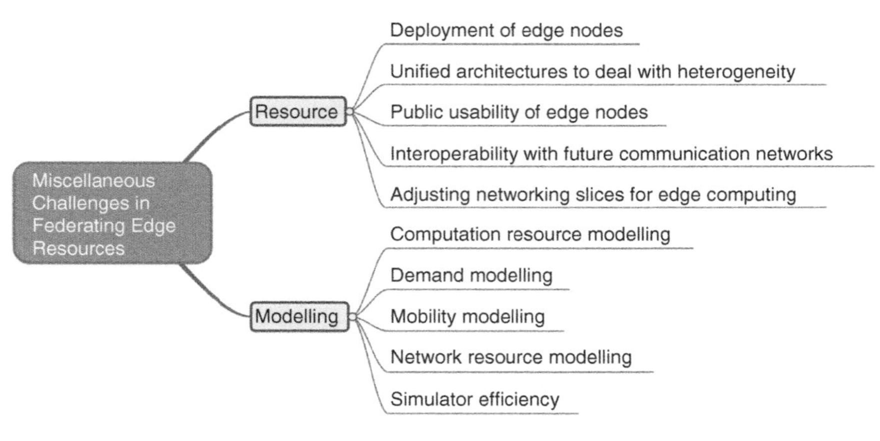

Figure 2.3 Resource and modeling challenges in federating edge resources.

architectures and test beds that validate these architectures, edge computing has not yet been publicly adopted and we have yet to see large-scale practical implementations of these systems. We present five resource challenges that will need to be addressed before an edge layer can become a practical reality.

2.4.1.1 Defined Edge Nodes

The first resource challenge is related to *deploying edge nodes.* It is still not clear whether edge nodes are likely to be: (i) traffic routing nodes, such as routers, switches, gateways, and mobile base stations that integrate general-purpose computing via CPUs on them; (ii) dedicated computing nodes with low-power compute devices on which general purpose computing can be achieved, such as micro clouds; or (iii) a hybrid of the former.

In the retail market, products that enable general-purpose computing on traffic routing nodes are available. For example, Internet gateways that are edge enabled are currently available in the market.[1] Additionally, there is ongoing research that aims to use micro cloud data centers at the edge of the network.[2] It seems that there is a business use case for either option, but how the latter will coexist with traffic routing nodes is yet to be determined. Furthermore, the migration to the former may take a long time since existing traffic routing nodes will need to be upgraded.

2.4.1.2 Unified Architectures to Account for Heterogeneity

The second resource challenge is related to *developing unified architectures to account for heterogeneity.* Bringing different types of edge-based nodes with varying performance and compute resources as a coherent single layer or multiple layers can be challenging from a software, middleware, and hardware perspective. Given the wide variety of edge computing options proposed ranging from small home routers to micro cloud installations, federating them will require the development of unified interoperable standards across all these nodes. This is unprecedented and will be unlike the standards that have been used on the cloud where large collections of compute resources have the same underlying architecture. If this were the case, then applications and services would need to be executed in a manner oblivious to the underlying hardware. However, current research that enables this via virtualization or containerization is not suitable and is not available for all hardware architectures.

2.4.1.3 Public Usability of Edge Nodes

The third resource challenge is related to *public usability of edge nodes.* Regardless of how the edge layer is enabled, it is anticipated to be accessible both for

---

1 http://www.dell.com/uk/business/p/edge-gateway

2 http://www.dell.com/en-us/work/learn/rack-scale-infrastructure

bringing computation from clouds closer to the user devices and for servicing requests from user devices or processing data generated from a large collection of sensors before it is sent to the cloud. This raises several concerns:

1. How will edge nodes be audited?
2. Which interface will be used to make them publicly accessible?
3. Which billing model will be required?
4. Which security and privacy measures will need to be adopted on the edge?

These concerns are beyond the scope of this chapter. However, they will need to be addressed for obtaining publicly usable edge nodes.

2.4.1.4 Interoperability with Communication Networks

The fourth resource challenge is related to *interoperability with future communication networks*. In the context of edge computing systems, the network itself is a critical resource that defines the overall performance of an edge solution. Resource management strategies should consider the network resources as well as the computational resources for the efficient operation of the edge systems [14]. Initial edge computing proposals almost exhaustively employ WLAN technologies for accessing computational resources. However, this is likely to change, given the emergence of 5G. QoS provided at the level of tactile Internet makes 5G systems a strong alternative for edge access [52]. Considering the potential of edge systems, European Telecommunications Standards Institute (ETSI) started the multi-access edge computing (MEC) standardization with many contributors from the telecom industry [53]. MEC, in principle, is an edge computing architecture that is envisioned as an intrinsic component of 5G systems. With 5G practical deployments, whether edge computing services will rely on the MEC functionality or whether they will make use of high-bandwidth 5G network capabilities and position themselves as over-the-top (OTT) is not yet clear and will depend on parameters including cost and openness. The two options will dictate radically different positions for the federation of edge computing systems. ETSI-MEC with its inherent position within the 5G architecture will closely couple edge systems in general and their federation with the telecom operators' operation of the whole network.

2.4.1.5 Network Slices for Edge Systems

The fifth resource challenge is related to *adjusting network slices for edge systems*. Another important opportunity that is anticipated to be provided by future networks is slicing. Network slices are defined as logical networks overlaid on a physical or virtual network that can be created on demand with a set of parameters [52]. Slicing will allow network operators to cater to QoS specific to a service or a group of services.

Although, slicing is not an approach that is particular to edge computing, it will significantly affect the operation and the performance of edge systems. An end-to-end slice dedicated to each service on an edge server would be beneficial. However, this becomes challenging as the slices become fine-grained due to scalability and management challenges. In addition, slicing related to edge computing needs to consider the volume of interaction between edge and cloud servers. A simple approach would be to assign slices for every edge deployment and expect the edge orchestration system to assign additional resources within the edge system. In a federated setting, a finite capacity of resources may be assigned to a set of standalone edge systems and the overall slice of the individual edge system may be adjusted globally considering resource usage patterns.

2.4.2 The Modeling Challenge

Edge computing has paved way for a variety of technologies, such as MEC, mobile MCC, cloudlets, and fog computing [5]. These indicate that edge solutions can be obtained in multiple domains using different techniques. Given that there are no de facto standards and that there is an abundance of edge architectures emerging in the literature, tools for modeling and analyzing edge systems are required.

One option to model edge systems is to implement test beds that are specific to the requirements of a use case. Given the availability of open source tools to virtualize resources (computational and network), it would be feasible to develop testbeds for a research environment. For example, the Living Edge Lab[3] is an experimental testbed. However, setting up testbeds can be quite expensive. Additionally, for a complete performance analysis, testbeds and sometimes even real-world deployments may not lend themselves to repeatable and scalable experiments as could be obtained using simulators [50]. Therefore, simulators are employed to complement experimental testbeds for a thorough evaluation.

At the heart of a simulator is a complex mathematical model that captures the environment. Although simulators are advantageous, numerous modeling challenges will need to be addressed in designing an ideal (or even a reasonable) simulator [51]. An ideal simulation environment should incorporate programming APIs, management of configuration files, and UI dashboards for easy modeling with minimum manual effort. We anticipate that the same principles would apply to an edge simulator. In this section, we consider five specific modeling challenges that will need to be addressed for an ideal edge simulator.

2.4.2.1 Computational Resource Modeling

Like a cloud data center, an edge server will provide computational power to its users via virtualization techniques, such as VMs and containers [12].

3 http://openedgecomputing.org/lel.html

A simulation environment in this context should support the creation, resizing, and migration of virtual resources and model CPU, memory, and network resource consumption at different levels of granularity (process, application, and entire node). The model will need to capture the possibilities of using existing traffic routing nodes, dedicated nodes, or a combination of these.

2.4.2.2 Demand Modeling

To be able to model the load on an edge computing system, the demand on the edge resource due to an individual user (or a collection of users) will need to be modeled. Accounting for the heterogeneity of mobile devices and the traffic generated by a variety of applications is complex. End-user devices or a cloud server may offload compute on to edge servers and this will need to be accounted for in a demand model. The distribution of demand and the inter-arrival times of traffic on the edge will need to be considered. Profiles of the users and/or family of applications with predefined distributions would also be beneficial.

2.4.2.3 Mobility Modeling

Mobility is a key component that will need to be considered for accurately modeling the time-varying demands on the edge. The need for mobility arises in multiple use-cases. For example, a human with a wearable gadget moving from the coverage area of one edge server to another could result in the migration of the service from one edge server onto another or the replication of the service with the user data on another edge server. In such a use case, the simulation environment should allow for designing experiments that realistically captures mobility in a variety of forms.

2.4.2.4 Network Modeling

Performance and behavior patterns of the network are critical for the overall operation of an edge system. Accurate network delay modeling will not be easy due to dynamic workloads that operate using different network access technologies, such as Wi-Fi, Bluetooth and cellular networks. In contrast to legacy network simulators, an edge simulation tool should be able to scale rapidly network resources. This requirement arises due to the *slicing* approach described previously in which multiple network slices will need to be modeled in the network [52].

2.4.2.5 Simulator Efficiency

Simulators will need to be scalable, extensible to changing infrastructure requirements, and easy to use. Taking into account the upcoming Internet of Things and machine-to-machine communications, the time complexity of simulators accounting for federated edge resources should model the connections of a large number of devices and users.

2.5 Conclusions

Computational resources that are typically concentrated in cloud data centers are now proposed to become available at the edge of the network via edge computing architectures. Edge resources will be geographically distributed and they will need to be federated for a globally accessible edge layer that can service both data center and user device requests. The aim of this chapter is to highlight some of the challenges that will need to be addressed for federating geographically distributed edge resources. The chapter first presented the network and management related issues. Then the chapter considered how existing research reported in the literature addresses these challenges and provided a roadmap of future directions. Subsequently, we presented additional challenges related to resources and modeling for a federated edge. The key message of this chapter is that federating edge resources is not an easy task. Let alone the social and legal aspects in federating, the underlying technology that will facilitate public edge computing is still in its infancy and rapidly changing. There are a number of technological challenges related to networking, management, resource, and modeling that will need to be addressed for developing novel solutions to make the federated edge computing a reality.

References

1 B. Varghese, and R. Buyya. Next generation cloud computing: New trends and research directions. *Future Generation Computer Systems*, 79(3): 849–861, February 2018.

2 W. Shi, and S. Dustdar. The promise of edge computing. *Computer*, 49(5): 78–81, May 2016.

3 T. Taleb, K. Samdanis, B. Mada, H. Flinck, S. Dutta, and D. Sabella. On multi-access edge computing: A survey of the emerging 5G network edge architecture and orchestration. *IEEE Communications Surveys & Tutorials*, 19(3): 1657–1681, May 2017.

4 R. Vilalta, A. Mayoral, D. Pubill, R. Casellas, R. Martínez, J. Serra, and R. Muñoz. End-to-end SDN orchestration of IoT services using an SDN/NFV-enabled edge node. In *Proceedings of Optical Fiber Communications Conference and Exhibition*, Anaheim, CA, USA, March 20–24, 2016.

5 A. C. Baktir, A. Ozgovde, and C. Ersoy. How can edge computing benefit from software-defined networking: A survey, Use Cases & Future Directions. *IEEE Communications Surveys & Tutorials*, 19(4): 2359–2391, June 2017.

6 T. Q. Dinh, J. Tang, Q.D. La, and T.Q.S. Quek. Offloading in mobile edge computing: Task allocation and computational frequency scaling. *IEEE Transactions on Communications*, 65(8): 3571–3584, August, 2017.

7 L. F. Bittencourt, M. M. Lopes, I. Petri, and O. F. Rana. Towards virtual machine migration in fog computing. *10th International Conference on P2P, Parallel, Grid, Cloud and Internet Computing*, Krakow, Poland, November 4–6, 2015.

8 J. Xu, L. Chen, and S. Ren, Online learning for offloading and autoscaling in energy harvesting mobile edge computing. *IEEE Transactions on Cognitive Communications and Networking*, 3(3): 361–373, September 2017.

9 N. Apolónia, F. Freitag, L. Navarro, S. Girdzijauskas, and V. Vlassov. Gossip-based service monitoring platform for wireless edge cloud computing. In *Proceedings of the 14th International Conference on Networking, Sensing and Control*, Calabria, Italy, May 16–18, 2017.

10 M. Satyanarayanan. Edge computing: Vision and challenges. *IEEE Internet of Things Journal*, 3(5): 637–646, June 2016.

11 N. Wang, B. Varghese, M. Matthaiou, and D. S. Nikolopoulos. ENORM: A framework for edge node resource management. *IEEE Transactions on Services Computing*, PP(99): 1–1, September 2017.

12 B. Varghese, N. Wang, S. Barbhuiya, P. Kilpatrick, and D. S. Nikolopoulos. Challenges and opportunities in edge computing. In *Proceedings of the International Conference on Smart cloud*, New York, USA, November 18–20, 2016.

13 Z. Hao, E. Novak, S. Yi, and Q. Li. Challenges and software architecture for fog computing. *IEEE Internet Computing*, 21(2): 44–53, March 2011.

14 C. Sonmez, A. Ozgovde, and C. Ersoy. EdgeCloudSim: An environment for performance evaluation of edge computing systems. In *Proceedings of the 2nd International Conference on Fog and Mobile Edge Computing*. Valencia, Spain, May 8–11, 2017.

15 S. Yi, C. Li, and Q. Li. A survey of fog computing: Concepts, applications and issues. In *Proceedings of the Workshop on Mobile Big Data*. Hangzhou, China, June 22–25, 2015.

16 L. M. Vaquero and L. Rodero-Merino. Finding your way in the fog: Towards a comprehensive definition of fog computing. *SIGCOMM Computer Communication Review*, 44(5): 27–32, October 2014.

17 I. Stojmenovic, S. Wen, X. Huang, and H. Luan. An overview of fog computing and its security issues, *Concurrency and Computation: Practice and Experience*, 28(10): 2991–3005, April 2015.

18 A. C. Baktir, A. Ozgovde, and C. Ersoy. Enabling service-centric networks for cloudlets using SDN. in *Proceedings of the 15th International Symposium on Integrated Network and Service Management*, Lisbon, Portugal, May 8–12, 2017.

19 H. Farhady, H. Lee, and A. Nakao. Software-Defined Networking: A survey, *Computer Networks*, 81(C): 79–95, December 2014.

20 R. Jain, and S. Paul. Network virtualization and software defined networking for cloud computing: A survey. *IEEE Communications Magazine*, 51(11): 24–31, 2013.

21 M. Jammal, T. Singh, A. Shami, R. Asal, and Y. Li. Software defined networking: State of the art and research challenges, *Computer Networks*, 72: 74–98, 2014.

22 V. R. Tadinada. Software defined networking: Redefining the future of Internet in IoT and cloud era. In *Proceedings of the 4th International Conference on Future Internet of Things and cloud*, Barcelona, Spain, August 22–24, 2014.

23 Open Networking Foundation. OpenFlow Switch Specification Version 1.5.1, https://www.opennetworking.org/images/stories/downloads/sdn-resources/onf-specifications/openflow/. Accessed December 2017.

24 S. Tomovic, M. Pejanovic-Djurisic, and I. Radusinovic. SDN based mobile networks: Concepts and benefits. *Wireless Personal Communications*, 78(3): 1629–1644, July 2014.

25 X. N. Nguyen, D. Saucez, C. Barakat, and T. Turletti. Rules placement problem in OpenFlow networks: A survey. *IEEE Communications Surveys & Tutorials*, 18(2): 1273–1286, December 2016.

26 G. Luo, S. Jia, Z. Liu, K. Zhu, and L. Zhang. sdnMAC: A software defined networking based MAC protocol in VANETs. In *Proceedings of the 24th International Symposium on Quality of Service*, Beijing, China, June 20–21, 2016.

27 Geni, http://groups.geni.net/geni/wiki/OpenFlowDiscoveryProtocol/. Accessed on 14 March, 2018.

28 B. Pfaff, J. Pettit, T. Koponen, E. J. Jackson, A. Zhou, J. Rajahalme, J. Gross, A. Wang, J. Stringer, P. Shelar, K. Amidon, and M. Casado. 2015. The design and implementation of open vSwitch. In *Proceedings of the 12th USENIX Conference on Networked Systems Design and Implementation*, Berkeley, CA, USA, May 7–8, 2015.

29 R. Mijumbi, J. Serrat, J. Rubio-Loyola, N. Bouten, F. De Turck, and S. Latré. Dynamic resource management in SDN-based virtualized networks, in *Proceedings of the 10th International Conference on Network and Service Management*, Rio de Janeiro, Brazil, November 17–21, 2014.

30 J. Bailey, and S. Stuart. Faucet: Deploying SDN in the enterprise, *ACM Queue*, 14(5): 54–68, November 2016.

31 C. Puliafito, E. Mingozzi, and G. Anastasi. Fog computing for the Internet of Mobile Things: Issues and challenges, in *Proceedings of the 3rd International Conference on Smart Computing*, Hong Kong, China, May 29–31, 2017.

32 A. Mendiola, J. Astorga, E. Jacob, and M. Higuero. A survey on the contributions of Software-Defined Networking to Traffic Engineering, *IEEE Communications Surveys & Tutorials*, 19(2), 918–953, November 2016.

33 N. B. Truong, G. M. Lee, and Y. Ghamri-Doudane. Software defined networking-based vehicular ad hoc network with fog computing, in *Proceedings of IFIP/IEEE International Symposium on Integrated Network Management*, Ottawa, ON, Canada, May 11–15, 2015.

34 K. Bakshi, Considerations for software defined networking, SDN): Approaches and use cases, in *Proceedings of IEEE Aerospace Conference*, Big Sky, MT, USA, March 2–9, 2013.

35 Open Networking Foundation. SDN Definition, https://www.opennetworking.org/sdn-resources/sdn-definition. Accessed on November 2017.

36 C. J. Bernardos, A. De La Oliva, P. Serrano, A. Banchs, L. M. Contreras, H. Jin, and J. C. Zúñiga. An architecture for software defined wireless networking, *IEEE Wireless Communications*, 21(3), 52–61, June 2014.

37 Open Networking Foundation. Northbound Interfaces, https://www.opennetworking.org/images/stories/downloads/working-groups/charter-nbi.pdf, Accessed on: 14 March, 2018.

38 Open Networking Foundation. Special Report: OpenFlow and SDN - State of the union, https://www.opennetworking.org/images/stories/downloads/sdn-resources/special-reports/Special-Report-OpenFlow-and-SDN-State-of-the-Union-B.pdf. Accessed on 14 March, 2018.

39 B. Amento, B. Balasubramanian, R. J. Hall, K. Joshi, G. Jung, and K. H. Purdy. FocusStack: Orchestrating edge clouds using location-based focus of attention. In *Proceedings of IEEE/ACM Symposium on Edge Computing*, Washington, DC, USA, October 27–28, 2016.

40 P. Liu, D. Willis, and S. Banerjee. ParaDrop: Enabling lightweight multi-tenancy at the network's extreme edge. In *Proceedings of IEEE/ACM Symposium on edge Computing*, Washington, DC, USA, October 27–28, 2016.

41 B. Varghese, N. Wang, J. Li, and D. S. Nikolopoulos. Edge-as-a-service: Towards distributed cloud architectures. In *Proceedings of the 46th International Conference on Parallel Computing*, Bristol, United Kingdom, August 14–17, 2017.

42 S. Nastic, H. L. Truong, and S. Dustdar. A middleware infrastructure for utility-based provisioning of IoT cloud systems. In *Proceedings of IEEE/ACM Symposium on edge Computing*, Washington, DC, USA, October 27–28, 2016.

43 V. K. Vavilapalli, A. C. Murthy, C. Douglas, S. Agarwal, M. Konar, R. Evans, T. Graves, J. Lowe, H. Shah, S. Seth, B. Saha, C. Curino, O. O'Malley, S. Radia, B. Reed, and E. Baldeschwieler. Apache Hadoop YARN: Yet another resource negotiator, in *Proceedings of the 4th Annual Symposium on cloud Computing*, Santa Clara, California, October 1–3, 2013.

44 B. Hindman, A. Konwinski, M. Zaharia, A. Ghodsi, A. D. Joseph, R. Katz, S. Shenker, and I. Stoica. Mesos: A platform for fine-grained resource

sharing in the data center, in *Proceedings of the 8th USENIX Conference on Networked Systems Design and Implementation*, Berkeley, CA, USA, March 30–April 01, 2011.

45 C. Clark, K. Fraser, S. Hand, J. G. Hansen, E. Jul, C. Limpach, I. Pratt, and A. Warfield. Live migration of virtual machines. In *Proceedings of the 2nd conference on Symposium on Networked Systems Design & Implementation*, Berkeley, CA, USA, May 2–4, 2005.

46 S. Wang, R. Urgaonkar, M. Zafer, T. He, K. Chan, and K. K. Leung. Dynamic service migration in mobile edge-clouds, *IFIP Networking Conference*, 91(C): 205–228, September 2015.

47 F. Callegati, and W. Cerroni. Live migration of virtualized edge networks: Analytical modelling and performance evaluation, in *Proceedings of the IEEE SDN for Future Networks and Services*, Trento, Italy, November 11–13, 2013.

48 D. Darsena, G. Gelli, A. Manzalini, F. Melito, and F. Verde. Live migration of virtual machines among edge networks viaWAN links. In *Proceedings of the 22nd Future Network & Mobile Summit*, Lisbon, Portugal, July 3–5, 2013.

49 S. Shekhar, and A. Gokhale. Dynamic resource management across cloud-edge resources for performance-sensitive applications. In *Proceedings of the 17th IEEE/ACM International Symposium on Cluster, Cloud and Grid Computing*, Madrid, Spain, May 14–17, 2017.

50 G. D'Angelo, S. Ferretti, and V. Ghini. Modelling the Internet of Things: A simulation perspective. In *Proceedings of the International Conference on High Performance Computing Simulation*, Genoa, Italy, July 17–21, 2017.

51 G. Kecskemeti, G. Casale, D. N. Jha, J. Lyon, and R. Ranjan. Modelling and simulation challenges in Internet of Things, *IEEE Cloud Computing*, 4(1): 62–69, January 2017.

52 X. Foukas, G. Patounas, A. Elmokashfi, and M.K. Marina. Network slicing in 5G: Survey and challenges, *IEEE Communications Magazine*, 55(5): 94–100, May 2017.

53 Y.C. Hu, M. Patel, D. Sabella, N. Sprecher, and V. Young. Mobile edge computing—A key technology towards 5G, *ETSI White Paper*, 11(11):1–16, September 2015.

3

Integrating IoT + Fog + Cloud Infrastructures: System Modeling and Research Challenges

Guto Leoni Santos, Matheus Ferreira, Leylane Ferreira, Judith Kelner, Djamel Sadok, Edison Albuquerque, Theo Lynn, and Patricia Takako Endo

3.1 Introduction

There is widespread recognition, from academia, industry, and policymakers, that social media, cloud computing, big data, and associated analytics, mobile technologies, and the Internet of Things (IoT) are transforming how society operates and interacts with technology and each other [1, 2]. Often referred to as the Internet of Everything (IoE) or "Third IT Platform," these technologies presage a future of greater inter-dependencies between people, devices, and the infrastructure that supports these relationships. Cisco estimates that there are IoE, with estimates of 8–10 billion connected today [3].

While cloud computing has been a key enabling technology for the IoT, a small increase in the percentage of connected or cyber-physical objects represents dramatic change in the feature space of computing and a potential tsunami of computation and hyper-connectivity, which today's infrastructure will struggle to accommodate at historic levels of quality of service (QoS). Large-scale distributed control systems, geo-distributed applications, time-dependent mobile applications, and applications that require very low and predictable latency or interoperability between service providers are just some of the IoT application categories that existing cloud infrastructures are not well-equipped to manage at a hyperscale [4]. Traditional cloud computing architectures were simply not designed with an IoT, characterized by extreme geographic distribution, heterogeneity and dynamism, in mind. As such, a novel approach is required to meet the requirements of IoT including transversal requirements (scalability, interoperability, flexibility, reliability, efficiency, availability, and security) as well as cloud-to-thing (C2T)-specific computation, storage and communication needs [5].

Fog and Edge Computing: Principles and Paradigms, First Edition.
Edited by Rajkumar Buyya and Satish Narayana Srirama.

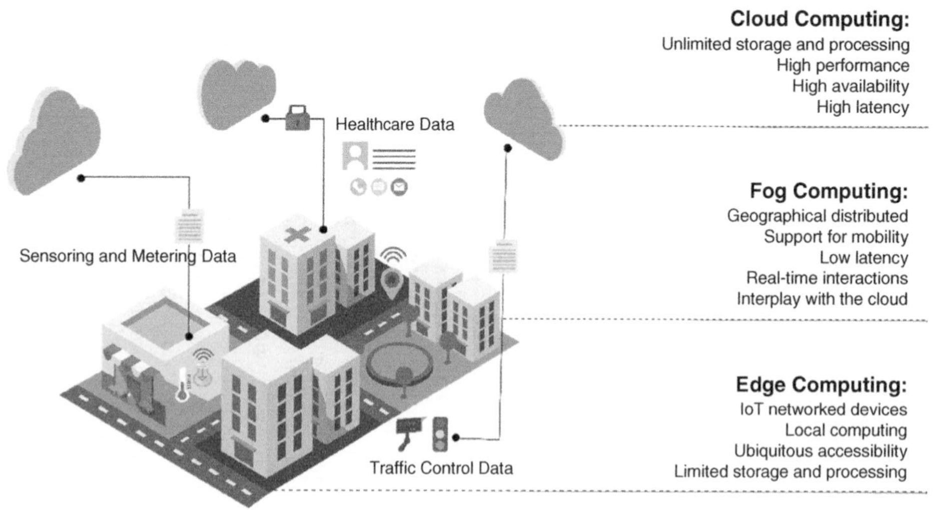

Figure 3.1 Integration of IoT devices with fog and cloud computing.

In response to the need for a new intermediary layer along the C2T continuum, fog computing has emerged as a computing paradigm situated between the cloud and connected or smart end-devices where intermediary compute elements (fog nodes) provide data management and/or communications services to facilitate the execution of relevant IoT applications [6], as shown in Figure 3.1. The ambition for fog computing is greater support for interoperability between service providers, real-time processing and analytics, mobility, geographic distribution, and different device or fog node form factors, and as a result the achievement of QoS expectations [4]. Despite these advantages, fog computing adds a layer of complexity that operators across the C2T continuum need to account for, not least resource orchestration and management [4, 7]. Fog computing represents both an opportunity to exploit but also a risk to mitigate. Not only do failures in the cloud and end points need to be considered but the potentiality and impact of failures across the entire C2T continuum.

In this chapter, we review the literature with regarding to the use of modeling techniques to represent and evaluate an integrated C2T system comprising cloud computing, fog computing, and the IoT (C2F2T). The remainder of this chapter is organized as follows. The next section describes the methodology adopted to guide this literature review and provides a descriptive analysis of the final works selected for use in this chapter. Section 4.3 presents an analysis of existing system modeling techniques used in cloud computing, fog computing, and IoT research against four categories – analytical models, Petri Net models, integer linear programming, and other approaches. Section 4.4 discusses the main scenarios modeled in extant research while section 4.5 discusses the metrics used in evaluation. The chapter concludes with a discussion of research challenges and future directions for research.

3.2 Methodology

The objective of this systematic literature review is to present an overview of academic literature on (i) the use of modeling techniques to represent and evaluate an integrated C2F2T system; (ii) the main scenarios modeled; and (iii) the metrics used to evaluate models. In general, the literature review follows the methodology outlined in [8] and illustrated in Figure 3.2.

While [9] suggests authors aim for complete coverage, such coverage is not feasible. Thus, we limit this review to the computer science discipline and only publication outlets featured in three repositories, namely: Science Direct, IEEE Xplore, and ACM Digital Library. The literature search was limited to modeling, cloud computing, and the IoT. We did not include "fog" in the string, as it was deemed likely that papers on integrated C2F2T featuring fog computing would need to feature cloud computing and IoT keywords. As such,

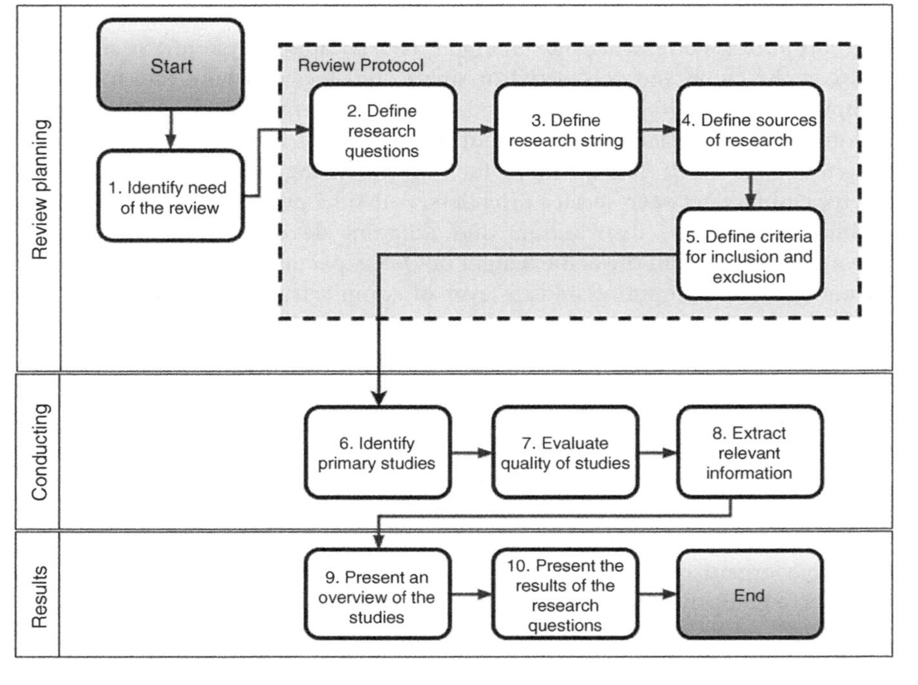

Figure 3.2 Systematic review steps. Adapted from [8].

the search was limited to publications resulting from the following search expression: model AND (cloud OR cloud computing) AND (IoT OR Internet of Things).

Our initial search yielded 1,857 publications from 2013 to 2017. We considered conference and journal papers, but books, PhD theses, and industry publications were excluded. Abstracts were further scrutinized and a final list was produced with 23 relevant articles. Papers were primarily omitted on the grounds that (i) their main focus was not concerned with an integrated C2T system; (ii) the papers were concerned with business modeling; (iii) architectures were misrepresented as models; or (iv) there is a lack of models or modeling techniques. Full texts for the final 23 articles were evaluated to validate that the articles meets the literature search criteria. Table 3.1 presents the number of articles in the initial search and the final selection; while Table 3.2 shows the list of articles selected by year and publication source.

As can be seen in Table 3.2, the largest number of papers was published in 2016 (11), followed by 2017 (5). As cloud computing, IoT, and fog computing are all relatively new fields, high-ranking outlets dealing specifically with the topic are scarce, and those that do exist may not be affiliated with IEEE or may require longer turnaround times for acceptance. Given the increasing number

Table 3.1 Summary of systematic search results.

Repository	Science Direct	IEEE Explore	ACM Digital Library	Total
Initial search	1,244	426	187	1,857
Final selection	10	12	1	23

of conference papers since 2013, one would expect a greater number of journal articles in the coming years.

3.3 Integrated C2F2T Literature by Modeling Technique

In this section, we present an analysis of the modeling techniques used in the sample of papers identified in the conducting phase. A wide range of techniques was identified in the papers analyzed. The analysis presented in Figure 3.3 suggests that analytical models followed by Petri Nets and Integer Linear Programs are the most common techniques used for modeling an integrated C2F2T system.

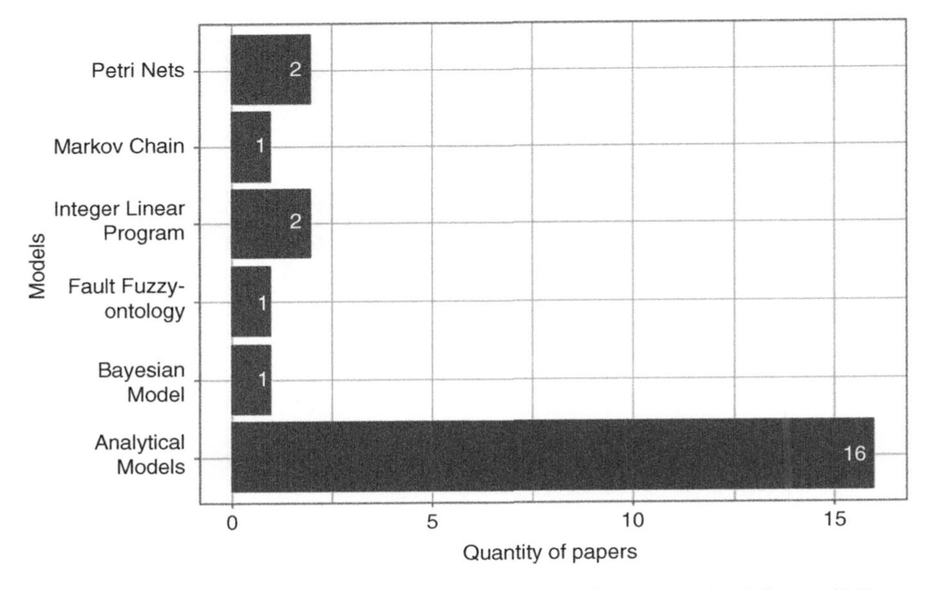

Figure 3.3 The most approaches used to model the integration among cloud, fog, and IoT.

Table 3.2 Articles about modelling of IoT, fog, and cloud integration.

Year/ Source	IEEE	Science Direct	ACM
2013		High-performance scheduling model for multisensory gateway of cloud sensor system-based smart-living [10] QoS-aware computational method for IoT composite service [11]	
Total	**0**	**2**	**0**
2014	Energy efficient and quality-driven continuous sensor management for mobile IoT applications [12]	A fault fuzzy-ontology for large-scale fault-tolerant wireless sensor networks [13]	
Total	**1**	**1**	**0**
2015	Virtualization framework for energy efficient IoT networks [14] Opacity in IoT with cloud computing [15] Reliability modeling of service-oriented Internet of Things [16]		
Total	**3**	**0**	**0**
2016	Query processing for the IoT: Coupling of device energy consumption and cloud infrastructure billing [17] A location-based interactive model for IoT and cloud (IoT-cloud) [19] An IoT-based system for collision detection on guardrails [21]	System modeling and performance evaluation of a three-tier cloud of things [18] Collaborative building of behavioral models based on IoT [20] Deviation-based neighborhood model for context-aware QoS prediction of cloud and IoT services [22]	

	Towards distributed service allocation in fog-to-cloud (F2C) scenarios [23]	Event prediction in an IoT environment using naïve Bayesian models [24]	
	Interconnecting fog computing and microgrids for greening IoT [25]	Mobile crowdsensing as a service: A platform for applications on top of sensing clouds [26]	
	Theoretical modeling of fog computing: a green computing paradigm to support IoT applications [27]		
Total	**6**	**5**	**0**
2017	Application-aware resource provisioning in a heterogeneous IoT [28]	Wearable IoT data stream traceability in a distributed health information system [29]	A novel distributed latency-aware data processing in fog computing-enabled IoT networks [30]
	Leveraging renewable energy in edge clouds for data stream analysis in IoT [31]	Incentive mechanism for computation offloading using edge computing: A Stackelberg game approach [32]	
Total	**2**	**2**	**1**

3.3.1 Analytical Models

Analytical models are mathematical models that have a closed-form solution, i.e. the solution to the equations used to describe changes in a given system can be expressed as a mathematical analytic function. In general, analytical models can be used to predict computing resource requirements related to workload behavior, content, and volume changes, and to measure effects of hardware and software changes [33]. However, most of analytical models rely on approximations, and it is important to realize how these impact on any given models' results. It can be observed that papers using analytical models dominate C2F2T literature featuring modeling techniques. In this systematic review, we found 16 articles using analytical models.

The authors in [18] define the architecture, where each physical and virtual component of layers are described as a vector of associated features. A set of equations is defined to calculate metrics such as power consumed and time latency of scenarios.

In [30], the authors consider the number of gateways (devices that receive data from IoT devices and forward to cloud infrastructure or fog devices) present in an architecture, the total data received for each gateway, and the time spent by each gateway to process the data, in their modeling. Equations are proposed to represent the available buffer and the occupancy efficiency of gateway buffers. Depending on the available space on the gateways, the data are transferred to higher layers, increasing the delay for data processing. Several equations are defined for the calculation of the delay and the authors propose an optimization of the gateways' efficiency, improving the occupancy and the response time efficiency for all gateways in the system.

The authors in [20] propose an analytical model to represent a scheduling mechanism in IoT environment. An equation is defined to represent the addition of a load to be processed in a device. This process takes into account several variables, such as processor load, free memory, and free bandwidth, among others.

The work presented in [22] proposes a method to predict the QoS of IoT services. The model is based on a neighborhood collaborative filter and allows an efficient global optimization scheme. A set of equations is defined to predict correctly the QoS of a service, based on aspects such as latency, response time, and user network condition.

A scheduling model is proposed in [10] to manage sensor applications by a gateway. The requirement resources of applications are taken into account in scheduling problem. Equation (3.1) shows the problem formulated by [10]. In a scenario with n applications $A = \{a_i(R_i, r_i, w_i(t), s_i(t))$, where R_i is a resource requirement vector, $r_i \in [0, 1]$ is a priority, $w_i(t)$ is a required work status, and $s_i(t)$ is an actual working status of a_i.

$$min.\varphi = \sum_{i=1}^{n} \int_{0}^{T} r_i w_i(t)[1 - s_i(t)]dt$$

$$s.t.(1) \sum_{i=1}^{n} u_i s_i(t) \leq \alpha_c u_c$$

$$(2) \sum_{i=1}^{n} c_i s_i(t) \leq \alpha_m c_m$$

$$(3) \sum_{i=1}^{n} b_i^I s_i(t) \leq \alpha_I b_I$$

$$(4) \sum_{i=1}^{n} b_i^O s_i(t) \leq \alpha_O b_O$$

Equation 3.1 Scheme proposed in [10].

The integral calculates the total waiting time of the tasks, where T is evaluation time. u_c, c_m, b_I and b_O are, respectively, total CPU utilization, total memory capacity, input bandwidth and output bandwidth the gateway can offer. and the vector $[\alpha_c, \alpha_m, -\alpha_I, \alpha_O]$ represents the overloading factors for the CPU, memory, input and output bandwidth, respectively, and indicates that the system can run smoothly to some extent.

In [17], a set of analytic expressions are proposed for representing the expected energy consumption of devices, as well as a cloud billing method for a group of devices on an IoT aggregator. Considering n devices, Equation (3.2) computes the energy consumption of each device over the monitoring period, T. Where $E[\Psi_e]$ is a query data volume of devices, g_e is consumption rate (joule-per-bit), and i_e is "idle" energy consumption by each device. The integral of the second term represents the expected energy consumption of a device in idle mode. $c_e E[\Psi_e]$ is a threshold where the application activates "idle" mode, and $P_e(\omega_e)$ represent the probability density function of Ψ_e.

$$E_{\exp} = E[\Psi_e]g_e + i_e \int_{0}^{c_e E[\Psi_e]} (c_e E[\Psi_e] - \omega_e) P_e(\omega_e) d\omega_e$$

Equation 3.2 Energy consumption model presented in [17].

Equation (3.3) shows the calculation of expected cloud billing cost, taking into account n aggregated query volumes from n devices. $E[\Psi_b]g_b$ represents the transfer/storage costs, the first integral represents the idle billing cost, and the second integral corresponds to the active billing cost. i_b is the cost of transferring one bit (dollars-per-query-bit), and c_b is a coupling point that define the expect billing cost.

In [25], the utilization of a strategy using fog computing with microgrids to reduce the energy consumption of IoT applications is considered. Two

$$B_{\exp} = E[\Psi_b]g_b + i_b \int_0^{c_b} (c_b - \omega_b)P(\omega_b)d\omega_b + p_b \int_{c_b}^{\infty} (\omega_b - c_b)P(\omega_b)d\omega_b$$

Equation 3.3 Billing model presented in [15].

equations are proposed to evaluate energy consumption. Equation (3.4) calculates the energy consumed by an IoT service using cloud computing. This equation takes into account the energy consumed by IoT gateways when receiving data from IoT devices and sensors (E_{GW-r}), the energy consumed by IoT gateways to transmit data to the cloud data center (E_{GW-t}), the energy consumed by transport network between IoT gateways and cloud (E_{net}), and the energy consumed by components of data center (E_{DC}).

$$E_{IoT-cloud} = E_{GW-r} + E_{GW-t} + E_{net} + E_{DC}$$

Equation 3.4 Calculate of energy consumption between IoT and cloud infrastructure [25].

$$E_{IoT-fog} = E_{GW-r} + E_{GW-c} + \beta(E_{GW-t} + E_{net} + E_{DC})$$

Equation 3.5 Calculate of energy consumption between IoT and fog infrastructure [25].

Equation (3.5) calculates the energy consumption of communication between the IoT and the fog. This equation takes into account the same components of previous equation plus two other components: E_{GW-c}, the energy consumed by IoT gateways for local computation and processing, and β, a ratio of the number of updates from the fog to the cloud for synchronization.

In [16], an analytical modeling for estimating the reliability of an IoT scenario is proposed in a smart home context. An algorithm is proposed to estimate the reliability of the IoT service, which is formed by n subsystems. The calculation of the IoT system reliability is defined in Equation (3.6). It considers the availability of the all k programs running on the virtual machines (P_{pr}), the availability of the f input files for the programs (P_f), and the reliability of each subsystem (ISR), i.e. the reliability of the VM being executed.

$$R_s(t_b) = \prod_{i=1}^{N} ISR_i\ ISR_i \times \prod_{i=1}^{f} P_f(i) \times \prod_{i=1}^{k} P_{pr}(i)$$

Equation 3.6 Reliability equation presented in [16].

In [15], the authors presented a model to evaluate the security level of C2T systems. The focus of the model is the flow of information, where an initial state of the system is defined, and a set of operations are performed. Thus, after performing these actions, the system reaches new states.

Works presented in [11, 23], and [32] use analytical models in optimization problems. In [23], authors used a knapsack problem (MKP) to find the optimal service allocation in C2F2T scenarios. For this, they consider a number of application aspects: load balancing, delay and energy consumption. So, the service allocation is defined as an MKP problem, where the objective is threefold: minimizing the energy consumption by devices, minimizing the overload in terms of processing capacity, and minimize the overall allocation of services in infrastructure. In [32], authors address the interactions among cloud operator and IoT service provide as optimization problem. They formulated analytically the problem and maximize the utilities of cloud service with the purpose of obtaining optimal payment and computation offloading. In [11], authors propose analytical modeling to represent QoS of IoT composite services, taking into account such metrics as availability, reliability, and response time. An optimization algorithm is proposed to find optimal cost with QoS constraints.

In [27] and [31], authors use analytical models to compare two layers of architectures proposed. In [31], analytical models are used to decide if offloading computing will be processed in IoT devices or in cloud, taking into account the desired QoS and energy level available in IoT devices. By contrast, in [27], analytical models are proposed to compare the performance between the fog architecture against traditional cloud computing. Authors consider several aspects, such as location of devices, operation mode, hardware details, and type of events.

In [12] and [19], authors present analytical models to represent mobile nodes connected to cloud computing. In both articles, the proposed models consider the movement of the devices that are connected to the cloud. In [19], the model details the architecture in a selection of components: wireless sensor network (WSN), the cloud infrastructure, applications, and mobile users. While in [12], authors consider that the mobile devices are connected to the cells to send data to the cloud.

3.3.2 Petri Net Models

According to [34], a Petri Net is a well-known model to represent systems with respect to evaluate performance and dependability. To solve a Petri Net, one can use two options: (i) analytic solution by using Markov chain (in which case all transitions must follow exponential distributions); or (ii) simulations using theory of discrete event simulation. Although Markov chains are also indicated to represent the availability of a system, Petri Nets allows a more fine-grained representation of the system, utilizing Markovian and non-Markovian distributions, and represents the system behavior with a fewer number of states [34]. We found two works that used Petri Net models: [26] and [29].

In [26], a mobile crowdsensing framework, integrating mobile devices into services hosted in the cloud is proposed. In order to demonstrate that the proposed framework is better than the standard mobile crowdsensing architectures, two Petri Net models are proposed. Figure 3.4 illustrates these two models. Figure 3.4a shows the proposed framework, while Figure 3.4b shows a common mobile crowdsensing architecture.

The places represent four possible states of a contributing node: contributing node availability (*Av/ NAv*) and position on the interest area (*In/Out*). The transitions represent the probability of the devices entering or exiting these states. The framework model is more complex as extra modules were added, such as contributor enrollment and churn management.

In [29], the problem of managing the traceability of data in C2T scenario is evaluated. A Petri Net model is proposed (Figure 3.5) to map and match device data to users that assists tracking a transparent data trace route, and possible detection of data compromises.

Here, the Petri Net represents the behavior of a proposed wearable IoT architecture. The places represent the different sources where data are generated or

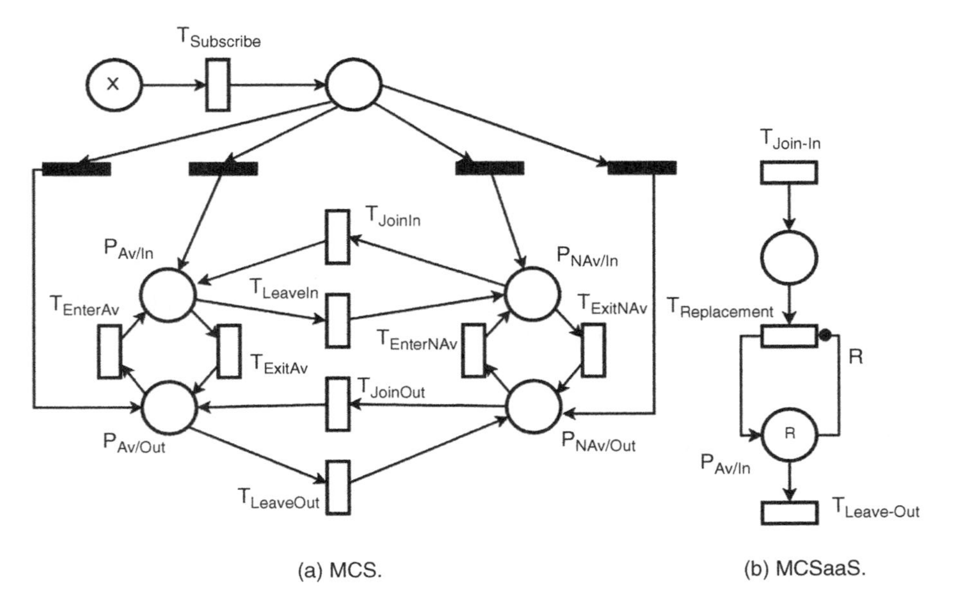

Figure 3.4 Petri Net model proposed in [26]. © Elsevier. Reproduced with the permission of Elsevier.

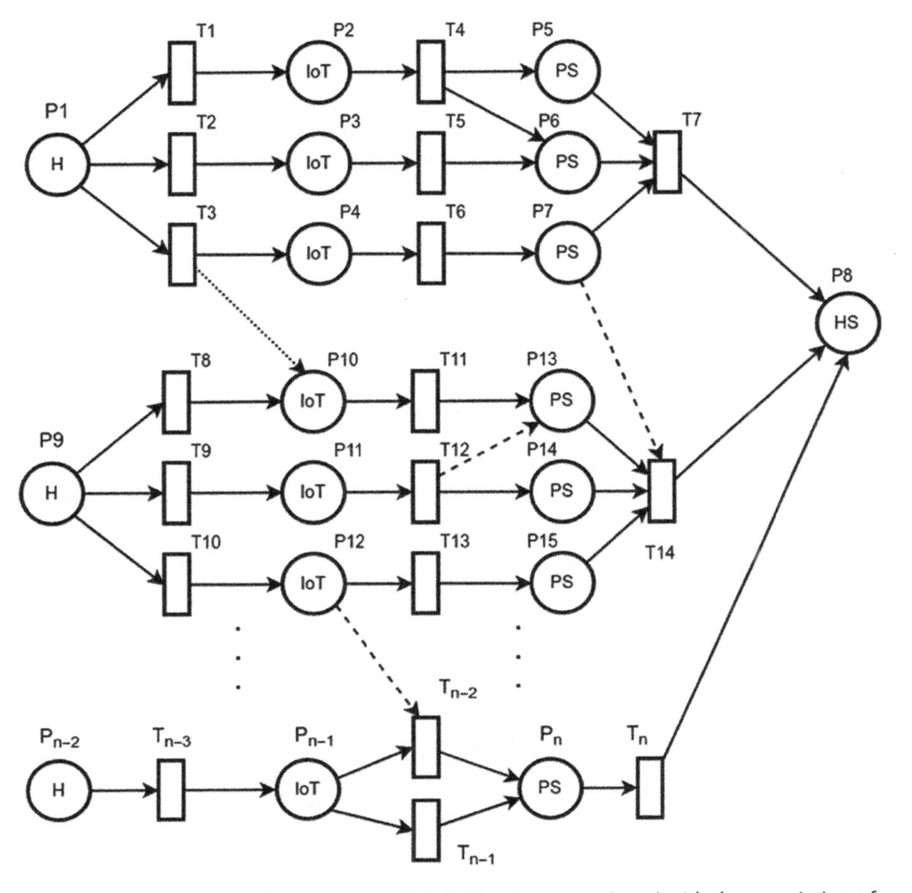

Figure 3.5 Petri Net model presented in [29]. © Elsevier. Reproduced with the permission of Elsevier.

collected. Transitions represent events that may occur, such as new medical data readings, for example vitals, from the wearable IoT device.

3.3.3 Integer Linear Programming

Some optimization problems can be modeled using integer linear programming (ILP). Here, the objective function and all the constraints are expressed as linear functions [35]. However, if the problem involves continuous and discrete variables, a mixed-integer linear programming (MILP) approach can be used to solve it. In this systematic review, two works modeled their problems using ILP and MILP, [28] and [14], respectively.

In [28], an ILP model is proposed for calculating the financial cost of a fog-cloud architecture located in a metropolitan area network (MAN). The authors represent application profiles and characteristics of each node in the MAN as two vectors. The ILP model minimizes the operational cost necessary to support the traffic in network topology while satisfy the application constraints. Equation (3.7) is the objective function that needs to be minimized. $Cost_t$ represents the total cost, $Cost_p$ is the processing cost, $Cost_s$ is the storage cost, $Cost_u$ are the total upstream and downstream costs respectively, and $Cost_c$ is the total MAN link capacity.

$$Cost_t = Cost_p + Cost_s + Cost_u + Cost_d + Cost_c$$

Equation 3.7 Objective function used in ILP model presented in [28].

Authors in [14] model the energy consumption of an IoT architecture by using MILP approach. This architecture is composed of mini clouds. The objective of the model is to minimize the energy consumption, where this consumption is composed of traffic-induced energy consumption and processing-induced energy consumption.

3.3.4 Other Approaches

According to [36], Markov chains model a sequence of random variables, which correspond to the system states, in which a state at one time depends only on the state in the previous time. Markov chains are being widely applied as statistical models of real-world problems. One article was identified in this category [21]. In this work, the authors proposed a Markov chain model with the goal of representing the energy consumption of an IoT-based system for collision detection on guardrails. This system is composed of a WSN, where gateways collect the sensors information and send it to the cloud. The Markov chain model is presented in Figure 3.6. Each state of the Markov chain represents a state of the system, with an associated energy consumption level. It is possible to estimate the energy consumption of the system by calculating the probability of a given state materializing.

Another modeling approach that can be used to represent the C2F2T integration is a probability-based one. While probability is useful to express the likelihood of an occurrence of an event, Bayesian probability represents a conditional measure of uncertainty associated with the occurrence of a given event, considering available information and prior beliefs [37]. The authors in [24] propose a Bayesian model for predicting events that may occur in an IoT application that is connected with the cloud. For this, the model calculates the probability of future events occurring based on historical event data. In addition, the authors assume that the occurrence of an event in the application

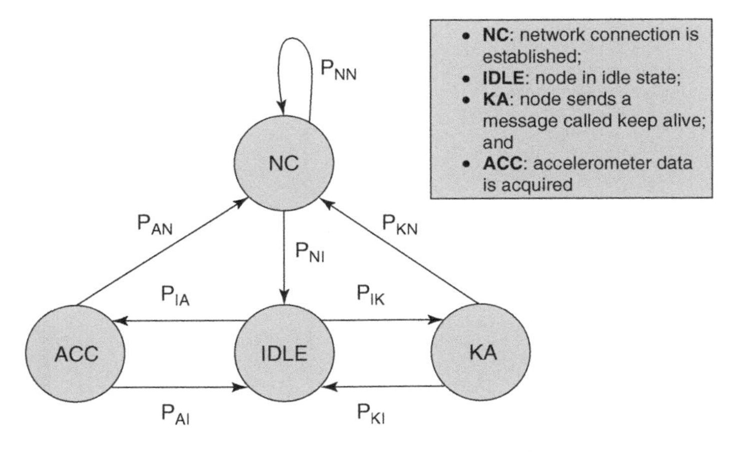

Figure 3.6 Markov model presented in [21]. Adapted from [21].

may imply the occurrence of several chain events. Prediction of flight delays resulting from problems in airplane equipment is used as a use case scenario. Thus, the Bayesian model calculates the conditional probabilities of such events occurring.

Beyond the approaches presented previously, different techniques can be combined to achieve a goal. For instance, in [17], a fault fuzzy-ontology is proposed for analyzing faults in WSN scenarios. In this work, the authors combine an ontology with fuzzy logic, arguing that ontology is appropriate to describe fault, error, and failure domains of systems, while fuzzy logic is a good approach for fault diagnosis. In this way, this approach provides a representation of heterogeneity of faults that allows one understand failures from different perspectives, e.g., applications, devices, and communications. The schema proposed allows us to detect and categorize the faults that occur in WSN. Figure 3.7 illustrates the fuzzy-ontology to detect faults in a hardware system. From right to left, the first layer of the fuzzy-ontology model represents a selection of fault possibilities that can occur in a wireless sensor network. The second layer represents the fault categories. The subsequent layers represent the propagation of fault in wireless sensor network.

3.4 Integrated C2F2T Literature by Use-Case Scenarios

In this section, we describe the use case scenarios modeled in the identified papers e.g. resource management, smart cities, WSN, health and other/generic. These are summarized in Table 3.3.

A number of studies present models to represent applications in health use case scenarios. Authors in [15] use a medical application as a case study for

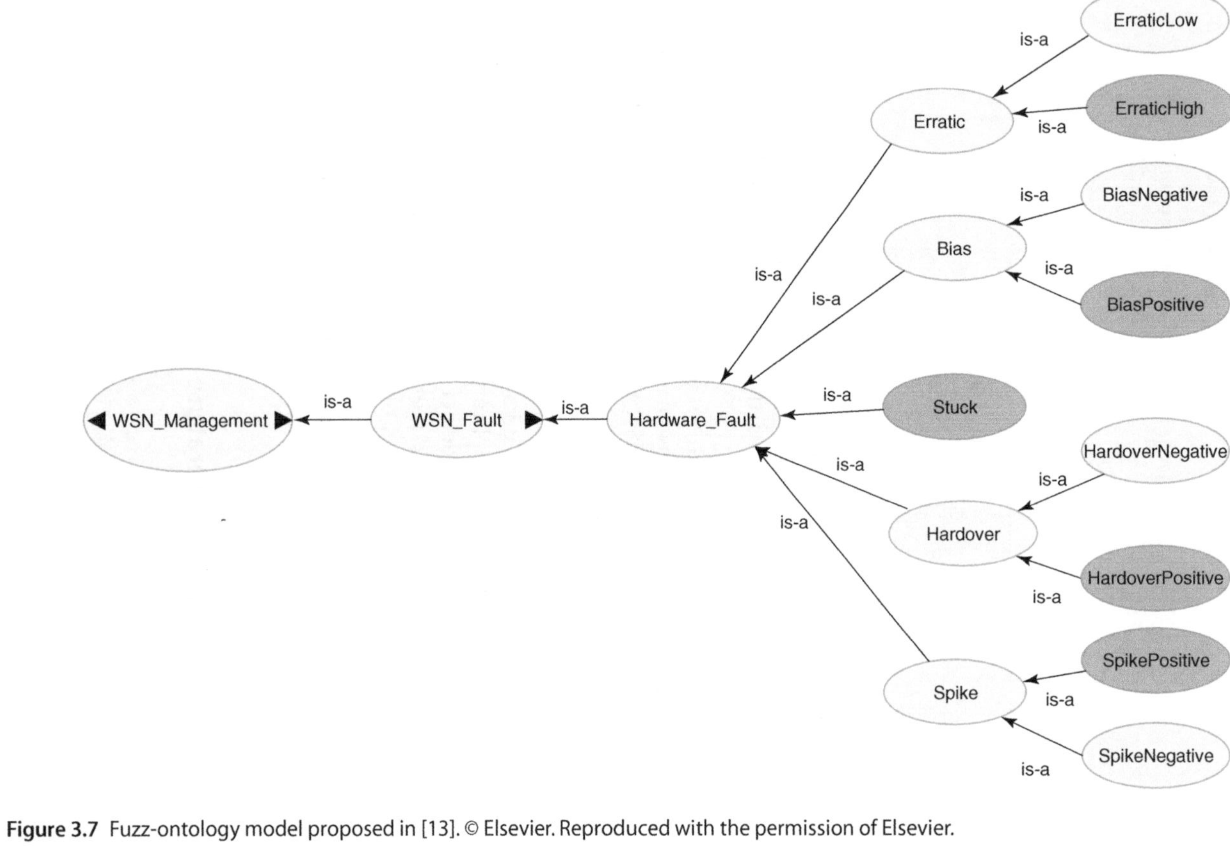

Figure 3.7 Fuzz-ontology model proposed in [13]. © Elsevier. Reproduced with the permission of Elsevier.

their proposed model that analyzes the security of the information flow in IoT systems integrated with cloud infrastructures. In [20], authors propose a framework that enables multiple applications to share IoT computational devices for health monitoring. The use case scenario in [29] is a wearable IoT architecture for healthcare systems.

Use case scenarios that address the WSN context can be found in [17] and [19]. In [13], authors propose a fault fuzzy-ontology that can be used to verify fault tolerance in large-scale WSN using service-oriented applications while [19] proposes a WSN model for sensing as a service that integrates IoT and cloud infrastructures.

The literature widely identifies cases used in smart cities. Authors in [26] and [12] explore a crowdsensing application. The former uses a smart traffic application to evaluate an architecture proposed for mobile crowdsensing-as-a-service [24]. The latter presents an air quality sensor application using mobile sensors and devices as inputs to the proposed model [12]. In [31], authors present a vehicle-to-cloud monitoring application that interacts with edge computing; their analytic model defines where the data will be processed – e.g. in the edge or in the cloud infrastructure. They use a video-streaming application to validate their model. The work presented in [16] is a fire alarm system. Authors evaluate it by using an IoT reliability model. The authors in [24] propose an airplane monitoring scenario, where data generated by C2T application are analyzed through a model in order to

Table 3.3 Scenarios presented in articles.

Scenarios	Applications	Articles
Generic	Generic IoT/fog/cloud applications	[12, 18, 30]
	Edge offloading computing	[32]
Health	Health monitoring	[20, 29]
	Medical research	[15]
WSN	Sensing as a service	[13]
	Fault tolerance	[19]
Smart cities	Mobile crowd sensing as a service	[26]
	Smart traffic	[26, 31]
	Smart living	[10, 16, 20]
	Airplane monitoring	[24]
Resource management	Resource allocation	[23, 28]
	Energy efficiency	[14, 17, 21, 25]
	Quality of service (QoS)	[11, 22]

estimate flights delay. Finally, in [10], authors use a smart living space in their scheduling model with gateways to access cloud resources.

In the resource management category, works were identified that analyze energy consumption. Authors in [17] and [14] study energy efficiency of IoT devices in a cloud-IoT context. Renna et al. [17] analyzes the relation between energy consumption and cloud infrastructure billing cost. Benazzouz and I. Parissis [13] presents a model to improve the energy efficiency of the IoT devices. In [21], authors use a model to analyze the energy consumption of a WSN. Similarly, [25] studies energy consumption but, in this case, the authors analyze how the integration of IoT and fog can reduce this consumption. Papers that address resource allocation in a C2F2T context were also identified, i.e. [23] and [28]. Finally, [11] and [22] discuss QoS in IoT and cloud-IoT environments, respectively. Ming and Yan [11] utilize a mathematical model to calculate the QoS of a set of IoT services (IoT sensor network). In [11], the authors propose a neighborhood model that makes QoS predictions for IoT and cloud services.

Some works do not describe a specific use case scenario or application for their models. For instance, Li et al. [18] presents a three-tier model that address the C2F2T scenario; and Desikan, Srinivasan, and Murthy [30] proposes a model to improve the response time of an IoT system that uses gateways to communicate to the cloud, using fog computing as an intermediary platform. Similarly, the work in [27] is a generic fog architecture and [32] proposes an architecture focused on offloading computing to the edge to improve the mobile user's experience.

3.5 Integrated C2F2T Literature by Metrics

In this section, we explore the sample of papers identified by the evaluation metrics (Table 3.4) used to derive insights in to the main concerns relating to C2F2T scenarios. We do not describe the metrics addressed in some of the papers in this section. Benazzouz and Parissis [13] do not evaluate the model proposed; while in [20, 22], and [24], the authors only evaluate the efficiency of the proposed models.

3.5.1 Energy Consumption

By design, most IoT devices are limited by power source, typically a battery, which comprises application performance. Unsurprisingly, energy consumption features in most (10) of the articles examined. Authors in [18] propose a model to represent a C2F2T architecture. They evaluated the energy consumption in each of the application layers (IoT, fog, and cloud), identified

Table 3.4 Metrics observed in articles.

Metric	Variations	Articles
Energy consumption	Devices power consumption	[14, 17, 18, 25, 27, 31],
	Energy efficiency	[19]
	Percentage energy saved	[12, 23]
Performance	Latency	[18, 27, 28]
	Effective arrival rate	[30]
	Average system response	[30]
	System effectiveness	[26, 30]
	Cache performance	[10]
Resource consumption	Usage of devices	[23, 32]
	Average buffer occupancy	[30]
Costs	Itemized costs	[28]
	Operational costs	[17]
QoS	Quality of pattern recognition in images	[31]
	Response time	[11]
	Reliability	[11, 16, 29]
Security	Opacity	[15]

the main sources of consumption of each, and verified the energy consumption in relation to the increase in the number of devices in the architecture.

In [17], the energy consumption of devices in an IoT infrastructure are presented in a different way. The model represents the device energy cost in three modes: active, idle, and when the application switches from idle to active state. In [12], the authors also address the energy consumption of idle and active devices sending data to a cloud environment, but in this work, the energy consumption for transmitting, receiving, listening, sensing, and computing are taken into account. In [21], the authors consider the operational mode of devices, however, in this work the energy consumption depends on the distance between the communicating nodes and the number of bits that will be transmitted. In addition, the authors take into account the medium where the communication will occur – for instance, open space.

In [12, 14, 27], the authors consider that energy consumption is directly related to the traffic between hierarchical devices layers. In [14], the energy consumption is also impacted by processing-induced VMs located in the networking elements at the upper three layers. In [12], the movement of devices is considered as an additional variable.

Jalali et al. [25] evaluate the energy consumption with respect to two perspectives: the synchronization between IoT devices and the fog, and between IoT devices and the cloud. Moreover, they consider how the data flow of applications affect energy consumption.

The work presented in [31] assesses the energy consumption of fog devices and cloud equipment. The fog devices are more similar to the IoT devices with respect to computational capacity. This paper is relatively unique in that it presents a renewable energy source where the fog devices are powered by photovoltaic panels. They evaluate the consumption of energy produced for photovoltaic panels, and other energy sources, such as batteries.

In [23], Souza et al. try to minimize the excessive energy consumption of C2F2T architectures to find on optimal service allocation.

3.5.2 Performance

Another metric that has been evaluated in many articles is application performance. In [18] and [27], the system performance is evaluated with respect to the latency of the application. This latency can be divided into (i) processing latencies and (ii) transmission latencies. The processing latency is the time taken by an application to process all tasks. The transmission latency is the communication delay to send a unit of data to the destination. In [19], the packet delivery latency is evaluated. This is calculated by the number of hops to the destination, the sleep interval of a node, and the time to transmit a data packet. In [28], the delay of processing applications requests is defined analytically as a combination of computational complexity of devices and the average flow size of application.

In [10], Lyu et al. evaluate the performance of a scheduling mechanism. For this, they assess the scalability of the scheduling and caching solver and add several sensors connected to cloud servers. In addition, the average waiting time and throughput applications are analyzed.

In [26], a mobile crowdsensing scenario is used to evaluate the performance and the effectiveness of the system using a Petri Net model proposed.

In [30], the authors examine the performance of C2F2T applications using a variety of means. The response time of system is defined as the time elapsed from the data was generated until it is processed by a gateway. Efficiency processing is formulated mathematically and takes account of the time to process the application data, the occupancy buffer, and the response time of all gateways.

3.5.3 Resource Consumption

Due of the limited capabilities of IoT devices, careful attention is required when allocating tasks on these devices. An analysis of this workload allows assigning tasks more optimally, taking into account the resources currently available. In addition, because fog and cloud computing offer additional capabilities for IoT devices, it is necessary to examine where such workloads should be processed.

In [23], authors aim to obtain the optimal allocation of services on the available resources, taking into account resources of devices available in C2F2T infrastructures. They present the advantages of using fog computing and evaluate the number of allocated resources and the usage of resource devices.

In [32], authors assess the level consumption in C2T scenario. They evaluated the consumption about two aspects: cloud infrastructure consumption and IoT devices consumption.

Due to the limited computing and storage capacity of IoT devices, [30] evaluates the effectiveness of gateways by examining the buffer occupancy of the gateways, where the arrival of data to the gateways was represented by a queuing system.

3.5.4 Cost

Applications that require a lot of cloud computing resources can dramatically increase the cost to the service provider and ultimately the consumer. Obviously, there is overlap between energy consumption studies and cost studies.

Renna et al. [17] examine the billing costs when computational resources are reserved to process queries uploaded by IoT applications. Therefore, billing costs are directly proportional to the expected query volume generated by IoT devices.

Sturzinger et al. [28] attempt to minimize the operational cost of provisioning IoT traffic to the cloud, while satisfying all application constraints. The total cost is composed of the sum of the cost of all devices including as processing, storage, upstream and downstream cost, local and global traffic, and the link capacity cost.

3.5.5 Quality of Service

In IoT scenarios, many factors can impact on QoS such as network-related delays, available computational resources, and the number of IoT devices consuming resources.

In [11], composite services are evaluated in an IoT environment. In order to evaluate the QoS, the authors consider it is necessary to divide composite services into simpler more granular services, and then evaluate the QoS of each one separately.

The use case presented in [31] analyses of videostreams in the cloud generated by vehicles on a road. Authors represent QoS as the detection accuracy of an object in video images. They vary the resolution of video in order to decrease the CPU, memory and bandwidth consumption. In addition, considering the scenario where there are $2n+1$ cars, the accuracy is defined as the probability of a result (object detected in image) appears exceeds $n + 1$ times among $2n + 1$ results.

In [16], reliability of the IoT system depends on a variety of factors, including the availability of the service, the availability of the input files to service, and the reliability of each sub-system that composes the overall IoT system. In [29], reliability was defined as the ability of the traceability model to detect and prevent attacks on applications.

3.5.6 Security

Only one article evaluated security issues. In [15], the authors evaluate the concept of opacity in data flow in the C2T context. Opacity is a uniform approach to describe security properties expressed as predicates. The authors used analytical models to verify the opacity in the medical research application scenario.

3.6 Future Research Directions

Fog and cloud computing address a number of problems encountered in the IoT; however, they also increase management complexity. Despite fog and cloud computing offering greater availability and resilience, they can also be viewed as vulnerabilities or potential points of failure. As such, in addition to device/end point failure, we now also need to pay attention to fog node and cloud infrastructure failures. While cloud and fog integration is relatively well known and shares common technologies, the integration/extension with IoT is a nontrivial task, mostly due to massive device heterogeneity and service requirements.

As presented in previous sections, many studies examined have proposed computational models to understand how IoT, fog, and cloud infrastructures can be integrated in order to improve the overall system performance and availability. Those works consider a wide range of applications and scenarios, modeling, and analysis regarding the reduction of application energy consumption resulting from greater C2F2T integration. Future research may address the use of gateways to distribute/balance requests to be processed in the cloud infrastructure or by fog devices.

Another area for future research concerns failure management and, in particular, the minimization of failure, whether on the device or fog node, on application availability. Some applications have high criticality, such as those in the connected health space e.g. health monitoring. In these case, any downtime can lead, in the worst case, to death. Here, the main goal is to identify the bottlenecks in this integrated system and propose strategies to minimize the application downtime, prevent failure, and guide investment decision-making.

Given the complexity of the resource pool in the C2T continuum, resource management is a fruitful area for future research, as requests can be allocated locally, in the fog, or in one or more clouds. There is a wide range of data for informing resource allocation including device location, user information, application throughput and scalability, to name but a few. As the IoT feature space becomes more standardized per use case, more fine-grained data can be used to inform and model user and device behavior to inform these decisions. More sophisticated QoS mechanisms can then be assigned application priorities and resources can be allocated appropriately.

The cloud-fog-IoT space is complex, and such complexity is exacerbated by lack of standardization and extreme heterogeneity. From a modeling perspective, this results in much more computationally complex models. State-based models, such as Markov chains and Petri Nets, grow exponentially with the size of the model and can suffer from the so-called state-space explosion problem [38]. As such, one needs to pay attention to the models' scalability and run-time performance. Moreover, the complexity and, to some extent, the uncertainty resulting from a rapidly changing environment and the chain of service provision in the cloud-fog-IoT space can result in significant challenges in validating models against real scenarios. Researchers need to consider how best to improve the effectiveness and accuracy of models from a methodological perspective. This may require additional effort e.g. prototyping to some other methodology to validate the model's performance accuracy.

3.7 Conclusions

According to Cisco,[1] approximately 20 billion objects will be connected in 2020. These devices generate a massive volume of data at high velocity and varying formats, and require additional processing and storage capacity. In this aspect, cloud computing provides "unlimited" capacity to IoT devices based on the pay-as-you-go model, and one completes the other. However, this integration is complex to manage for many reasons, not least security, performance, communication delay, QoS, and so on. Fog computing adds a layer between cloud and IoT to solve problems related to communication, since the fog devices are located geographically closer to the IoT devices.

A wide variety of applications depend on C2F2T integration. Smart cities, WSN, e-health, traffic management, and smart buildings are some of the use case scenarios featured in the literature examined. These applications have their own domain-specific requirements and associated limitations that need to be evaluated on a case-by-case basis, such as data security and integrity, availability, reliability, real-time data, etc. Models have a useful role to play in the

1 https://www.cisco.com/c/dam/en_us/about/ac79/docs/innov/IoT_IBSG_0411FINAL.pdf

evaluation of many of the components, variables, and aspects of integrated C2TF2 systems.

In this chapter, we presented a systematic literature review about models that represent C2F2T integrations. We analyzed other relevant aspects including typical scenarios and metrics evaluated in the articles. We identified that the most used modeling techniques in the articles were analytical models, with 16 articles, followed by Petri Net (2 articles), and ILP (2 articles). In itself, this suggests a greater need for exploration of the C2T topic through a wider range of methodological lenses. Similarly, our descriptive analytics suggest that energy consumption is the topic of most concern based on the number of articles focusing on this particular unit of measurement. While the literature examined covers a relatively small number of techniques and units of measurement, there is greater variability in the use case scenarios as illustrated in the previous paragraph. Our descriptive analytics reflect a substantial opportunity for future academic research through the identification of studies whose uniqueness can be designed from the interstices of C2F2T components and systems, modeling approaches, units of measurement or metrics, and use case scenarios.

Some aspects were not considered in the articles analyzed. The absence of a significant number of studies regarding application availability in C2F2T scenarios is noteworthy. For example, unavailability of a health monitoring system or healthcare application, for even a short period, can have unacceptable outcomes. Furthermore, greater examination is needed regarding the impact of each C2F2T layer (cloud, fog, and IoT) on the overall availability of the application is needed. Such studies will lead to novel strategies to improve the availability of applications in the C2F2T scenarios, which in themselves will require further evaluation.

The IoT ecosystem is a significant economic and societal opportunity whose evolution may lead to the Internet of Everything, where people, processes, things, data, and networks are all connected and interconnected. It is a vision for the future that is not without challenges. System modeling can play an important role in understanding and optimizing C2F2T systems and accelerating the evolution and maturity of the IoT for everyone's benefit.

Acknowledgments

This work is partially funded by the European Union's Horizon 2020 and FP7 Research and Innovation Programmes through RECAP (http://www.recap-project.eu) under Grant Agreement No. 732667.

The authors would like CAPES (Coordenação de Aperfeiçoamento de Pessoal de Nível Superior) and CNPq (Conselho Nacional de Desenvolvimento Científico e Tecnológico) for the support.

References

1 F. Gens, TOP 10 PREDICTIONS. IDC Predictions 2012: Competing on the 3rd Platform. https://www.virtustream.com/sites/default/files/IDCTOP10Predictions2012.pdf. March, 2018.

2 S. Aguzzi, D. Bradshaw, M. Canning, M. Cansfield, P. Carter, G. Cattaneo, S. Gusmeroli, G. Micheletti, D. Rotondi, R. Stevens. Definition of a Research and Innovation Policy Leveraging Cloud Computing and IoT Combination – Digital Agenda for Europe – European Commission. https://ec.europa.eu/digital-agenda/en/news/definition-research-and-innovation-policy-leveraging-clou1d09-0computing-and-iot-combination. March, 2018.

3 J. Bradley, J. Barbier, D. Handler. Embracing the Internet of Everything to Capture Your Share of $4.4 Trillion. Cisco IBSG Group. http://www.cisco.com/web/about/ac79/docs/innov/IoE_Economy.pdf. March, 2018.

4 F. Bonomi, R. Milito, P. Natarajan, et al. Fog computing: A platform for Internet of Things and analytics, *Big Data and Internet of Things: A Roadmap for Smart Environments*. Springer, Switzerland, 2014.

5 A. Botta, W. De Donato, V. Persico, and A. Pescapé. Integration of cloud computing and internet of things: A survey. *Future Generation Computer Systems,* 56: 684–700, March 2016.

6 L. Iorga, L. Feldman, R. Barton, M. Martin, N. Goren, C. Mahmoudi. The NIST definition of fog computing – draft, NIST Special Publication 800 (191), 2017.

7 P-O. Östberg, J. Byrne, P. Casari, P. Eardley, A. Fernandez Anta, J. Forsman, J. Kennedy, T. Le Duc, M. Noya Marino, R. Loomba, M.A. Lopez Pena, J. Lopez Veiga, T. Lynn, V. Mancuso, S. Svorobej, A. Torneus, S. Wesner, P. Willis, and J. Domaschka. Reliable capacity provisioning for distributed cloud/edge/fog computing applications. European Conference on Networks and Communications (EuCNC), Oulu, Finland, June 12–15, 2017.

8 FQB da Silva, M. Suassuna, A. César C. França, et al. Replication of empirical studies in software engineering research: a systematic mapping study. *Empirical Software Engineering*, 19(3): 501–557, June 2014.

9 F. Rowe. What literature review is not: diversity, boundaries and recommendations. *European Journal of Information Systems,* 23(3): 241–255, May 2014.

10 Y. Lyu, F. Yan, Y. Chen, et al, High-performance scheduling model for multisensor gateway of cloud sensor system-based smart-living. *18th International Conference on Information Fusion (Fusion)* 21: 42–56, January 2015.

11 Z. Ming and M. A. Yan. QoS-aware computational method for IoT composite service. *The Journal of China Universities of Posts and Telecommunications* 20 (2013): 35–39.

12 L. Skorin-Kapov, K. Pripuzic, M. Marjanovic, et al, Energy efficient and quality-driven continuous sensor management for mobile IoT applications, *Collaborative Computing: Networking, Applications and Worksharing (CollaborateCom), 2014 International Conference on,* Miami, Florida, October 22–25, 2014.

13 Y Benazzouz and I. Parissis, A fault fuzzy-ontology for large scale fault-tolerant wireless sensor networks, *Procedia Computer Science* 35 (September 2014): 203–212.

14 Zaineb T. Al-Azez, et al. Virtualization framework for energy efficient IoT networks. *4th IEEE International Conference on Cloud Networking (CloudNet),* Niagra Falls, Canada, October 5–7, 2015.

15 W. Zeng, K. Maciej, and P. Watson. *Opacity in Internet of Things with Cloud Computing. University of Newcastle Upon Tyne,* Newcastle upon Tyne University Computing Science, Newcastle, England, 2015.

16 R. K. Behera, K. Ranjit Kumar, K.R. Hemant, K. Reddy, and D.S. Roy. Reliability modelling of service-oriented Internet of Things. *Infocom Technologies and Optimization (ICRITO)(Trends and Future Directions), 2015 4th International Conference on,* Noida, India, September 2–4, 2015.

17 F. Renna, J. Doyle, V. Giotsas, Y. Andreopoulos. Query processing for the Internet-of-Things: Coupling of device energy consumption and cloud infrastructure billing. *2016 IEEE First International Conference on Internet-of-Things Design and Implementation (IoTDI).* Berlin, Germany, April 4–8, 2016.

18 W. Li, I. Santos, F.C. Delicato, P.F. Pires, L. Pirmez, W. Wei, H. Song, A. Zomaya, S. Khan. System modelling and performance evaluation of a three-tier cloud of things. *Future Generation Computer Systems* 70 (2017): 104–125.

19 T. Dinh, K. Younghan, and L. Hyukjoon. A location-based interactive model for Internet of Things and cloud (IoT-cloud). *Ubiquitous and Future Networks (ICUFN), 2016 Eighth International Conference on,* Vienna, Austria, July 5–8, 2016.

20 J.F. Colom, H. Mora, D. Gil, M.T. Signes-Pont. Collaborative building of behavioural models based on internet of things, *Computers & Electrical Engineering,* 58: 385–396, February 2017.

21 T. Gomes, D. Fernandes, M. Ekpanyapong, J. Cabral. An IoT-based system for collision detection on guardrails. 2016 IEEE International Conference on Industrial Technology (ICIT), Taipei, Tawan, May 14–17, 2016.

22 H. Wu, K. Yue, C. H. Hsu, Y. Zhao, B. Zhang, G. Zhang. Deviation-based neighborhood model for context-aware QoS prediction of cloud and IoT services. *Future Generation Computer Systems* 76: 550–560, November 2017.

23 V. B. Souza, X Masip-Bruin, E. Marin-Tordera, W. Ramirez, and S. Sanchez. Towards Distributed Service Allocation in Fog-to-Cloud (F2C) Scenarios, *Global Communications Conference (GLOBECOM), 2016 IEEE*, Washington, DC, USA, December 4–8, 2016.

24 B. Karakostas. Event prediction in an IoT Environment using Naïve Bayesian models. *Procedia Computer Science,* 83: 11–17, 2016.

25 F. Jalali, A. Vishwanath, J. de Hoog, F. Suits. Interconnecting fog computing and microgrids for greening IoT. *IEEE Innovative Smart Grid Technologies-Asia (ISGT-Asia),* Melbourne, Australia, 28 November–1 December, 2016.

26 G. Merlino, S. Arkoulis, S. Distefano, C. Papagianni, A. Puliafito, and S. Papavassiliou. Mobile crowdsensing as a service: a platform for applications on top of sensing clouds. *Future Generation Computer Systems,* 56: 623–639, March 2016.

27 S. Sarkar and M. Sudip. Theoretical modelling of fog computing: a green computing paradigm to support IoT applications. *IET Networks* 5.2: 23–29, March 2016.

28 E. Sturzinger, T. Massimo, and M. Biswanath. Application-aware resource provisioning in a heterogeneous Internet of Things. *IEEE 21th International Conference on Optical Network Design and Modeling(ONDM), Budapest, Hungary,* May 15–18, 2017.

29 R. K. Lomotey, J. Pry, and S. Sriramoju. Wearable IoT data stream traceability in a distributed health information system. *Pervasive and Mobile Computing,* 40: 692–707, September 2017.

30 K. E. Desikan, M. Srinivasan, and C. Murthy. A Novel Distributed Latency-Aware Data Processing in Fog Computing-Enabled IoT Networks. In *Proceedings of the ACM Workshop on Distributed Information Processing in Wireless Networks,* Chennai, India, July 10–14, 2017.

31 Y. Li, A.C. Orgerie, I. Rodero, M. Parashar, and J.-M. Menaud. Leveraging Renewable Energy in Edge Clouds for Data Stream Analysis in IoT. In *Proceedings of the 17th IEEE/ACM International Symposium on Cluster, Cloud and Grid Computing,* Madrid, Spain, May 14–17, 2017.

32 Y. Liu, C Xu, Y. Zhan, Z. Liu, J. Guan, and H. Zhang. Incentive mechanism for computation offloading using edge computing: a Stackelberg game approach. *Computer Networks* 129: 399–409, 2017.

33 Gregory V. Caliri. Introduction to Analytical Modeling, *Int. CMG Conference,* Orlando, USA, December 10–15, 2000.

34 F. Bause and P. S. Kritzinger. *Stochastic Petri Nets,* Springer Verlag, Germany, 2002.

35 E. Oki. *Linear Programming and Algorithms for Communication Networks: A Practical Guide to Network Design, Control, and Management*, CRC Press, USA, 2012.
36 W. Ching, X. Huang, Michael K. Ng, and T.-K. Siu. *Markov Chains: Models, Algorithms and Applications*, Springer, USA, 2006.
37 T. Ando. *Bayesian Model Selection and Statistical Modeling*, CRC Press, USA, 2010.
38 E.M. Clarke, W. Klieber, M. Novacek, P. Zuliani. Model checking and the state explosion problem. *Tools for Practical Software Verification*, Germany, 2011.

4

Management and Orchestration of Network Slices in 5G, Fog, Edge, and Clouds

Adel Nadjaran Toosi, Redowan Mahmud, Qinghua Chi, and Rajkumar Buyya

4.1 Introduction

The major digital transformation happening all around the world these days has introduced a wide variety of applications and services ranging from smart cities and vehicle-to-vehicle (V2V) communication to virtual reality (VR)/augmented reality (AR) and remote medical surgery. Design and implementation of a network that can simultaneously provide the essential connectivity and performance requirements of all these applications with a single set of network functions not only is massively complex but also is prohibitively expensive. The 5G infrastructure public–private partnership (5G-PPP) has identified various use case families of enhanced mobile broadband (eMBB), massive machine-type communications (mMTC), and ultra-reliable low-latency communication (uRLLC) or critical communications that would simultaneously run and share the 5G physical multi-service network [1]. These applications essentially have very different quality of service (QoS) requirements and transmission characteristics. For instance, video-on-demand streaming applications in eMMB category require very high bandwidth and transmit a large amount of content. By contrast, mMTC applications, such as the Internet of Things (IoT), typically have a multitude of low throughput devices. The differences between these use cases show that the *one-size-fits-all* approach of the traditional networks does not satisfy different requirements of all these vertical services.

A cost-efficient solution toward meeting these requirements is slicing physical network into multiple isolated logical networks. Similar to server virtualization technology successfully used in cloud-computing era, network slicing intends to build a form of virtualization that partitions a shared physical network infrastructure into multiple end-to-end level logical networks allowing for traffic grouping and tenants' traffic isolation. Network slicing is considered as the critical enabler of the 5G network where vertical service

Fog and Edge Computing: Principles and Paradigms, First Edition.
Edited by Rajkumar Buyya and Satish Narayana Srirama.

providers can flexibly deploy their applications and services based on the requirements of their service. In other words, network slicing provides a *network-as-a-service* (NaaS) model, which allows service providers to build and set up their own networking infrastructure according to their demands and customize it for diverse and sophisticated scenarios.

Software-defined networking (SDN) and network function virtualization (NFV) can serve as building blocks of network slicing by facilitating network programmability and virtualization. SDN is a promising approach to computer networking that separates the tightly coupled control and data planes of traditional networking devices. Thanks to this separation, SDN can provide a logically centralized view of the network in a single point of management to run network control functions. NFV is another trend in networking gaining momentum quickly, with the aim of transferring network functions from proprietary hardware to software-based applications executing on general-purpose hardware. NFV intends to reduce the cost and increase the elasticity of network functions by building virtual network functions (VNFs) that are connected or chained together to build communication services.

With this in mind, in this chapter, we aim to review the state-of-the-art literature on network slicing in 5G, edge/fog, and cloud computing, and identify the spectrum challenges and obstacles that must be addressed to achieve the ultimate realization of this concept. We begin with a brief introduction of 5G, edge/fog, and clouds and their interplay. Then, we outline the 5G vision for network slicing and identify a generic framework for 5G network slicing. We then review research and projects related to network slicing in cloud computing context, while we focus on SDN and NFV technologies. Further, we explore network slicing advances in emerging fog and edge cloud computing. This leads us to identify the key unresolved challenges of network slicing within these platforms. Concerning this review, we discuss the gaps and trends toward the realization of network slicing vision in fog and edge and software-defined cloud computing. Finally, we conclude the chapter.

Table 4.1 lists acronyms and abbreviations referenced throughout the chapter.

4.2 Background

4.2.1 5G

The renovation of telecommunications standards is a continuous process. Practicing this, 5th generation mobile network or 5th generation wireless system, commonly called 5G, has been proposed as the next telecommunications standards beyond the current 4G/IMT advanced standards [2]. The wireless networking architecture of 5G follows 802.11ac IEEE wireless networking

Table 4.1 Acronyms and abbreviations.

5G	5th generation mobile networks or 5th generation wireless systems
AR	augmented reality
BBU	baseband unit
CRAN	cloud radio access network
eMBB	enhanced mobile broadband
FRAN	fog radio access network
IoT	Internet of Things
MEC	mobile edge computing
mMTC	massive machine-type communications
NaaS	network-as-a-service
NAT	network address translation
NFaaS	network function as a service
NFV	network function virtualization
QoS	quality of service
RRH	remote radio head
SDC	software-defined clouds
SDN	software-defined networking
SFC	service function chaining
SLA	service level agreement
uRLLC	ultra-reliable low-latency communication
V2V	vehicle to vehicle
VM	virtual machine
VNF	virtualized network function
VPN	virtual private network
VR	virtual reality

criterion and operates on millimeter wave bands. It can encapsulate extremely high frequency (EHF) from 30 to 300 gigahertz (GHz) that ultimately offers higher data capacity and low latency communication [3].

The formalization of 5G is still in its early stages and is expected to be mature by 2020. However, the main intentions of 5G include enabling Gbps data rate in a real network with least round-trip latency and offering long-term communication among the large number of connected devices through high-fault tolerant networking architecture [1]. Also, it targets improving the energy usage both for the network and the connected devices. Moreover, it is anticipated that 5G will be more flexible, dynamic, and manageable compared to the previous generations [4].

4.2.2 Cloud Computing

Cloud computing is expected to be an inseparable part of 5G services for providing an excellent backend for applications running on the accessing devices. During the last decade, cloud has evolved into a successful computing paradigm for delivering on-demand services over the Internet. The cloud data centers adopted virtualization technology for efficient management of resources and services. Advances in server virtualization contributed to the cost-efficient management of computing resources in the cloud data centers.

Recently, the virtualization notion in cloud data centers, thanks to the advances in SDN and NFV, has extended to all resources, including compute, storage, and networks, which formed the concept of software defined clouds (SDC) [5]. SDC aims to utilize the advances in areas of cloud computing, system virtualization, SDN, and NFV to enhance resource management in data centers. In addition, cloud is regarded as the foundation block for *cloud radio access network (CRAN)*, an emerging cellular framework that aims to meet ever-growing end-user demand on 5G. In CRAN, the traditional base stations are split into radio and baseband parts. The radio part resides in the base station in the form of the remote radio head (RRH) unit and the baseband part is placed to cloud for creating a centralized and virtualized baseband unit (BBU) pool for different base stations.

4.2.3 Mobile Edge Computing (MEC)

Among the user proximate computing paradigms, MEC is considered as one of the key enablers of 5G. Unlike CRAN [6], in MEC, base stations and access points are equipped with edge servers that take care of 5G-related issues at the edge network. MEC facilitates a computationally enriched distributed RAN architecture upon the LTE-based networking. Ongoing research on MEC targets real-time context awareness [7], dynamic computation offloading [8], energy efficiency [9], and multi-media caching [10] for 5G networking.

4.2.4 Edge and Fog Computing

Edge and fog computing are coined to complement the remote cloud to meet the service demands of a geographically distributed large number of IoT devices. In edge computing, the embedded computation capabilities of IoT devices or local resources accessed via ad-hoc networking are used to process IoT data. Usually, an edge computing paradigm is well suited to perform light computational tasks and does not probe the global Internet unless intervention of remote (core) cloud is required. However, not all the IoT devices are computationally enabled, or local edge resources are computational-enriched to execute different large-scale IoT applications simultaneously. In this case, executing latency sensitive IoT applications at remote cloud can degrade the QoS significantly [11]. Moreover, a huge amount of the IoT workload sent to remote cloud can flood the Internet and congest the network. In response to these challenges, fog computing offers infrastructure and software services through distributed fog nodes to execute IoT applications within the network [12].

In fog computing, traditional networking devices such as routers, switches, set-top boxes, and proxy servers, along with dedicated nano-servers and micro-data centers, can act as fog nodes and create wide area cloud-like services both in an independent or clustered manner [13]. Mobile edge servers or cloudlets [14] can also be regarded as fog nodes to conduct their respective jobs in fog-enabled MCC and MEC. In some cases, edge and fog computing are used interchangeably, although, in a broader perspective, edge is considered as a subset of fog computing [15]. However, in edge and fog computing, the integration of 5G has already been discussed in terms of bandwidth management during computing instance migration [16] and SDN-enabled IoT resource discovery [17]. The concept of fog radio access network (FRAN) [18] is also getting attention from both academia and industry where fog resources are used to create BBU pool for the base stations.

Working principle of these computing paradigms largely depends on virtualization techniques. The alignment of 5G with different computing paradigms can also be analyzed through the interplay between network and resource virtualization techniques. Network slicing is one of the key features of 5G network virtualization. Computing paradigms can also extend the vision of 5G network slicing into data center and fog nodes. By the latter, we mean that the vision of network slicing can be applied to the shared data center network infrastructure and fog networks to provide an end-to-end logical network for applications by establishing a full-stack virtualized environment. This form of network slicing can also be expanded beyond a data center network into multiclouds or even cluster of fog nodes [19]. Whatever the extension may be, this creates a new set of challenges to the network, including wide area network (WAN) segments, cloud data centers (DCs), and fog resources.

4.3 Network Slicing in 5G

In recent years, industries and academia have undertaken numerous research initiatives to explore different aspects of 5G. Network architecture and its associated physical and MAC layer management are among the prime focuses of current 5G research. The impact of 5G in different real-world applications, sustainability, and quality expectations is also gaining predominance in the research arena. However, among the ongoing research in 5G, network slicing is drawing more attractions since this distinctive feature of 5G aims at supporting diverse requirements at the finest granularity over a shared network infrastructure [20, 21].

Network slicing in 5G refers to sharing a physical network's resources to multiple virtual networks. More precisely, network slices are regarded as a set of virtualized networks on the top of a physical network [22]. The network slices can be allocated to specific applications/services, use cases or business models to meet their requirements. Each network slice can be operated independently with its own virtual resources, topology, data traffic flow, management policies, and protocols. Network slicing usually requires implementation in an end-to-end manner to support coexistence of heterogeneous systems [23].

The network slicing paves the way for customized connectivity among a high number of interconnected end-to-end devices. It enhances network automation and leverages the full capacity of SDN and NFV. Also, it helps to make the traditional networking architecture scalable according to the context. Since network slicing shares a common underlying infrastructure to multiple virtualized networks, it is considered as one of the most cost-effective ways to use network resources and reduce both capital and operational expenses [24]. Besides, it ensures that the reliability and limitations (congestion, security issues) of one slice do not affect the others. Network slicing assists isolation and protection of data, control and management plane that enforce security within the network. Moreover, network slicing can be extended to multiple computing paradigms such as edge [25], fog [13], and cloud that eventually improves their interoperability and helps to bring services closer to the end user with less service-level agreement (SLA) violations [26].

Apart from the benefits, the network slicing in current 5G context is subjected to diversified challenges, however. Resource provisioning among multiple virtual networks is difficult to achieve since each virtual network has a different level of resource affinity and it can be changed with the course of time. Besides, mobility management and wireless resource virtualization can intensify the network slicing problems in 5G. End-to-end slice orchestration and management can also make network slicing complicated. Recent research in 5G network slicing mainly focuses on addressing the challenges through efficient network slicing frameworks. Extending the literature [26, 27], we depicted a generic framework for 5G network slicing in Figure 4.1 The framework consists of three main layers: *infrastructure layer*, *network function layer*, and *service layer*.

4.3.1 Infrastructure Layer

The infrastructure layer defines the actual physical network architecture. It can be expanded from edge cloud to remote cloud through radio access network and the core network. Different software defined techniques are encapsulated to facilitate resource abstraction within the core network and the radio access network. Besides, in this layer, several policies are conducted to deploy, control, manage, and orchestrate the underlying infrastructure. This layer allocates

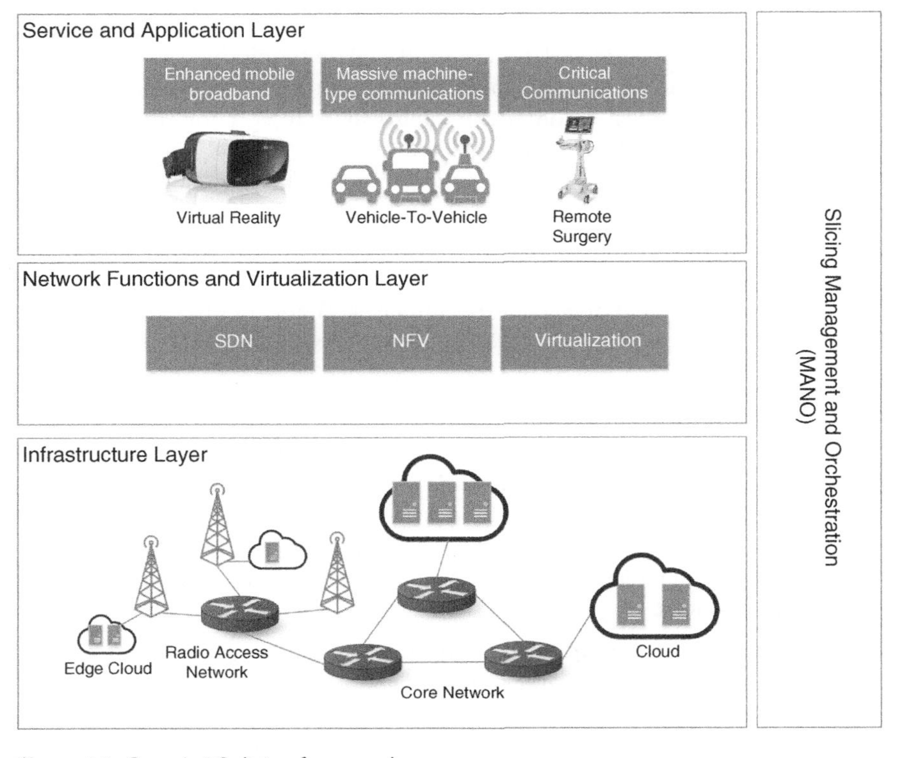

Figure 4.1 Generic 5G slicing framework.

resources (compute, storage, bandwidth, etc.) to network slices in such way that upper layers can get access to handle them according to the context.

4.3.2 Network Function and Virtualization Layer

The network function and virtualization layer executes all the required operations to manage the virtual resources and network function's life cycle. It also facilitates optimal placement of network slices to virtual resources and chaining of multiple slices so that they can meet specific requirements of a particular service or application. SDN, NFV, and different virtualization techniques are considered as the significant technical aspect of this layer. This layer explicitly manages the functionality of core and local radio access network. It can handle both coarse-grained and fine-grained network functions efficiently.

4.3.3 Service and Application Layer

The service and application layer can be composed by connected vehicles, virtual reality appliances, mobile devices, etc. having a specific use case or business model and represent certain utility expectations from the networking infrastructure and the network functions. Based on requirements or high-level description of the service or applications, virtualized network functions are mapped to physical resources in such way that SLA for the respective application or service does not get violated.

4.3.4 Slicing Management and Orchestration (MANO)

The functionality of the above layers are explicitly monitored and managed by the slicing management and orchestration layer. There are three main tasks in this layer:

1. Create virtual network instances upon the physical network by using the functionality of the infrastructure layer.
2. Map network functions to virtualized network instances to build a service chain with the association of network function and virtualization layer.
3. Maintain communication between service/application and the network slicing framework to manage the life cycle of virtual network instances and dynamically adapt or scale the virtualized resources according to the changing context.

The logical framework of 5G network slicing is still evolving. Retaining the basic structure, extension of this framework to handle the future dynamics of network slicing can be a potential approach to further standardization of 5G.

According to Huawei, a high-level perspective of 5G network [28], Cloud-Native network architecture for 5G has four characteristics:

1. It provides cloud data center–based architecture and logically independent network slicing on the network infrastructure to support different application scenarios.
2. It uses Cloud-RAN[1] to build radio access networks (RAN) to provide a substantial number of connections and implement 5G required on-demand deployments of RAN functions.
3. It provides simpler core network architecture and provides on-demand configuration of network functions via user and control plane separation, unified database management, and component-based functions.
4. In an automatic manner, it implements network slicing service to reduce operating expenses.

---

1 CLOUD-RAN (CRAN) is a centralized architecture for radio access network (RAN) in which the radio transceivers are separated from the digital baseband processors. This means that operators can centralize multiple base band units in one location. This simplifies the amount of equipment needed at each individual cell site. Ultimately, the network functions in this architecture become virtualized in the Cloud.

In the following section, we intend to review the state-of-the-art related work on network slice management happening in cloud computing literature. Our survey in this area can help researcher to apply advances and innovation in 5G and clouds reciprocally.

4.4 Network Slicing in Software-Defined Clouds

Virtualization technology has been the cornerstone of resource management and optimization in cloud data centers for the last decade. Many research proposals have been expressed for VM placement and virtual machine (VM) migration to improve utilization and efficiency of both physical and virtual servers [29]. In this section, we focus on the state-of-the-art network-aware VM/VNF management in line with the aim of the report, i.e., network slicing management for SDCs. Figure 4.2 illustrates our proposed taxonomy of network-aware VM/VNF management in SDCS. Our taxonomy classifies existing works based on the objective of the research, the approach used to

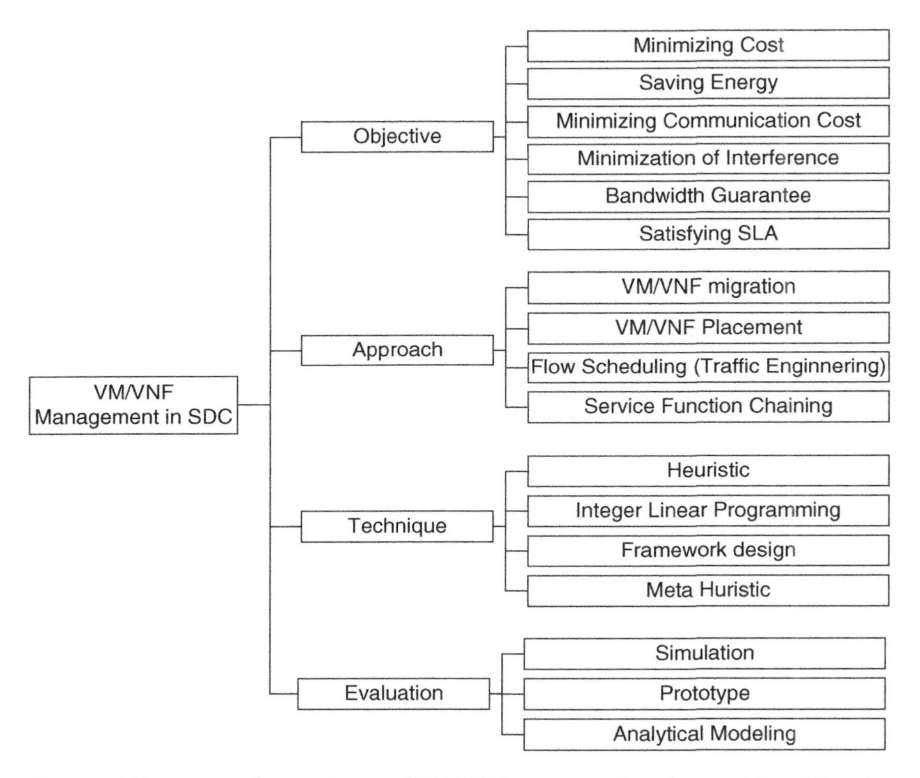

Figure 4.2 Taxonomy of network-aware VM/VNF Management in software-defined Clouds

address the problem, the exploited optimization technique, and finally, the evaluation technique used to validate the approach. In the remaining parts of this section, we cover network slicing from three different perspectives and map them to the proposed taxonomy: Network-aware VM management, network-aware VM migration, and VNF management.

4.4.1 Network-Aware Virtual Machines Management

Cziva et al. [29] present an orchestration framework to exploit time-based network information to live migrate VMs and minimize the network cost. Wang et al. [30] propose a VM placement mechanism to reduce the number of hops between communicating VMs, save energy, and balance the network load. Remedy [31] relies on SDN to monitor the state of the network and estimate the cost of VM migration. Their technique detects congested links and migrates VMs to remove congestion on those links.

Jiang et al. [32] worked on joint VM placement and network routing problem of data centers to minimize network cost in real-time. They proposed an online algorithm to optimize the VM placement and data traffic routing with dynamically adapting traffic loads. VMPlanner [33] also optimizes VM placement and network routing. The solution includes VM grouping that consolidates VMs with high inter-group traffic, VM group placement within a rack, and traffic consolidation to minimize the rack traffic. Jin et al. [34] studied joint host-network optimization problem. The problem is formulated as an integer linear problem that combines VM placement and routing problem. Cui et al. [35] explore the joint policy-aware and network-aware VM migration problem and present a VM management to reduce network-wide communication cost in data center networks while considering the policies regarding the network functions and middleboxes. Table 4.2 summarizes the research projects on network-aware VM management.

4.4.2 Network-Aware Virtual Machine Migration Planning

A large body of literature has focused on improving the efficiency of VM migration mechanism [36]. Bari et al. [37] propose a method for finding an efficient migration plan. They try to find a sequence of migrations to move a group of VMs to their final destinations while migration time is minimized. In their method, they monitor residual bandwidth available on the links between source and destination after performing each step in the sequence. Similarly, Ghorbani et al. [38] propose an algorithm to generate an ordered list of VMs to migrate and a set of forwarding flow changes. They concentrate on imposing bandwidth guarantees on the links to ensure that link capacity is not violated during the migration. The VM migration planning problem is also tackled by Li et al. [39] where they address the workload-aware migration problem

Table 4.2 Network-aware virtual machines management.

Project	Objectives	Approach/Technique	Evaluation
Cziva et al. [29]	Minimization of the network communication cost	VM migration – Framework Design	Prototype
Wang et al. [30]	Reducing the number of hops between communicating VMs and network power consumption	VM placement – Heuristic	Simulation
Remedy [31]	Removing congestion in the network	VM migration – Framework Design	Simulation
Jiang et al. [32]	Minimization of the network communication cost	VM Placement and Migration – Heuristic (Markov approximation)	Simulation
VMPlanner [33]	Reducing network power consumption	VM placement and traffic flow routing - Heuristic	Simulation
PLAN [35]	Minimization of the network communication cost while meeting network policy requirements	VM Placement - Heuristic	Prototype/ Simulation

and propose methods for selection of candidate virtual machines, destination hosts, and sequence for migration. All these studies focus on the migration order of a group of VMs while taking into account network cost. Xu et al. [40] propose an interference-aware VM live migration plan called *iAware* that minimizes both migration and co-location interference among VMs. Table 4.3 summarizes the research projects on VM migration planning.

4.4.3 Virtual Network Functions Management

NFV is an emerging paradigm where network functions such as firewalls, network address translation (NAT), and virtual private networks (VPNs) are virtualized and divided up into multiple building blocks called virtualized network functions (VNFs). VNFs are often chained together and build service function chains (SFC) to deliver a required network functionality. Han et al. [41] present a comprehensive survey of key challenges and technical requirements of NFV where they present an architectural framework for NFV. They focus on the efficient instantiation, placement and migration of VNFs, and network performance.

Table 4.3 Virtual machine migration planning.

Project	Objectives	Approach/Technique	Evaluation
Bari et al. [37]	Finding sequence of migrations while migration time is minimized	VM migration – Heuristic	Simulation
Ghorbani et al. [38]	Finding sequence of migrations while imposing bandwidth guarantees	VM migration – Heuristic	Simulation
Li et al. [39]	Finding sequence of migrations and destination hosts to balance the load	VM migration – Heuristic	Simulation
iAware [40]	Minimization of migration and co-location interference among VMs	VM migration – Heuristic	Prototype/ Simulation

VNF-P is a model proposed by Moens and Turck [42] for efficient placement of VNFs. They propose a NFV burst scenario in a hybrid scenario in which the base demand for network function service is handled by physical resources while the extra load is handled by virtual service instances. Cloud4NFV [43] is a platform following the NFV standards by the European Telecommunications Standards Institute (ETSI) to build network function as a service using a cloud platform. Its VNF Orchestrator exposes RESTful APIs, allowing VNF deployment. A cloud platform such as OpenStack supports management of virtual infrastructure at the background. vConductor [44] is another NFV management system proposed by Shen et al. for the end-to-end virtual network services. vConductor has simple graphical user interfaces (GUIs) for automatic provisioning of virtual network services and supports the management of VNFs and existing physical network functions. Yoshida et al. [45] proposed as part of vConductor using virtual machines (VMs) for building NFV infrastructure in the presence of conflicting objectives that involve stakeholders such as users, cloud providers, and telecommunication network operators.

Service chain is a series of VMs hosting VNFs in a designated order with a flow going through them sequentially to provide desired network functionality. Tabular VM migration (TVM) proposed by [46] aims at reducing the number of hops (network elements) in service chains of network functions in cloud data centers. They use VM migration to reduce the number of hops the flow should traverse to satisfy SLAs. SLA-driven ordered variable-width

Table 4.4 Virtual network functions management projects.

Project	Objectives	Approach/Technique
VNF-P	Handling burst in network services demand while minimizing the number of servers	Resource allocation - Integer linear programming (ILP)
Cloud4NFV	Providing network function as a service	Service provisioning – Framework design
vConductor	Virtual network services provisioning and management	Service provisioning – Framework design
MORSA	Multi objective placement of virtual services	Placement - Multi-objective genetic algorithm
TVM	Reducing number of hops in service chain	VNF migration - heuristic
SOVWin	Increasing user requests acceptance rate and minimization of SLA violation	VNF placement - heuristic
Clayman et al.	Providing automatic placement of the virtual nodes	VNF placement - heuristic
T-NOVA	Building a marketplace for VNF	Marketplace – framework design
UNIFY	Automated, dynamic service creation and service function chaining	Service provisioning– framework design

windowing (SOVWin) is a heuristic proposed by Pai et al. [47] to address the same problem, however, using initial static placement. Similarly, an orchestrator for the automated placement of VNFs across the resources proposed by Clayman et al. [48].

The EU-funded T-NOVA project [49] aims to realize the NFaaS concept. It has designed and implemented integrated management and orchestrator platforms for the automated provisioning, management, monitoring, and optimization of VNFs. UNIFY [50] is another EU-funded FP7 project aimed at supporting automated, dynamic service creation based on a fine-granular SFC model, SDN, and cloud virtualization techniques. For more details on SFC, interested readers are referred to the literature survey by Medhat et al. [51]. Table 4.4 summarizes the state of the art projects on VNF management.

4.5 Network Slicing Management in Edge and Fog

Fog computing is a new trend in cloud computing that attempts to address the quality of service requirements of applications requiring real-time and low-latency processing. While fog acts as a middle layer between edge and core clouds to serve applications close to the data source, core cloud data centers provide massive data storage, heavy-duty computation, or wide area connectivity for the application.

One of the key visions of fog computing is to add compute capabilities or general-purpose computing to edge network devices such as mobile base stations, gateways, and routers. On the other hand, SDN and NFV play key roles in prospective solutions to facilitate efficient management and orchestration of network services. Despite natural synergy and affinity between these technologies, significant research does not exist on the integration of fog/edge computing and SDN/NFV, as both are still in their infancy. In our view, intraction between SDN/NFV and fog/edge computing is crucial for emerging applications in IoT, 5G, and stream analytics. However, the scope and requirements of such interaction are still an open problem. In the following, we provide an overview of the state-of-the-art within this context.

Lingen et al. [52] define a model-driven and service-centric architecture that addresses technical challenges of integrating NFV, fog, and 5G/MEC. They introduce an open architecture based on NFV MANO proposed by the European Telecommunications Standards Institute (ETSI) and aligned with the OpenFog Consortium (OFC) reference architecture[2] that offers uniform management of IoT services spanning through cloud to the edge. A two-layer abstraction model along with IoT-specific modules and enhanced NFV MANO architecture is proposed to integerate cloud, network, and fog. As a pilot study, they presented two use cases for physical security of fog nodes and sensor telemetry through street cabinets in the city of Barcelona.

Truong et al. [53] are among the earliest who have proposed an SDN-based architecture to support fog computing. They have identified required components and specified their roles in the system. They also showed how their system can provide services in the context of vehicular adhoc networks (VANETs). They showed benefits of their proposed architecture using two use-cases in data streaming and lane-change assistance services. In their proposed architecture, the central network view by the SDN controller is utilized to manage resources and services and optimize their migration and replication.

Bruschi et al. [54] propose a network slicing scheme for supporting multidomain fog/cloud services. They propose SDN-based network slicing scheme to build an overlay network for geographically distributed Internet services using non-overlapping OpenFlow rules. Their experimental results show that the number of unicast forwarding rules installed in the overlay network significantly drops compared to the fully meshed and OpenStack cases.

2 OpenFog Consortium, https://www.openfogconsortium.org/

Inspired by Open Network Operating System (ONOS)[3] SDN controller, Choi et al. [55] propose a fog operating system architecture called *FogOS* for IoT services. They identified four main challenges of fog computing:

1. *Scalability* for handling significant number of IoT devices,
2. *Complex inter-networking* caused by diverse forms of connectivity, e.g., various radio access technologies,
3. *Dynamics and adaptation* in topology and quality of service (QoS) requirements, and
4. *Diversity and heterogeneity* in communications, sensors, storage, and computing powers, etc.

Based on these challenges, their proposed architecture consists of four main components:

1. Service and device abstraction
2. Resource management
3. Application management
4. Edge resource: registration, ID/addressing, and control interface

They also demonstrate a preliminary proof-of-concept demonstration of their system for a drone-based surveillance service.

In a recent work, Diro et al. [56] propose a mixed SDN and fog architecture that gives priority to critical network flows while taking into account fairness among other flows in the fog-to-things communication to satisfy QoS requirements of heterogeneous IoT applications. They intend to satisfy QoS and performance measures such as packet delay, lost packets, and maximized throughput. Results show that their proposed method can serve critical and urgent flows more efficiently while allocating network slices to other flow classes.

4.6 Future Research Directions

In this section, we discuss open issues in software-defined clouds and edge computing environments along with future directions.

4.6.1 Software-Defined Clouds

Our survey on network slicing management and orchestration in SDC shows that the community very well recognizes the problem of joint provisioning of hosts and network resources. In the earlier research, a vast amount of attention has been given to solutions for the optimization of cost/energy focusing only

3 ONOS, https://onosproject.org/

on either host [57] or network [58], not both. However, it is essential for the management component of the system to take into account both network and host cost at the same time. Otherwise, optimization of one can exacerbate the situation for the other.

To address this issue, many research proposals have also focused on the joint host and network resource management. However, most of the proposed approaches suffer from high computational complexity, or they are not optimal. Therefore, it is important to develop algorithms that manage joint hosts and network resource provisioning and scheduling. In joint host and network resource management and orchestration, two conditions must be satisfied: finding the minimum subset of hosts and network resources that can handle a given workload and meeting SLA and users' QoS requirements (e.g., latency). The problem of joint host and network resource provisioning becomes more sophisticated when SDC supports VNF and SFC.

SFC is a hot topic, attaining a significant amount of attention by the community. However, little attention has been paid to VNF placement while meeting the QoS requirements of the applications. PLAN [35] intends to minimize the network communication costs while meeting network policy requirements. However, it only considers traditional middleboxes, and it does not take into account the option of VNF migration. Therefore, one of the areas requires more attention and development of novel optimization techniques is the management and orchestration of SFCs. This has to be done in a way that the placement and migration of VNFs are optimized while SLA violation and cost/energy are maximized.

Network-aware virtual machines management is a well-studied area. However, the majority of works in this context consider VM migration and VM placement to optimize network costs. The traffic engineering and dynamic flow scheduling combined with migration and placement of VMs also provide a promising direction for the minimization of network communication cost. For example, SDN, management, and orchestration modules of the system can be used to install flow entries on the switches of the shortest path with the lowest utilization to redirect VM migration traffic to an appropriate path.

The analytical modeling of SDCs has not been investigated intensely in the literature. Therefore, research is warranted that focuses on building a model based on priority networks that can be used for analysis of the SDCs network and validation of results from experiments conducted via simulation.

Auto-scaling of VNFs is another area that requires more in-depth attention by the community. VNFs providing networking functions for the applications are subject to performance variation due to different factors such as the load of the service or overloaded underlying hosts. Therefore, development of auto-scaling mechanisms that monitor the performance of the VMs hosting VNFs and adaptively adds or remove VMs to satisfy the SLA requirements of the applications is of paramount importance for management and orchestration of network slices. In fact, efficient placement of VNFs [59] on hosts near to the service component producing data streams or users generating requests minimizes latency and reduces the overall network cost. However, placement on a more powerful node far in the network might improve processing time [60]. Existing solutions mostly focus on either scaling without placement or placement without scaling. Moreover, auto-scaling techniques of VNFs, they typically focus on auto-scaling of a single network service (e.g., firewall), while in practice auto-scaling of VNFs must be performed in accordance with SFCs. In this context, node, and link capacity limits must be considered, and the solution must maximize the benefit gained from existing hardware using techniques such as dynamic pathing. Therefore, one of the promising avenues for future research on auto-scaling of VNFs is to explore the optimal dynamic resource allocation and placement.

4.6.2 Edge and Fog Computing

In both edge and fog computing, the integration of 5G so far has been discussed within a very narrow scope. Although 5G network resource management and resource discovery in edge/fog computing have been investigated, many other challenging issues in this area are still unexplored. Mobility-aware service management in 5G enabled fog computing and forwarding large amount of data from one fog node to another in real-time overcoming communication overhead can be very difficult to ensure. In addition, due to decentralized orchestration and heterogeneity among fog nodes, modeling, management and provisioning of 5G network resources are not as straightforward as other computing paradigms.

Moreover, compared to mobile edge servers, cloudlets and cloud datacenters, the number of fog nodes and their probability of failure are very high. In this case, implementation of SDN (one of the foundation blocks of 5G) in fog computing can get obstructed significantly. On the other hand, fog computing enables traditional networking devices to process incoming data and due to 5G, this data amount can be significantly huge. In such scenario, adding more resources in traditional networking devices will be very costly, less secured and hinders their inherent functionalities like routing, packet forwarding, etc. which in consequence affect the basic commitments of 5G network and NFV.

Nonetheless, fog infrastructures can be owned by different providers that can significantly resist developing a generalized pricing policy for 5G-enabled fog computing. Prioritized network slicing for forwarding latency-sensitive IoT data can also complicate 5G enabled fog computing. Opportunistic scheduling and reservation of virtual network resources is tough to implement in fog as it deals with a large number of IoT devices, and their data sensing frequency can change with the course of time. Load balancing on different virtual networks and QoS can degrade significantly unless efficient monitoring is imposed. Since fog computing is a distributed computing paradigm, centralized monitoring of network resources can intensify the problem. In this case, distributed monitoring can be an efficient solution, although it can fail to reflect the whole network context in a body. Extensive research is required to solve this issue. Besides, in promoting fault tolerance of 5G-enabled fog computing, topology-aware application placement, dynamic fault detection, and reactive management can play a significant role, which is subjected to uneven characteristics of the fog nodes.

4.7 Conclusions

In this chapter, we investigated research proposals for the management and orchestration of network slices in different platforms. We discussed emerging technologies such as software-defined networking SDN and NFV. We explored the vision of 5G for network slicing and discussed some of the ongoing projects and studies in this area. We surveyed state-of-the-art approaches to network slicing in software-defined clouds and application of this vision to the cloud computing context. We disscussed state-of-the-art literature on network slices in emerging fog/edge computing. Finally, we identified gaps in this context and provided future directions toward the notion of network slicing.

Acknowledgments

This work is supported through Huawei Innovation Research Program (HIRP). We also thank Wei Zhou for his comments and support for the work.

References

1 J. G. Andrews, S. Buzzi, W. Choi, S. V. Hanly, A. Lozano, A. C. K. Soong, and J. C. Zhang. What Will 5G Be? *IEEE Journal on Selected Areas in Communications* 32(6): 1065–1082, 2014.
2 D. Ott, N. Himayat, and S. Talwar. 5G: Transforming the User Wireless Experience. *Towards 5G: Applications, Requirements and Candidate Technologies*, R. Vannithamby, and S. Talwar (eds.). Wiley Press, Hoboken, NJ, USA, Jan. 2017.
3 J. Zhang, X. Ge, Q. Li, M. Guizani, and Y. Zhang. 5G millimeter-wave antenna array: Design and challenges. *IEEE Wireless Communications* 24(2): 106–112, 2017.
4 S. Chen and J. Zhao. The Requirements, Challenges, and Technologies for 5G of terrestrial mobile telecommunication. *IEEE Communication Magazine* 52(5): 36–43, 2014.

5 R. Buyya, R. N. Calheiros, J. Son, A.V. Dastjerdi, and Y. Yoon. Software-defined cloud computing: Architectural elements and open challenges. In *Proceedings of the 3rd International Conference on Advances in Computing, Communications and Informatics (ICACCI'14)*, pp. 1–12, New Delhi, India, Sept. 24–27, 2014.

6 M. Afrin, M.A. Razzaque, I. Anjum, et al. Tradeoff between user quality-of-experience and service provider profit in 5G cloud radio access network. *Sustainability* 9(11): 2127, 2017.

7 S. Nunna, A. Kousaridas, M. Ibrahim, M.M. Hassan, and A. Alamri. Enabling real-time context-aware collaboration through 5G and mobile edge computing. In *Proceedings of the 12th International Conference on Information Technology-New Generations (ITNG'15)*, pp. 601-605, Las Vegas, USA, April 13–15, 2015.

8 I. Ketykó, L. Kecskés, C. Nemes, and L. Farkas. Multi-user computation offloading as multiple knapsack problem for 5G mobile edge computing. In *Proceedings of the 25th European Conference on Networks and Communications (EuCNC'16)*, pp. 225–229, Athens, Greece, June 27–30, 2016.

9 K. Zhang, Y. Mao, S. Leng, Q. Zhao, L. Li, X. Peng, L. Pan, S. Maharjan and Y. Zhang. Energy-efficient offloading for mobile edge computing in 5G heterogeneous networks. *IEEE Access* 4: 5896–5907, 2016.

10 C. Ge, N. Wang, S. Skillman, G. Foster and Y. Cao. QoE-driven DASH video caching and adaptation at 5G mobile edge. In *Proceedings of the 3rd ACM Conference on Information-Centric Networking*, pp. 237–242, Kyoto, Japan, Sept. 26–28, 2016.

11 M. Afrin, R. Mahmud, and M.A. Razzaque. Real time detection of speed breakers and warning system for on-road drivers. In *Proceedings of the IEEE International WIE Conference on Electrical and Computer Engineering (WIECON-ECE'15)*, pp. 495-498, Dhaka, Bangladesh, Dec. 19–20, 2015.

12 A. V. Dastjerdi and R. Buyya. Fog computing: Helping the Internet of Things realize its potential. *Computer. IEEE Computer,* 49(8): 112–116, 2016.

13 F. Bonomi, R. Milito, J. Zhu, and S. Addepalli. Fog computing and its role in the internet of things. In *Proceedings of the first edition of the MCC workshop on Mobile Cloud computing* (MCC'12), pp. 13–16, Helsinki, Finland, Aug. 17, 2012.

14 R. Mahmud, M. Afrin, M. A. Razzaque, M. M. Hassan, A. Alelaiwi and M. A. AlRubaian. Maximizing quality of experience through context-aware mobile application scheduling in Cloudlet infrastructure. *Software: Practice and Experience,* 46(11): 1525–1545, 2016.

15 R. Mahmud, K. Ramamohanarao, and R. Buyya. Fog computing: A taxonomy, survey and future directions. Internet of Everything: Algorithms, Methodologies, Technologies and Perspectives. Di Martino

Beniamino, Yang Laurence, Kuan-Ching Li, and Esposito Antonio (eds.), ISBN 978-981-10-5861-5, Springer, Singapore, Oct. 2017.

16 D. Amendola, N. Cordeschi, and E. Baccarelli. Bandwidth management VMs live migration in wireless fog computing for 5G networks. In *Proceedings of the 5th IEEE International Conference on Cloud Networking* (Cloudnet'16), pp. 21–26, Pisa, Italy, Oct. 3–5, 2016.

17 M. Afrin, R. Mahmud. Software Defined Network-based Scalable Resource Discovery for Internet of Things. *EAI Endorsed Transaction on Scalable Information Systems* 4(14): e4, 2017.

18 M. Peng, S. Yan, K. Zhang, and C. Wang. Fog-computing-based radio access networks: issues and challenges. *IEEE Network,* 30(4): 46–53, 2016.

19 R. Mahmud, F. L. Koch, and R. Buyya. Cloud-fog interoperability in IoT-enabled healthcare solutions. In *Proceedings of the 19th International Conference on Distributed Computing and Networking* (ICDCN'18), pp. 1–10, Varanasi, India, Jan. 4–7, 2018.

20 T. D. P. Perera, D. N. K. Jayakody, S. De, and M. A. Ivanov. A Survey on Simultaneous Wireless Information and Power Transfer. *Journal of Physics: Conference Series,* 803(1): 012113, 2017.

21 P. Pirinen. A brief overview of 5G research activities. In *Proceedings of the 1st International Conference on 5G for Ubiquitous Connectivity* (5GU'14), pp. 17–22, Akaslompolo, Finland, November 26–28, 2014.

22 A. Nakao, P. Du, Y. Kiriha, et al. End-to-end network slicing for 5G mobile networks. *Journal of Information Processing* 2 (2017): 153–163.

23 K. Samdanis, S. Wright, A. Banchs, F. Granelli, A. A. Gebremariam, T. Taleb, and M. Bagaa. 5G Network Slicing: Part 1–Concepts, Principales, and Architectures [Guest Editorial]. *IEEE Communications Magazine,* 55(5) (2017): 70–71.

24 S. Sharma, R. Miller, and A. Francini. A cloud-native approach to 5G network slicing. *IEEE Communications Magazine,* 55(8): 120–127, 2017.

25 W. Shi, J. Cao, Q. Zhang, Y. Li, and L. Xu. Edge computing: vision and challenges. *IEEE Internet of Things Journal,* 3(5): 637–646, 2016.

26 X. Foukas, G. Patounas, A. Elmokashfi, and M. K. Marina. Network Slicing in 5G: Survey and Challenges. *IEEE Communications Magazine,* 55(5): 94–100, 2017.

27 X. Li, M. Samaka, H. A. Chan, D. Bhamare, L. Gupta, C. Guo, and R. Jain. Network slicing for 5G: Challenges and opportunities., *IEEE Internet Computing,* 21(5): 20–27, 2017.

28 Huawei Technologies' white paper. 5G Network Architecture A High-Level Perspective, http://www.huawei.com/minisite/hwmbbf16/insights/5G-Nework-Architecture-Whitepaper-en.pdf (Last visit: Mar, 2018).

29 R. Cziva, S. Jouët, D. Stapleton, F.P. Tso and D.P. Pezaros. SDN-Based Virtual Machine Management for Cloud Data Centers. *IEEE Transactions on Network and Service Management,* 13(2): 212–225, 2016.

30 S.H. Wang, P.P. W. Huang, C.H.P. Wen, and L. C. Wang. EQVMP: Energy-efficient and QoS-aware virtual machine placement for software defined datacenter networks. In *Proceedings of the International Conference on Information Networking* (ICOIN'14), pp. 220–225, Phuket, Thailand, Feb. 10–12, 2014.

31 V. Mann, A. Gupta, P. Dutta, A. Vishnoi, P. Bhattacharya, R. Poddar, and A. Iyer. Remedy: Network-aware steady state VM management for data centers. In *Proceedings of the 11th international IFIP TC 6 conference on Networking* (IFIP'12), pp. 190–204, Prague, Czech Republic, May 21–25, 2012.

32 J. W. Jiang, T. Lan, S. Ha, M. Chen, and M. Chiang. Joint VM placement and routing for data center traffic engineering. In *Proceedings of the IEEE International Conference on Computer Communications* (INFOCOM'12), pp. 2876–2880, Orlando, USA, March 25–30, 2012.

33 W. Fang, X. Liang, S. Li, L. Chiaraviglio, N. Xiong. VMPlanner: Optimizing virtual machine placement and traffic flow routing to reduce network power costs in Cloud data centers. *Computer Networks* 57(1): 179–196, 2013.

34 H. Jin, T. Cheocherngngarn, D. Levy, A. Smith, D. Pan, J. Liu, and N. Pissinou. Joint host-network optimization for energy-efficient data center networking. In *Proceedings of the 27th IEEE International Symposium on Parallel and Distributed Processing* (IPDPS'13), pp. 623–634, Boston, USA, May 20–24, 2013.

35 L. Cui, F.P. Tso, D.P. Pezaros, W. Jia, and W. Zhao. PLAN: Joint policy- and network-aware VM management for cloud data centers. *IEEE Transactions on Parallel and Distributed Systems,* 28(4):1163–1175, 2017.

36 W. Voorsluys, J. Broberg, S. Venugopal, and R. Buyya. Cost of virtual machine live migration in clouds: a performance evaluation. In *Proceedings of the 1st International Conference on Cloud Computing* (CloudCom'09), pp. 254–265, Beijing, China, Dec. 1–4, 2009.

37 M.F. Bari, M.F. Zhani, Q. Zhang, R. Ahmed, and R. Boutaba. CQNCR: Optimal VM migration planning in cloud data centers. In *Proceedings of the IFIP Networking Conference,* pp. 1–9, Trondheim, Norway, June 2–4, 2014.

38 S. Ghorbani, and M. Caesar. Walk the line: consistent network updates with bandwidth guarantees. In *Proceedings of the 1st workshop on Hot topics in software defined networks* (HotSDN'12), pp. 67–72, Helsinki, Finland, Aug. 13, 2012.

39 X. Li, Q. He, J. Chen, and T. Yin. Informed live migration strategies of virtual machines for cluster load balancing. In *Proceedings of the 8th IFIP international conference on Network and parallel computing* (NPC'11), pp. 111–122, Changsha, China, Oct. 21–23, 2001.

40 F. Xu, F. Liu, L. Liu, H. Jin, B. Li, and B. Li. iAware: Making Live Migration of Virtual Machines Interference-Aware in the Cloud. *IEEE Transactions on Computers,* 63(12): 3012–3025, 2014.

41 B. Han, V. Gopalakrishnan, L. Ji, and S. Lee. Network function virtualization: Challenges and opportunities for innovations. *IEEE Communications Magazine,* 53(2): 90–97, 2015.
42 H. Moens and F. D. Turck. VNF-P: A model for efficient placement of virtualized network functions. In *Proceedings of the 10th International Conference on Network and Service Management* (CNSM'14), pp. 418–423, Rio de Janeiro, Brazil, Nov. 17–21, 2014.
43 J. Soares, M. Dias, J. Carapinha, B. Parreira, and S. Sargento. Cloud4NFV: A platform for virtual network functions. In *Proceedings of the 3rd IEEE International Conference on Cloud Networking* (CloudNet'14), pp. 288–293, Luxembourg, Oct. 8–10, 2014.
44 W. Shen, M. Yoshida, T. Kawabata, et al. vConductor: An NFV management solution for realizing end-to-end virtual network services. In *Proceedings of the 16th Asia-Pacific Network Operations and Management Symposium* (APNOMS'14), pp. 1–6, Hsinchu, Taiwan, Sept.17–19, 2014.
45 M. Yoshida, W. Shen, T. Kawabata, K. Minato, and W. Imajuku. MORSA: A multi-objective resource scheduling algorithm for NFV infrastructure. In *Proceedings of the 16th Asia-Pacific Network Operations and Management Symposium* (APNOMS'14), pp. 1–6, Hsinchu, Taiwan, Sept. 17–19, 2014.
46 Y. F. Wu, Y. L. Su and C. H. P. Wen. TVM: Tabular VM migration for reducing hop violations of service chains in cloud datacenters. In *Proceedings of the IEEE International Conference on Communications* (ICC'17), pp. 1–6, Paris, France, May 21–25, 2017.
47 Y.-M. Pai, C.H.P. Wen and L.-P. Tung. SLA-driven ordered variable-width windowing for service-chain deployment in SDN datacenters. In *Proceedings of the International Conference on Information Networking* (ICOIN'17), pp. 167–172, Da Nang, Vietnam, Jan. 11–13, 2017
48 S. Clayman, E. Maini, A. Galis, A. Manzalini, and N. Mazzocca. The dynamic placement of virtual network functions. In *Proceedings of the IEEE Network Operations and Management Symposium* (NOMS'14), pp. 1–9, Krakow, Poland, May 5–9, 2014.
49 G. Xilouris, E. Trouva, F. Lobillo, J.M. Soares, J. Carapinha, M.J. McGrath, G. Gardikis, P. Paglierani, E. Pallis, L. Zuccaro, Y. Rebahi, and A. Koutis. T-NOVA: A marketplace for virtualized network functions. In *Proceedings of the European Conference on Networks and Communications* (EuCNC'14), pp. 1–5, Bologna, Italy, June 23–26, 2014.
50 B. Sonkoly, R. Szabo, D. Jocha, J. Czentye, M. Kind and F. J. Westphal. UNIFYing cloud and carrier network resources: an architectural view. In *Proceedings of the IEEE Global Communications Conference* (GLOBECOM'15), pp. 1–7, San Diego, USA, Dec. 6–10, 2015.
51 A. M. Medhat, T. Taleb, A. Elmangoush, G. A. Carella, S. Covaci and T. Magedanz. Service function chaining in next generation networks: state of

the art and research challenges. *IEEE Communications Magazine,* 55(2): 216–223, 2017.

52 F. van Lingen, M. Yannuzzi, A. Jain, R. Irons-Mclean, O. Lluch, D. Carrera, J. L. Perez, A. Gutierrez, D. Montero, J. Marti, R. Maso, and A. J. P. Rodriguez. The unavoidable convergence of NFV, 5G, and fog: A model-driven approach to bridge cloud and edge. *IEEE Communications Magazine,* 55 (8): 28–35, 2017.

53 N.B. Truong, G.M. Lee, and Y. Ghamri-Doudane. Software defined networking-based vehicular adhoc network with fog computing. In *Proceedings of the IFIP/IEEE International Symposium on Integrated Network Management* (IM'15), pp. 1202–1207, Ottawa, Canada, May 11–15, 2015.

54 R. Bruschi, F. Davoli, P. Lago, and J.F. Pajo. A scalable SDN slicing scheme for multi-domain fog/cloud services. In *Proceedings of the IEEE Conference on Network Softwarization* (NetSoft'17), pp. 1-6, Bologna, Italy, July 3–7, 2017.

55 N. Choi, D. Kim, S. J. Lee, and Y. Yi. A fog operating system for user-oriented IoT services: Challenges and research directions. *IEEE Communications Magazine,* 55(8): 44–51, 2017.

56 A.A. Diro, H.T. Reda, and N. Chilamkurti. Differential flow space allocation scheme in SDN based fog computing for IoT applications. *Journal of Ambient Intelligence and Humanized Computing*, DOI: 10.1007/s12652-017-0677-z.

57 A. Beloglazov, J. Abawajy, R. Buyya. Energy-aware resource allocation heuristics for efficient management of data centers for cloud computing. *Future Generation Computer Systems,* 28(5): 755–768, 2012.

58 B. Heller, S. Seetharaman, P. Mahadevan, Y. Yiakoumis, P. Sharma, S. Banerjee, and N. McKeown. ElasticTree: Saving energy in data center networks. In *Proceedings of the 7th USENIX conference on Networked systems design and implementation* (NSDI'10), pp. 249–264, San Jose, USA, April 28–30, 2010.

59 A. Fischer, J.F. Botero, M.T. Beck, H. de Meer, and X. Hesselbach. Virtual network embedding: A survey. *IEEE Communications Surveys & Tutorials,* 15(4):1888–1906, 2013.

60 S. Dräxler, H. Karl, and Z.A. Mann. Joint optimization of scaling and placement of virtual network services. In *Proceedings of the 17th IEEE/ACM International Symposium on Cluster, Cloud and Grid Computing* (CCGrid '17), pp. 365–370, Madrid, Spain, May 14–17, 2017.

5

Optimization Problems in Fog and Edge Computing

Zoltán Ádám Mann

5.1 Introduction

Fog / edge computing arises through the increasing convergence and integration of several – traditionally distinct – disciplines: cloud computing on one hand, mobile computing and the Internet of Things (IoT) on the other hand, and advanced networking technologies as a glue between them. The main idea is to combine the strengths of these technologies to provide the necessary compute power to end-user applications in a cost-effective and secure way, with low latencies. Thus, fog / edge computing brings significant benefits to all of the underlying fields.

The notions of fog computing and edge computing are somewhat vaguely defined in the literature and have largely overlapping meaning [1]. In this chapter, we use the terms "fog computing" and "edge computing" interchangeably to refer to an architecture combining cloud computing with resources on the network edge and end-user devices.

In cloud computing, there has been an evolution for several years from centralized architectures (one or few large data centers) toward increasing decentralization (several smaller data centers), which is still continuing, and fog / edge computing is a natural next step on this evolution trajectory [2]. Geographically distributed data centers lead to decreased latency for applications involving distributed data sources and sinks (e.g., users or sensors / actuators), since each data source / sink can be served by a nearby data center. Other benefits include improved fault tolerance as well as access to green energy sources of limited capacity [3].

From the point of view of mobile computing and IoT, the devices' limited computational capacity and limited battery life span are major challenges [4]. By offloading resource-intensive compute tasks to more powerful nodes – such as servers in a data center or compute resources at the network edge – the range of possible applications can be widened significantly [5].

Fog and Edge Computing: Principles and Paradigms, First Edition.
Edited by Rajkumar Buyya and Satish Narayana Srirama.

Optimization plays a crucial role in fog computing. For example, minimizing latency and energy consumption are just as important as maximizing security and reliability. Because of the high complexity of typical fog deployments (many different types of devices, with many different types of interactions) and their dynamic nature (mobile devices coming and going, devices or network connections failing permanently or temporarily etc.), it has become virtually impossible to ensure the best solution by design. Rather, the best solution should be determined using appropriate optimization techniques.

For this purpose, it is vital to define the relevant optimization problem(s) carefully and precisely. Indeed, the used problem formulation can have dramatic consequences on the practical applicability of the approach (e.g., omitting an important constraint may lead to solutions that cannot be applied in practice), as well as on its computational complexity.

Research on fog computing is still in its infancy. Some specific optimization problems have been defined, but in an ad hoc manner, independently from each other. As a result, it is difficult to compare or combine different approaches, because they usually address different variants or facets of the same problem and such subtle differences are often not apparent. (Earlier, we have witnessed a similar situation in cloud computing research as well [6, 7]). In addition, the quality and level of detail of existing problem formulations is quite heterogeneous.

Therefore, the aim of this chapter is to propose a generic conceptual framework for optimization problems in fog computing, based on consistent, well-defined, and formalized notation for constraints and optimization objectives. Using a taxonomy of problem formulations, their relationships will become clear, also highlighting the gaps that necessitate further research. With this standard reference, we hope to contribute significantly to the maturation of this field of research.

5.2 Background / Related Work

The concept of fog computing was introduced by Cisco in 2012 as a means to extend cloud computing capabilities to the network edge, thus enabling more advanced applications [8]. Since then, an increasing number of research papers have been published on fog computing. This is exemplified by Figure 5.1, which shows the development of the number of papers and number of citations in fog computing, available in the Scopus database[1] on 7th December 2017. The used search query was "TITLE-ABS-KEY ("fog computing")", meaning that the phrase "fog computing" must occur in the title, the abstract, or the keywords of the paper.

1 https://www.scopus.com

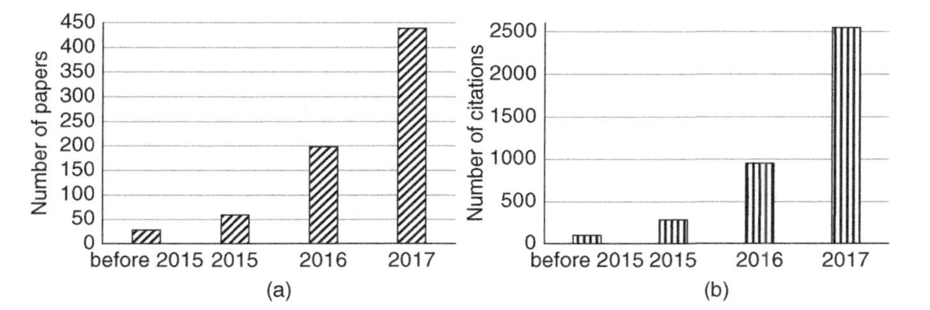

Figure 5.1 Number of (a) papers and (b) citations in fog computing.

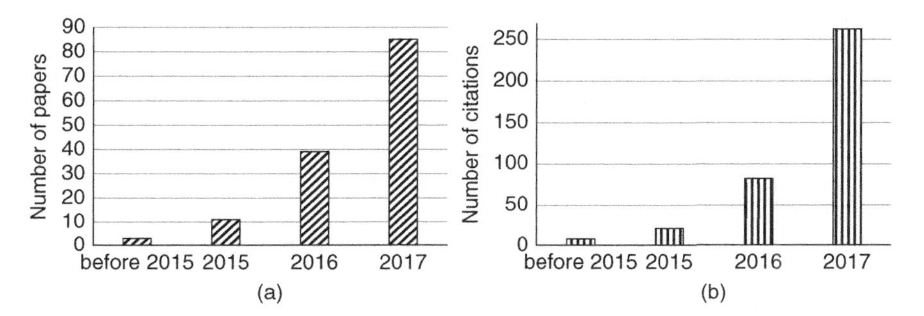

Figure 5.2 Number of (a) papers and (b) citations about optimization in fog computing.

Several of those papers describe technologies, architectures, and applications in a fog computing setting. However, the number of papers that deal with optimization in fog computing is also quickly rising. This is demonstrated by Figure 5.2, which shows the number of papers and number of citations obtained from the Scopus database on 7th December 2017, with the search query "TITLE-ABS-KEY ("fog computing") AND TITLE-ABS-KEY (optim*)", meaning that both the phrase "fog computing" and a word starting with "optim" (like optimal, optimized, or optimization) must occur in the title, the abstract, or the keywords of the paper.

Later in Section 5.9, when the essential characteristics of optimization problems in fog computing have already been defined, we will show how existing literature on optimization in fog computing can be classified.

5.3 Preliminaries

Before delving into optimization problems and optimization approaches in fog computing, we describe some essential properties and notions of optimization in general.

An optimization problem is generally defined by the following [9]:

- a list of variables $\overline{x} = (x_1, \ldots, x_n)$.
- the domain – i.e., the set of valid values – of each variable; the domain of variable x_i is denoted by D_i.
- a list of constraints $(C_1, \ldots, C_m)$; constraint C_j relates to some variables $x_{j_1}, \ldots, x_{j_k}$ and defines the valid tuples for those variables in the form of a set $R_j \subseteq D_{j_1} \times \cdots \times D_{j_k}$.
- an objective function $f : D_1 \times \cdots \times D_n \rightarrow \mathbb{R}$.

The problem then consists of finding appropriate values $v_1, \ldots, v_n$ for the variables, such that all of the following holds:

(1) $v_i \in D_i$ for each $i = 1, \ldots, n$.
(2) for any constraint C_j relating to variables $x_{j_1}, \ldots, x_{j_k}$, it holds that $(v_{j_1}, \ldots, v_{j_k}) \in R_j$.
(3) $f(v_1, \ldots, v_n)$ is maximum among all $(v_1, \ldots, v_n)$ tuples that satisfy (1) and (2).

A tuple $(v_1, \ldots, v_n)$ that satisfies (1) and (2) is called a solution of the problem. Thus, the goal is to find the solution with highest f value. At least, this is the case for maximization problems (as defined above). For a minimization problem, the goal is to find the solution with lowest f value, which is equivalent to finding the solution that maximizes the objective function $f' = -f$. In case of minimization problems, the objective function is often called cost function because it represents some – real or fictive – cost that needs to be minimized.

It is important to differentiate between a practical problem in engineering – e.g., minimization of power consumption in fog computing – and a formally defined optimization problem as outlined above. Deriving a formalized optimization problem from a practical problem is a nontrivial process, in which the variables, their domains, the constraints, and the objective function must be defined. In particular, there are usually many different ways to formalize a practical problem, leading to different formal optimization problems. Formalizing the problem is also a process of abstraction, in which some nonessential details are suppressed or some simplifying assumptions are made. Different formalizations of the same practical problem may exhibit different characteristics – for example, in terms of computational complexity. Therefore, the decisions made during problem formalization have a high impact. Problem formalization implies finding the most appropriate trade-off between the generality and applicability of the formalized problem on one hand and its simplicity, clarity, and computational tractability on the other hand. This requires expertise and an iterative approach in which different ways of formalizing the problem are evaluated.

It should be mentioned that some papers jump from an informal problem description directly to devising some algorithm, without formally defining the problem first. This, however, has the disadvantage of prohibiting precise reasoning about the problem itself, e.g., about its computational complexity or its similarity with known other problems that could lead to the adoption of existing algorithms.

In the above definition of a general optimization problem, it was assumed that there is a single real-valued objective function. However, in several practical problems, there are multiple objectives and the difficulty of the problem often lies in balancing between conflicting objectives. Let the objective functions be $f_1, \dots, f_q$, where the aim is to maximize all of them. Since there is generally no solution that maximizes all of the objective functions simultaneously, some modification is necessary to obtain a well-defined optimization problem. The most common approaches for that are the following [10]:

- Adding lower bounds to all but one of the objective functions and maximizing the last one. That means adding constraints of the form $f_s(v_1, \dots, v_n) \geq l_s$, where l_s is an appropriate constant, for all $s = 1, \dots, q-1$, and maximizing $f_q(v_1, \dots, v_n)$.
- Scalarizing all objective functions into a single combined objective function $f_{combined}(v_1, \dots, v_n) = F(f_1(v_1, \dots, v_n), \dots, f_q(v_1, \dots, v_n))$. Common choices for the function F are product and weighted sum.
- Looking for Pareto-optimal solutions. A solution $(v_1, \dots, v_n)$ dominates another solution $(v'_1, \dots, v'_n)$, if $f_s(v_1, \dots, v_n) \geq f_s(v'_1, \dots, v'_n)$ holds for all $s = 1, \dots, q$, and $f_s(v_1, \dots, v_n) > f_s(v'_1, \dots, v'_n)$ holds for at least one value of s, i.e., $(v_1, \dots, v_n)$ is at least as good as $(v'_1, \dots, v'_n)$ regarding each objective and it is strictly better regarding at least one objective. A solution is called Pareto-optimal if it is not dominated by any other solution. In other words, a Pareto-optimal solution can only be improved with regard to an objective if it is worsened regarding some other objective. Different Pareto-optimal solutions of a problem represent different trade-offs between the objectives, but all of them are optimal in the above sense.

5.4 The Case for Optimization in Fog Computing

The fundamental motivation for the developments leading to fog computing are strongly related to some important quality attributes that should be improved. As explained earlier, fog computing can be seen as an extension of cloud computing towards the network edge, with the aim of providing lower latencies for latency-critical applications within end devices. In other words, the optimization objective of minimizing latency is a major driving force behind fog computing [11].

On the other hand, from the point of view of end devices, fog computing promises significantly increased compute capabilities, enabling the execution of compute-intensive tasks quickly and without major impact on energy consumption of the device. Therefore, optimization relating to execution time and energy consumption are also fundamental aspects of fog computing.

As we will see shortly in Section 5.6, several other optimization objectives are relevant to fog computing as well. Moreover, there are nontrivial interactions, sometimes also conflicts, among the different objectives. Hence, it is important to systematically study the different aspects of optimization in fog computing.

5.5 Formal Modeling Framework for Fog Computing

Before discussing individual optimization objectives, it is useful to define a generic framework for modeling different variants of the problem.

As shown in Figure 5.3, fog computing can be represented by a hierarchical three-layer model [12]. Higher layers represent higher computational capacity, but at the same time also higher distance – and thus higher latency – from the end devices. On the highest layer is the cloud with its virtually unlimited, high-performance, and cost- and energy-efficient resources. The middle layer consists of a set of edge resources: machines offering compute services near the network edge, e.g. in base stations, routers, or small, geographically distributed data centers of telecommunication providers. The edge resources are all connected to the cloud. Finally, the lowest layer contains the end devices like mobile phones or IoT devices. Each end device is connected to one of the edge resources.

More formally, let c denote the cloud, E the set of edge resources, D_e the set of end devices connected to edge resource $e \in E$, and $D = \bigcup_{e \in E} D_e$ the set of all end devices. The set of all resources is $R = \{c\} \cup E \cup D$. Each resource $r \in R$ is associated with a compute capacity $a(r) \in \mathbb{R}^+$ and a compute speed $s(r) \in \mathbb{R}^+$. Moreover, each resource has some power consumption, which depends on its computational load. Specifically, the power consumption of resource r increases by $w(r) \in \mathbb{R}^+$ for every instruction to be carried out by r.

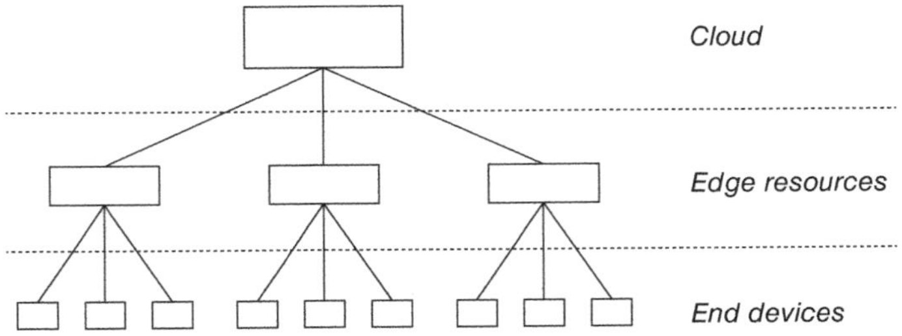

Figure 5.3 Three-layer model of fog computing.

Table 5.1 Notation overview.

Notation	Explanation
c	cloud
E	set of edge resources
D_e	set of end devices connected to edge resource $e \in E$
R	set of all resources
$a(r)$	compute capacity of resource $r \in R$
$s(r)$	compute speed of resource $r \in R$
$w(r)$	marginal energy consumption of resource $r \in R$
L	set of all links between resources
$t(l)$	latency of link $l \in L$
$b(l)$	bandwidth of link $l \in L$
$w(l)$	marginal energy consumption of link $l \in L$

The set of links between resources is $L = \{ce : e \in E\} \cup \{ed : e \in E, d \in D_e\}$. Each link $l \in L$ is associated with a latency $t(l) \in \mathbb{R}^+$ and a bandwidth $b(l) \in \mathbb{R}^+$. Moreover, transmitting one more byte of data over link l increases power consumption by $w(l) \in \mathbb{R}^+$. Table 5.1 gives an overview of the used notation.

5.6 Metrics

As already mentioned, there are several metrics that need to be optimized in a fog computing system. Depending on the specific optimization problem variant, these metrics may indeed be optimization objectives, but they can also be used as constraints. For example, one problem variant may look at a real-time application, in which overall execution time needs to be constrained by an upper bound, while energy consumption should be minimized. In another application, the finite battery capacity of a mobile device may be the bottleneck, so that energy consumption should be constrained by an upper bound, while execution time should be minimized.

Independently from the specific application – and hence, problem variant– some metrics play an important role in fog computing. These metrics are reviewed next.

5.6.1 Performance

There are several performance-related metrics, like execution time, latency, and throughput. Generally, performance is related to the amount of time needed

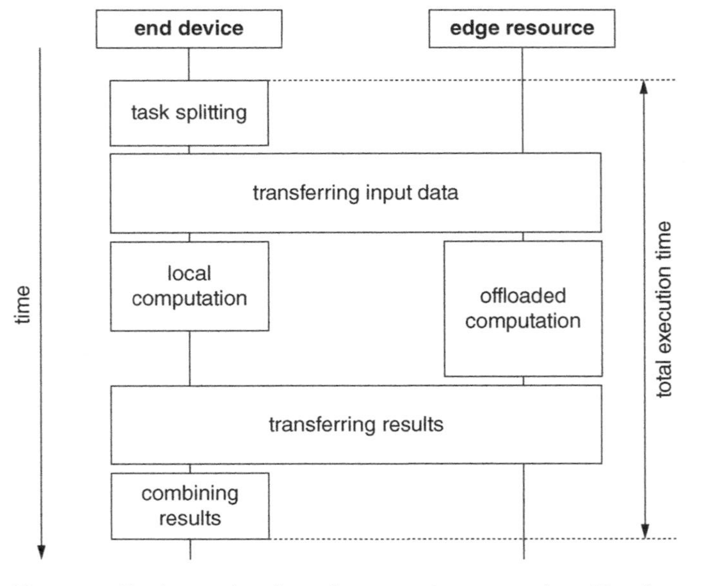

Figure 5.4 Total execution time of an example computation offloading scenario.

to accomplish a certain task. In a fog computing setting it is important to note that accomplishing a task usually involves multiple resources, often on different levels of the reference model of Figure 5.3. Hence, the completion time of the task may depend on the computation time of multiple resources, plus the time for data transfer between the resources. Some of these steps might be made in parallel (e.g., multiple devices can perform computations in parallel), whereas others must be made one after the other (e.g., the results of a computation can only be transferred once they have been computed). The total execution time depends on the critical path of compute and transfer steps. For instance, if a computation is partly done in an end device and partly offloaded from the end device to an edge resource, this may lead to a situation such as the one depicted in Figure 5.4, in which the total execution time is determined by the sum of multiple computation and data transfer steps.

5.6.2 Resource Usage

Especially in the lower layers of the reference model of Figure 5.3, the economical use of the scarce resources is vital. This particularly applies to end devices, which typically have very limited CPU and memory capacity. Edge resources typically offer higher capacities, but also those capacities can be limited, given that edge resources may include machines like routers that do not offer exhaustive computational capabilities.

To some extent, CPU usage can be traded off with execution time, i.e., overbooking the CPU may lead to a situation where the application is still running, but more slowly. This may be acceptable for some applications, but not for time-critical ones. Moreover, memory poses a harder constraint on resource consumption, since overbooking the memory may lead to more serious problems like application failure.

Beyond CPU and memory, also network bandwidth can be a scarce resource, both between end devices and edge resources and between edge resources and the cloud. Hence, also the use of network bandwidth may have to be either minimized or constrained by an upper bound.

It is important to note that, in contrast to performance, which is a global metric spanning multiple resources, resource consumption needs to be considered at each network node and link separately.

5.6.3 Energy Consumption

Energy can also be seen as a scarce resource, but it is quite different from the other resource types already considered. Energy is consumed by all resources as well as the network. Even idle resources and unused network elements consume energy, but their energy consumption increases with usage. Generally, assuming that the power consumption of a resource depends linearly on its CPU load is a good approximation [13]. It is important, though, to highlight the difference between power consumption and energy consumption, since energy consumption also depends on the amount of time during which power is consumed. Thus, it is, for instance, beneficial in terms of overall energy consumption to move a compute task from one resource to a significantly faster one, even if the faster machine has slightly higher power consumption.

Energy consumption is important on each layer of the fog, but in different ways. For end devices, battery power is often a bottleneck, and thus preserving it as much as possible is a primary concern. Edge resources are typically not battery-powered; hence, their energy consumption is less important. For the cloud, energy consumption is again very important, but because of its financial implications: electric power is a major cost driver in cloud data centers. Finally, also the overall energy consumption of the whole fog system is important because of its environmental impact.

5.6.4 Financial Costs

As already mentioned, energy consumption has implications on financial costs. But also other aspects influence costs. For example, the use of the cloud or edge infrastructure may incur costs. These costs can be fixed or usage-based, or some combination thereof. Similarly, the use of the network for transferring data may incur costs.

5.6.5 Further Quality Attributes

All aspects covered so far are easily quantifiable. However, they are not sufficient to guarantee a high quality of experience for users. For this, quality attributes like reliability [14], security [12], and privacy [16] also need to be taken into account, which are harder to quantify.

Traditionally, such quality attributes are not captured by optimization problems, but, rather, addressed with appropriate architectural or technical solutions. For instance, reliability may be achieved by creating redundancy in the architecture, security may be achieved by using appropriate cryptographic techniques for encryption, while privacy may be achieved by applying anonymization of personal data. Nevertheless, there are several ways to address quality attributes during optimization of a fog system, as shown by the following representative examples:

- To increase reliability, it is beneficial to let multiple resources perform the same critical computations in parallel, so that the result is available even if some of the resources stop working or become unreachable, and also to compare the results with each other to filter out flawed results. The higher the number of resources used in parallel, the higher level of reliability can be achieved this way. Therefore, the number of resources used in parallel is an important optimization objective that should be maximized [15].
- Both security and privacy concerns may be mitigated by preferring trusted resources. Using existing techniques to quantify trust, for instance based on reputation scores [16], the usage of trusted resources becomes an optimization objective, in which trust levels of the used resources should be maximized.
- Co-location of computational tasks belonging to different users / tenants may increase the likelihood of tenant-on-tenant attacks. Therefore, minimizing the number of tenants whose tasks are co-located is an optimization objective that helps to keep security and privacy risks at an acceptably low level.
- Co-location of tasks belonging to the same user decreases the need for exchanging data over the network, which in turn decreases the likelihood of eavesdropping, man-in-the-middle, and other network-based attacks. Hence, minimizing the number of resources used also helps in decreasing risks related to information security.

It is important to note that the above optimization objectives relating to quality attributes typically conflict with other optimization objectives relating to costs, performance, etc. For example, increasing redundancy may be beneficial for improving reliability but at the same time it can lead to higher costs. Similarly, preferring service providers with high reputation is advantageous from the point of view of security, but may also lead to higher costs. Constraining co-location options may improve privacy, but may lead to worse performance or higher energy consumption, and so on. This is one of the main reasons why it is beneficial to include also quality attributes in optimization problems, because this enables explicit reasoning about the optimal trade-off between the conflicting objectives.

5.7 Optimization Opportunities along the Fog Architecture

Optimization problems in fog computing can be classified according to which layer(s) of the three-layer fog model (cf. Figure 5.3) is/are involved.

In principle, it is possible that only one layer is involved. This, however, is typically not regarded as fog computing. For example, if only the cloud layer is involved, then we have a pure cloud optimization problem. Likewise, if only end devices are involved, then the problem would not be in the realm of fog computing, but rather – depending on the kinds of devices and their interconnections – in mobile computing, IoT, wireless sensor networks etc.

Therefore, real fog computing problems involve at least two layers. This consideration leads to the following classification of optimization problems in fog computing:

- Problems involving the cloud and the edge resources. This is a meaningful setting, which allows for example to optimize overall energy consumption of cloud and edge resources, subject to capacity and latency constraints [17]. This setup shows some similarity to distributed cloud computing; a potential difference is that the number of edge resources can be several orders of magnitude higher than the number of data centers in a distributed cloud.
- Problems involving edge resources and end devices. The collaboration of end devices with edge resources (e.g., offloading computations) is a typical fog computing problem, and because of the limited resources of end devices, optimization plays a vital role in such cases. An often studied special case of this problem setup is when a single edge resource is considered together with the end devices that it serves [18]. However, the more general case in which multiple edge resources – together with the end devices that they serve – are considered has also received attention [19]. The latter leads to more complex optimization problems, but has the advantage to balance computational load among multiple edge resources.
- In principle, all three layers can be optimized together. This, however, is seldom studied, probably because of the difficulties of such optimization. The difficulties relate on one hand to the computational complexity of large-scale optimization problems involving decision variables for all fog resources. On the other hand, many different technical issues would have to be integrated

into a single optimization problem to capture the different optimization concerns of the cloud, the edge resources, and the end devices, which is challenging in itself. In addition, changes to the cloud, the edge resources, and the end devices are typically made by different stakeholders on different time scales, which is also a rationale for independent optimization of the different fog layers.

In each of the fog layers, optimization may target the distribution of data, code, tasks, or a combination of these. In data-related optimization, decisions have to be made about which pieces of data are stored and processed where in the fog architecture. In code-related optimization, program code can be deployed on multiple resources and the goal is to find the optimal placement of the program code. Finally, in task-related optimization, the aim is to find the optimal split of tasks among multiple resources.

Finally, it should be noted that the distributed nature of fog computing systems may make it necessary to perform optimization in a distributed fashion. Ideally, the locally optimal decisions of the participating autonomous resources should lead to a globally optimal behavior [20].

5.8 Optimization Opportunities along the Service Life Cycle

Just like cloud computing, fog computing is characterized by the provision and consumption of services. By looking at the different optimization opportunities at the different stages of the service life cycle, one can differentiate between the following options:

- **Design-time optimization**. When a fog service is designed, exact information about the end devices to be served is typically not available. Hence, optimization will be constrained mostly to the cloud and edge layers of the architecture, where more information may already be available at design time. Concerning the end devices, optimization is constrained to questions dealing with types of devices (as opposed to device instances, which will be known only during run time).
- **Deployment-time optimization**. When the deployment of the service on specific resources is planned, the available information of the resources can be used to make further optimization decisions. For example, the exact capacity of the edge resources to be used may become available at this time, so that the split of tasks between the cloud and the edge resources can be (re-)optimized.
- **Run-time optimization**. Although some aspects of a fog system may be optimized in advance (i.e., during design time or deployment time), many important aspects become clear only when the system is running and

used. Examples include the specific end devices with their parameters (e.g., compute capacity) and the compute tasks that the end devices want to offload to the edge resources. These aspects are vital for making sound optimization decisions. Moreover, these aspects keep changing during the operation of the system. As a consequence, much of the system operation needs to be optimized during run time. This requires continuous monitoring of important system parameters, analysis of whether the system still operates with acceptable effectiveness and efficiency, and re-optimization whenever necessary [20].

As can be seen, run-time optimization plays a very important role in the optimization of fog computing systems. This has some important consequences. First, the time available for executing an optimization algorithm during run time is seriously limited, thus the adopted optimization algorithms have to be fast. Second, run-time optimization is usually not about laying out a system from scratch, but rather, about adapting an existing setup. This implies, in particular, that the costs associated with changes to the system have to be taken into account.

5.9 Toward a Taxonomy of Optimization Problems in Fog Computing

The different aspects of optimization covered so far can form the basis to devise a taxonomy of optimization problems in fog computing. In the following, we illustrate this by means of classifying some representative publications taken from the literature along the presented dimensions.

As a first example, Table 5.2 shows the classification of the work of Do et al. [21]. This paper considers a video streaming service, consisting of a central cloud data center and a huge number of geographically distributed edge resources (called fog computing nodes or FCNs in the paper), which are to provide end devices with streaming video. The aim is to determine the data rate of video streaming for each edge resource, taking into account the different utility provided by different data rates at different edge resources, data center energy consumption, and the workload capacity of the data center. The paper proposes a distributed iterative improvement algorithm inspired by the ADMM (Alternating Direction Method of Multipliers) method.

As another example, Table 5.3 shows the classification of the work of Sardellitti et al. [22] according to the presented dimensions. That paper considers the computation offloading problem in a mobile edge computing (MEC) setting, where some mobile end devices offload some compute tasks to a nearby edge resource. For each compute task of each end device, it can be decided whether or not it should be offloaded, and in case of offloading, which radio

Table 5.2 Classification of the work of Do et al. [21] according to the presented dimensions.

Paper:	Do et al.: A proximal algorithm for joint resource allocation and minimizing carbon footprint in geo-distributed fog computing [21]
Context / domain:	Video-streaming service with a central cloud serving distributed edge resources which, in turn, serve end devices
Considered metrics:	• "Utility" (weighted data rate of the edge resources) • Compute capacity of the cloud data center • Energy consumption of the cloud data center
Considered layer / resources:	• Cloud • Edge resources
Phase in life cycle:	Design / deployment time
Optimization algorithm:	Distributed iterative improvement

Table 5.3 Classification of the work of Sardellitti et al. [22] according to the presented dimensions.

Paper:	Sardellitti et al.: Joint optimization of radio and computational resources for multicell mobile-edge computing [22]
Context / domain:	Computation offloading from mobile end devices to an edge resource
Considered metrics:	• Energy consumption of the end devices • Total time to transfer and execute offloaded tasks • Amount of compute power of edge resource occupied by offloaded tasks of the devices
Considered layer / resources:	• Edge resource • End devices
Phase in life cycle:	Run time
Optimization algorithm:	Iterative heuristic using successive convex approximation

channel should be used for the communication. The optimization problem is formed in terms of energy consumption and latency. The paper first formulates the problem for a single end device, which can be solved explicitly in closed form. However, for several end devices with potentially interfering communication, the problem becomes much tougher (in particular, nonconvex), which the authors solved by means of an appropriate heuristic.

Finally, Table 5.4 describes the work of Mushunuri et al. [23], which addresses the problem of finding the optimal work distribution among cooperating robots. The robots (end devices) offload their compute tasks to a server

Table 5.4 Classification of the work of Mushunuri et al. [23] according to the presented dimensions.

Paper:	Mushunuri et al.: Resource optimization in fog enabled IoT deployments [23]
Context / domain:	Cooperating mobile robots sharing compute tasks
Considered metrics:	• Communication cost between end devices and edge resource • Battery power of end devices • CPU capacity of end devices and edge resource
Considered layer / resources:	• Edge resource • End devices
Phase in life cycle:	Run time
Optimization algorithm:	Nonlinear optimization with the COBYLA (Constrained Optimization by Linear Approximations) algorithm within the NLOpt library

(edge resource), which distributes it among the end devices and itself. It is assumed that compute tasks can be split arbitrarily. The optimization, carried out at run time by the edge resource, takes into account the communication costs, battery status, and compute capacities of the devices, and uses an off-the-shelf nonlinear optimization package.

As can be seen from these three examples that cover different optimization problems within fog computing, the presented aspects can be applied successfully to classify the approaches from the literature and capture the characteristics that are relevant for optimization.

5.10 Optimization Techniques

The three examples presented in Section 5.9 show that the optimization techniques adopted in fog computing optimization problems are quite heterogeneous. The following characteristics seem to be quite common, though:

- Adoption of nonlinear, sometimes even nonconvex optimization techniques
- Usage of heuristics (as opposed to exact algorithms) to derive – potentially suboptimal – results to hard problems with limited computational effort
- Usage of distributed algorithms, accounting for the distributed resources and the distributed knowledge in fog computing

In the future, with the maturation of the field, a consolidation of the used methods may take place. However, since the considered problem variants are also manifold, we expect the field to continue to require several different types of algorithmic techniques, including exact algorithms, heuristics, as well as hybrid approaches [24].

5.11 Future Research Directions

Fog computing is still in its early days, with optimization taking an ever more important role in it. Accordingly, there are several areas where significant future research is needed:

- **Co-optimization**. One of the key challenges in optimizing fog computing systems is that several different technical systems and sub-systems must be tuned to achieve an overall optimal, or at least good enough configuration. This includes on the one hand the different devices making up a fog system and on the other hand the different technical aspects like networking, computation, volatile memory and persistent storage, sensors and actuators etc. Optimizing all those aspects together, or finding good ways to decompose this huge optimization problem into sub-problems that can be solved mostly independently, remains an important challenge for future research.
- **Balancing multiple optimization objectives**. Another important characteristic of optimization in fog computing is that multiple, often conflicting optimization objectives must be considered simultaneously. Current practices to handle multicriteria optimization in fog computing – e.g., using the weighted sum of the different optimization objectives – are simple and may lead to good results in several cases, but may lead to implausible solutions in extreme situations, hindering the practical adoption of such approaches. Finding more robust ways of incorporating multiple optimization objectives thus remains an important future research direction.
- **Algorithmic techniques**. So far, optimization algorithms have been selected largely arbitrarily, often based primarily on authors' previous experience with different techniques. With the maturation of the field, the community should develop a better understanding of which algorithmic approaches work well for which problem variants.
- **Evaluation of optimization algorithms**. Existing approaches were evaluated in rather ad hoc ways. Before methods can be transferred from research into practice, it is vital to evaluate the applicability of the proposed algorithms in a sound, thorough, and repeatable manner. This requires the definition of benchmark problems with publicly available problem instances, consensus in the community on evaluation methodologies and test environments, development of reliable and realistic simulators, and unbiased comparison of competing approaches under realistic – also

including extreme – situations. Also, theoretical methods to prove algorithm properties in a rigorous way will be necessary.

5.12 Conclusions

In this chapter, we have presented a review of optimization problems in fog computing. In particular, we have explained why optimization plays a vital role in fog computing and why it is important to define optimization problems unambiguously, preferably using a formal problem model. The most important aspects of optimization in fog computing have been reviewed according to multiple dimensions: the metrics that serve as optimization objectives or as constraints, the considered layers within the fog architecture, and the relevant phase in the service life cycle. These dimensions also lend themselves to form a taxonomy, which can be used to classify existing or future problem variants.

We have also argued that there are several important directions for future research, including the improved handling of multiple optimization objectives, the co-optimization of multiple technical aspects, better understanding of which algorithmic techniques work best for which problem variant, and devising disciplined evaluation methodologies.

Acknowledgments

The work of Z. Á. Mann has been supported by the Hungarian Scientific Research Fund (Grant Nr. OTKA 108947) and the European Union's Horizon 2020 research and innovation program under grant 731678 (RestAssured).

References

1 L. M. Vaquero, L. Rodero-Merino. Finding your way in the fog: Towards a comprehensive definition of fog computing. *ACM SIGCOMM Computer Communication Review,* 44(5): 27–32, October 2014.
2 B. Varghese and R. Buyya. Next generation cloud computing: New trends and research directions. *Future Generation Computer Systems,* 79(3): 849–861, February 2018.
3 E. Ahvar, S. Ahvar, Z. Á. Mann, et al. CACEV: A cost and carbon emission-efficient virtual machine placement method for green distributed clouds. In *IEEE International Conference on Services Computing,* pp. 275–282, IEEE, 2016.
4 A. V. Dastjerdi, R. Buyya. Fog computing: Helping the Internet of Things realize its potential, *Computer,* 49(8): 112–116, 2016.

5 K. Kumar, Y.-H. Lu. Cloud computing for mobile users: Can offloading computation save energy? *Computer,* 43(4): 51–56, April 2010.
6 Z. Á. Mann. Modeling the virtual machine allocation problem. In *Proceedings of the International Conference on Mathematical Methods, Mathematical Models and Simulation in Science and Engineering,* pp. 102–106, 2015.
7 Z. Á. Mann. Allocation of virtual machines in cloud data centers – A survey of problem models and optimization algorithms. *ACM Computing Surveys,* 48(1): article 11, September 2015.
8 F. Bonomi, R. Milito, J. Zhu, S. Addepalli. Fog computing and its role in the Internet of Things. In *Proceedings of the 1st ACM Mobile Cloud Computing Workshop,* pp. 13–15, 2012.
9 Z. Á. Mann, *Optimization in computer engineering – Theory and applications.* Scientific Research Publishing, 2011.
10 R. T. Marler, J. S. Arora. Survey of multi-objective optimization methods for engineering. *Structural and Multidisciplinary Optimization,* 26(6): 369–395, April 2004.
11 S. Soo, C. Chang, S. W. Loke, S. N. Srirama. Proactive mobile fog computing using work stealing: Data processing at the edge. *International Journal of Mobile Computing and Multimedia Communications,* 8(4): 1–19, 2017.
12 I. Stojmenovic and S. Wen. The fog computing paradigm: Scenarios and security issues. In *Proceedings of the 2014 Federated Conference on Computer Science and Information Systems (FedCSIS),* pp. 1–8, 2014.
13 S. Rivoire, P. Ranganathan, C. Kozyrakis. A comparison of high-level full-system power models. In *Proceedings of the 2008 Conference on Power Aware Computing and Systems (HotPower '08),* article 3, 2008.
14 H. Madsen, B. Burtschy, G. Albeanu, F. Popentiu-Vladicescu. Reliability in the utility computing era: Towards reliable fog computing. *20th International Conference on Systems, Signals and Image Processing,* pp. 43–46, 2013.
15 I. Kocsis, Z. Á. Mann, D. Zilahi. Optimised deployment of critical applications in infrastructure-as-a-service clouds. *International Journal of Cloud Computing,* 6(4): 342–362, 2017.
16 S. Yi, Z. Qin, Q. Li. Security and privacy issues of fog computing: A survey. *International Conference on Wireless Algorithms, Systems, and Applications,* pp. 685–695, 2015.
17 R. Deng, R. Lu, C. Lai, and T.H. Luan. Towards power consumption-delay tradeoff by workload allocation in cloud-fog computing. *IEEE International Conference on Communications,* pp. 3909–3914, 2015.
18 X. Chen, L. Jiao, W. Li, and X. Fu. Efficient multi-user computation offloading for mobile-edge cloud computing. *IEEE/ACM Transactions on Networking,* 24(5): 2795–2808, 2016.

19 J. Oueis, E. C. Strinati, S. Barbarossa. The fog balancing: Load distribution for small cell cloud computing. *81st IEEE Vehicular Technology Conference,* 2015.

20 J. O. Kephart, D. M. Chess. The vision of autonomic computing. *Computer,* 36(1): 41–50, 2003.

21 C. T. Do, N. H. Tran, C. Pham, M. G. R. Alam, J. H. Son, and C. S. Hong. A proximal algorithm for joint resource allocation and minimizing carbon footprint in geo-distributed fog computing. *International Conference on Information Networking,* pp. 324–329, IEEE, 2015.

22 S. Sardellitti, G. Scutari, S. Barbarossa. Joint optimization of radio and computational resources for multicell mobile-edge computing. *IEEE Transactions on Signal and Information Processing over Networks,* 1(2): 89–103, 2015.

23 V. Mushunuri, A. Kattepur, H. K. Rath, and A. Simha. Resource optimization in fog enabled IoT deployments. *2nd International Conference on Fog and Mobile Edge Computing,* pp. 6–13, 2017.

24 D. Bartók and Z. Á. Mann. A branch-and-bound approach to virtual machine placement. In *Proceedings of the 3rd HPI Cloud Symposium "Operating the Cloud,"* pp. 49–63, 2015.

Part II

Middlewares

6

Middleware for Fog and Edge Computing: Design Issues

Madhurima Pore, Vinaya Chakati, Ayan Banerjee, and Sandeep K. S. Gupta

6.1 Introduction

Edge computing and fog computing have combined in a way to facilitate a wide variety of applications that involve human interactions, which are geographically distributed and have stringent real-time performance requirements. The Internet of Things (IoT) or Internet of Everything (IoT) has introduced edge devices that now obtain information from user environment and need to respond intelligently to changes in real time. The scale of an application has increased from mere single mobile device to a large number of edge devices that are geographically distributed and change locations dynamically. Even though cloud support can be used in processing the data generated by the edge devices, the delay incurred in communication to cloud devices is excessively more than the real-time constraints of some of the latency sensitive applications. With such a large scale of data being generated in geographically distributed locations, sending the code toward the data is in some cases more efficient than processing in the cloud. Fog computing introduced computation solution in the form of fog devices, cloudlets, and mobile edge computing (MEC), which provide computation services in the network edge. It can also meet the real-time requirements of such applications.

Apart from the components that pertain to application logic, there are large design components that perform the underlying task of managing the network, computation, and resources of fog and edge architecture (FEA). Due to the dynamically changing context in the edge devices, underlying algorithms for managing the execution and processing data become complex. In addition to the processing and data communication of the distributed application, the control data and algorithm decisions incur excess resources when they are executed on the edge devices. In this chapter, we discuss different aspects of design of middleware for FEA.

Fog and Edge Computing: Principles and Paradigms, First Edition.
Edited by Rajkumar Buyya and Satish Narayana Srirama.

6.2 Need for Fog and Edge Computing Middleware

Fog and edge computing are gaining acceptance due to high availability, low latency and low cost. Domains such as smart cities, virtual reality and entertainment, vehicular systems use edge processing for real-time operations [1, 2]. The efficient design of middleware enables the realization of full potential of fog and edge infrastructures. Middleware handles different tasks such as communication management, network management, task scheduling, mobility management, and security management, thereby reducing the complexity of the distributed mobile application design.

Middleware design of fog/edge infrastructure is challenging because of the stringent application requirements such as (i) availability of context on the sensing devices; (ii) cost of data transfer and processing in different tiers of FEA; (iii) limitations on number of edge devices present and dynamic changes in context and mobility of the devices; and (iv) strict latency constraints. Including the dynamically changing context of a user and capturing the user interactions patterns can essentially enable intelligent and informed execution of the applications.

In this chapter, we present state-of-the-art middleware for fog and edge infrastructures and propose an architecture of middleware that supports distributed mobile applications with specific requirements of applications. Proposed middleware primarily focuses on application-aware task scheduling and data acquisition.

6.3 Design Goals

A varied class of mobile applications can utilize FEA middleware. Requirements for emerging applications can be summarized as following:

1. Newer distributed applications increasingly demand a large number of resources and low latency to meet the real-time response constraints. While the use of cloud has eased the implementation of large-scale distributed mobile applications, in many newer applications, the strict real-time response may not be feasible unless processing is done near the edge.
2. Geo-distributed edge applications such as monitoring oil plant and electricity grid management are geographically distributed. Processing enormous sensor generated data streams in real-time constraints require a large processing facility, but also incur huge communication infrastructure or bandwidth. Edge infrastructure can reduce the communication overhead involved in large data streams.
3. Large-scale distributed management applications such as connected railways and smart grids involve processing huge data in real time to provide

control towards reliable operation. Increasing real-time monitoring and analytical processing in edge can adapt the system itself to dynamic faults and changes.
4. Smart and connected applications such as real-time traffic monitoring and connected vehicles can leverage local edge infrastructure for fast and real-time updates and response related to locally sensed data.

Even though the FEA can support different types of applications, common functionalities that are required in such applications can be provided by the middleware. Following are design goals of FEA middleware.

6.3.1 Ad-Hoc Device Discovery

Data sources in fog/edge may belong to a wide category of devices ranging from IoT sensors, mobile devices to fixed sensors. The data are processed locally or sent to fog/edge devices for further processing. A channel of communication needs to be set up between the requesting devices and ad-hoc discovered devices that perform the application task of acquisition and processing. Once a communication channel is set up, it allows dynamically changing set of devices to join and participate. Given the dynamic nature of participating edge devices that acquire and process the data, the device discovery allows setup of a communication layer to enable further communication between devices.

6.3.2 Run-Time Execution Environment

The middleware provides a platform that executes the application task remotely on the edge devices. Functionality includes code download, remote execution in the edge devices, and delivery of results such that they are available to the requesting device.

6.3.3 Minimal Task Disruption

Task disruption during execution affects the reliability of execution of FEA task. Often it results in reinitialization of the task or undesirable/unavailable results. Device usage patterns, mobility, and network disconnections may cause unexpected changes in the context of the device. This may render the device inappropriate for continuing the execution of sensing or computation task. Anticipatory techniques can be used to minimize the interruption in tasks, thereby promoting intelligent scheduling decisions.

6.3.4 Overhead of Operational Parameters

Establishing communication between ad-hoc edge devices, selection of candidate edge devices, distribution of FEA tasks between multiple edge devices, and managing remote execution in a sequence of FEA tasks incurs additional usage of bandwidth and energy consumption on the edge devices. As these resources are expensive, minimizing these operational parameters is an important aspect of middleware operations. Additionally, several devices may enforce usage limit on their resources that are available for sharing.

6.3.5 Context-Aware Adaptive Design

Innovative contexts such as the mental state of [3–5] and user activity [6] are now used in mobile applications for sensing useful data. For successful execution in FEA, dynamic changes in the context of the devices as well its environment require the middleware to adapt to these changes. Self-adaptive services can enhance its operations and improve the FEA quality of service.

6.3.6 Quality of Service

Quality of service (QoS) of an architecture is highly dependent on the application. Many edge/fog applications use multidimensional data for achieving specific goals. Acquiring and processing such huge sensor data within real-time constraints is a requirement for these applications. Real-time response is an important QoS measure. Other application specific QoS parameters can be the relevance of the acquired data, its correctness, and uninterrupted data acquisition.

6.4 State-of-the-Art Middleware Infrastructures

Applications in fog and edge computing are discussed in some of the recent works. Real-time data streaming applications include traffic monitoring Waze [7, 8], smart traffic light systems [9], real-time replay in the stadium [10], and video analytics [12, 13]. Real-time applications that process the requests for emergency rescue in disaster and searching for missing persons [10, 11]. Application of geographically distributed systems such as wind farms [9] and smart vehicle-to-vehicle systems [14] are becoming popular in fog and edge computing. Common requirements of these applications present a need for middleware to support easy design and development of such applications.

Middleware features such as security, mobility, context awareness and data analytics addressed in recent research are shown in Table 6.1. Popular IoT platforms such as GoogleFit [15, 16] have a cloud-based IoT middleware for smartphones. Provisioning of sensing services on the mobile devices is discussed in M-Sense [17]. Service oriented middleware like GSN [18–20] are proposed for processing data in the distributed environment. Further, Carrega et al. propose microservices-based middleware using a user interface [20].

Table 6.1 Middleware features in fog and edge architectures.

	Devices	Security	Mobility support	Context awareness	Data analytics	Optimized selection of devices
FemtoCloud [21]	Mobile	N	N	Y	Y	Y
Nakamura et al. [22]	Mobile and sensor	N	N	N	Y	N
Aazam et al. [27]	Fog, MEC, Cloud	Y	N	N	Y	Y
Bonomi et al. [9]	Fog, Cloud	N	Y	N	Y	Y
Verbelen et al. [28]	Mobile Cloudlet	N	N	N	N	Y
Cloudaware [26]	Cloudlet	N	Y	Y	N	Y
Hyrax [10]	Cloud	N	Y	N	N	Y
Grewe et al. [14]	MEC	Y	Y	N	Y	Y
Carrega et al. [20]	MEC, Fog	Y	Y	N	Y	Y
Piro et al. [16]	Cloud	Y	Y	Y	Y	Y

In FemtoCloud system, the mobile devices in the edge can be configured to provide the services to requesting devices [21]. "Process on Our Own (PO3)" concept where a data stream generated on each device is processed on itself is proposed by Nakamura et al. [22]. CoTWare middleware proposed by Jaroodi et al. suggests a novel way of integrating things, fog devices, and cloud by using cloud-hosted services to manage processing of IoT data in fog resources [23]. MobiPADs [24] and MobiCon [25] are context-aware middleware solutions that enable adaptive design in mobile applications by reconfiguring the services with respect to dynamic changes in context for edge devices. Recently proposed CloudAware [26] is an example of adaptive middleware for constantly changing context such as connectivity for cloudlet.

6.5 System Model

FEA includes devices that can be broadly classified into five types, as shown in Figure 6.1. Mobile devices connected with sensors and actuators are the nearest devices to the users. As the processing devices move away from the edge, the latency of the communication increases. Also, the availability of resources for processing and data storage increases toward the cloud. The components of FEA are discussed in detail in the following sections.

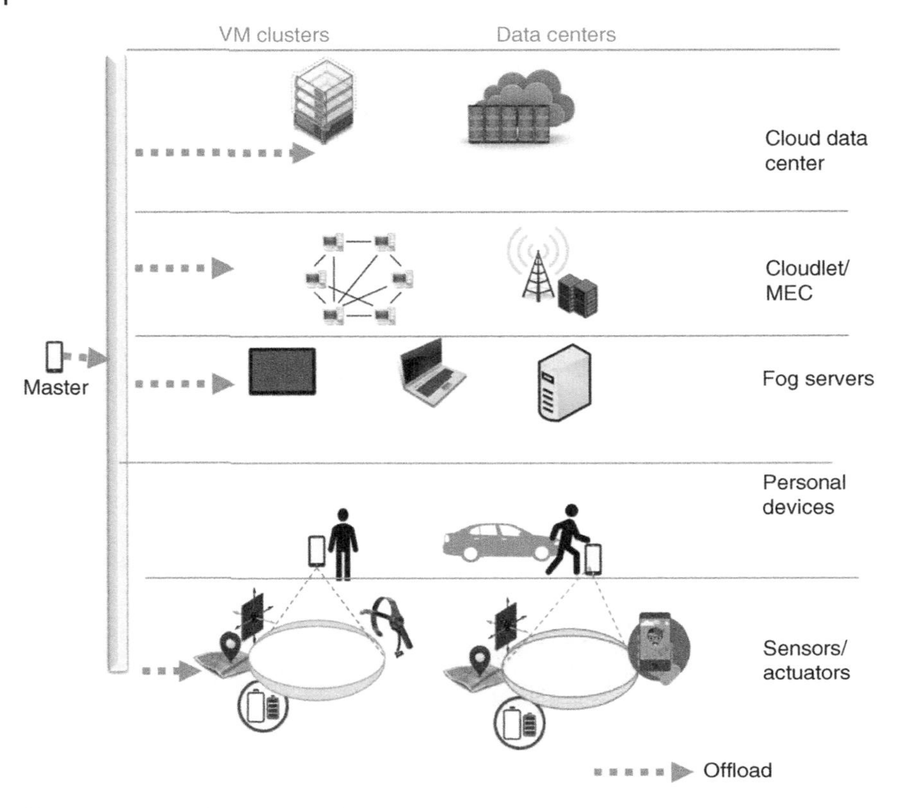

Figure 6.1 Fog and edge computing devices.

6.5.1 Embedded Sensors or Actuators

Embedded sensors and actuators are installed in physical structures or deployed on a human body. The sensors are responsible for obtaining the environmental or physiological signals that are processed by the system, while the actuators execute the actions initiated by the system. Built-in networking capabilities allow these devices to communicate to the nearby devices. They may also have a limited computing capability.

6.5.2 Personal Devices

Personal devices, or smartphones, inherently demonstrate mobility as they are owned by human users. These devices connect with the embedded sensors and actuators. They often act as an intermediate data hub or computation platform, and/or provide a communication link to servers. A part of their resources may be shared to execute the fog and edge distributed applications.

6.5.3 Fog Servers

Fog servers are computationally more powerful than the personal mobile devices.

As these devices are closer to the edge, they provide a cheaper option for offload with respect to communication costs. These nodes exist between the edge devices and cloud. They can be used to process data and also act as intermediary storage. Further, communication to other edge devices can be achieved through peer-to-peer (P2P) or device-to-device (D2D) techniques such as WiFi, Bluetooth, and WiFi Direct.

6.5.4 Cloudlets

Cloudlets were proposed by Satyanarayan et al. [29] as a small-scale dedicated set of servers that have high bandwidth Internet connectivity but are situated close to the edge. They are also known as a data center in a box. Another implementation of edge computation is offered by telecom companies that bring the compute resources in the base station of mobile towers. They are known as mobile edge computing (MEC) servers.

6.5.5 Cloud Servers

Cloud servers have the most computational, communication, and storage capability in the hierarchy. The cloud servers are usually associated with a pay-as-you-go model. They can easily scale the number of VMs according to the request.

6.6 Proposed Architecture

Fog and edge computing applications include the following: (i) batch processing that needs large-scale data acquisition and distributed processing; (ii) quick-response application that needs a response in real time; and (iii) stream applications that require processing of a continuous data stream in real time [11, 12].

Such applications exist in different domains such as healthcare, emergency rescue and response systems, traffic management, vehicle-to-vehicle systems, and environment monitoring. Due to huge processing requirements, such applications need a large distributed architecture that processes data in multiple tiers. Lower tier near the edge perform filtering, preprocessing, and extraction of useful information while the edge and fog servers are used for processing and analytics. FEA (Figure 6.2) mainly consists of middleware services that are common to FEA applications, as discussed below.

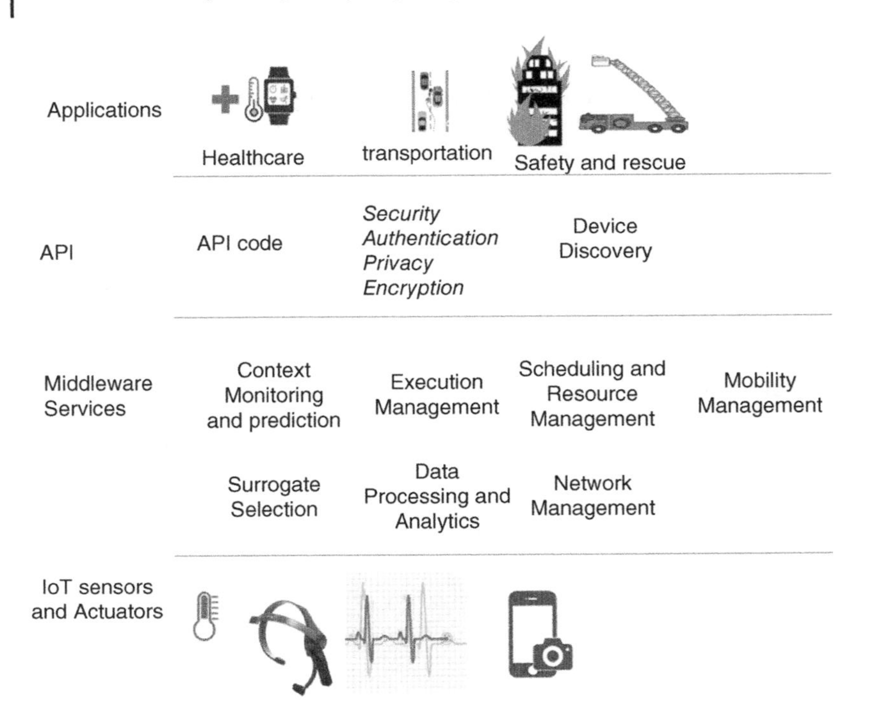

Figure 6.2 Fog and edge computing architecture.

6.6.1 API Code

Services common to fog and edge applications can be designed as an API. The API is then integrated into an app, enabling the design of different functionalities in the middleware with simple and easy to use functions.

6.6.2 Security

Security in edge devices is essential to ensure access to authorized users and establish a secure communication channel for communication of user data.

6.6.2.1 Authentication

Data ownership and protecting access to the private information is very important to the edge participants. Many "pay-as-you-go" services necessitate authentication to prevent any unwanted access. Public key infrastructure-based systems have been proposed for user and device authentications for distributed key management [30]. Authentication of VM instances and migration of VMs that has been used in cloud [31] can be adopted for VMs in MEC and fog servers. Authentication as a service is proposed by Mukherjee et al. to enable authentication in the participating fog nodes [32]. Ibrahim proposed a lightweight mutual authentication scheme for roaming fog nodes [33] using one long-lived master secret key. This algorithm is efficient and lightweight for limited resource devices such as sensors and actuator and doesn't require the devices to re-register.

6.6.2.2 Privacy

Data privacy is very important with respect to handling data from user devices. The main challenge in FEA is to ensure the privacy of devices that exhibit mobility in the edge. Though the sensors and edge devices have limited resources, the fog nodes can provide the necessary encryption capabilities for edge processing [34]. Existing works propose anonymization techniques or pseudo anonymization through which user identity is protected. A lightweight privacy preserving scheme using Chinese remainder theorem is proposed by Lu et al. [35]. Laplacian mechanism query model is proposed by Wang et al. [36] that deals with privacy preservation of location-aware services. Policy-based access control mechanisms for fog computing is proposed by Dsouza et al. in [37].

6.6.2.3 Encryption

Many existing studies propose the use of data encryption [34]. Recent work proposes crypto processors and encryption functions [38]. However, using encryption increases the computation, energy usage, and time incurred.

6.6.3 Device Discovery

Device discovery allows the new devices to participate and leave the network as they become available in the network. Many researchers use MQTT as the standard publish-subscribe message API [39], which is a lightweight messaging protocol, designed for constrained devices and low-bandwidth, high-latency, or unreliable networks. Fog and edge-distributed middleware can also use pub-sub as a service from a third party such as Nearby Message [40], or PubNub [41] that may be integrated into the middleware. Additionally, these services may provide security, scalability, and reliability for message exchange.

6.6.4 Middleware

Components of middleware commonly used in fog and edge applications are discussed in the following subsections.

6.6.4.1 Context Monitoring and Prediction

FEA can adapt to dynamic changes in the user environment using the context-aware design of middleware. This may involve continuous monitoring of relevant context and adaptive actions that are based on changes in the context. Also, recent research shows that several human-dependent contexts have patterns. These patterns can be learned to intelligently manage the operations between multiple devices. Several techniques such as time series, stochastic, or machine learning can be used to model and predict the human-mobile contextual changes [42, 43].

6.6.4.2 Selection of Participating Devices

FEA employs devices from the environment that can sense and/or process the data acquired in the FEA applications. Selection of the surrogate device can be based on different policies designed in the middleware. Research shows several policies such as fairness-based selection, game theoretic [8], context optimization [44], and resource optimization approaches that are used in surrogate selection. Participating users are selected based on different criterion ranging from simple user context such as the location of the device to selection based on the reputation of user task completion history [45]. Following are different surrogate selection techniques.

Energy-Aware Selection. Remaining battery is critical to every mobile user and it determines the amount of resources that the device owner may share. Selection of surrogates is a trade-off between the quality of information gathered and the remaining battery on the device with an incentive budget [46].

Delay Tolerance-Based Selection. Real-time applications and streaming data application require the processing to be completed in a given time constraint [12]. Performance-based selection of surrogates is proposed in incentivized schemes by Petri et al. [47].

Context-Aware Selection. Context-aware functionality is used in many mobile applications. Applications are designed to adapt themselves based on changes in context on the mobile device or that of the user. Recently proposed context-aware recruitment scheme focused on improving the mobile selection based on context requirements of the application [44]. In applications like crowdsensing apart from individual context prediction, large-scale activity prediction such as proposed in [48] is now becoming useful. Change in location of mobile users can be modeled using different techniques such as random waypoint model and statistical models such as Markov [49, 50]. Spatiotemporal model of user location is proposed by Wang et al. [51].

6.6.4.3 Data Analytics

FEA introduces the idea of processing near the edge. Extensive analytics may be involved in an application that is processing across different layers in the architecture. Some of the analytics tasks can also be used to extract the essential information from the raw data obtained on the user devices. This not only reduces the processing requirements centrally but also reduces the communication costs. Data analytics module on the user device can be used to send essential data towards a central server [52]. Cloud server/data center may be used to aggregate information and process high-level data analytics tasks. Bonomi et al. [9] discuss processing data analytics tasks in multiple use cases in a fog/edge environment.

6.6.4.4 Scheduling and Resource Management

This engine works continuously to monitor the incoming tasks and their assignment using the surrogate selection policy. It monitors the availability of resources in different layers such as the availability of new, incoming user devices as well as tenant resources such as VMs that process data in fog devices and the cloud.

6.6.4.5 Network Management

FEA uses the multitier network to distribute the fog and edge applications. It may use software-defined networking or virtual network topology in multitenant resources in fog and cloud devices. User devices are usually connected using point-to-point network topologies that may either use TCP socket – WiFi connection, WiFi direct, or Bluetooth communication. This module is also responsible for monitoring connection and triggering the connection resume procedures for a lost connection.

6.6.4.6 Execution Management

This module facilitates the application specific code functionality to execute on the edge and fog nodes. Existing work in fog computing proposed the use of a virtual environment [28] or use of private OS stack provided by CISCO iox [53]. Virtualization with migration support on mobile devices is proposed by Bellavista et al. [54]. In some research, the code offload techniques such as DEX compositions in android [55] or .NET may be used. Other works propose plug-in based designs that are downloaded and integrated into the app in runtime [56].

6.6.4.7 Mobility Management

MEC supports mobile edge devices that are constantly on the move. In such cases, the data and the middleware services follow the devices. The idea is commonly known as Follow me Cloud (FMC) [57] and uses Locator/ID separation (LISP) protocol.

6.6.5 Sensor/Actuators

The sensors handle the important task of obtaining real-time data from the environment and user's surrounding. The information obtained through sensors is used in several forms. Sensor data may be acquired in the FEA application itself. It can also be used to evaluate and extract context information of the device user. In more complex applications, the closed-loop information is acquired and analyzed and further used for taking real-time actions using the actuators.

6.7 Case Study Example

This section describes an example of a perpetrator tracking application that can be designed through middleware in Section 6.5. This is a mobile application that performs real-time tracking of perpetrators through video surveillance using surrogate mobile phones available in the vicinity:

- **Device discovery**. One of the devices initiates the perpetrator tracking application by sending a request on the publish subscribe channel. Participating devices respond to the request and communication channel is established for further communication.
- **Context monitoring**. The location is the main context that is acquired using GPS data on the mobile device. Accelerometer data enables to obtain accelerometer variance context of mobile users. Accelerometer variance helps to obtain images/video from mobile devices that are not moving, thus reducing the motion-related distortions in the acquired image data. The orientation of mobile device enables prediction of the potential location of the perpetrator.
- **Data analytics**. Instead of sending all the image data from the mobile devices for perpetrator recognition, only images that have faces is sent. A face detection algorithm runs on the mobile device that eliminates images that do not contain face/s. In the fog server, face images are input to face recognition application that detects if perpetrator is found in the input images.
- **Mobility support**. As the perpetrator moves from one location to another, the set of devices selected to run the application change. Moreover, the devices selected need to be stationary and moving mobile devices are not being used.
- **Network management**. This application involves a point-to-point connection with other devices that are connected using WiFi. Also, the mobile devices connect to fog server over WiFi.
- **Execution management**. Mobile code of face detection is offloaded to the mobile device while the web server application that performs face recognition application is set up on the fog.

- **Scheduling**. In runtime, application requirement i.e. GPS location of perpetrator changes as the perpetrator is on the move. Scheduling module is responsible for matching the search location with the candidate device locations. Other considerations in the optimized selection are minimally moving devices, the orientation of devices, availability of battery and communication bandwidth on the mobile devices.
- **Security**. Authentication of new devices is performed in a fog server. Data encryption is performed while communicating data to the fog.

6.8 Future Research Directions

Different aspects of middleware can be explored in future research in order to improve the performance of mobile distributed applications.

6.8.1 Human Involvement and Context Awareness

The FEA may involve context-based decisions in several aspects of middleware, such as choice of participating devices, activation triggers of distributed applications, and anticipatory scheduling based on historical context data. Increasingly, more context-aware control and management of fog and edge devices will be used to intelligently schedule the distributed applications. The existing works primarily focus on location tracking of the users [51] and several other useful contexts [58] for executing an application. Several other contexts such as user activity patterns, prediction of user environments, and device usage patterns may be explored to improve the execution of the distributed application. Anticipatory context-aware computing is studied for mobile domain [59], but its adoption in a collaborative edge environment is yet to be seen.

6.8.2 Mobility

Mobile edge nodes need to provide services as requested by the application as they move from one location to another. The standard methods of network virtualization and VM migration exist. Research involves managing costs in VM migration, consideration of the mobility changes in edge devices, intermittent connectivity, and task partitioning within time constraints. In a mobile environment like vehicle-to-vehicle systems focuses on the prefetching, data caching and migration of services [14]. Future work in MEC must address how to guarantee service continuity in highly dynamic scenarios.

6.8.3 Secure and Reliable Execution

Participating nodes can include many private devices, as well as resources with ownership of telecom companies or companies that provide cloud computing services. With a wide variety of devices involved, establishing and maintaining a secured channel of communication that is lightweight enough to execute on the personal mobile devices is a future research area. Methods such as encryption of data and key-based authentication can incur excessive energy and computation on edge devices that have limited resources. Another area of possible research is the secure offload of application tasks to other edge devices for outsourcing.

6.8.4 Management and Scheduling of Tasks

Traditionally, VM management includes migration and replication that easily allows the transfer of the soft state of the application from one node to another. Such techniques are currently being proposed in fog resources [60]. However, heterogeneity in networks and devices in FEA is a barrier for migration. A dynamically changing set of suitable devices to execute application requires seamless handover of tasks. Methods, checkpoints, and offload mechanisms need to be explored to make this handover seamless as well as time and resource efficient. The devices must be able to make real-time decisions regarding whether to execute the task in the cloud or in the edge. Also, they need to account for the overhead of management and other security features such as encryption for optimal task scheduling.

6.8.5 Modularity for Distributed Execution

Modular software components should exist vertically in different layers FEA. Different platforms such as network virtualization and software-defined networking (SDN) are being explored [61] for orchestrating distributed execution in edge resources. Standard protocols like OpenFlow are promising options for virtual network design across cloud, fog, and edge devices. Recent works propose the use of dockers with migration support for edge processing [11].

6.8.6 Billing and Service-Level Agreement (SLA)

Existing research demonstrates SLA for VM in edge devices [62] for MEC architectures or MCC [63], which are static resources. However, in the case of edge devices associated guarantees with the payment model is not studied. As such, the fog nodes and other edge participants incur energy and bandwidth as well as provide access to several resources and private data while executing distributed applications. Designing a payment model for such fog and edge node services is yet to be explored.

6.8.7 Scalability

Some of the existing works propose service-oriented middleware approach for scalability of edge devices [20]. For data acquisition and processing from a large number of edge devices that might be geographically distributed, the middleware design can be distributed with decentralized processing and decision making [64]. Hierarchical clustering may be an approach for designing these systems to attain the specific goal of edge applications such as real-time responsiveness.

6.9 Conclusions

In this chapter, we discussed the changes introduced by edge and fog devices in the distributed computing. Fog and edge architecture can now support mobile sensing applications that are real time, latency sensitive, and geographically distributed. The dynamically changing set of fog and edge participating devices also need more design support to execute such distributed applications. We discussed middleware architecture and existing works that deal with different aspects of fog and edge computing, such as context awareness, mobility support, selection of edge participants apart from the network, and computation management. Broadly speaking, these architecture aspects can be adapted to MEC, fog, and cloudlet implementation of FEA. We also highlighted some of the newer areas of research that will improve the design of FEA in the future.

References

1 V. Dastjerdi and R. Buyya. Fog computing: Helping the Internet of Things realize its potential, *Computer*, 49(8): 112–116, 2016.
2 E. Koukoumidis, M. Margaret, and P. Li-Shiuan. Leveraging smartphone cameras for collaborative road advisories. *Transactions on Mobile Computing*, 11(5): 707–723, 2012.
3 K. S. Oskooyee, A. Banerjee, and S.K.S Gupta. Neuro movie theatre: A real-time internet-of-people. In *16th International Workshop on Mobile Computing Systems and Applications*, Santa Fe, NM, February, 2015.
4 M. Pore, K. Sadeghi, V. Chakati, A. Banerjee, and S.K.S. Gupta. Enabling real-time collaborative brain-mobile interactive applications on volunteer mobile devices. In *Proceedings of the 2nd Intl. Workshop on Hot topics in Wireless*, Paris, France, September 2015.
5 K. Sadeghi, A. Banerjee, J. Sohankar, and S.K.S. Gupta. SafeDrive: An autonomous driver safety application in aware cities. In *PerCom Workshops*, Sydney, Australia, 14 March 2016.
6 X. Bao and R.R. Choudhury. Movi: mobile phone based video highlights via collaborative sensing. In *8th International Conference on Mobile Systems, Applications, and Services*, San Francisco, California, USA, June 15, 2010.

7 D. Hardawar. Driving app Waze builds its own Siri for hands-free voice control. VentureBeat, 2012.

8 Y. Liu, C. Xu, Y. Zhan, Z. Liu, J. Guan, and H. Zhang. Incentive mechanism for computation offloading using edge computing: A Stackelberg game approach. *Computer Networks*, 129(2): 339–409, 2017.

9 F. Bonomi, R. Milito, P. Natarajan, and J. Zhu. Fog computing: A platform for Internet of Things and analytics. In *Big Data and Internet of Things: A Roadmap for Smart Environments, Studies in Computational Intelligence*, 546: 169–186. Springer International Publishing, Cham, Switzerland, March 13, 2014.

10 J. Rodrigues, Eduardo R.B. Marques, L.M.B. Lopes, and F. Silva. Towards a Middleware for Mobile Edge-Cloud Applications. In *Proceeding of MECC*, Las Vegas, NV, USA, December 11, 2017.

11 P. Bellavista, S. Chessa, L. Foschini, L. Gioia, and M. Girolami. Human-enabled edge computing: exploiting the crowd as a dynamic extension of mobile edge computing. *IEEE Communications Magazine*, 56(1): 145–155, January 12, 2018.

12 S. Yi, Z. Hao, Q. Zhang, Q. Zhang, W. Shi, and Q. Li. LAVEA: Latency-Aware Video Analytics on Edge Computing Platform. In *37th International Conference on Distributed Computing Systems (ICDCS)*, Atlanta GA, July 17, 2017.

13 G. Ananthanarayanan, P. Bahl, P. Bodík, K. Chintalapudi, M. Philipose, L. Ravindranath, and S. Sinha. Real-Time Video Analytics: The Killer App for Edge Computing, *Computer*, 50(10): 58–67, October 3, 2017.

14 D. Grewe, M. Wagner, M. Arumaithurai, I. Psaras, and D. Kutscher. Information-centric mobile edge computing for connected vehicle environments: Challenges and research directions. In *Proceedings of the Workshop on Mobile Edge Communications*, Los Angeles, CA, USA, August 21, 2017.

15 Google, GoogleFit, https://www.google.com/fit/, January 15, 2018.

16 G. Piro, M. Amadeo, G. Boggia, C. Campolo, L. A. Grieco, A. Molinaro, and G. Ruggeri. Gazing into the crystal ball: when the Future Internet meets the Mobile Clouds, *Transactions on Cloud Computing*, 55(7): 173–179, 2017.

17 C. Chang, S. N. Srirama, and M. Liyanage. A Service-Oriented Mobile Cloud Middleware Framework for Provisioning Mobile Sensing as a Service. In *21st International Conference on Parallel and Distributed Systems (ICPADS)*, Melbourne, VIC, Australia, January 18, 2016.

18 K. Aberer. Global Sensor Network, LSIR, http://lsir.epfl.ch/research/current/gsn/, January 18, 2018.

19 W. Botta, W. D. Donato, V. Persico, and A. Pescapé. Integration of cloud computing and internet of things: a survey. *Future Generation Computer Systems*, 56: 684–700, 2016.

20 Carrega, M. Repetto, P. Gouvas, and A. Zafeiropoulos. A Middleware for Mobile Edge Computing. *IEEE Cloud Computing*, 4(4): 26–37, October 12, 2017.

21 K. Habak, M. Ammar, K.A. Harras, and E. Zegura. Femto Clouds: Leveraging Mobile Devices to Provide Cloud Service at the Edge. In *8th International Conference on Cloud Computing (CLOUD)*, New York, NY, USA, August 20, 2015.

22 Y. Nakamura, H. Suwa, Y. Arakawa, H. Yamaguchi, and K. Yasumoto. Middleware for Proximity Distributed Real-Time Processing of IoT Data Flows. In *36th International Conference on Distributed Computing Systems (ICDCS)*, Nara, Japan, August 11, 2016.

23 J. Al-Jaroodi, N. Mohamed, I. Jawhar, and S. Mahmoud. CoTWare: A Cloud of Things Middleware. In *37th International Conference on Distributed Computing Systems Workshops (ICDCSW)*, Atlanta, GA, USA, July 17, 2017.

24 S.-N. Chuang and A. T. Chan. MobiPADS: a reflective middleware for context-aware mobile computing. *IEEE Transactions on Software Engineering* 29(12), 2003: 1072–1085.

25 Y. Lee, Y. Ju, C. Min, J. Yu, and J. Song. MobiCon: Mobile context monitoring platform: Incorporating context-awareness to smartphone-centric personal sensor networks. In *9th annual IEEE Conference on Communications Society Conf. on Sensor, Mesh and Ad Hoc Communications and Networks (SECON)*, Seoul, South Korea, August 23, 2012.

26 G. Orsini, D. Bade, and W. Lamersdorf. CloudAware: A Context-Adaptive Middleware for Mobile Edge and Cloud Computing Applications. In *1st International Workshops on Foundations and Applications of Self* Systems (FAS*W)*, Augsburg, Germany, December 19, 2016.

27 M. Aazam and E.-N. Huh. Fog computing: The cloud-IoT/IoE middleware paradigm. *Potentials*, 35(3): 40–44, May–June 2016.

28 T. Verbelen, S. Pieter, F.D. Turck, and D. Bart. Adaptive application configuration and distribution in mobile cloudlet middleware. *MOBILWARE*, LNICST 65: 178–191, 2012.

29 M. Satyanarayanan, P. Bahl, R. Caceres et al. The Case for VM-Based Cloudlets in Mobile Computing, in *Pervasive Computing* 8(4), October 6, 2009.

30 Y.W. Law, P. Marimuthu, K. Gina, and L. Anthony. WAKE: Key management scheme for wide-area measurement systems in smart grid. *IEEE Communications Magazine*, 51(1): 34–41, January 04, 2013.

31 R. Chandramouli, I. Michaela, and S. Chokhani. Cryptographic key management issues and challenges in cloud services. In *Secure Cloud Computing*, Springer, New York, NY, USA, December 7, 2013.

32 M. Mukherjee, R. Matam, L. Shu, L. Maglaras, M. A. Ferrag, N. Choudhury, and V. Kumar. Security and privacy in fog computing: Challenges. *In Access,* 5: 19293–19304, September 6, 2017.

33 M.H. Ibrahim. Octopus: An edge-fog mutual authentication scheme. *International Journal of Network Security,* 18(6): 1089–1101, November 2016.

34 A. Alrawais, A. Alhothaily, C. Hu, and X. Cheng. Fog computing for the Internet of Things: Security and privacy issues. *Internet Computing,* 21(2): 34–42, March 1, 2017.

35 R. Lu, K. Heung, A.H. Lashkari, and A. A. Ghorbani. A Lightweight Privacy-Preserving Data Aggregation Scheme for Fog Computing-Enhanced IoT. *Access* 5, March 02, 2017): 3302–3312.

36 T. Wang, J. Zeng, M.Z.A. Bhuiyan, H. Tian, Y. Cai, Y. Chen, and B. Zhong. Trajectory Privacy Preservation based on a Fog Structure in Cloud Location Services. *IEEE Access,* 5: 7692–7701, May 3, 2017.

37 Dsouza, G.-J. Ahn, and M. Taguinod. Policy-driven security management for fog computing: Preliminary framework and a case study. In *15th International Conference on Information Reuse and Integration (IRI),* Redwood City, CA, USA, March 2, 2015.

38 R.A. Popa, C.M.S. Redfield, N. Zeldovich, and H. Balakrishnan. Cryptdb: Processing queries on an encrypted database. *Communications of ACM* 55(9), September, 2012): 103–111.

39 Stanford-Clark and A. Nipper. Message Queuing Telemetry Transport, http://mqtt.org/, January 21, 2018.

40 Google, Nearby Connections API, https://developers.google.com/nearby/messages/android/pub-sub. Accessed January 17, 2018.

41 PubNub, Realtime Messaging, PubNub, https://www.pubnub.com/. Accessed January 18, 2018.

42 J.H. Rosa, J.L.V. Barbosa, and G.D. Ribeiro, ORACON: An adaptive model for context prediction. *Expert Systems with Applications,* 45: 56–70, March 1, 2016.

43 S. Sigg, S. Haseloff, and K. David. An alignment approach for context prediction tasks in ubicomp environments. *IEEE Pervasive Computing,* 9(4): 90–97, February 5, 2011.

44 Hassan, P.D. Haghighi, and P.P. Jayaraman. Context-Aware Recruitment Scheme for Opportunistic Mobile Crowdsensing. In *21st International Conference on Parallel and Distributed Systems,* Melbourne, VIC, Australia, January 18, 2016.

45 X. Liu, M. Lu, B.C. Ooi, Y. Shen, S. Wu, and M. Zhang. CDAS: a crowdsourcing data analytics system, *VLDB Endowment,* 5(10): 1040–1051, 2012.

46 L. Harold, B. Zhang, X. Su, J. Ma, W. Wang, and K.K. Leung. Energy-aware participant selection for smartphone-enabled mobile crowd sensing. *IEEE Systems Journal,* 11(3): 1435–1446, 2017.

47 O. Petri, F. Rana, J. Bignell, S. Nepal, and N. Auluck. Incentivising resource sharing in edge computing applications. In *International Conference on the Economics of Grids, Clouds, Systems, and Services*, October 7, 2017.

48 Y. Zhang, C. Min, M. Shiwen, L. Hu, and V.C.M. Leung. CAP: Community activity prediction based on big data analysis. *IEEE Network*, 28(4): 52–57, July 24, 2014.

49 S. Reddy, D. Estrin, and M. Srivastava. Recruitment framework for participatory sensing data collections. In *Proceedings of the 8th international conference on Pervasive Computing*. Lecture Notes in Computer Science, 6030, Springer, Berlin, Heidelberg, 2010.

50 A. Banerjee and S. K.S Gupta. Analysis of smart mobile applications for healthcare under dynamic context changes. *Transactions on Mobile Computing*, 14(5): 904–919, 2015.

51 L. Wang, Z. Daqing, W. Yasha, C. Chao, H. Xiao, and M. S. Abdallah. Sparse mobile crowdsensing: challenges and opportunities, in *IEEE Communications Magazine*, 54(7): 161–167, July 2016.

52 W. Sherchan, P. P. Jayaraman, S. Krishnaswamy, A. Zaslavsky, S. Loke, and A. Sinha. Using on-the-move mining for mobile crowdsensing. In *13th International Conference on Mobile Data Management (MDM)*, Bengaluru, Karnataka, India, November 12, 2012.

53 CISCO. IOx and Fog Applications. CISCO, https://www.cisco.com/c/en/us/solutions/internet-of-things/iot-fog-applications.html, January 21 2018.

54 P. Bellavista, A. Zanni, and M. Solimando. A migration-enhanced edge computing support for mobile devices in hostile environments, in *13th International Wireless Communications and Mobile Computing Conference (IWCMC)*, Valencia, Spain July 20, 2017.

55 Z. Ying, H. Gang, L. Xuanzhe, Z. Wei, M. Hong, and Y. Shunxiang. Refactoring Android Java code for on-demand computation offloading. In *International conference on object-oriented programming systems languages and applications*. Tucson AZ, USA, October 19, 2012.

56 P. P. Jayaraman, C. Perera, D. Georgakopoulos, and A. Zaslavsky. Efficient opportunistic sensing using mobile collaborative platform mosden. In *Collaborative Computing: Networking, Applications and Worksharing (Collaboratecom)*, Austin, TX, USA, December 12, 2013.

57 A. Ksentini, T. Taleb, and F. Messaoudi. A LISP-Based Implementation of Follow Me Cloud. *Access* 2 (September 24): 1340–1347, 2014.

58 P. Perera, P. Jayaraman, A. Zaslavsky, D. Georgakopoulos, and P. Christen. Mosden: An Internet of Things middleware for resource constrained mobile devices. In *47th Hawaii International Conference in System Sciences (HICSS)*, Waikoloa, HI, USA, March 10, 2014.

59 V. Pejovic and M. Musolesi. Anticipatory mobile computing: A survey of the state of the art and research challenges. *ACM Computing Surveys (CSUR)*, 47(3) (April), 2015.

60 T. Taleb, S. Dutta, A. Ksentin, M. Iqbal, and H. Flinck. Mobile edge computing potential in making cities smarter. *IEEE Communications Magazine*, 5(3) (March 13): 38–43, 2017.
61 C. Baktir, A. Ozgovde, and C. Ersoy. How can edge computing benefit from software-defined networking: A survey, use cases, and future directions. *Communications Surveys & Tutorial*, 19(4) (June): 2359–2391, 2017.
62 T. Katsalis, G. Papaioannou, N. Nikaein, and L. Tassiulas. SLA-driven VM Scheduling in Mobile Edge Computing. In *9th International Conference on Cloud Computing (CLOUD)*, San Francisco, CA, USA, January 19, 2017.
63 M. Al-Ayyoub, Y. Jararweh, L. Tawalbeh, E. Benkhelifa, and A. Basalamah. Power optimization of large scale mobile cloud computing systems. In *3rd International Conference on Future Internet of Things and Cloud*, Rome, Italy, October 26, 2015.
64 Y. Jararweh, L. Tawalbeh, F. Ababneh, A. Khreishah, and F. Dosari. Scalable cloudlet-based mobile computing model. In *Procedia Computer Science*, 34: 434–441, 2014.

7

A Lightweight Container Middleware for Edge Cloud Architectures

David von Leon, Lorenzo Miori, Julian Sanin, Nabil El Ioini, Sven Helmer, and Claus Pahl

7.1 Introduction

In typical cloud applications, most of the data processing is done on the back end and the clients are relatively thin. Integrating Internet-of-Things (IoT) devices and sensors into such an environment in a straightforward manner causes several problems. If billions of new devices start shipping data into the cloud, this will have a major impact on the flow of network traffic. Also, certain applications require real-time behavior (e.g. self-driving cars) and cannot afford to wait for data, which may arrive late due to network delays. Finally, users may also not want to send sensitive or private data into the cloud, losing control over it (this is especially important for healthcare applications). Consequently, cloud computing is moving away from large, centralized structures toward multicloud environments. The integration of cloud and sensor-based IoT environments results in edge cloud or fog computing [1, 2], in which a substantial part of the data processing takes place on the IoT devices themselves. Rather than moving data from the IoT to the cloud, the computation moves to the edge of the cloud [3].

However, when running these kinds of workloads on such an infrastructure, we are confronted with different issues: the deployed devices are constrained in terms of computational power, storage capability, reliable connectivity, power supply, and other resources. For a start, we need solutions that are lightweight enough to be run on resource-constrained devices. Nevertheless, we still aim to develop visualized solutions providing scalability, flexibility, and multi-tenancy. We address flexibility and multi-tenancy via containerization. Containers form the basis of a middleware platform that suits the needs of platform-as-a-service (PaaS) clouds [4, 5], where application packaging and orchestration are key issues [6–8]. We address scalability by proposing to cluster small devices to boost and share their computational power and

Fog and Edge Computing: Principles and Paradigms, First Edition.
Edited by Rajkumar Buyya and Satish Narayana Srirama.

other resources. Using Raspberry Pi (RPi) clusters as a proof-of-concept, we demonstrate how this can be achieved [9].

In edge cloud environments, additional requirements include cost-efficiency, low power consumption, and robustness. In a sense, our solution should not just be lighweight in terms of the software platform but also in terms of the hardware platform. We show how the additional requirements can be met by implementing containers on clusters of single-board devices like RPis [10–12]. The lightweight hardware and software architecture we envision allows us to build applications based on multicloud platforms on a range of nodes from data centers to small devices.

Edge cloud systems are also subject to security concerns. Data, software, and hardware might join or leave a system at any time, requiring all to be identified and their traces to be tracked. Traceability and auditability also apply for the orchestration aspects. By looking at blockchain technologies, we explore a conceptual architecture that manages security concerns for IoT fog and edge architecture (FEA) using blockchain mechanisms.

Our chapter is organized as follows. We first introduce architecture requirements and review technologies and architectures for edge clouds. We then discuss core principles of an edge cloud reference architecture that is based on containers as the packaging and distribution mechanism. We specifically present different options for storage, orchestration, and cluster management for distributed clusters in edge cloud environments. To this end, we report on experimental results with Raspberry Pi clusters to validate the proposed architectural solution. The settings included are: (i) an own-build storage and cluster orchestration; (ii) an OpenStack storage solution; (iii) Docker container orchestration; and (iv) IoT/sensor integration. We also include practical concerns such as installation and management efforts, since lightweight edge clusters are meant to be run in remote areas.

7.2 Background/Related Work

We identify some principles and requirements for a reference architecture for edge cloud computing, illustrating these principles and requirements with a use case.

7.2.1 Edge Cloud Architectures

Supporting the management of data collections, including their pre-processing and further distribution, via computational and storage resources in the case of edge computing with integrated IoT objects is different to traditional cloud computing architectures. It is facilitated by smaller devices spread across a distributed network and due to these smaller device sizes results in different

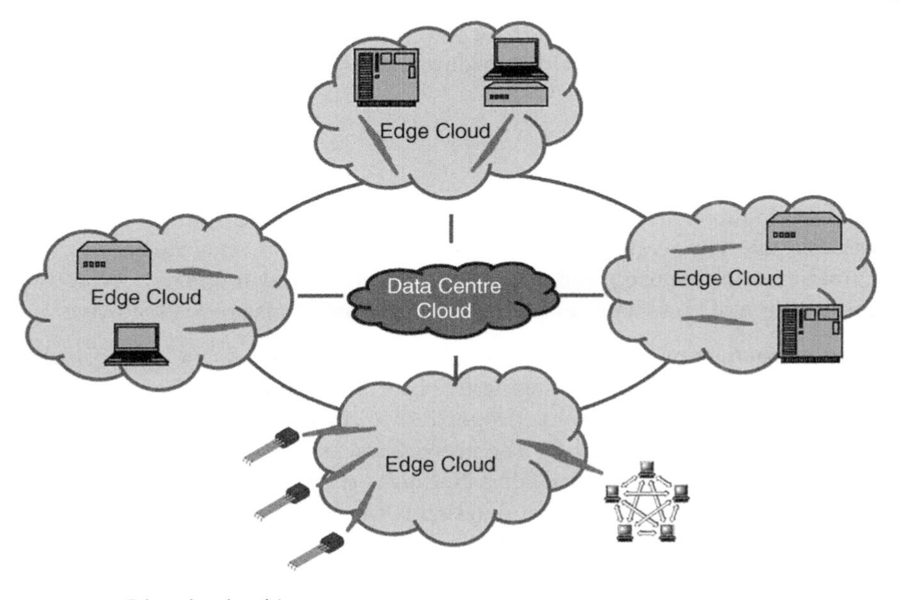

Figure 7.1 Edge cloud architecture.

resource restrictions, which in turn requires some form of lightweightness [13]. Still, virtualization is a suitable mechanism for edge cloud architectures [14, 15], as recent work on software-defined networks (SDNs) shows. The compute and storage resources can be managed, i.e., packaged, deployed and orchestrated, by platform services. Figure 7.1 shows that we need to provide compute, storage, and network resources between end devices and traditional data centers, including data transfer support between virtualized resources.

Other requirements that emerge are location awareness, low latency, and software mobility support to manage cloud end points with rich (virtualized) services. A specific requirement is continued configuration and update – this particularly applies to service management. What is needed is a development layer that allows for provision and manage applications on these edge architectures. We propose to adjust the abstraction level for edge cloud management at a typical PaaS layer.

We propose an edge cloud reference architecture based on containers as the packaging and distribution mechanism, addressing a number of concerns. This includes application construction and orchestration and resource scheduling as typical distributed systems services. Furthermore, we need an orchestration model for an (edge) cloud-native architecture where applications are deployed on provided platform services. For this type of architecture, we need a combination of lightweight technologies – single-board devices as lightweight hardware combined with containers as a lightweight software platform.

Container technologies can take care of the application management. They are specifically useful for constrained resources in edge computing clusters.

7.2.2 A Use Case

Consider the following use case. Modern ski resorts operate extensive IoT-cloud infrastructures. Sensors gather a variety of data in this setting – on weather (air temperature/humidity, sun intensity), on snow quality (snow humidity, temperature), and on people (location and numbers). With the combination of these data sources, two sample functions are the following:

- **Snow management**. Snow groomers (snow cats) are heavy-duty vehicles that rely on sensor data (ranging from tilt sensors in the vehicle and GPS location all the way to snow properties) to provide an economic solution in terms of time needed for the preparation of slopes, while at the same time allowing a near-optimal distribution of snow. This is a real-time application where cloud-based computation is not feasible (due to unavailability of suitable connectivity). As a consequence, local data processing for all data collection, analysis, and reaction is needed.
- **People management**. Through mobile phone apps, skiers can get recommendations regarding snow quality and possible overcrowding at lifts and on the slopes. A mobile phone app can use the cloud as an intermediary to receive data from. For performance, however, the application architecture would benefit from data pre-processing at the sensor location to reduce the data traffic between local devices and cloud services.

Performance is a critical concern that can be addressed by providing local computation. This avoids high data volumes to be transferred into centralized clouds. Local processing of data, particularly for the snow management where data sources and actions resulting through the snow groomers happen in the same place, is beneficial, but needs to be facilitated through robust technologies that can operate in remote areas under difficult environmental conditions. Clusters of single-board computers such as Raspberry Pis are a suitable, robust technology.

Another critical concern is flexibility. The application would benefit from flexible platform management with different platform and application services deployed at different times in different locations to facilitate short-term and long-term change [16]. For instance, a sensor responsible for people management during daytime could support snow management during the night. Containers are suitable but require good orchestration support. Two orchestration patterns emerge that illustrate this point. The first pattern is about fully localized processing in clusters (organized around individual slopes with their profile): full computation on board and locally between snow groomers is required, facilitated by the deployment of analysis, but also decision making and actuation features, all as containers. The second pattern is about data pre-processing for people management: reducing data volume in transfer to the cloud is the aim. Analytics services packaged as containers that filter and aggregate data need to be deployed on selected edge nodes.

7.2.3 Related Work

Container technologies that provide lightweight virtualization are a viable option to hypervisors, as demonstrated in [17]. This lightweightness is a benefit for smaller devices due to their limitations as a result of the reduced size and capabilities.

Bellavista and Zanni [18] have investigated an infrastructure based on Raspberry Pis to host Docker container. Their work also confirms the suitability of single-board devices. Work carried out at the University of Glasgow [19] also uses Raspberry Pis for edge cloud computing. The work there is driven by lessons learned from practical applications of RPis in real-world settings. In addition to the results presented there, we have added here a comparative evaluation of different cluster-based architectures based on architecture-relevant observations regarding installation, performance, cost, power, and security.

If we want to consider a middleware platform for a cluster architecture of smaller devices, be that in constrained or mobile environments, the functional scope of a middleware layer needs to be suitably adapted [20]:

- Robustness is a requirement that needs to be facilitated through fault tolerance mechanisms that deal with failure of connections and nodes. Flexible orchestration and load balancing are such functions.
- Security is another requirement, here relevant in the form of identity management in unsecured environments. Other security concerns such as data provenance or smart contracts accompanying orchestration instructions are also relevant. De Coninck et al. [21] also approach this problem from a middleware perspective. Dupont et al. [22] look at container migration to enhance the flexibility, which is an important concern in IoT settings.

7.3 Clusters for Lightweight Edge Clouds

In the following, we explain how to build platforms that are lightweight, in terms of software and hardware.

7.3.1 Lightweight Software – Containerization

Containerization allows a lightweight virtualization through the construction of containers as application packages from individual images (generally retrieved from an image repository). This addresses performance and portability weaknesses of current cloud solutions. Given the overall importance of the cloud, a consolidating view on current activities is important. Many container solutions build on top of Linux LXC techniques, providing kernel mechanisms such as namespaces and cgroups to isolate operating system processes. Docker, which is basically an extension of LXC, is the most popular container platform at the moment [23].

Orchestration is about constructing and managing a possibly distributed assembly of container-based software applications. Container orchestration is not only about the initial deployment, starting and stopping of containers, but also about the management of the multicontainers as a single entity, concerning availability, scaling, and networking of the containers, and moving them between servers. In this way, edge cloud-based container construction is a form of orchestration within the distributed cloud environment. However, the management solution for containers provided by cluster management solutions needs to be combined with development and architecture support. A multi-PaaS based on container clusters can serve as a solution for managing distributed software applications in the cloud, but this technology still faces challenges. These include a lack of suitable formal descriptions or user-defined metadata for containers beyond image tagging with simple IDs. Description mechanisms need to be extended to clusters of containers and their orchestration as well [24]. The topology of distributed container architectures must be more explicitly specified and its deployment and execution orchestrated [25]. So far, there is no accepted solution for these orchestration challenges.

Docker has started to develop its own orchestration solution (Swarm) and Kubernetes is another relevant project, but a more comprehensive solution that would address the orchestration of complex application stacks could involve Docker orchestration based on the topology-based service orchestration standard TOSCA [26]. The latter is done by the Cloudify PaaS, which supports TOSCA.

In Figure 7.2, we illustrate an orchestration plan for the use case from the previous section. A container host selects either the people management or the snow management as the required node configuration. For instance, the people management architecture could be upgraded to a more local processing mode that includes analysis and storage locally. The orchestration engine takes care of the deployment of the containers at the right time. Container clusters need network support. Usually, containers are visible on the network using the shared host's address. In Kubernetes, each group of containers (called pods) receives its own unique IP address that is reachable from any other pod in the cluster, whether co-located on the same physical machine or not. This needs to be supported by network virtualization with specific routing features.

Container cluster management also needs data storage support. Managing containers in Kubernetes clusters cause challenges due to the need of

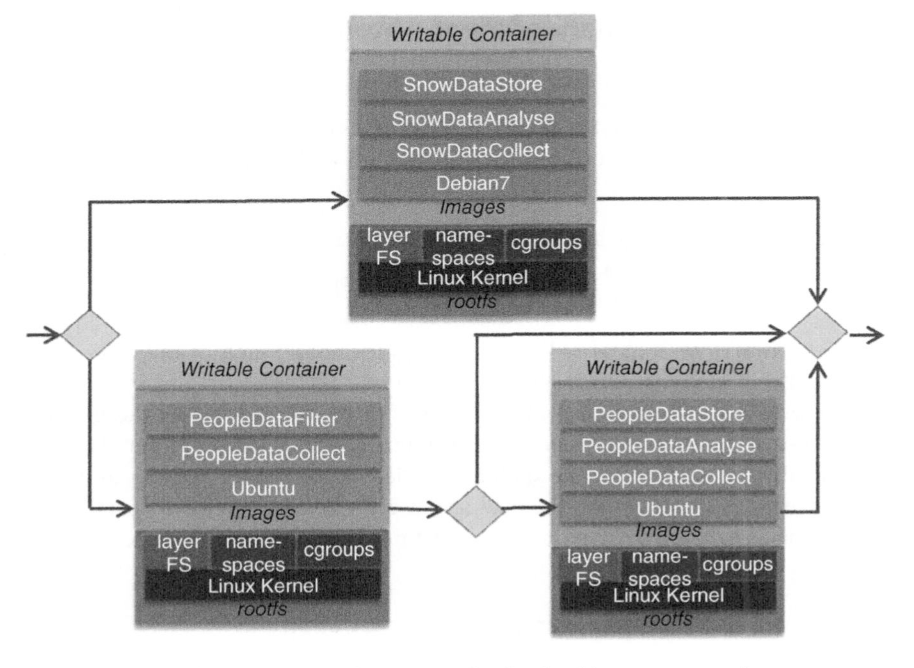

Figure 7.2 Simplified container orchestration plan for the ski resort case study.

Kubernetes pods to co-locate with their data: a container needs to be linked to a storage volume that follows it to any physical machine.

7.3.2 Lightweight Hardware – Raspberry Pi Clusters

We focus on Raspberry Pis as the hardware device infrastructure and illustrate orchestration for Raspberry Pi clusters in edge cloud environments. These small single-board computers create both opportunities and challenges. Creating and managing clusters are typical PaaS functions, including setting up and configuring hardware and system software, monitoring and maintaining the system, up to hosting container-based applications.

A Raspberry Pi (RPi) is relatively cheap (around $30, depending on the version) and has a low power consumption, which makes it possible to create an affordable and energy-efficient cluster that is particularly suitable for environments for which high-tech installations are not possible. Since a single RPi lacks computing power, in general we cannot run computationally intensive software. On the other hand, this limitation can be overcome by combining several RPis into a cluster. This allows platforms to be better configured and customized so that they are at the same time robust against failure through their cluster architecture.

7.4 Architecture Management – Storage and Orchestration

Raspberry Pi clusters are the hardware basis for our middleware platform. In order to explore different options for this, we look at different implementations types:

1. An own-build storage and cluster orchestration
2. An OpenStack storage implementation
3. A Docker container orchestration

Furthermore, we look at IoT/sensor integration. For each of the three core architectural patterns, we describe the concrete architecture and implementation work, which we will evaluate later. The aim is to address the general suitability of the proposed architectures in terms of performance, but also take specifically practical concerns such as physical maintenance, power, and cost into account. Consequently, the evaluation criteria are as follows: installation and management effort, cost, power consumption, and performance.

7.4.1 Own–Build Cluster Storage and Orchestration

7.4.1.1 Own–Build Cluster Storage and Orchestration Architecture

As demonstrated in (Abrahamsson, 2013), our Raspberry Pi cluster can be configured with up to 300 nodes (using RPi 1 with a lower spec compared to RPi 2 and 3 strengthens the case for lightweightness). The core of an RPi 1 is a single board with an integrated circuit with an ARM 700 MHz processor (CPU), a Broadcom VideoCore graphics processor (GPU) and 256 or 512 MB of RAM. Also provided is an SD card slot for storage and I/O units for USB, Ethernet, audio, video, and HDMI. Power support is enabled using a micro-USB connector. Raspbian is a version of the widely used Linux distribution Debian, optimized for the ARMv6 architecture, that serves as the operating system. We use a Debian 7 image to support core middleware services such as storage and cluster management. In [27], we have investigated basic storage and cluster management for an RPi cluster management solution.

The topology of our cluster is a star network. In this configuration, one switch acts as the core and other switches link the core to the RPIs. A master node and an uplink to the Internet are connected to the core switch. In addition to deploying existing cluster management tools such as Swarm or Kubernetes, we also built our own dedicated tool, covering important features for a dynamic edge cloud environment, such as low-level configuration, monitoring, and maintenance of the cluster as an architectural option. This approach gives flexibility for monitoring the joining and leaving of nodes to and from the cluster, with the master handling the (de)registration process.

Table 7.1 Speed and power consumption of the Raspberry Pi cluster. Adapted from [28].

Device	Page/Sec	Power
RPi	17	3W
Kirkwood	25	13W
MK802	39	4W
Atom 330	174	35W
G620	805	45W

7.4.1.2 Use Case and Experimentation

A key objective is the suitability of an RPi for running standard application in terms of performance and power. In previously conducted experiments, a sample file with a size of 64.9 KB was used. An RPi (model B) was compared to different other processor configurations: a 1.2 GHz Marvell Kirkwood, a 1 GHz MK802, a 1.6 GHz Intel Atom 330, and a 2.6 GHz dual core G620 Pentium. All tested systems had a wired 1 GB Ethernet connection (which the Raspberry Pi could not utilize fully, only including a 10/100 Mbit Ethernet card). As a benchmark, ApachBench2 was used. The test involved a total of 1000 requests, with 10 of them running concurrently. In Table 7.1, page/sec as performance measure and power consumptions (in Watts) are summarized.

Table 7.1 shows that RPis are suitable for most sensor integration and data processing requirements. They are suitable in an environment where robustness is required and that is subject to power supply problems.

7.4.2 OpenStack Storage

7.4.2.1 Storage Management Architecture

In Miori [29], we have investigated OpenStack Swift as a distributed storage device for our setting, porting OpenStack Swift onto RPis. By using a fully fledged platform such as OpenStack Swift, we can substantially extend our self-built storage approach. The challenges are adopting an open-source solution that is meant for significantly larger devices.

Swift is useful for distributing storage clusters. Using a network storage system can improve the cluster performance in a common filesystem. In our own implementation of Swift, we used a four-bay network attached storage (NAS) from QNAP Systems, but can now demonstrate that a more resource-demanding OpenStack Swift is a feasible option. The Swift cluster provides a solution for storing objects such as application data as well as system data. Data are replicated and distributed among different nodes: we considered different topologies and configurations. While we showed that this is technically feasible, the performance of OpenStack Swift performance is a key downside that would need further optimization to become a practically relevant solution.

7.4.2.2 Use Case and Experimentation

In order to evaluate the Swift-based storage, we have run several benchmarks based on YCSB and ssbench.

- **Single node installation**. Here, a significant bottleneck around data uploads emerges. This means that a single server cannot handle all the traffic, resulting in either cache (memcached) or container server failures.
- **Clustered file storage**. Here, a real-world case study has been carried out using the ownCloud cloud storage system. A middleware layer on a Raspberry cluster was configured and benchmarked. We can demonstrate the utility of this option by running ownCloud on top that facilitates a (virtualized) storage service across the cluster. Performance is not great, but is acceptable.

In the implementation, we used a FUSE (filesystem-in-userspace) module called cloudfuse. This connects to a Swift cluster and manages content as in traditional directory-based filesystems. An ownCloud instance accesses the Swift cluster via cloudfuse. The application GUI loads sufficiently fast. File listing is slower, but is still acceptable. The significant limitations here stem from cloudfuse. Some operations, like renaming folders, are not possible. Neither is it always sufficiently efficient. This could be remedied by a direct implementation, or possibly by improving the built-in Swift support.

Apart from performance, scalability remains a key concern. We can demonstrate that the addition of more Raspberry Pi predictably results in better performances, i.e., that Swift is scalable. Though based on hardware configuration limitations, we cannot confirm if the trend is linear.

The next concern is costs. The cluster costs are acceptable (see Table 7.2 for a pricing of some cluster configurations). The PoE (Power over Ethernet) add-on boards and PoE managed switches that we used are not specific to the project and could easily be replaced by a cheaper solution that involves a separate power supply unit and a simple unmanaged switch without having a negative impact on the system's performance. In comparing our configuration with other architectures, modern gateway servers (e.g., Dell Gateway 5000 series) would be higher, all hardware included.

7.4.3 Docker Orchestration

Docker and Kubernetes as the most prominent examples of lightweight virtualization mechanisms via containers have been successfully placed on Raspberry

Table 7.2 Approximate costs of the Raspberry Pi cluster.

Component	Price	Units	Total
Raspberry Pi	35 €	7	245 €
PoE module	45 €	7	315 €
Cat.5e SFTP Cable	3 €	7	21 €
Aruba 2530 8 PoE+	320 €	1	320 €
Total			901

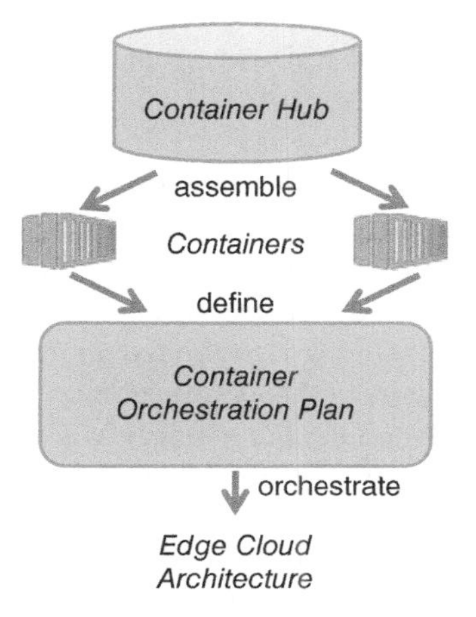

Figure 7.3 Overall orchestration flow.

Pis [23]. This demonstrates the feasibility of running container clusters on clustered RPi architectures.

As our focus is on edge cloud architectures, we investigate the core components for a middleware platform for edge clouds. Figure 7.3 describes a complete orchestration flow for containers on edge cloud architectures. The first step is the construction of a container from individual images, for instance from an open repository of images such as a container hub. Different containers for a specific processing problem are composed, which forms an orchestration plan. This plan defines the edge cloud topology on which the orchestration is enacted. This orchestration mechanism realizes central components of an edge PaaS-oriented middleware platform. Containerization helps to overcome limitations of the earlier two solutions we discussed.

7.4.3.1 Docker Orchestration Architecture

With RPis, we can facilitate in a cost-effective way an intermediate layer for local edge data processing. Its advantages are reliability, low-energy consumption, low-cost devices that are still capable of performing data-intensive computations.

Implementation – Hardware and Operating System. As in the earlier architectural patterns, we constructed clusters composed of a number Raspberry Pis. Installation and power management were initial concerns. Technically, the devices were connected to a switch via cables for signal processing and power supply to the devices. Each unit was equipped with a PoE module, connected to the Raspberry Pi. By replicating the GPIO interface, further modules can be connected easily. A network connection is setup by connecting the switch through an Ethernet port. The switch can be configured to connect to an existing DHCP (dynamic host configuration protocol) server that can distribute network configuration parameters like IP (internet protocol) addresses. Furthermore, via virtual LANs subnets can be created. As the operating system, Hypriot OS, a Debian distribution, was chosen. The distribution already contains Docker software.

Swarm Cluster Architecture and Security. The cluster nodes have different roles. A selected node becomes the user gateway into the cluster. This is initialized by creating Docker Machines on the gateway node. Then both the OS and the Docker daemon are configured on all Raspberry Pis cluster nodes. Docker Machines can manage remote hosts by sending the commands from the Docker client to the Docker demon on the remote machine over a secured connection. When the first Docker Machine is created, in order to create a trusted network, new TLS protocol certificates are created on the local machine and then copied to the remote machines. In order to address security concerns, we replaced the default authentication, which is considered insecure, by a public-key authentication during the cluster setup process. This way, we avoid a password-based authentication. We enhanced security by requiring the SSH daemon on the remote machine to only accept public-key authentication.

We used Docker Swarm for cluster management: normal nodes run one container that identifies them as a swarm node. The swarm managers deploy an additional dedicated container that provides the management interface. Swarm managers can be configured to support fault tolerance through redundancy (running as replicas in that case). There are mechanisms to avoid inconsistencies in the swarm that could lead to potential misbehavior. If several swarm managers exist, they can also share their knowledge about the swarm by communicating information from a nonleading manager to the one in charge.

Service Discovery. Information about a swarm, images, and how they can be reached must be shared. In a multihost network, we can use, for instance, a key-value store that keeps information about network state (e.g., discovery, networks, endpoints and IP addresses).

We used Consul as a key-value store for our implementation, which supports redundant swarm managers and works without a continuous Internet connection, which is important for our need to support intermittent connectivity. For fault tolerance, it can be replicated. Consul selects a lead node in a cluster and manages information distribution across the nodes.

Swarm Handling. In a properly set-up swarm configuration of Docker Machines, the nodes communicate their presence to both Consul server and swarm manager. Users can interact with the swarm manager and each Docker Machine separately. Docker-specific environment variables can be requested from a Docker Machine for this. A Docker client tunnels into the manager and executes remote commands over there. Users can obtain swarm information and execute swarm tasks (e.g., launching a new container) in this way. The manager then deploys it according to any given constraints following the selected swarm strategy.

7.4.3.2 Docker Evaluation – Installation, Performance, Power

This evaluation section looks at the key concerns of performance and power consumption. Furthermore, we will also address practical concerns such as installation effort. The evaluation of the project focuses on the complexity to build and handle it and its costs, before concentrating on the performance and power consumption [11].

Installation Effort and Costs. Assembling the hardware for the Raspberry Pi cluster does not require special tools or advanced skills. This makes the architecture suitable to be installed and managed in remote areas without expert support available. Once running, handling the cluster is straightforward. Interacting with clusters does not differ from single Docker installations. One drawback is the reliance on the ARM architecture, where images are not always available, causing the need for them to be created on demand.

Performance. We stress-tested the swarm manager by deploying larger numbers of containers (with a fixed image) over a given period of time. We measured

- Time to deploy the images
- Launch time for containers

In the test, we deployed 250 containers on the swarm with five requests at a time. To determine the efficiency of the Raspberry Pi cluster, both the time to execute the analysis and the power consumption are measured and put into perspective with a virtual machine cluster on a desktop computer and a single Raspberry Pi.

The tests were run on a desktop PC, which was a 64-bit Intel Core 2 Quad Q9550 @2.83GHz Windows 10 machine with 8GB Ram and a 256GB SSD.

If we compare the RPi setup with a normal VM configuration in Table 7.3, there is less performance for the Raspberry Pi cluster. This is a consequence of the limits of the single board architecture. A particular problem is the I/O

Table 7.3 Time comparison – listing the overall, the mean, and the maximal time of container.

	Launching	Idle	Load
Raspberry Pi cluster	228s	2137ms	9256ms
Single Raspberry Pi node	510s	5025ms	14115ms
Virtual machine cluster	49s	472ms	1553ms
Single virtual machine node	125s	1238ms	3568ms

of the micro SD card slot, which is slow in terms of reading and writing, with a maximum of 22MB/s and 20MB/s, respectively, for the two operations. This can be partially explained by a network connectivity of only 10/100Mbit/s.

Power. The observations for power consumption are presented in Tables 7.4 and 7.5. With 26W (2.8W per unit) under load, as shown in Table 7.4, the modest power consumption of the Raspberry Pi cluster puts its moderate performance that we noted above into perspective. Table 7.5 details the consumption in two situations (idle and under load).

With still-acceptable performance and suitability from the installation and operations perspective (including power consumption), the suitability for an environment with limitations that requires robustness can be assumed.

Table 7.4 Comparison of the power consumption while idling and under load.

	Idle	Load
Raspberry Pi cluster	22.5W	25-26W
Single Raspberry Pi node	2.4W	2.8W
Virtual machine cluster	85-90W	128-132W
Single virtual machine node	85-90W	110-114W

Table 7.5 Power consumption of the Raspberry Pi cluster while idling and under load.

	Idle	Load
Single node	2.4W	2.7W
All nodes	16W	17-18W
Switch	5W	8W
Complete system	22.5W	25-26W

7.5 IoT Integration

Apart from looking at the suitability of the three different architectural options, we also need to analyze the suitability of the proposed solutions for IoT applications with sensor integration. In order to demonstrate this, we refer to a medical application. For this healthcare application, we integrated health status sensing devices into a Raspberry Pi infrastructure.

Protocols exist for bridging between the sensor world and Internet-enabled technologies such as MQTT, making the installation and management work easy. Our experiments, however, have demonstrated the need for dedicated power management. Some sensors required considerable energy and caused overheating. Thus, solutions to prevent overheating and reduce consumption are needed.

7.6 Security Management for Edge Cloud Architectures

IoT/edge computing networks are distributed environments in which we cannot assume that sensor owners, network, and device providers trust each other. In order to guarantee a secure edge cloud computing architecture [30] with reliable and secure orchestration activities, we need to consider several aspects:

- Things (sensors, devices, software) might join, leave, and rejoin the network, so we need to be able to identify them.
- Data are generated and communicated, making it necessary to trace this by providing provenance and making sure that data have not been tampered with.
- Dynamic and local architectural management decisions, e.g. changing or updating software for maintenance or in emergency situations, need to be agreed upon by the relevant participants.

In this section, we explore the suitability of blockchain technology for providing a security platform that addresses the above concerns for edge architectures. Blockchains enable a form of distributed software architectures, where agreement on shared state for decentralized and transactional data can be established across a network of untrusted participants – as it is in the case in edge clouds. This approach avoids relying on central trusted integration points, which quickly become single points of failure. Edge platforms built on blockchains can take advantage of common blockchain properties such as data immutability, integrity, fair access, transparency, and nonrepudiation of transactions.

The key aim is to manage trust locally in lightweight edge clusters with low computational capabilities and limited connectivity. Blockchain technologies can be applied to identity management, data provenance, and transaction processing. For orchestration management, we can employ advanced blockchain concepts such as smart contracts.

7.6.1 Security Requirements and Blockchain Principles

Blockchain technology is a solution for untrusted environments that lack a central authority or trusted third party: many security-related problems can be addressed using the decentralized, autonomous, and trusted capabilities of blockchains. Additionally, blockchains are tamper-proof, distributed, and shared databases where all participants can append and read transactions but no one has full control over it. Every added transaction is digitally signed and timestamped. This means that all operations can be traced back and their provenance can be determined [31].

The security model implemented by blockchains ensures data integrity using consensus-driven mechanisms to enable the verification of all the transactions in the network, which makes all records easily auditable. This is particularly important since it allows tracking all sources of insecure transactions in the network (e.g., vulnerable IoT devices) [32]. A blockchain can also strengthen the security of edge components in terms of identity management and access control and prevent data manipulation.

The principles of blockchains can be summarized as follows:

- A *transaction* is a signed piece of information created by a node in the network, which is then broadcast to the rest of the network. The transactions are digitally signed to maintain integrity and enforce nonrepudiation.
- A *block* is a collection of transactions that are appended to the chain. A newly created block is validated by checking the validity of all transactions contained within.
- A *blockchain* is a list of all the created and validated blocks that make up the network. The chain is shared between all the nodes in the network. Each newly created and validated block is linked to the previous block in the chain with a hash value generated by applying a hashing algorithm over its content. This allows the chain to maintain nonrepudiation.
- *Public keys* act as addresses. Participants in the network use their *private keys* to sign their transactions.
- A block is appended to the existing blockchain using a specific *consensus method* and respective coordination protocol. Consensus is driven by collected self-interest.
- Three types of blockchain platforms can be identified: (i) permissionless, where anyone can have a copy of the database and join the network both for reading and writing; (ii) permissioned, where access to the network is controlled by a preselected set of participants; and (iii) private, where the participants are added and validated by a central organization.

One of the (more recent) key concepts that has been introduced in blockchains is a smart contract, which is a piece of executable code residing on the blockchain that gets executed if a specific agreement (condition) is met. Smart contracts are not processed until their invoking transactions are included in a new block. Blocks impose an order on transactions, thus resolving nondeterminism that might otherwise affect their execution results. Blockchain contracts increase the autonomy of the edge/IoT devices by allowing them to sign agreements directly with any interested party that respects the contract requirements.

7.6.2 A Blockchain-Based Security Architecture

However, blockchains cannot be considered the silver bullet to all security issues in edge/IoT devices, especially due to massive data replication, performance and scalability. This is a challenge in the constrained environment we are working in. Blockchain technology has been applied for transactional processing before, but the novelty here is the application to lightweight IoT architectures, as shown in Figure 7.4:

- Application of consensus methods and protocols in localized clusters of edge devices to manage trust between the participating Edge/IoT devices.
- Smart contracts define orchestration decisions in the architecture.

In such environments, IoT/edge endpoints are generally what we call sleepy nodes, meaning that they are not online all the time to save battery life. This constrains them to only have intermittent Internet connectivity, especially when they are deployed in remote locations. We propose to use blockchains to

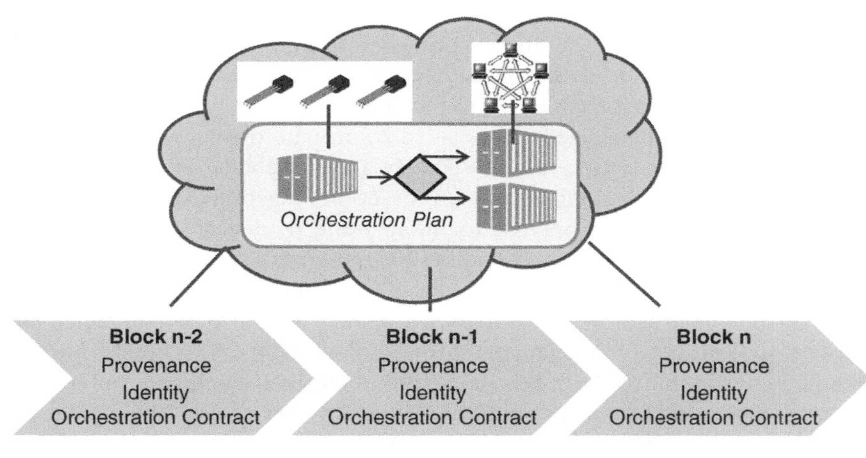

Figure 7.4 Blockchain-based IoT orchestration and security management.

manage security (trust, identity) in distributed autonomous clusters [10, 12]. As a starting point, we use permissioned blockchains with brokers, since they achieve higher performance in terms of block mining time and reduce transaction validation time and cost. We recommend using partially centralized/decentralized settings with permissioned blockchains with permissions for fine-grained operations on the transaction level (e.g., permission to create assets). In an implementation, we can consider both permissioned blockchains with permissioned miners (write) and also permissionless normal nodes (read). Additionally, not all IoT/edge endpoints need to behave as full blockchain nodes. Rather, they would act as lightweight nodes that access a blockchain to retrieve instructions or identity-related information (e.g., who has access to sensor data). For instance, each of the IoT endpoints, when connected to the network, would receive a receipt proof of payment (proof of payment is a receipt proving that a specific party has the necessary credentials to access a certain resource) that states which devices to trust and interact with. Then, when the endpoint receives a request, it needs to be signed by one of the trusted devices. A verifier is a third party that provides information about the external world. When the validation of a transaction depends on the external state, the verifier is requested to check the external state and to provide the result to the validator (miner), which then validates the condition. A verifier can be implemented as a server outside the blockchain, and has the permission to sign transactions using its own key pair on demand.

With respect to concerns such as cost efficiency, performance, and flexibility, a crucial point is choosing what data and computation should be placed on-chain and what should be kept off-chain.

Basing our architecture on container-based orchestration, software becomes another artefact that is subject to identity and authorization concerns, since edge computing is essentially based on the idea to bring software to the edge (to process data locally) rather than to bring data to the cloud center. Device and container orchestration involving the deployed software can be implemented within a smart contract transaction of the blockchain.

Blockchains use specialized protocols to coordinate the consensus process. The protocol configuration affects security and scalability. Different strategies have been used to confirm that a transaction is securely appended to the blockchain, for instance to prevent double spending in blockchains like bitcoin. An option is to wait for a certain number (X) of blocks to have been generated after the transaction is included into the blockchain. We will also investigate mechanisms such as checkpointing with respect to the best suitability for trusted orchestration management through blockchains. The option here is to add a checkpoint to the blockchain, so that all the participants can accept the transactions up to the checkpoint as valid and irreversible. Checkpointing relies on an entity trusted by the community to define the checkpoint (see discussion on architectural options with trusted broker), while traditional

X-block confirmation can be decided by the developers of the applications using blockchain.

Consensus protocols can be configured to improve scalability in terms of transaction processing rate (sample sizes are 1 to 8MB). Larger sizes can include more transactions into a block and thus increase maximum throughput. Another configuration change would be to adjust mining difficulty to shorten the time required to generate a block, thus reducing latency and increasing throughput (but a shorter inter-block time would lead to an increased frequency of forks).

7.6.3 Integrated Blockchain-Based Orchestration

We singled out data provenance, data integrity, identity management, and orchestration as important concerns in our framework. Based on the outline architecture from Figure 7.4, we detail now how blockchains are integrated into our framework. The starting point is the W3C PROV standard (https://www.w3.org/TR/prov-overview/). According to the PROV standard, provenance is information about entities, activities, and people involved in producing, in our case, data. This provenance data aids the assessment of quality, reliability or trustworthiness in the data production (See Figure 7.5.) The goal of PROV is to enable the representation and interchange of provenance information using common formats such as XML.

Provenance records describe the provenance of entities, which in our case are data objects. An entity's provenance can refer to other entities, e.g. compiled sensor data to the original records. Activities create and change entities, often making use of previously existing entities to achieve this. They are dynamic parts, here the processing components. The two fundamental activities are generation and usage of entities, which are represented by relationships in the model. Activities are carried out on behalf of agents that also act as owners of entities, i.e. are responsible for the processing. An agent takes some degree of responsibility for the activity taking place. Actors in our case are orchestrators in charge of deploying software and managing infrastructure.

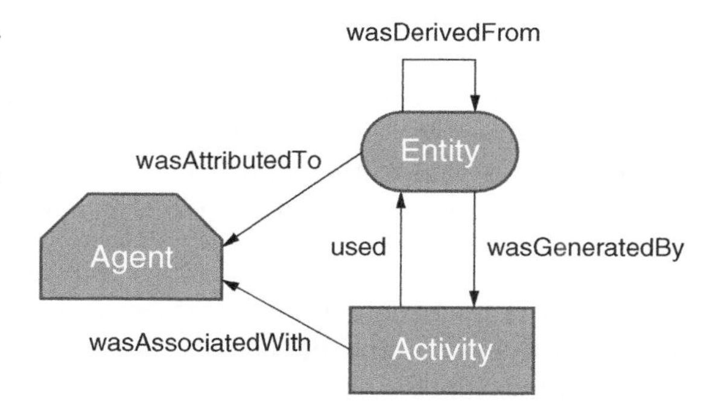

Figure 7.5 Provenance model. Adapted from W3C. "PROV Model Primer," April 30, 2013. © 2013 World Wide Web Consortium, (MIT, ERCIM, Keio, Beihang). https://www.w3.org/TR/2013/NOTE-prov-primer-20130430/.

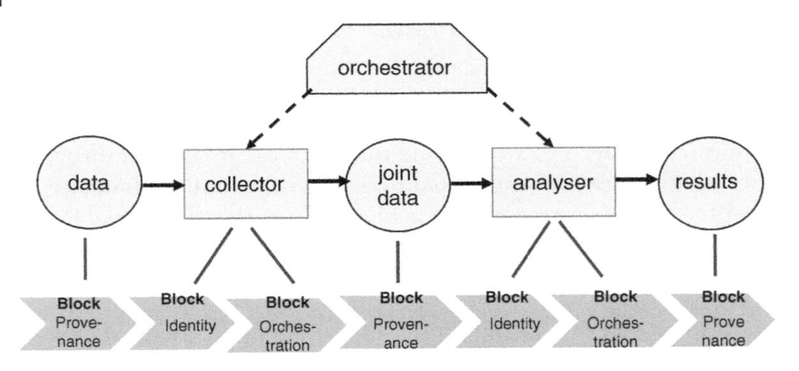

Figure 7.6 Blockchain-based tracking of an orchestration plan.

We can expand this idea by considering the provenance of an agent. In our case, the orchestrator is also a container, though one with a management rather than an application role. In order to make provenance assertions about an agent in PROV, the agent must then be declared explicitly both as an agent and as an entity.

In the schematic example in Figure 7.6, the orchestrator is the agent that orchestrates, i.e. deploys the collector and analyzer containers. This effectively forms a contract between orchestrator and nodes, whereby the nodes are contracted to carry out the collection and analyzer activities:

- The collector USES sensor data and GENERATES the joint data.
- The analyzer USES the joint data and GENERATES the results.

This sequence of activities forms an orchestration plan. This plan is enacted based on the blockchain smart contract concept, requiring the contracted activity to

- Obtain permissions (credentials) to retrieve the data (USES)
- Create output entities (GENERATE) as an obligation defined in the contract.

A smart contract is defined through a program that defines the implementation of the work to be done. It includes the obligations to be carried out, the benefits (in terms of SLAs), and the penalties for not achieving the obligations. Generally, fees paid to the contractor and possible penalties to compensate the contract issuer shall be neglected here. Each step based on the contract is recorded in the blockchain:

- The generation of data through a provenance entry: what, by whom, when.
- The creation of a credentials object defining, based on the identity of the processing component, the authorized activities.

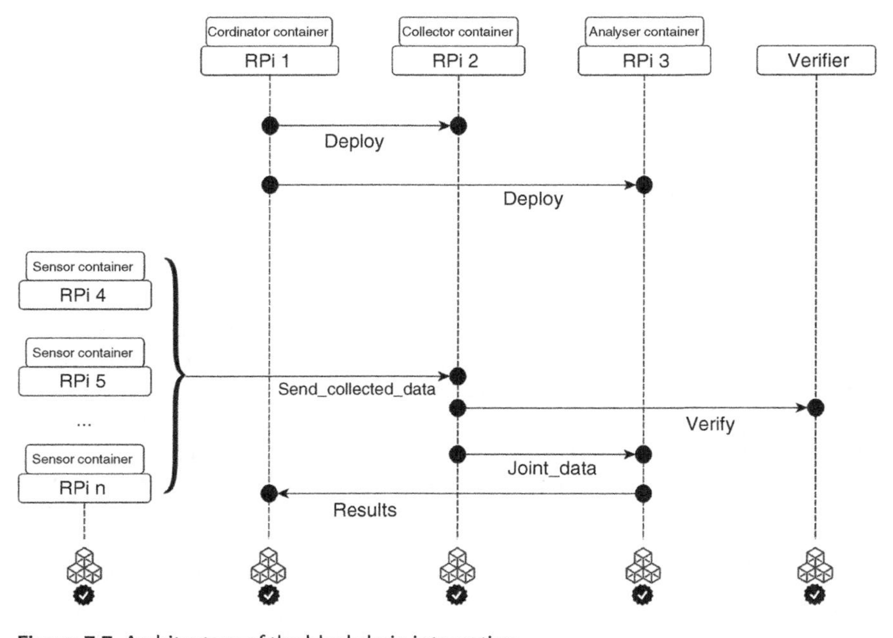

Figure 7.7 Architecture of the blockchain integration.

- The formalized contract between the orchestrator and the activity node. The obligations formalized include in the IoT edge context data-oriented activities such as storage, filtering and analysis, and container-oriented activities such as deploying or redeploying (updating) a container.

Figure 7.7 shows the full architecture of the system, including the interactions between all the components. All transactions are recorded in the blockchain to guarantee data provenance. Additionally, the identity of all components (e.g., containers, verifier) is stored to ensure identity. The transactions are executed by invoking the appropriate smart contract. For instance, when a sensor container collects data, it invokes the *send_collected_data* smart contract defined by the collector container by passing a signed hash of the collected data. At this point the collected container checks the identity of the sensor container (e.g., signature) and the integrity of the data (e.g., the hash of the data), and then downloads the data in order to process it.

7.7 Future Research Directions

We identified some limitations and have discussed concerns that require further work such as the security aspect where we explored blockchain technologies for provenance and identity management.

In the cloud context, some existing PaaS platforms have started to address limitations in the orchestration and DevOps. Some observations shall clarify this:

- Containers: Containers are now widely adopted for PaaS clouds.
- DevOps: Development and operations integration is still at an early stage, particularly if complex orchestrations on distributed topologies are considered that need to be managed in an integrated DevOps-style pipeline.

As a first concern, architecting applications for container-based clustered environments in a DevOps style is addressed by microservice-style software architecting [33–36]. Microservices are small, self-contained, and independently deployable architectural units that find their counterparts at deployment level in the form of containers.

For cluster management, more work is needed in addition to (static) architectural concerns. The question arises to which extent their distribution reaches an edge consisting of small devices and embedded systems and what the platform technology for that might be [31, 38]. A sample question is whether devices that run small Linux distributions such as the Debian-based DSL (which requires about 50MB storage) can support container host and cluster management. Significant improvements are still required to reliably support data and network management. Orchestration, the way it is realized in cluster solutions at the moment, is ultimately not adequate and requires further improvements, among them performance management. What is needed are controllers [39, 40] that manage performance, workload, and failure in these unreliable contexts, while, for instance, fault-tolerance [41] or performance management [42] has been addressed for the cloud, edge and fog adaptation are still needed.

Another concern that needs more attention is security. We have discussed blockchain technology for provenance management and other security concerns as a possible solution, but here more implementation and empirical evaluation work are needed.

Our ultimate aim is an edge cloud PaaS. We have implemented, experimented with, and evaluated some core ingredients of such an edge cloud PaaS, demonstrating that containers are the most suitable technology platform to reach this goal. Currently, cloud management platforms are still at an earlier stage than the container platforms they build on. Some recent third-generation PaaS support a build-your-own-PaaS idea, while being lightweight at the same time. We believe that the next development step could be a fourth-generation PaaS in the form of an edge cloud PaaS bridging the gap between IoT and cloud technology.

7.8 Conclusions

Edge cloud computing environments are distributed to bring specific services to the users away from centralized computing infrastructures [43], shifting computation from heavyweight data center clouds to more lightweight resources. Consequently, we require more lightweight virtualization mechanisms on these lightweight devices and have identified the need to orchestrate the deployment of services in this environment as key challenges. We looked at requirements for a platform (PaaS) middleware solution that specifically supports application service packaging and orchestration as a key PaaS concern.

We presented and evaluated different cluster management architectural options, including the recently emerging container technology, an open-source cloud solution (OpenStack), and an own-build solution to analyze the suitability of these options for edge clouds built on single-board lightweight device clusters. Our observations and evaluations support the current trend toward container technology as the most suitable option. Container technology is better suited than the other options to migrate and apply cloud PaaS technology toward distributed heterogeneous clouds through lightweightness and interoperability as key properties.

References

1 A. Chandra, J. Weissman, and B. Heintz. Decentralized Edge Clouds. *IEEE Internet Computing*, 2013.

2 F. Bonomi, R. Milito, J. Zhu, and S. Addepalli. Fog computing and its role in the internet of things. *Workshop Mobile Cloud Computing*, 2012.

3 N. Kratzke. A lightweight virtualization cluster reference architecture derived from Open Source PaaS platforms. *Open Journal of Mobile Computing and Cloud Computing*, 1: 2, 2014.

4 O. Gass, H. Meth, and A. Maedche. PaaS characteristics for productive software development: An evaluation framework. *IEEE Internet Computing*, 18(1): 56–64, 2014.

5 C. Pahl and H. Xiong. Migration to PaaS clouds – Migration process and architectural concerns. *International Symposium on the Maintenance and Evolution of Service-Oriented and Cloud-Based Systems*, 2013.

6 C. Pahl, A. Brogi, J. Soldani, and P. Jamshidi. Cloud container technologies: a state-of-the-art review. *IEEE Transactions on Cloud Computing*, 2017.

7 C. Pahl and B. Lee. Containers and clusters for edge cloud architectures – a technology review. *Intl Conf on Future Internet of Things and Cloud*, 2015.

8 C. Pahl. Containerization and the PaaS Cloud. *IEEE Cloud Computing*, 2015.

9 C. Pahl, S. Helmer, L. Miori, J. Sanin, and B. Lee. A container-based edge cloud PaaS architecture based on Raspberry Pi clusters. *IEEE Intl Conference on Future Internet of Things and Cloud Workshops*, 2016.

10 C. Pahl, N. El Ioini, and S. Helmer. A decision framework for blockchain platforms for IoT and edge computing. *International Conference on Internet of Things, Big Data and Security*, 2018.

11 D. von Leon, L. Miori, J. Sanin, N. El Ioini, S. Helmer, and C. Pahl. A performance exploration of architectural options for a middleware for decentralised lightweight edge cloud architectures. *International Conference on Internet of Things, Big Data and Security*, 2018.

12 C. Pahl, N. El Ioini, and S. Helmer. An Architecture Pattern for Trusted Orchestration in IoT Edge Clouds. *Third IEEE International Conference on Fog and Mobile Edge Computing FMEC*, 2018.

13 J. Zhu, D.S. Chan, M.S. Prabhu, P. Natarajan, H. Hu, and F. Bonomi. Improving web sites performance using edge servers in fog computing architecture. *Intl Symp on Service Oriented System Engineering*, 2013.

14 A. Manzalini, R. Minerva, F. Callegati, W. Cerroni, and A. Campi. Clouds of virtual machines in edge networks. *IEEE Communications*, 2013.

15 C. Pahl, P. Jamshidi, and O. Zimmermann. Architectural principles for cloud software. *ACM Transactions on Internet Technology*, 2018.

16 C. Pahl, P. Jamshidi, and D. Weyns. Cloud architecture continuity: Change models and change rules for sustainable cloud software architectures. *Journal of Software: Evolution and Process*, 29(2): 2017.

17 S. Soltesz, H. Potzl, M.E. Fiuczynski, A. Bavier, and L. Peterson. Container-based operating system virtualization: a scalable, high-performance alternative to hypervisors, *ACM SIGOPS Operating Syst Review*, 41(3): 275–287, 2007.

18 P. Bellavista and A. Zanni. Feasibility of fog computing deployment based on docker containerization over Raspberry Pi. *International Conference on Distributed Computing and Networking*, 2017.

19 P. Tso, D. White, S. Jouet, J. Singer, and D. Pezaros. The Glasgow Raspberry Pi cloud: A scale model for cloud computing infrastructures. *Intl. Workshop on Resource Management of Cloud Computing*, 2013.

20 S. Qanbari, F. Li, and S. Dustdar. Toward portable cloud manufacturing services, *IEEE Internet Computing*, 18(6): 77–80, 2014.

21 E. De Coninck, S. Bohez, S. Leroux, T. Verbelen, B. Vankeirsbilck, B. Dhoedt, and P. Simoens. Middleware platform for distributed applications incorporating robots, sensors and the cloud. *Intl Conf on Cloud Networking*, 2016.

22 C. Dupont, R. Giaffreda, and L. Capra. Edge computing in IoT context: Horizontal and vertical Linux container migration. *Global Internet of Things Summit*, 2017.

23 J. Turnbull. *The Docker Book*, 2014.

24 V. Andrikopoulos, S. Gomez Saez, F. Leymann, and J. Wettinger. Optimal distribution of applications in the cloud. *Adv Inf Syst* Eng: 75–90, 2014.

25 P. Jamshidi, M. Ghafari, A. Ahmad, and J. Wettinger. A framework for classifying and comparing architecture-centric software evolution research. *European Conference on Software Maintenance and Reengineering*, 2013.

26 T. Binz, U. Breitenbücher, F. Haupt, O. Kopp, F. Leymann, A. Nowak, and S. Wagner. OpenTOSCA – a runtime for TOSCA-based cloud applications, *Service-Oriented Computing*: 692–695, 2013.

27 P. Abrahamsson, S. Helmer, N. Phaphoom, L. Nicolodi, N. Preda, L. Miori, M. Angriman, Juha Rikkilä, Xiaofeng Wang, Karim Hamily, Sara Bugoloni. Affordable and energy-efficient cloud computing clusters: The Bolzano Raspberry Pi Cloud Cluster Experiment. *IEEE 5th Intl Conference on Cloud Computing Technology and Science*, 2013.

28 R. van der Hoeven. "Raspberry pi performance," http://freedomboxblog.nl/raspberry-pi-performance/, 2013.

29 L. Miori. Deployment and evaluation of a middleware layer on the Raspberry Pi cluster. BSc thesis, University of Bozen-Bolzano, 2014.

30 C.A. Ardagna, R. Asal, E. Damiani, T. Dimitrakos, N. El Ioini, and C. Pahl. Certification-based cloud adaptation. *IEEE Transactions on Services Computing*, 2018.

31 A. Dorri, S. Salil Kanhere, and R. Jurdak. Towards an Optimized BlockChain for IoT, Intl Conf on IoT Design and Implementation, 2017.

32 N. Kshetri. Can Blockchain Strengthen the Internet of Things? *IT Professional*, 19(4): 68–72, 2017.

33 P. Jamshidi, C. Pahl, N.C. Mendonça, J. Lewis, and S. Tilkov. Microservices – The Journey So Far and Challenges Ahead. *IEEE Software*, May/June 2018.

34 R. Heinrich, A. van Hoorn, H. Knoche, F. Li, L.E. Lwakatare, C. Pahl, S. Schulte, and J. Wettinger. Performance engineering for microservices: research challenges and directions. In *Proceedings of the 8th ACM/SPEC on International Conference on Performance Engineering Companion*, 2017.

35 C.M. Aderaldo, N.C. Mendonça, C. Pahl, and P. Jamshidi. Benchmark requirements for microservices architecture research. In *Proceedings of the 1st International Workshop on Establishing the Community-Wide Infrastructure for Architecture-Based Software Engineering*, 2017.

36 D. Taibi, V. Lenarduzzi, and C. Pahl. Processes, motivations, and issues for migrating to microservices architectures: an empirical investigation. *IEEE Cloud Computing*, 4(5): 22–32, 2017.

37 A. Gember, A Krishnamurthy, S. St. John, et al. Stratos: A network-aware orchestration layer for middleboxes in the cloud. Duke University, Tech Report, 2013.

38 T.H. Noor, Q.Z. Sheng, A.H.H. Ngu, R. Grandl, X. Gao, A. Anand, T. Benson, A. Akella, and V. Sekar. Analysis of Web-Scale Cloud Services. *IEEE Internet Computing*, 18(4): 55–61, 2014.

39 P. Jamshidi, A. Sharifloo, C. Pahl, H. Arabnejad, A. Metzger, and G. Estrada. Fuzzy self-learning controllers for elasticity management in dynamic cloud architectures, *Intl ACM Conference on Quality of Software Architectures*, 2016.

40 P. Jamshidi, A.M. Sharifloo, C. Pahl, A. Metzger, and G. Estrada. Self-learning cloud controllers: Fuzzy q-learning for knowledge evolution. *International Conference on Cloud and Autonomic Computing ICCAC*, pages 208–211, 2015.

41 H. Arabnejad, C. Pahl, G. Estrada, A. Samir, and F. Fowley. A Fuzzy Load Balancer for Adaptive Fault Tolerance Management in Cloud Platforms, *European Conference on Service-Oriented and Cloud Computing (CCGRID)*: 109–124, 2017.

42 H. Arabnejad, C. Pahl, P. Jamshidi, and G. Estrada. A comparison of reinforcement learning techniques for fuzzy cloud auto-scaling, *17th IEEE/ACM International Symposium on Cluster, Cloud and Grid Computing (CCGRID)*: 64–73, IEEE, 2017.

43 S. Helmer, C. Pahl, J. Sanin, L. Miori, S. Brocanelli, F. Cardano, D. Gadler, D. Morandini, A. Piccoli, S. Salam, A.M. Sharear, A. Ventura, P. Abrahamsson, and T.D. Oyetoyan. Bringing the cloud to rural and remote areas via cloudlets. ACM Annual Symposium on Computing for Development, 2016.

8

Data Management in Fog Computing

Tina Samizadeh Nikoui, Amir Masoud Rahmani, and Hooman Tabarsaied

8.1 Introduction

Fog computing plays an important role in a huge and real-time data management system for Internet of Things (IoT). IoT is a popular topic; however, as a new one, it has its own challenges in handling the huge amount of data and providing on-time response are some of them. The high growth rate of data generation in the IoT ecosystem is a considerable issue. It was stated that in 2012, 2500 petabytes of data were created per day [1]. In [2] it was mentioned that 25,000 records were generated per second in a health application with 30 million users. Pramanik et al. [3] outlined that by a fast growing rate, in the near future health-related data will be in scale of zettabytes. In smart cities, the amount of data is even more, while Qin et al. [1] noted that 1 million/second records may be produced in smart cities. One exabyte of data is generated per year by US smart grid and approximately 2.4 petabytes of data are generated per month by US Library of Congress [4].

The processing time in the cloud and delay of transferring cause the latency that affects performance, and that latency is unacceptable in IoT applications like e-Health, because late feedback about a suspicious or emergency situation may endanger someone's life.

The sensors and end devices periodically generate row data that include useless, noisy, or repetitive records, but transferring huge amount of data leads to increased errors, packet loss, and high probability of data congestion. In addition, processing and storing the repetitive or noisy data waste the resource with no gain. So interactional applications with large scale of data generation must decrease end-to-end delay and achieve real-time data processing and analytics. Therefore, there is a need to do some local processing. Because of resource constraints and lack of aggregated data in each of IoT devices, however, they are not capable of processing and storing generated data.

Bringing the storage, processing, and network close to the end-devices in fog computing paradigm is considered a proper solution. There are many

Fog and Edge Computing: Principles and Paradigms, First Edition.
Edited by Rajkumar Buyya and Satish Narayana Srirama.

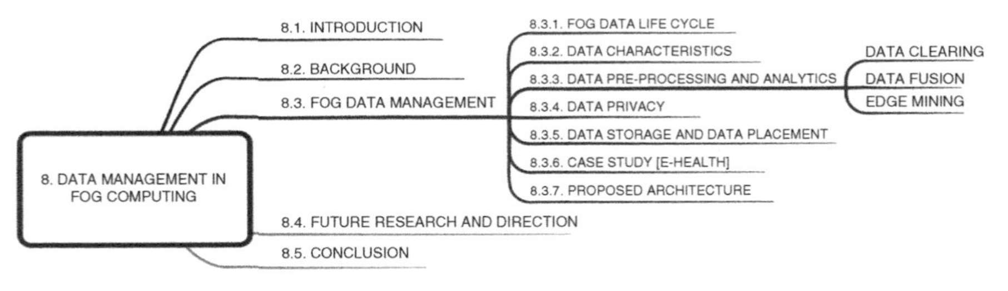

Figure 8.1 Structure of data management in fog computing.

definitions for fog computing. Qin defined it as "a highly virtualized platform that provides compute, storage, and networking services between end devices and traditional cloud computer data-centers, typically, but not exclusively located at the edge of network" [1]. In this way, only on-site processing and storage would be possible. However, fog computing has other benefits, such as better privacy by providing encryption and decryption, data integration [5], dependability, and load balancing.

A schematic structure of the contents in this chapter is shown in Figure 8.1. This chapter focused on data management in fog computing and proposed a conceptual architecture. In section 8.2, we review fog data management and highlight the management issues; additionally, a number of studies that are done on fog data management are presented. The main concepts of fog data management and the proposed architecture are presented in section 8.3. Future research and directions are presented in section 8.4. We summarize the main contents that are discussed in this chapter in section 8.5.

8.2 Background

Fog plays a role as a mediator between devices and cloud; it is responsible for temporary data storage, some preliminary processing, and analytics. By this way after data generation by IoT devices, fog does some preliminary process and may store data for a while; these data are consumed by the cloud applications, proper feedback is generated by fog or cloud, and they are returned to a device. A three-layer fog diagram with a data view is depicted in Figure 8.2.

This chapter addresses surveys and paper on fog computing architecture. Typical fog computing architecture has three basic layers: device layer (or physical layer), fog layer (or edge network layer), and cloud layer [6]. A reference architectural model for fog computing is addressed in [7]. In [8] with the aim of data acquisition and management in fog computing paradigms, a three-layer architecture was provided. It was composed of IoT sensor nodes, gateways, and IoT middleware. Data management, processing, virtualization, and service provisioning are done in the fog layer. In [9] a fog-based schema was proposed

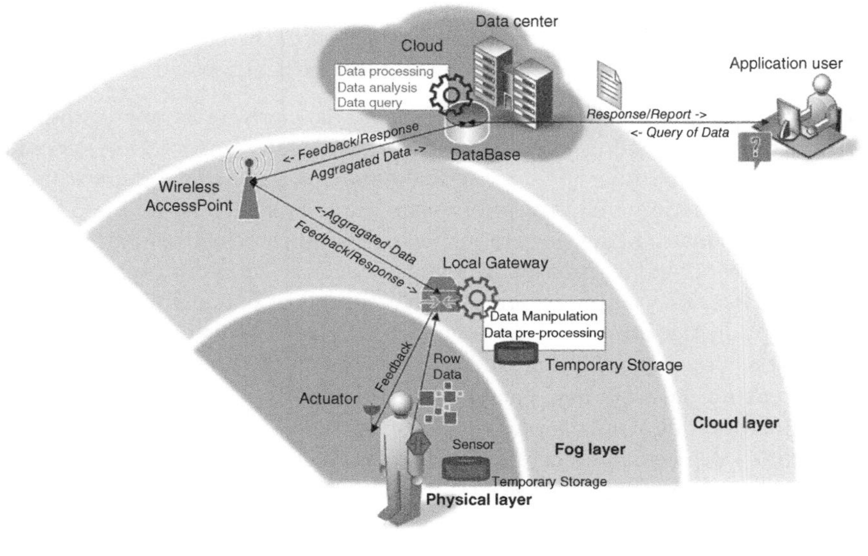

Figure 8.2 Basic data management diagram in fog computing.

for data analytics. The authors proposed a fog-based architecture with a vertical and three horizontal layers, for crowdsensing applications. In [10] a programming framework was provided to define the processing model on streams of data. Another fog computing multitier framework for data analysis was proposed in [11].

A data-centered platform for fog computing was proposed in [12]. Fog servers, fog edge nodes, and foglets are introduced as fog elements. Fog servers are responsible for interaction between the fog and cloud. The other entity focuses on data processing, storage, and communication. Foglet is a software agent that plays a middleware role and interacts between fog servers and other fog edge nodes. It is also used for monitoring, controlling, and maintaining.

Fog data management is about handling the data and its related concepts such as data aggregation approaches, data filtering techniques, data placement, providing data privacy, etc. Based on the three-layer fog architecture, which is shown in Figure 8.2, sensory and collected data as a part of the system are sent to the upper layer and should be managed properly. As before mentioned, end-to-end latency and network traffic are two of main motivations to use fog computing. Local data management yields benefits such as better efficiency, more privacy, and so on. The main advantages of data management in fog computing are described as following:

- **Increasing efficiency**. Local processing on data and elimination of corrupted, repetitive, or unneeded data in fog layer reduces the network load and increase the network efficiency. Because the transferred data to the cloud must be processed, stored, and analyzed in the cloud, by decreasing the amount of data, cloud processing and storage needs would also decrease.

- **Increasing the level of privacy**. Ensuring data privacy is one of the IoT and cloud computing challenges. In IoT systems, sensors may generate and transfer sensitive and confidential data, but transferring them without any manipulation and encryption bears the risk of disclosure. In addition, resource constraint devices cannot handle complicated mathematical operations. The privacy-preserving mechanisms such as encrypting algorithms in end-devices may be impossible. Therefore, privacy, data manipulations and encryption algorithms can be done in a fog layer. Nevertheless, protection of fog devices is another issue that will be investigated further.
- **Increasing data quality**. Quality of data would be increased though the elimination of low-quality data such as repetitive, corrupted, or noisy data and the integration of received data in a fog layer.
- **Decreasing the end-to-end latency**. Because of the nature of networks, existence of delay is obvious and inevitable, so response time must account for issues such as network delay and processing time when gathering feedback from cloud in IoT scenarios. Putting data pre-processing close to the devices in fog layer will minimize the end-to-end delay.
- **Increasing dependability**. System dependability is about the ability of a system to provide the service as is expected. Refer to the definition of dependability which is provided in ISO/IEC/IEEE 24765 [13], three main aspects of dependability are reliability, availability, and maintainability. Fog devices and the local network can cover the possible failure of cloud networks and provide local data processing, therefore, the availability and reliability of a system would be increased.
- **Decreasing cost**. Local data processing and data compressing in a fog layer reduce the cost of network usage, cloud processing, and storage. However, the cost of fog devices should be considered, and there must be a trade-off.

8.3 Fog Data Management

Data life cycle and fog data characteristics as the essential concepts of fog data management are elaborated in this section, as well as the other important issues in fog computing such as data cleaning, data fusion, data analysis, privacy concerns, and fog data storage. In addition, a case study of employing fog computing in e-health application is described. Finally, the proposed architecture is presented.

8.3.1 Fog Data Life Cycle

The fog data life cycle consists of several steps that start from data acquisition in the device layer where data is generated, continue with processing and storing in upper layers and sending feedback to the device layer, and finally end with execution of commands in the device layer. As is shown in Figure 8.3, we consider five main steps: *data acquisition, lightweight processing, processing and analysis, study feedback,* and *command execution.* In the following, the main steps are elaborated.

8.3.1.1 Data Acquisition

Data from different types of end devices should be acquired. It must be sent to upper layers. To this end, a sink node or local gateway node may exist to gather data, or sensors can send data directly to the fog.

8.3.1.2 Lightweight Processing

This step provides lightweight data manipulation and local data processing on the collected data. Lightweight processing may include data aggregation, data filtering, and elimination of unnecessary or repetitive data, data cleaning, compression/decompression, or some lightweight data analysis and pattern extraction. As data may be stored for a while in fog devices, the last period's data would be accessible locally, so more feasibility for data pre-processing will be provided. The aggregated data will be transferred to the cloud via the network. In addition, the feedback as response data should be transferred to the device. Also, as the feedback is received from cloud layer and sent to the device layer, there may be a need for data decompression, data decryption, doing some format change on the received data, etc. These types of change must be supported by the fog layer.

8.3.1.3 Processing and Analysis

Received data may be stored permanently in the cloud layer, and it is processed based on predefined requirements. In addition, the application users may access data to get reports or data analysis. Different types of analysis on stored data may be applied to obtain valuable information and knowledge, and these types of processing and analysis are almost in the scale of big data. They need big data platforms and technologies such as HDFS for storage and map-reduce for processing [3]. More information about big data concepts and analytics is provided in [14].

8.3.1.4 Sending Feedback

Based on data processing and analyzing, feedbacks such as proper commands or decisions are generated and sent to the fog layer.

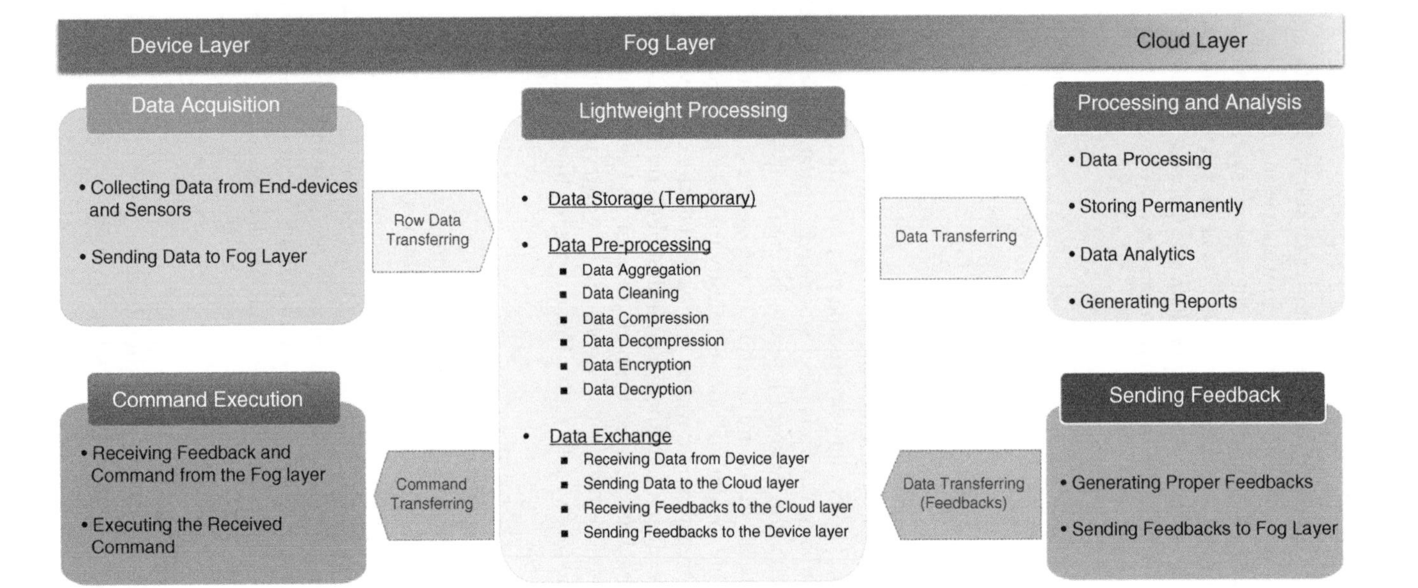

Figure 8.3 Data life cycle in fog computing.

8.3.1.5 Command Execution

Actuators must run the proper action based on the received data. In this way, proper feedback and responses are applied to the environment.

8.3.2 Data Characteristics

Reviewing data characteristics is necessary to define and refine data quality and integration standards and to handle the related challenges properly in the data management process. Data quality refers to how much data characteristics are suitable and can comply with consumer requirements.

Some of the main IoT data characteristics are mentioned in [15], which are as follow: uncertainties, erroneous, noisy data, voluminous and distributed, smooth variation, continuous, correlation, periodicity, and Markovian behavior. They reviewed accuracy, confidence, completeness, data volume, and timeliness as data quality dimensions. Three other additional data quality dimensions are ease of access, access security, and interpretability.

Also in [1] IoT data characteristics were categorized into three categories: data generation, data quality, and data interoperability. The IoT data quality characteristics include uncertainty, redundancy, ambiguity, and inconsistency. In the traditional way, after data are captured from different devices, they are stored for further steps. After the data are gathered in storage, batch processing is applied. By increasing in data generation speed and data volumes, new data analytics requirements are raised. One of them is stream processing, which is applied on a continuous and ongoing stream of data. Management of main IoT data characteristics and the related issues can be done in the fog data management process to fulfill the requirements. In the following, these characteristics are reviewed:

- **Heterogeneity**. Distributed heterogeneous end devices generate data in different formats. Generated data may be diversely varying in terms of structure or format [16, 17].
- **Inaccuracy**. Inaccuracy or uncertainties of the sensed data refer to the sensing precision, accuracy, or misreading of data [1, 15, 17, 18].
- **Weak semantics**. As mentioned before, the collected raw data that may be heterogeneous in terms of data formats, data structure, data source, etc. must be processed and managed. Using the concepts of semantic web and injecting some information and extra data to the raw data make the data readable and understandable for machines. Nevertheless, most of the collected data from the environment has weak semantics [1, 16, 17].
- **Velocity**. Data generation rates and sampling frequencies are varying in different types of end devices [1].
- **Redundancy**. Repetitive data that are sent by one or more end devices lead to redundancy in the collected data [1].

- **Scalability**. Large numbers of end devices and high data sampling rate that may exist in different scenarios may lead to generation of a huge amount of data [1].
- **Inconsistency**. Low precision or misreading in the sensed data may cause inconsistency in the gathered data [1].

8.3.3 Data Pre-Processing and Analytics

In this section, three of the main data pre-processing and analytics concepts that play important roles in fog data management are reviewed. They are data cleaning, data fusion, and edge mining.

8.3.3.1 Data Cleaning

Because of the mentioned characteristics, sensory data are not fully reliable, which is unpleasant for further processing and decision-making. Jeffery et al. stated, "Dirty data" refer to missed readings and unreliable readings [19]. Cleaning mechanisms can be applied on the collected data in fog layers to reduce the effect of dirty and unreliable data, and to increase the quality of them. Data cleaning approaches can be divided in two categories: declarative data cleaning and model-based data cleaning [18]. In the following, each of them is reviewed briefly:

- **Declarative data cleaning**. High-level declarative queries such as CQL (continuous query language) are used to define the sensor values constraints. In this way, the user can express the queries and control the system easily via the provided interface. Extensible sensor stream processing (ESP) [18], is an example of this type. It is a declarative-based and pipelined framework for sensor data cleaning for use in pervasive applications.
- **Model-based data cleaning**. Anomalies are detected by comparison of raw values with the inferred values that are resulted as the most probable values based on selected models. The model-based approaches also have subcategories such as regression models. These include polynomial regression and Chebyshev regression [18, 20], probabilistic models such as Kalman filter [18], and outliner detection models [21].

8.3.3.2 Data Fusion

Data fusion refers to the elimination of redundant and ambiguous data and integration of data, and can be done in the fog layer as one of the data management tasks to increase the accuracy and efficiency. In [22], data fusion was defined as "multilevel, multifaceted process handling the automatic detection, association, correlation, estimation, and combination of data and information from several sources." Data fusion models can be categorized into three particular categories: data-based model, activity-based model, and role-based model [23].

Khaleghi et al. [24] categorizes data fusion frameworks into four classes based on data-related aspects: (i) imperfect data fusion framework; (ii) correlated data fusion framework; (iii) inconsistent data fusion framework; and (iv) disparate data fusion framework. The first category is related to data imperfection, which is one the main data challenges and may be caused by impreciseness, incompleteness, vagueness, or uncertainty [25]. The second category is related to dependency of data. The last two categories are about the conflict and diversity on data. There are some famous data fusion techniques and models, such as Intelligent Cycle (IC) [23] and Joint Directors of Laboratories (JDL) [23, 24].

8.3.3.3 Edge Mining

Fog computing can be effective for local analytics and stream processing to reduce the volume of data. Edge mining refers to utilize mining approaches on row data that are produced by devices in the edge of the network (fog layer). In this way, the size of the transferred data will be reduced and better energy savings can be achieved. Gaura et al. [26] stated that edge mining can be defined as "processing of sensory data near or at the point at which it is sensed, in order to convert it from a raw signal to contextually relevant information." General Spanish Inquisition Protocol (G-SIP) is one of the edge mining algorithms; it has three instantiations, which are Linear Spanish Inquisition Protocol (L-SIP), ClassAct, and Bare Necessities (BN). L-SIP is a lightweight algorithm for local data compression and aims to reduce data size through the state estimation and improve storage and responsiveness. In this model, end devices and fog devices use a predefined and shared model for state calculation and prediction. In case of unexpected data, data would be sent to the fog devices.

Based on [26], L-SIP, ClassAct, and BN reduce the packet transmission by 95%, 99.6%, and 99.98% respectively. Collaborative edge mining is another extension of edge mining that was proposed in [27] to reduce the transferred data size.

8.3.4 Data Privacy

Privacy preserving and protection of data against unauthorized access are considered as one of the fog computing functionalities to keeps malicious and unauthorized end devices out of the system. However, due to the mobility of devices in some kinds of applications such as smart-transportation, the authentication phase must consider the mobility and dynamic nature of the network.

Position is a sensitive data point that can represent the owner's location, and it should be protected, as location privacy is considered as one the data protection issues in the fog data privacy. It was addressed in [28] in secure positioning protocol. The authors defined correctness, positioning security, and location privacy as three properties that the proposed protocol must satisfy to be secure.

Providing the privacy of data aggregation was addressed in [5], which proposed a privacy-preserving data aggregation schema for fog enhanced IoT. In the proposed approach, Chinese remainder theorem, one-way hash, and homomorphic Paillier were used for fault-tolerant data aggregation from hybrid IoT devices, authentication, and detection of false data injection in the fog layer.

8.3.5 Data Storage and Data Placement

Data storage and data placement are the other issues that must be handled in fog data management. Based on the predefined policy, data may be discarded after pre-processing or may be stored for temporary in the fog devices for further processing or aggregation purpose. It should be noted that in addition to storage and memory constraints, for the sake of decreasing the end-to-end latency and providing real-time response time, storage should have low-latency, cache, and cache management techniques.

Also, making decisions about the duration and volume of stored data is very dependent to application type and infrastructure capabilities. Another issue concerns efficient placing of gathered data in fog storages based on node characteristics, geographical zone, and type of application, because data placement strategy affects service latencies. Naas et al. proposed using iFogStore to reduce the latency, taking into account fog device characteristics as well as heterogeneity and location [29]. Sharing of data by different data consumers, dynamic location of data consumers, and the capacity constraints of fog devices are considered in iFogStore. In addition, to reduce the overall latency, it considers the storing and retrieval times.

To provide the real-time decision-making, in [30] based on the basic three-layer architecture, a storage management architecture in edge (fog) computing was proposed. In the edge (fog) layer, the architecture has six components to provide storage, and data management mechanism in a storage constrained system: monitoring, data preparation, adaptive algorithm, specification list, storage, and mediator component. The other two layers are cloud layer and gathering layer (device layer). The first is responsible for storing the historical data and the latter generates the row data.

8.3.6 e-Health Case Study

For clarifying the effect of fog data management, the role of fog data management in e-Health as one of IoT applications is investigated. e-Health applications aim to help take care of the elderly and patients. There have been a lot of studies on e-Health in the last decade, such as [30–32]. In [33], benefits of healthcare systems are described. Some of the main benefits are ease of use, reduction of cost, more considerable availability, and services. Healthcare applications such as ECG devices may generate several GBs of data in a day. Transferring and processing it means that it conserves network bandwidth, storage, and processing cycles [33]. Healthcare solutions can be used for monitoring, controlling, or prediction of emergency conditions based on the captured data.

In emergency conditions, fog computing performs faster than such processes being performed on the cloud layer. In comparison with nursing care, e-health application can monitor and control patients 24/7 and at lower cost. A number of papers have been published on electronic healthcare systems, such as [35] and [36]. This kind of application remotely monitors health status by controlling some parameters and data such as blood glucose, blood pressure, heartbeat, electroencephalography, electrocardiography, motion, and location data. These sensors transmit collected data to the local gateway (fog device) in short intervals (e.g. every one minute). These data are temporarily stored in the local storage. In addition, these data are pre-processed in terms of emergency conditions, such as blood pressure higher than 140/90 mmHg, or blood glucose above than 400 mg/dl. Therefore, in the event of an emergency, the necessary actions will be taken immediately through fog devices. For example, as the sequence diagram shows in Figure 8.4, if the measured blood glucose is above 400 mg/dl, injection must be done through the inclusion bracelet, or in the case of high blood pressure above 140/90 mmHg, notifications are sent to emergency services.

Some kinds of data compression or encryption can be done in the fog layer before sending data to the cloud layer in order to increase efficacy and privacy. After preliminary data aggregation, manipulation and processing were done in fog devices. Then data were sent to the cloud layer in the predefined intervals or events. Data can be stored and processed in the cloud layer and the application users can access these data and receive the health reports.

8.3.7 Proposed Architecture

This section provides a conceptual architecture based on the three-layer model to handle data management issues. As shown in Figure 8.5, our proposed architecture consists of a device layer, fog layer, and cloud layer. Located sensors and actuators in the device layer interact with the physical environment – the sensors collect data and the actuators run the commands that are received from the fog.

The device layer sends the collected data to the fog layer and receives commands from it. The fog layer is divided into two sub-layers. The lower fog sub-layer, called the fog-device sub-layer, is responsible for controlling physical device routines, protocol interpretation, de-noising the received signals, authentication, and data storage. In addition, lightweight analysis and local decision-making that are based on the business of fog application are

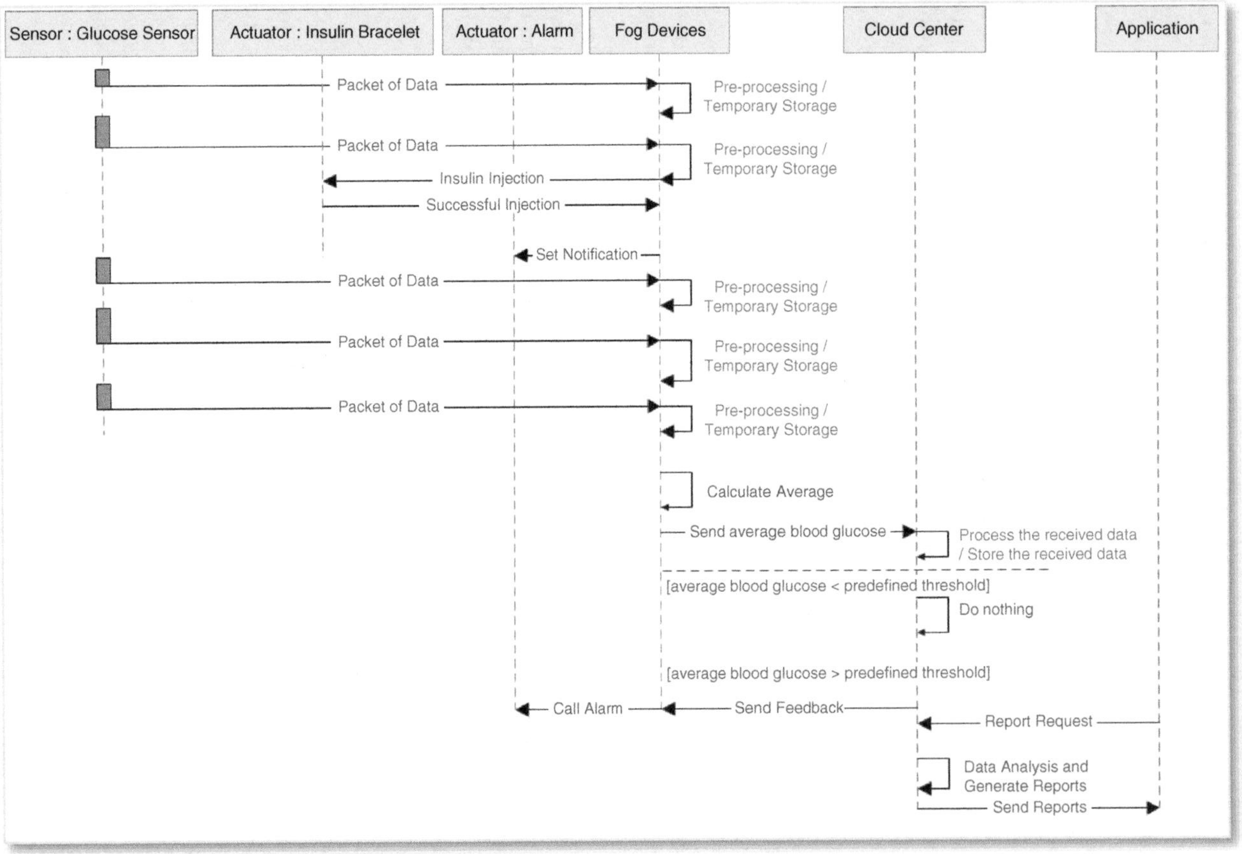

Figure 8.4 A simple sequence diagram of an e-health application.

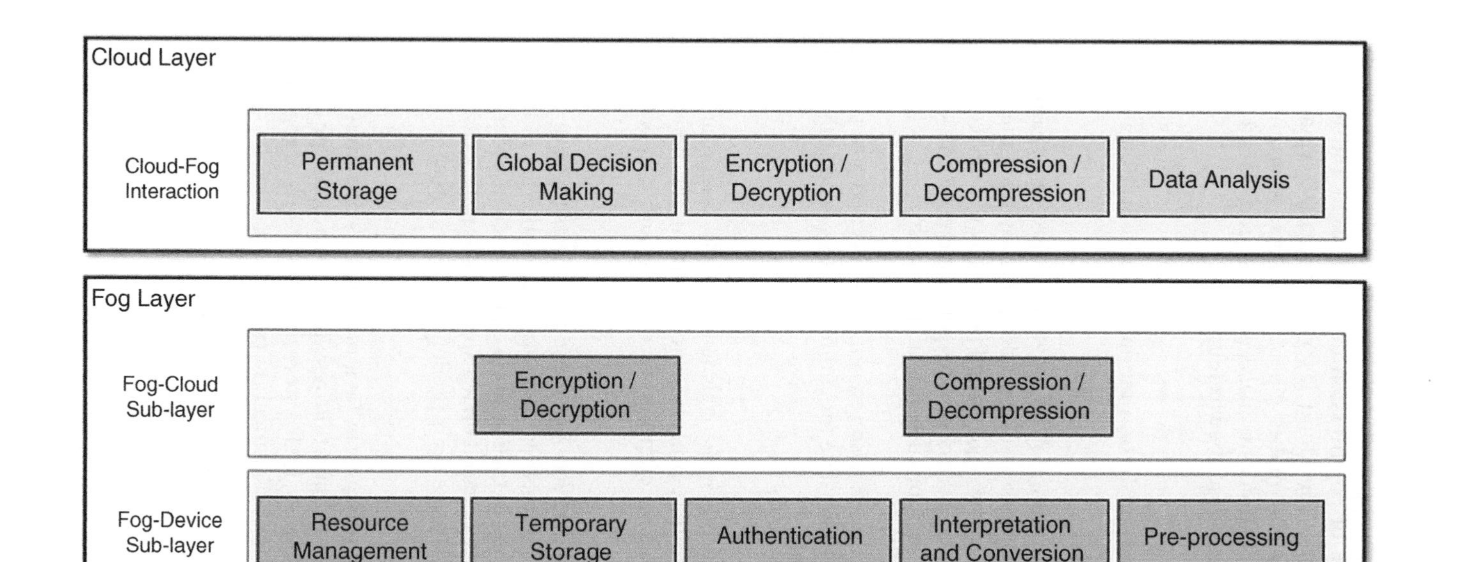

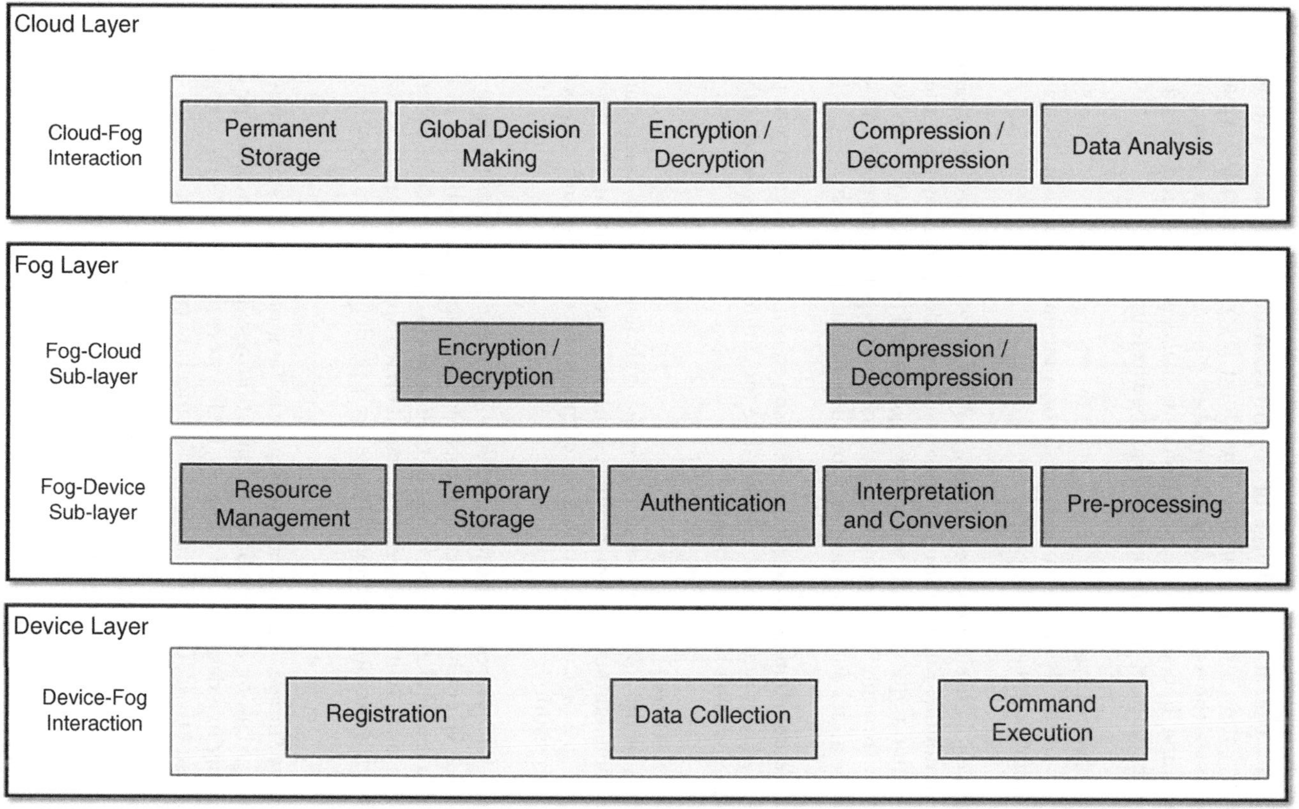

Figure 8.5 Proposed architecture.

located in this layer. The other sub-layer, the fog-cloud sub-layer, interacts with the cloud layer. It is in charge of compression/decompression and encryption/decryption on the packets. The cloud layer stores data permanently; it processes the received data and makes global decisions. Also, in terms of incoming query from applications, it analyses the stored data to send responses. Each of the modules is described as follows.

8.3.7.1 Device Layer

Modules of the device layer are registration module, data collection module, and command execution module.

Registration. Physical devices can join to the network or leave it dynamically via this module. Registration is necessary for sending and receiving messages. Registration requests as an initial message should be sent by the device to the fog layer. Through the registration process, devices get a unique ID and key that should be attached in the messages for authentication process.

Data Collection/Command Execution. Registered sensors collect data to transfer to the fog layer. Actuators are responsible for running the received commands from the fog layer by this module.

8.3.7.2 Fog Layer

Modules of the fog layer are resource management module, temporary storage module, authentication module, protocol interpretation and conversion module, pre-processing module, encryption/decryption, and compression/decompression.

Resource Management. The fog layer receives the join requests that are sent by the device layer, the resource management module queries the list of devices and adds the device specifications to the list in case of absence or deactivation of device, and it will be added. Also, based on fog applications policy, registered devices that are not sending message for a predefined period may be deactivated.

Temporary Storage. Temporary storage can be a module to store some of the incoming data or intermediate computation results, for example, for further processing in a database. In addition, it also stores specifications of registered devices and their IDs and keys for authentication process.

Authentication. The authentication module searches the list of registered devices on temporary storage to find related key and ID to authenticate the incoming messages.

Interpretation and Conversion. The communication of devices and fog may be received data, heterogeneity in terms of communication types between devices and the fog layer is possible, so data may be sent via different technologies such as Wi-Fi, Bluetooth, ZigBee, RFID, or etc. Therefore, Interpretation and conversion provides different protocols and conversion methods.

Pre-Processing. Storing the received data in the temporary storage, preparing and aggregating them are the responsibilities for the pre-processing module. Also some kinds of data processing may be needed on the received data, such as data cleaning, data fusion, edge mining, and quality improvement of received signals by data filtering or de-noising them, decision making in case of emergency situations by checking and comparing the received or collected data against the predefined threshold and conditions. Lightweight analyzing, feature extraction, pattern recognition, and decision making need more specific algorithms to be applied on data. However, selected approaches in level of the fog, must be simple and comply with the existed constraints.

Encryption/Decryption. To improve data privacy and protect sensitive data in the message, encryption and decryption algorithms are provided in this module.

Compression/Decompression. Compression techniques can be applied on the packet size of data to reduce overhead of networks by this module.

8.3.7.3 Cloud Layer

Modules of the cloud layer are permanent storage, global decision making, encryption/decryption, compression/decompression and data analysis modules. This section describes each of them.

Permanent Storage. This module receives data from different fog zones and stores the data in permanent storage. Depending on the fog application type, the size of permanent storage might vary from gigabytes to even petabytes. Thus, big data technologies are applied in this layer to store data.

Global Decision-Making. Received data are processed to send proper feedback to the lower layers and to store the needed and useful data in permanent storage. In case of receiving/sending compressed or encrypted messages from fog layer, decompression or decryption units are used, respectively. Data in motion and data stream processing, which were described previously, are the other issues raised in IoT and fog computing for some kinds of applications.

Encryption/Decryption, Compression/Decompression. As mentioned before, for privacy and efficiency concerns and to support encryption and compression, these modules must be located in both the cloud and fog layers. Encryption and compression approaches must be agreed by both of them.

Data Analysis. Gathering of data will be very valuable when it leads to knowledge, so data analysis, pattern recognition, and knowledge discovery from the heterogeneous data that are collected from different devices are considered an important stage in the data life cycle. This module is responsible for data analysis based on the application user requests for providing reports and getting global analysis on the gathered data. Depending on the size of stored data in data storage, data analysis approaches and technologies may vary. In the case of large-scale data storage, big data analytics technologies may be applied.

Based on the proposed architecture shown in Figure 8.5, interaction of components are presented in Figure 8.6.

8.4 Future Research and Direction

Despite the benefits of fog computing, new challenges arise that should be handled in future research to provide better and more efficient services to the fog users. This section addresses some key challenges that are related to data management issues.

8.4.1 Security

As was mentioned before, fog computing can increase the privacy via encryption and local processing of sensitive data. However, methods for maintaining the encryption keys and selection of proper encryption algorithms must be considered. As fog devices might not be properly secured, controlling and protecting distributed fog devices against different attacks and data leakage are both considerable security challenges. Other issues regard the structure and dynamic nature of fog computing, which different devices can join to a region or leave it, protection of fog devices against inaccurate and malicious data, and implementation of proper authentication methods.

8.4.2 Defining the Level of Data Computation and Storage

In comparison with end devices, fog devices have more computing and storage resources, but these are still not enough for complicated processing or permanent storage. Therefore, lightweight algorithms for data pre-processing based on a short time history must be provided. Also, determining the level of processing and storage in the fog devices based on existing constraints must be further studied.

8.5 Conclusions

To decrease the response time of real-time systems for IoT applications and handle the huge amount of data in IoT systems, the fog computing paradigm can be considered a good solution. In this chapter, we reviewed data

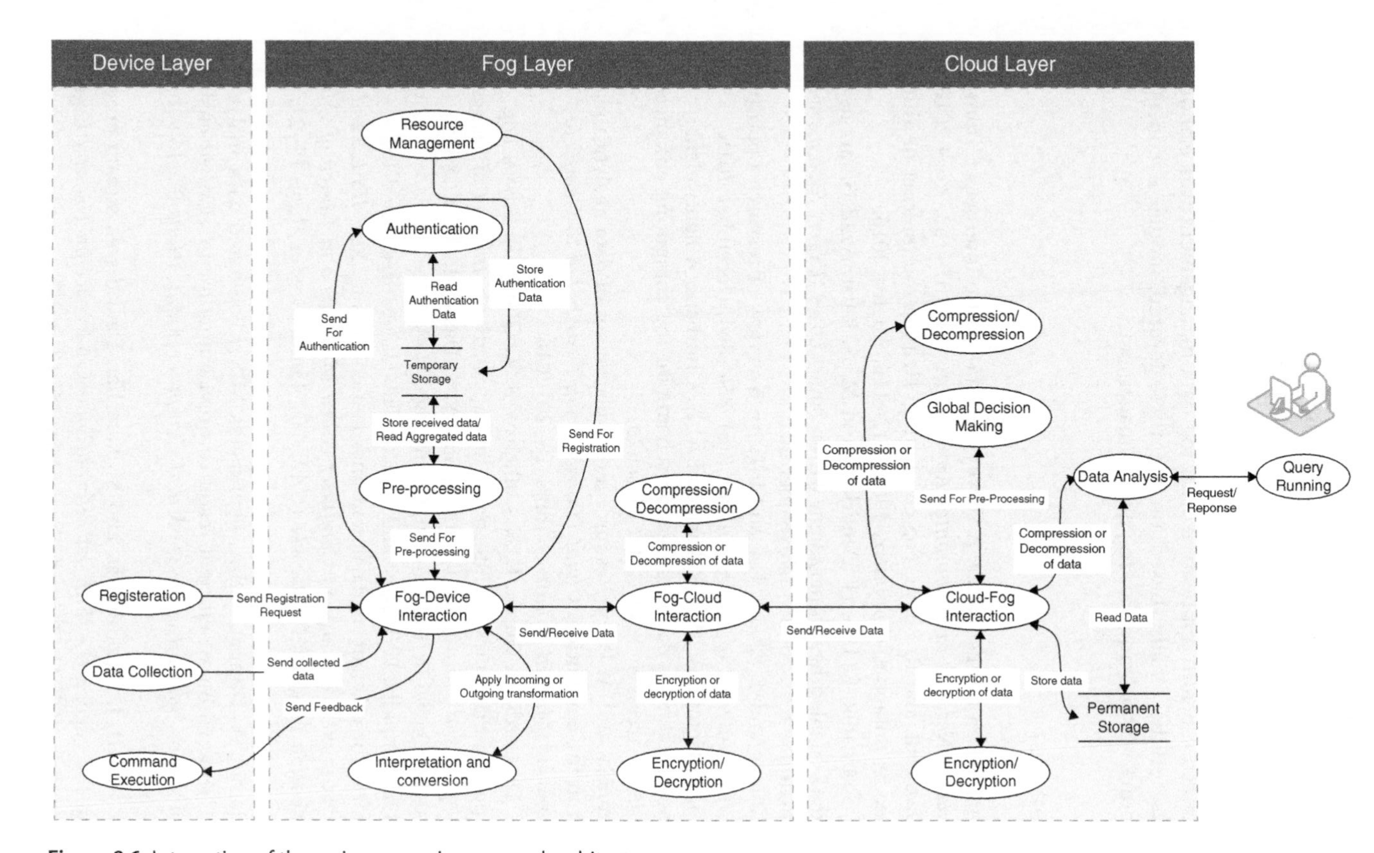

Figure 8.6 Interaction of the main process in proposed architecture.

management in fog computing, which plays an important and effective role in improving the quality of services for real-time IoT applications. We discussed the concept of fog data management, its main benefits, preliminary processes such as clearing mechanism, mining approaches and fusions, privacy issues, and data storage. To provide a better understanding of fog data management, an enhanced e-health application with fog data management was elaborated as a case study. Finally, based on the three-layered model, a conceptual architecture was provided for fog data management.

References

1 Y. Qin. When things matter: A survey on data-centric Internet of Things. *Journal of Network and Computer Applications,* 64: 137–153, April 2016.

2 A. Dastjerdi, and R. Buyya. Fog computing: Helping the Internet of Things realize its potential. *Computer,* 49(8): 112–116, August 2016.

3 M. I. Pramanik, R. Lau, H. Demirkan, and M. A. KalamAzad. Smart health: Big data enabled health paradigm within smart cities. *Expert Systems with Applications* 87: 370–383, November 2017.

4 M. Chiang and T. Zhang. Fog and IoT: An overview of research opportunities. *IEEE Internet of Things Journal,* 3(6): 854–864, December 2016.

5 R. Lu, K. Heung, A. H. Lashkari, and A. A. Ghorbani. A lightweight privacy-preserving data aggregation scheme for fog computing-enhanced IoT. *IEEE Access* 5: 3302–3312, March 2017.

6 M. Taneja, and A. Davy. Resource aware placement of data analytics platform in fog computing. *Cloud Futures: From Distributed to Complete Computing, Madrid, Spain,* October 18–20, 2016.

7 A. V. Dastjerdi, H. Gupta, R. N. Calheiros, S. K. Ghosh, R. Buyya. Fog computing: Principles, architectures, and applications. *Internet of Things: Principles and Paradigms,* R. Buyya, and A. V. Dastjerdi (Eds), ISBN: 978-0-12-805395-9, Todd Green, Cambridge, USA, 2016.

8 P. Charalampidis, E. Tragos, and A. Fragkiadakis. A fog-enabled IoT platform for efficient management and data collection. *2017 IEEE 22nd International Workshop on Computer Aided Modeling and Design of Communication Links and Networks (CAMAD),* Lund, Sweden, June 19–21, 2017.

9 A. Hamid, A. Diyanat, and A. Pourkhalili. MIST: Fog-based data analytics scheme with cost-efficient resource provisioning for IoT Crowdsensing Applications. *Journal of Network and Computer Applications* 82: 152–165, March 2017.

10 E. G. Renart, J. Diaz-Montes, and M. Parashar. Data-driven stream processing at the edge. *2017 IEEE 1st International Conference on Fog and Edge Computing (ICFEC),* Madrid, Spain, May 14–15, 2017.

11 J. He, J. Wei, K. Chen, Z. Tang, Y. Zhou, and Y. Zhang. Multi-tier fog computing with large-scale IoT data analytics for smart cities. *IEEE Internet of Things Journal*. Under publication, 2017.
12 J. Li, J. Jin, D. Yuan, M. Palaniswami, and K. Moessner. EHOPES: Data-centered fog platform for smart living. 2015 International Telecommunication Networks and Applications Conference (ITNAC), Sydney, Australia, November 18–20, 2015.
13 International Organization for Standardization. Systems and software engineering – Vocabulary. ISO/IEC/IEEE 24765:2010(E), December 2010.
14 R. Buyya, R. Calheiros, and A.V. Dastjerdi. *Big Data: Principles and Paradigms,* Todd Green, USA, 2016.
15 A. Karkouch, H. Mousannif, H. Al Moatassime, and T. Noel. Data quality in Internet of Things: A state-of-the-art survey. *Journal of Network and Computer Applications,* 73: 57–81, September 2016.
16 S. K. Sharma and X. Wang. Live data analytics with collaborative edge and cloud processing in wireless IoT networks. *IEEE Access,* 5: 4621–4635, March 2017.
17 M. Ma, P. Wang, and C. Chu. Data management for Internet of Things: Challenges, approaches and opportunities. *2013 IEEE and Internet of Things (iThings/CPSCom), IEEE International Conference on and IEEE Cyber, Physical and Social Computing Green Computing and Communications (GreenCom).* Beijing, China, August 20–23, 2013.
18 S. Sathe, T.G. Papaioannou, H. Jeung, and K. Aberer. A survey of model-based sensor data acquisition and management. *Managing and Mining Sensor Data,* C. C. Aggarwal (Eds.), Springer, Boston, MA, 2013.
19 S. R. Jeffery, G. Alonso, M. J. Franklin, W. Hong, and J. Widom. Declarative support for sensor data cleaning. *Pervasive Computing*. K.P. Fishkin, B. Schiele, P. Nixon, et al. (Eds.), 3968: 83–100. Springer, Berlin, Heidelberg, 2006.
20 N. Hung, H. Jeung, and K. Aberer. An evaluation of model-based approaches to sensor data compression. *IEEE Transactions on Knowledge and Data Engineering,* 25(11) (November): 2434–2447, 2012.
21 O. Ghorbel, A. Ayadi, K. Loukil, M.S. Bensaleh, and M. Abid. Classification data using outlier detection method in Wireless sensor networks. *2017 13th International Wireless Communications and Mobile Computing Conference (IWCMC),* Valencia, Spain, June 26–30, 2017.
22 F. E. White. *Data Fusion Lexicon.* Joint Directors of Laboratories, Technical Panel for C3, Data Fusion Sub-Panel, Naval Ocean Systems Center, San Diego, 1991.
23 M. M. Almasri and K. M. Elleithy. Data fusion models in WSNs: Comparison and analysis. *2014 Zone 1 Conference of the American Society for Engineering Education.* Bridgeport, USA, April 3–5, 2014.

24 B. Khaleghi, A. Khamis, F. O. Karray, and S.N. Razavi. Multisensor data fusion: A review of the state-of-the-art. *Information Fusion,* 14(1): 28–44, January 2013.

25 M.C. Florea, A.L. Jousselme, and E. Bosse. Fusion of Imperfect Information in the Unified Framework of Random Sets Theory. *Application to Target Identification.* Defence R&D Canada. Valcartier, Tech. Rep. ADA475342, 2007.

26 E.I. Gaura, J. Brusey, M. Allen, et al. Edge mining the Internet of Things. *IEEE Sensors Journal,* 13(10): 3816–3825, October 2013.

27 K. Bhargava, and S. Ivanov. Collaborative edge mining for predicting heat stress in dairy cattle. *2016 Wireless Days (WD).* Toulouse, France, March 23–25, 2016.

28 R. Yang, Q. Xu, M. H. Au, Z. Yu, H. Wang, and L. Zhou. Position based cryptography with location privacy: a step for fog computing. *Future Generation Computer Systems,* 78(2): 799–806, January 2018.

29 M. I. Naas, P. R. Parvedy, J. Boukhobza, J. Boukhobza, and L. Lemarchand. iFogStor: an IoT data placement strategy forF infrastructure. *2017 IEEE 1st International Conference on Fog and Edge Computing (ICFEC).* Madrid, Spain, May 14–15, 2017.

30 A.A. Rezaee, M. Yaghmaee, A. Rahmani, A.H. Mohajerzadeh. HOCA: Healthcare Aware Optimized Congestion Avoidance and control protocol for wireless sensor networks. *Journal of Network and Computer Applications,* 37: 216–228, January 2014.

31 A. A. Rezaee, M.Yaghmaee, A. Rahmani, and A. Mohajerzadeh. Optimized Congestion Management Protocol for Healthcare Wireless Sensor Networks. *Wireless Personal Communications,* 75(1): 11–34, March 2014.

32 S. M. Riazul Islam, D. Kwak, M.D.H. Kabir, M. Hossain, K.-S. Kwak. The Internet of Things for Health Care: A Comprehensive Survey. *IEEE Access* 3: 678–708, June 2015.

33 I. Lujic, V. De Maio, I. Brandic. Efficient edge storage management based on near real-time forecasts. *2017 IEEE 1st International Conference on Fog and Edge Computing (ICFEC).* Madrid, Spain, May 14–15, 2017.

34 B. Farahani, F. Firouzi, V. Chang, M. Badaroglu, N. Constant, and K. Mankodiya. Towards fog-driven IoT eHealth: Promises and challenges of IoT in medicine and healthcare. *Future Generation Computer Systems,* 78(2): 659–676, January 2018.

35 F. Alexander Kraemer, A. Eivind Braten, N. Tamkittikhun, and D. Palma. Fog computing in healthcare: A review and discussion. *IEEE Access,* 5: 9206–9222, May 2017.

36 B. Negash, T.N. Gia, A. Anzanpour, I. Azimi, M. Jiang, T. Westerlund, A.M. Rahmani, P. Liljeberg, and H. Tenhunen. Leveraging Fog Computing for Healthcare IoT. *Fog Computing in the Internet of Things.* A. Rahmani, P. Liljeberg, J.S. Preden, et al. (Eds.). Springer, Cham, 2018.

9

Predictive Analysis to Support Fog Application Deployment

Antonio Brogi, Stefano Forti, and Ahmad Ibrahim

9.1 Introduction

Connected devices are changing the way we live and work. In the next years, the Internet of Things (IoT) is expected to bring more and more intelligence around us, being embedded in or interacting with the objects that we will use daily. Self-driving cars, autonomous domotics systems, energy production plants, agricultural lands, supermarkets, healthcare, and embedded AI will more and more exploit devices and things that are an integral part of the Internet and of our existence without us being aware of them. CISCO foresees 50 billion connected entities (people, machines, and connected things) by 2020 [1], and estimates they will have generated around 600 zettabytes of information by that time, only 10% of which will be useful to some purpose [2]. Furthermore, cloud connection latencies are not adequate to host real-time tasks such as life-saving connected devices, augmented reality, or gaming [3]. In such a perspective, the need to provide processing power, storage, and networking capabilities to run IoT applications closer to sensors and actuators has been highlighted by various authors, such as [4, 5].

Fog computing [6] aims at selectively pushing computation closer to where data are produced, by exploiting a geographically distributed multitude of heterogeneous devices (e.g., gateways, micro-datacenters, embedded servers, personal devices) spanning the continuum from cloud to things. On one hand, this will enable low-latency responses to (and analysis of) sensed events, on the other hand, it will relax the need for (high) bandwidth availability from/to the cloud [7]. Overall, fog computing is expected to fruitfully extend the IoT+Cloud scenario, enabling quality-of-service- (QoS) and context-aware application deployments [5].

Modern applications usually consist of many independently deployable components – each with its hardware, software, and IoT requirements – that interact together distributedly. Such interactions may have stringent QoS

Fog and Edge Computing: Principles and Paradigms, First Edition.
Edited by Rajkumar Buyya and Satish Narayana Srirama.

requirements – typically on latency and bandwidth – to be fulfilled for the deployed application to work as expected [3]. If some application components (i.e., functionalities) are naturally suited to the cloud (e.g., service back-ends) and others are naturally suited to edge devices (e.g., industrial control loops), there are some applications for which functionality segmentation is not as straightforward (e.g., short- to medium-term analytics). Supporting deployment decision in the fog also requires comparing a multitude of offerings where providers can deploy applications to their infrastructure integrated with the cloud, with the IoT, with federated fog devices as well as with user-managed devices. Moreover, determining deployments of a multi-component application to a given fog infrastructure, while satisfying its functional and nonfunctional constraints, is an NP-hard problem [8].

As highlighted in [9], novel fog architectures call for modeling complex applications and infrastructures, based on accurate models of application deployment and behavior to predict their runtime performance, also relying on historical data [10]. Algorithms and methodologies are to be devised to help deciding how to map each application functionality (i.e., component) to a substrate of heterogeneously capable and variably available nodes [11].

All of this, including node mobility management, and taking into account possible variations in the QoS of communication links (fluctuating bandwidth, latency, and jitter over time) supporting component–component interactions as well as the possibility for deployed components to remotely interact with the IoT via proper interfaces [10]. Moreover, other orthogonal constraints such as QoS-assurance, operational cost targets, and administration or security policies should be considered when selecting candidate deployments.

Clearly, manually determining fog applications (re-)deployments is a time-consuming, error-prone, and costly operation, and deciding which deployments may perform better – without enacting them – is difficult. Modern enterprise IT is eager for tools that permit to virtually compare business scenarios and to design both pricing schemes and SLAs by exploiting *what-if* analyses [12] and predictive methodologies, abstracting unnecessary details.

In this chapter, we present an extended version of FogTorchΠ [13, 14], a prototype based on a model of the IoT+Fog+Cloud scenario to support application deployment in the Fog. FogTorchΠ permits to express processing capabilities, QoS attributes (viz., latency and bandwidth) and operational costs (i.e., costs of virtual instances, sensed data) of a fog infrastructure, along with processing and QoS requirements of an application. In short, FogTorchΠ:

1. Determines the deployments of an application over a fog infrastructure that meet all application (processing, IoT and QoS) requirements.
2. Predicts the *QoS-assurance* of such deployments against variations in the latency and bandwidth of communications links.

3. Returns an estimate of the *fog resource consumption* and a *monthly cost* of each deployment.

Overall, the current version of FogTorchΠ features (i) *QoS-awareness* to achieve latency reduction, bandwidth savings and to enforce business policies, (ii) *context-awareness* to suitably exploit both local and remote resources, and (iii) *cost-awareness* to determine the most cost-effective deployments among the eligible ones.

FogTorchΠ models the QoS of communication links by using probability distributions (based on historical data), describing variations in featured latency or bandwidth over time, depending on network conditions. To handle input probability distributions and to estimate the QoS assurance of different deployments, FogTorchΠ exploits the Monte Carlo method [17]. FogTorchΠ also exploits a novel cost model that extends existing pricing schemes for the cloud to fog computing scenarios, whilst introducing the possibility of integrating such schemes with financial costs that originate from the exploitation of IoT devices (sensing-as-a-service [15] subscriptions or data transfer costs) in the deployment of applications. We show and discuss how predictive tools like FogTorchΠ can help IT experts in deciding how to distribute application components over fog infrastructures in a QoS- and context-aware manner, also considering cost of fog application deployments.

The rest of this chapter is organized as follows. After introducing a motivating example of a smart building application (Section 9.2), we describe FogTorchΠ predictive models and algorithms (Section 9.3). Then, we present and discuss the results obtained by applying FogTorchΠ to the motivating example (Section 9.4). Afterward, some related work and a comparison with the iFogSim simulator are provided (Section 9.5). Finally, we highlight future research directions (Section 9.6) and draw some concluding remarks (Section 9.7).

9.2 Motivating Example: Smart Building

Consider a simple fog application (Figure 9.1) that manages fire alarm, heating and air conditioning systems, interior lighting, and security cameras of a smart building. The application consists of three microservices:

1. IoTController, interacting with the connected cyber-physical systems,
2. DataStorage, storing all sensed information for future use and employing machine learning techniques to update sense-act rules of the IoTController so to optimize heating and lighting management based on previous experience and/or on people behavior, and
3. Dashboard, aggregating and visualizing collected data and videos, as well as allowing users to interact with the system.

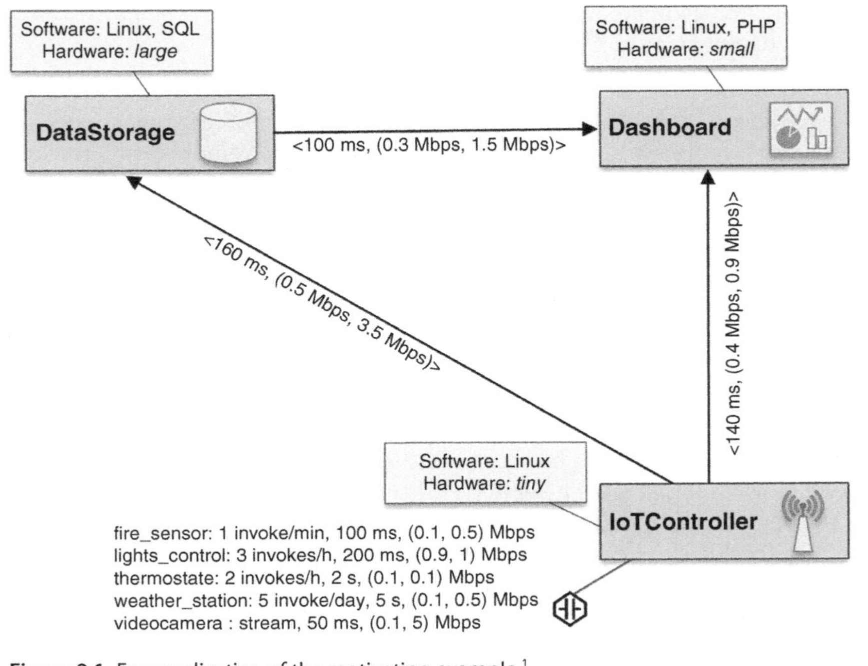

Figure 9.1 Fog application of the motivating example.[1]

Each microservice represents an independently deployable component of the application [16] and has hardware and software requirements in order to function properly (as indicated in the gray boxes associated with components in Figure 9.1). Hardware requirements are expressed in terms of the virtual machine (VM) types[2] listed in Table 9.1 and must be fulfilled by the VM that will host the deployed component.

Application components must cooperate so that well-defined levels of service are met by the application. Hence, communication links supporting component-component interactions should provide suitable end-to-end latency and bandwidth (e.g., the IoTController should reach the DataStorage within 160 ms and have at least 0.5 Mbps download and 3.5 Mbps upload free bandwidths). Component–things interactions have similar constraints, and also specify the sampling rates at which the IoTController is expected to query things at runtime (e.g., the IoTController should reach a fire_sensor, queried once per minute, within 100 ms having at least 0.1 Mbps download and 0.5 Mbps upload free bandwidths).

1 Links are labelled with the QoS required to support them in terms of latency and download/upload bandwidth. Arrows on the links indicate the upload direction.

2 Adapted from Openstack Mitaka flavors: https://docs.openstack.org/.

Table 9.1 Hardware specification for different VM types.

VM Type	vCPUs	RAM (GB)	HDD (GB)
tiny	1	1	10
small	1	2	20
medium	2	4	40
large	4	8	80
xlarge	8	16	160

Figure 9.2 shows the infrastructure – two cloud data centers, three fog nodes, and nine things – selected by the system integrators in charge of deploying the smart building application for one of their customers. The deployed application will have to exploit all the things connected to Fog 1 and the weather_station_3 at Fog 3. Furthermore, the customer owns Fog 2, what makes deploying components to that node cost-free.

All fog and cloud nodes are associated with pricing schemes, either to lease an instance of a certain VM type (e.g., a *tiny* instance at Cloud 2 costs €7 per month), or to build on-demand instances by selecting the required number of cores and the needed amount of RAM and HDD to support a given component.

Fog nodes offer software capabilities, along with consumable (i.e., RAM, HDD) and non-consumable (i.e., CPUs) hardware resources. Similarly, cloud nodes offer software capabilities, while cloud hardware is considered unbounded, assuming that, when needed, one can always purchase extra instances.

Finally, Table 9.2 lists the QoS of the end-to-end communication links supported by the infrastructure of Figure 9.2, connecting fog and cloud nodes. They are represented as probability distributions based on real data[3], to account for variations in the QoS they provide. Mobile communication links at Fog 2 initially feature a 3G Internet access. As per the current technical proposals (e.g., [6] and [10]), we assume fog and cloud nodes are able to access directly connected things as well as things at neighboring nodes via a specific middleware layer (through the associated communication links).

Planning to sell the deployed solution for €1,500 a month, the system integrators set the limit of the monthly deployment cost at €850. Also, the customer requires the application to be compliant with the specified QoS requirements at least 98% of the time. Then, interesting questions for the system integrators before the first deployment of the application are, for instance:

Q1(a) — *Is there any eligible deployment of the application reaching all the needed things at* Fog 1 *and* Fog 3, *and meeting the aforementioned financial*

3 Satellite: https://www.eolo.it, 3G/4G: https://www.agcom.it, VDSL: http://www.vodafone.it.

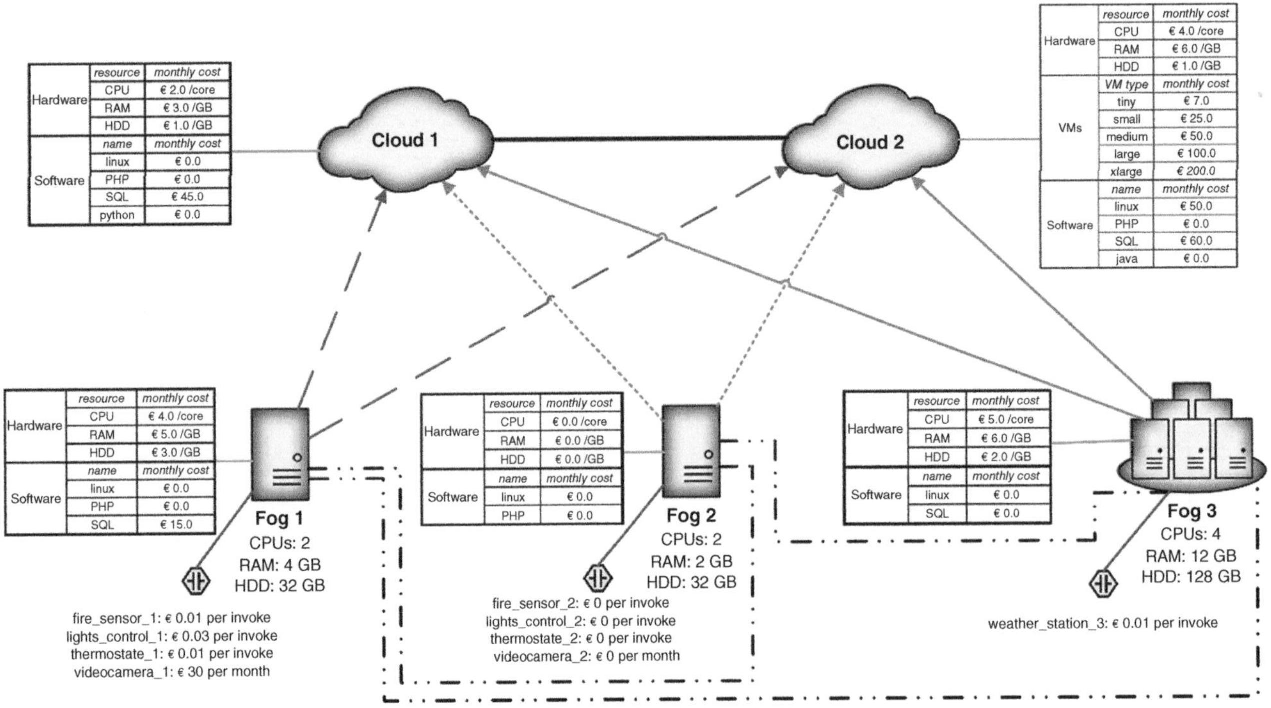

Figure 9.2 Fog infrastructure of the motivating example.[4]

Table 9.2 QoS profiles associated to the communication links.

Dash Type	Profile	Latency	Download	Upload
—— ——	Satellite 14M	40 ms	98%: 10.5 Mbps 2%: 0 Mbps	98%: 4.5 Mbps 2%: 0 Mbps
- - - - - - - - - -	3G	54 ms	99.6%: 9.61 Mbps 0.4%: 0 Mbps	99.6%: 2.89 Mbps 0.4%: 0 Mbps
	4G	53 ms	99.3%: 22.67 Mbps 0.7%: 0 Mbps	99.4%: 16.97 Mbps 0.6%: 0 Mbps
————	VDSL	60 ms	60 Mbps	6 Mbps
━━━━	Fiber	5 ms	1000 Mbps	1000 Mbps
— · · — ·	WLAN	15 ms	90%: 32 Mbps 10%: 16 Mbps	90%: 32 Mbps 10%: 16 Mbps

(at most € 850 per month) and QoS-assurance (at least 98% of the time) constraints?

Q1(b) — *Which eligible deployments minimize resource consumption in the fog layer so to permit future deployment of services and sales of virtual instances to other customers?*

Suppose also that with an extra monthly investment of €20, system integrators can exploit a 4G connection at Fog 2. Then:

Q2 — *Would there be any deployment that complies with all previous requirements and reduces financial cost and/or consumed fog resources if upgrading from 3G to 4G at* Fog 2?

In Section 9.4, we will show how FogTorchΠ can be exploited to obtain answers to all the above questions.

9.3 Predictive Analysis with FogTorchΠ

9.3.1 Modeling Applications and Infrastructures

FogTorchΠ [13, 14] is an open-source Java prototype[5] (based on the model presented in [8]) that determines eligible QoS-, context- and cost-aware multi-component application deployments to fog infrastructures.

4 Arrows on the links in Figure 9.2 indicate the upload direction.

5 Freely available at https://github.com/di-unipi-socc/FogTorchPI/tree/multithreaded/.

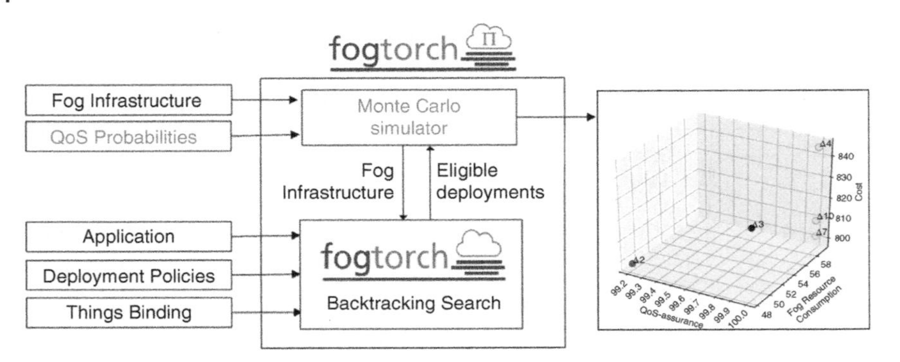

Figure 9.3 Bird's-eye view of FogTorchΠ.

FogTorchΠ input consists of the following:

1. **Infrastructure I.** The input includes a description of an *infrastructure* I specifying the IoT devices, the fog nodes, and the cloud data centers available for application deployment (each with its hardware and software capabilities), along with the probability distributions of the QoS (viz., latency, bandwidth) featured by the available (cloud-to-fog, fog-to-fog and fog-to-things) end-to-end communication links[6] and cost for purchasing sensed data and for cloud/fog virtual instances.
2. **Multicomponent application A.** This specifies all hardware (e.g., CPU, RAM, storage), software (e.g., OS, libraries, frameworks) and IoT requirements (e.g., which type of things to exploit) of each component of the application, and the QoS (i.e., latency and bandwidth) needed to adequately support component–component and component–thing interactions once the application has been deployed.
3. **Things binding ϑ.** This maps each IoT requirement of an application component to an actual thing available in I.
4. **Deployment policy δ.** The deployment policy white-lists the nodes where A components can be deployed[7] according to security or business-related constraints.

Figure 9.3 offers a bird's-eye view of FogTorchΠ, with the input to the tool on the left-hand side and its output on the right-hand side. In the next sections we will present the backtracking search exploited by FogTorchΠ to determine the *eligible deployments*, the models used to estimate the *fog resource consumption* and *cost* of such deployments, and the Monte Carlo method [17] employed

6 Actual implementations in fog landscapes can exploit data from monitoring tools (e.g., [51, 52]) to get updated information on the state of the infrastructure I.
7 When δ is not specified for a component γ of A, γ can be deployed to any compatible node in I.

to assess their *QoS-assurance* against variations in the latency and bandwidth featured by communication links.

9.3.2 Searching for Eligible Deployments

Based on the input described in Section 9.3.1, FogTorchΠ determines all the eligible deployments of the components of an application A to cloud or fog nodes in an infrastructure I.

An *eligible deployment* Δ maps each component γ of A to a cloud or fog node n in I so that:

1. n complies with the specified deployment policy δ (viz., $n \in \delta(\gamma)$) and it satisfies the hardware and software requirements of γ,
2. The things specified in the Things binding ϑ are all reachable from node n (either directly or through a remote end-to-end link),
3. The hardware resources of n are enough to deploy *all* components of A mapped to n, and
4. The component–component and component–thing interactions QoS requirements (on latency and bandwidth) are all satisfied.

To determine the eligible deployments of a given application A to a given infrastructure I (complying with both ϑ and δ), FogTorchΠ exploits a backtracking search, as illustrated in the algorithm of Figure 9.4. The preprocessing step (line 2) builds, for each software component $\gamma \in A$, the dictionary $K[\gamma]$ of fog and cloud nodes that satisfy conditions (1) and (2) for a deployment to be eligible, and also meet the latency requirements for the things requirements of component γ. If there exists even one component for which $K[\gamma]$ is empty, the algorithm immediately returns an empty set of deployments (lines 3–5). Overall, the preprocessing completes in $O(N)$ time, with N being the number of available cloud and fog nodes, having to check the capabilities of at most all fog and cloud nodes for each application component.

The call to the BACKTRACKSEARCH($D, \Delta, A, I, K, \vartheta$) procedure (line 7) inputs the result of the preprocessing and looks for eligible deployments. It visits a

```
1: procedure FINDDEPLOYMENTS(A, I, δ, ϑ)
2:     K ← PREPROCESS(A, I, δ, ϑ)
3:     if K = failure then
4:         return ∅                          ▷ ∃γ ∈ A s.t. K[γ] = ∅
5:     end if
6:     D ← ∅
7:     BACKTRACKSEARCH(D, ∅, A, I, K, ϑ)
8:     return D
9: end procedure
```

Figure 9.4 Pseudocode of the exhaustive search algorithm.

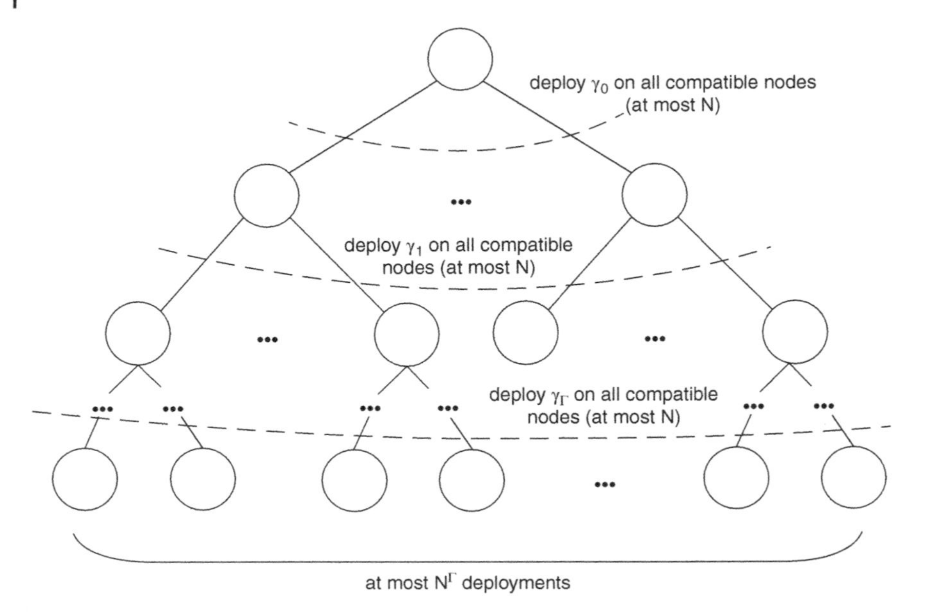

Figure 9.5 Search space to find eligible deployments of *A* to *I*.

(finite) search space tree having at most N nodes at each level and height equal to the number Γ of components of A. As sketched in Figure 9.5, each node in the search space represents a (partial) deployment Δ, where the number of deployed components corresponds to the level of the node. The root corresponds to an empty deployment, nodes at level i are partial deployments of i components, and leaves at level Γ contain complete eligible deployments. Edges from one node to another represent the action of deploying a component to some fog or cloud node. Thus, search completes in $O(N^\Gamma)$ time.

At each recursive call, BACKTRACKSEARCH (D, Δ, A, I, K, ϑ) first checks whether all components of A have been deployed by the currently attempted deployment and, if so, it adds the found deployment to set D (lines 2–3) as listed in Figure 9.6, and returns to the caller (line 4). Otherwise, it selects a component still to be deployed (SELECTUNDEPLOYEDCOMPONENT (Δ, A)) and it attempts to deploy it to a node chosen in $K[\gamma]$ (SELECTDEPLOYMENTNODE($K[\gamma]$, A). The ISELIGIBLE (Δ, γ, n, A, I, ϑ) procedure (line 8) checks conditions (3) and (4) for a deployment to be eligible and, when they hold, the DEPLOY (Δ, γ, n, A, I, ϑ) procedure (line 9) decreases the available hardware resources and bandwidths in the infrastructure, according to the new deployment association. UNDEPLOY (Δ, γ, n, I, A, ϑ) (line 11) performs the inverse operation of DEPLOY (Δ, γ, n, A, I, ϑ), releasing resources and freeing bandwidth when backtracking on a deployment association.

```
1: procedure BACKTRACKSEARCH(D, Δ, A, I, K, ϑ)
2:     if ISCOMPLETE(Δ) then
3:         ADD(Δ, D)
4:         return
5:     end if
6:     γ ← SELECTUNDEPLOYEDCOMPONENT(Δ, A);
7:     for all n ∈ SELECTDEPLOYMENTNODE(K[γ], A) do
8:         if ISELIGIBLE(Δ, γ, n, A, I, ϑ) then
9:             DEPLOY(Δ, γ, n, A, I, ϑ)
10:            BACKTRACKSEARCH(D, Δ, A, I, K, ϑ)
11:            UNDEPLOY(Δ, γ, n, I, A, ϑ)
12:        end if
13:    end for
14: end procedure
```

Figure 9.6 Pseudocode for the backtracking search.

9.3.3 Estimating Resource Consumption and Cost

Procedure FINDDEPLOYMENTS(A, I, δ, ϑ) computes an estimate of fog resource consumption and of the monthly cost[8] of each given deployment.

The *fog resource consumption* that is output by FogTorchΠ indicates the aggregated averaged percentage of consumed RAM and storage in the set of fog nodes[9] F, considering all deployed application components $\gamma \in A$. Overall, resource consumption is computed as the average

$$\frac{1}{2}\left(\frac{\sum_{\gamma \in A} RAM(\gamma)}{\sum_{f \in F} RAM(f)} + \frac{\sum_{\gamma \in A} HDD(\gamma)}{\sum_{f \in F} HDD(f)}\right)$$

where $RAM(\gamma)$, $HDD(\gamma)$ indicate the amount of resources needed by component γ, and $RAM(f)$, $HDD(f)$ are the total amount of such resources available at node f.

To compute the estimate of the monthly cost of deployment for application A to infrastructure I, we propose a novel cost model that extends to fog computing previous efforts in cloud cost modelling [18] and includes costs due to IoT [19], and software costs.

At any cloud or fog node n, our cost model considers that a hardware offering H can be either a *default VM* (Table 9.1) offered at a fixed monthly fee or an *on-demand VM* (built with an arbitrary amount of cores, RAM and HDD). Being R the set of resources considered when building on-demand VMs

8 Cost computation is performed *on-the-fly* during the search step, envisioning the possibility to exploit cost as a heuristic to lead the search algorithm toward a best-candidate deployment.
9 The actual implementation of FogTorchΠ permits to choose a subset of all the available fog nodes in I on which to compute fog resource consumption.

(viz., $R = \{CPU, RAM, HDD\}$), the estimated monthly cost for a hardware offering H at node n is:

$$p(H,n) = \begin{cases} c(H,n) & \text{if } H \text{ is a default } VM \\ \sum_{\rho \in R} [H.\rho \times c(\rho,n)] & \text{if } H \text{ is an on-demand } VM \end{cases}$$

where $c(H, n)$ is the monthly cost of a default VM offering H at fog or cloud node n, while $H.\rho$ indicates the amount of resources $\rho \in R$ used by[10] the on-demand VM represented by H, and $c(\rho, n)$ is the unit monthly cost at n for resource ρ.

Analogously, for any given cloud or fog node n, a software offering S can be either a predetermined software bundle or an on-demand subset of the software capabilities available at n (each sold separately). The estimated monthly cost for S at node n is:

$$p(S,n) = \begin{cases} c(S,n) & \text{if S is a bundle} \\ \sum_{s \in S} c(s,n) & \text{if S is on-demand} \end{cases}$$

where $c(S, n)$ is the cost for the software bundle S at node n, and $c(s, n)$ is the monthly cost of a single software s at n.

Finally, in sensing-as-a-service [15] scenarios, a thing offering T exploiting an actual thing t can be offered at a monthly subscription fee or through a pay-per-invocation mechanism. Then, the cost of offering T at thing t is:

$$p(T,t) = \begin{cases} c(T,t) & \text{if T is subscription based} \\ T.k \times c(t) & \text{if T is pay-per-invocation} \end{cases}$$

where $c(T, t)$ is the monthly subscription fee for T at t, while $T.k$ is the number of monthly invocations expected over t, and $c(t)$ is the cost per invocation at t (including thing usage and/or data transfer costs).

Assume that Δ is an eligible deployment for an application A to an infrastructure I, as introduced in Section 9.3. In addition, let $\gamma \in A$ be a component of the considered application A, and let $\gamma.\overline{\mathcal{H}}$, $\gamma.\overline{\Sigma}$ and $\gamma.\overline{\Theta}$ be its hardware, software and things requirements, respectively. Overall, the expected monthly cost for a given deployment Δ can be first approximated by combining the previous pricing schemes as:

$$cost(\Delta, \vartheta, A) = \sum_{\gamma \in A} \left[p\left(\gamma.\overline{\mathcal{H}}, \Delta(\gamma)\right) + p\left(\gamma.\overline{\Sigma}, \Delta(\gamma)\right) + \sum_{r \in \gamma \cdot \overline{\Theta}} p(r, \vartheta r) \right]$$

10 Capped by the maximum amount purchasable at any chosen cloud or fog node.

Although the above formula provides an estimate of the monthly cost for a given deployment, it does not feature a way to select the "best" offering to match the application requirements at the VM, software and IoT levels. Particularly, it may lead the choice always to on-demand and pay-per-invocation offerings when the application requirements do not match exactly default or bundled offerings, or when a cloud provider does not offer a particular VM type (e.g., starting its offerings from *medium*). This can lead to overestimating the monthly deployment cost.

For instance, consider the infrastructure of Figure 9.2 and the hardware requirements of a component to be deployed to Cloud 2, specified as $R = \{CPU : 1, RAM : 1GB, HDD : 20GB\}$. Since no exact matching exists between the requirements R and the offerings at Cloud 2, this first cost model would select an on-demand instance, and estimate its cost of €30.[11] However, Cloud 2 also provides a *small* instance that can satisfy the requirements at a (lower) cost of €25.

Since larger VM types always satisfy smaller hardware requirements, bundled software offerings may satisfy multiple software requirements at a lower price, and subscription-based thing offerings can be more or less convenient, depending on the number of invocations on a given thing, some policy must be used to choose the "best" offerings for each software, hardware and thing requirement of an application component. In what follows, we refine our cost model to also account for this.

A *requirement-to-offering matching policy* $p_m(r, n)$ matches hardware or software requirements r of a component ($r \in \{\gamma.\overline{\mathcal{H}}, \gamma.\overline{\Sigma}\}$) to the estimated monthly cost of the offering that will support them at cloud or fog node n, and a thing requirement $r \in \gamma.\overline{\Theta}$ to the estimated monthly cost of the offering that will support r at thing t.

Overall, this refined version of the cost model permits an estimation of the monthly cost of Δ including a cost-aware matching between application requirements and infrastructure offering (for hardware, software and IoT), chosen as per p_m. Hence:

$$cost(\Delta, \vartheta, A) = \sum_{\gamma \in A} \left[p_m\left[\gamma.\overline{\mathcal{H}}, \Delta(\gamma)\right] + p_m\left[\gamma.\overline{\Sigma}, \Delta(\gamma)\right] + \sum_{r \in \gamma \cdot \overline{\Theta}} p_m\left[r, \vartheta(r)\right] \right]$$

The current implementation of FogTorchΠ exploits a *best-fit lowest-cost* policy for choosing hardware, software and thing offerings. Indeed, it selects the cheapest between the first default VM (from *tiny* to *xlarge*) that can support $\gamma.\overline{\mathcal{H}}$ at node n and the on-demand offering built as per $\gamma.\overline{\mathcal{H}}$. Likewise, software requirements in $\gamma.\overline{\Sigma}$ are matched with the cheapest compatible version

11 €30 = 1 CPU x €4/core + 1 GB RAM x €6/GB + 20 GB HDD x €1/GB

```
 1: procedure MONTECARLO(A, I, ϑ, δ, n)
 2:     D ← ∅                         ▷ dictionary of ⟨Δ, counter⟩
 3:     parallel for n times
 4:         I_s ← SAMPLELINKSQOS(I)
 5:         E ← FINDDEPLOYMENTS(A, I_s, ϑ, δ)
 6:         D ← UNIONUPDATE(D, E)
 7:     end parallel for
 8:     for Δ ∈ keys(D) do
 9:         D[Δ] ← D[Δ]/n
10:     end for
11:     return D
12: end procedure
```

Figure 9.7 Pseudocode of the Monte Carlo simulation in FogTorchΠ.

available at n, and thing per invocation offer is compared to monthly subscription so to select the cheapest.[12]

Formally, the cost model used by FogTorchΠ can be expressed as:

$$p_m(\overline{\mathcal{H}}, n) = \min\{p(H, n)\} \forall H \in \{\text{default VMs, on-demand VM}\} \wedge H \models \overline{\mathcal{H}}$$
$$p_m(\overline{\Sigma}, n) = \min\{p(S, n)\} \forall S \in \{\text{on-demand, bundle}\} \wedge S \models \overline{\Sigma}$$
$$p_m(r, t) = \min\{p(T, t)\} \ \forall\ T \in \{\text{subscription, pay-per-invocation}\} \wedge T \models r$$

where $O \models R$ reads as offering O *satisfies requirements* R.

It is worth noting that the proposed cost model separates the cost of purchasing VMs from the cost of purchasing the software. This choice keeps the modelling general enough to include both IaaS and PaaS Cloud offerings. Furthermore, even if we referred to VMs as the only deployment unit for application components, the model can be easily extended so to include other types of virtual instances (e.g., containers).

9.3.4 Estimating QoS-Assurance

In addition to fog resource consumption and cost, FogTorchΠ outputs an estimate of the QoS-assurance of output deployments. FogTorchΠ exploits the algorithms described in Section 9.3.2 and parallel Monte Carlo simulations to estimate the QoS-assurance of output deployments, by aggregating the eligible deployments obtained when varying the QoS featured by the end-to-end communication links in I (as per the given probability distributions). Figure 9.7 lists the pseudocode of FogTorchΠ overall functioning.

First, an empty (thread-safe) dictionary D is created to contain key-value pairs $\langle\Delta$, `counter`$\rangle$, where the key (Δ) represents an eligible deployment and the value (`counter`) keeps track of how many times Δ will be generated during the Monte Carlo simulation (line 2). Then, the overall number n of Monte

---

12 Other policies are also possible such as, for instance, selecting the largest offering that can accommodate a component, or always increasing the component's requirements by some percentage (e.g., 10% before selecting the matching.

```
1: p ∈ [0, 1] ∧ q, q' ∈ Q
2: procedure SamplingFunction(p, q, q')
3:     r ← RandomDoubleInRange(0,1)
4:     if r ⩽ p then
5:         return q
6:     else
7:         return q'
8:     end if
9: end procedure
```

Figure 9.8 Bernoulli sampling function example.

Carlo runs is divided by the number w of available worker threads[13], each executing $n_w = \lceil n/w \rceil$ runs in a parallel *for* loop, modifying its own (local) copy of I (lines 4–6). At the beginning of each run of the simulation, each worker thread samples a state I_s of the infrastructure following the probability distributions of the QoS of the communication links in I (line 4).

The function FindDeployments(A, I_s, ϑ, δ) (line 5) is the exhaustive (backtracking) search of Section 9.3.2 to determine the set E of eligible deployments Δ of A to I_s, i.e. deployments of A that satisfy all processing and QoS requirements in that state of the infrastructure. The objective of this step is to look for eligible deployments, whilst dynamically simulating changes in the underlying network conditions. An example of sampling function that can be used to sample links QoS is shown in Figure 9.8, however, FogTorchΠ supports arbitrary probability distributions.

At the end of each run, the set E of eligible deployments of A to I_s is merged with D as shown in Figure 9.7. The function UnionUpdate (D, E) (line 6) updates D by adding deployments $\langle \Delta, 1 \rangle$ discovered during the last run ($\Delta \in E \setminus keys(D)$) and by incrementing the `counter` of those deployments that had already been found in a previous run ($\Delta \in E \cap keys(D)$).

After the parallel *for* loop is over, the output *QoS-assurance* of each deployment Δ is computed as the percentage of runs that generated Δ. Indeed, the more a deployment is generated during the simulation, the more it is likely to meet all desired QoS constraints in the actual infrastructure at varying QoS. Thus, at the end of the simulation ($n \geq 100,000$), the QoS-assurance of each deployment $\Delta \in keys(D)$ is computed by dividing the `counter` associated to Δ by n (lines 8–10), i.e. by estimating how likely each deployment is to meet QoS constraints of A, considering variations in the communication links as per historical behavior of I. Finally, dictionary D is returned (line 11).

In the next section, we describe the results of FogTorchΠ running over the smart building example of Section 9.2 to get answers to the questions of the system integrators.

13 The number of available worker threads can be set equal to the available physical or logical processors on the machine running FogTorchΠ.

9.4 Motivating Example (continued)

In this section, we exploit FogTorchΠ to address the questions raised by the system integrators in the smart building example of Section 9.2. FogTorchΠ outputs the set of eligible deployments along with their estimated QoS-assurance, fog resource consumption and monthly cost as per Section 9.3.

For question **Q1(a)** and **Q1(b)**:

Q1(a) — *Is there any eligible deployment of the application reaching all the needed things at* Fog 1 *and* Fog 3, *and meeting the aforementioned financial (at most € 850 per month) and QoS-assurance (at least 98% of the time) constraints?*

Q1(b) — *Which eligible deployments minimize resource consumption in the fog layer so as to permit future deployment of services and sales of virtual instances to other customers?*

FogTorchΠ outputs 11 eligible deployments (Δ1 — Δ11 in Table 9.3).

Table 9.3 Eligible deployments generated by FogTorchΠ for Q1 and Q2.[14]

Dep. ID	IoTController	DataStorage	Dashboard
Δ1	Fog 2	Fog 3	Cloud 2
Δ2	Fog 2	Fog 3	Cloud 1
Δ3	Fog 3	Fog 3	Cloud 1
Δ4	Fog 2	Fog 3	Fog 1
Δ5	Fog 1	Fog 3	Cloud 1
Δ6	Fog 3	Fog 3	Cloud 2
Δ7	Fog 3	Fog 3	Fog 2
Δ8	Fog 3	Fog 3	Fog 1
Δ9	Fog 1	Fog 3	Cloud 2
Δ10	Fog 1	Fog 3	Fog 2
Δ11	Fog 1	Fog 3	Fog 1
Δ12	Fog 2	Cloud 2	Fog 1
Δ13	Fog 2	Cloud 2	Cloud 1
Δ14	Fog 2	Cloud 2	Cloud 2
Δ15	Fog 2	Cloud 1	Cloud 2
Δ16	Fog 2	Cloud 1	Cloud 1
Δ17	Fog 2	Cloud 1	Fog 1

14 Results and Python code to generate 3D plots as in Figures. 13.4 and 13.5 are available at: https://github.com/di-unipi-socc/FogTorchPI/tree/multithreaded/results/SMARTBUILDING18.

It is worth recalling that we envision remote access to things connected to fog nodes from other cloud and fog nodes. In fact, some output deployments map components to nodes that do not directly connect to all the required things. For instance, in the case of $\Delta 1$, IoTController is deployed to Fog 2 but the required Things (fire_sensor_1, light_control_1, thermostate_1, video_camera_1, weather_station_3) are attached to Fog 1 and Fog 3, still being reachable with suitable latency and bandwidth.

Figure 9.9 only shows the five output deployments that satisfy the QoS and budget constraints imposed by the system integrators. $\Delta 3$, $\Delta 4$, $\Delta 7$ and $\Delta 10$ all feature 100% QoS-assurance. Among them, $\Delta 7$ is the cheapest in terms of cost, consuming as much fog resources as $\Delta 4$ and $\Delta 10$, although more with respect to $\Delta 3$. On the other hand, $\Delta 2$, still showing QoS-assurance above 98% and consuming as much fog resources as $\Delta 3$, can be a good compromise at the cheapest monthly cost of € 800, what answers question **Q1(b)**.

Finally, to answer question **Q2**:

Q2 — *Would there be any deployment that complies with all previous requirements and reduces financial cost and/or consumed fog resources if upgrading from 3G to 4G at* Fog 2?

we change the Internet access at Fog 2 from 3G to 4G. This increases the monthly expenses by € 20. Running FogTorchΠ now reveals six new eligible deployments ($\Delta 12$ — $\Delta 17$) in addition to the previous output. Among those, only $\Delta 16$ turns out to meet also the QoS and budget constraints that the system integrators require (Figure 9.10). Interestingly, $\Delta 16$ costs € 70 less than the best candidate for **Q1(b)** ($\Delta 2$), whilst sensibly reducing fog resource consumption. Hence, overall, the change from 3G to 4G would lead to an estimated monthly saving of € 50 with $\Delta 16$ with respect to $\Delta 2$.

The current FogTorchΠ prototype leaves to system integrators the final choice for a particular deployment, permitting them to freely select the "best" trade-off among QoS-assurance, resource consumption and cost. Indeed, the analysis of application-specific requirements (along with data on infrastructure behavior) can lead decision toward different segmentations of an application from the IoT to the cloud, trying to determine the best trade-off among metrics that describe likely runtime behavior of a deployment and make it possible to evaluate changes in the infrastructure (or in the application) before their actual implementation (*what-if* analysis [12]).

9.5 Related Work

9.5.1 Cloud Application Deployment Support

The problem of deciding how to deploy multicomponent applications has been thoroughly studied in the cloud scenario. Projects like SeaClouds [20],

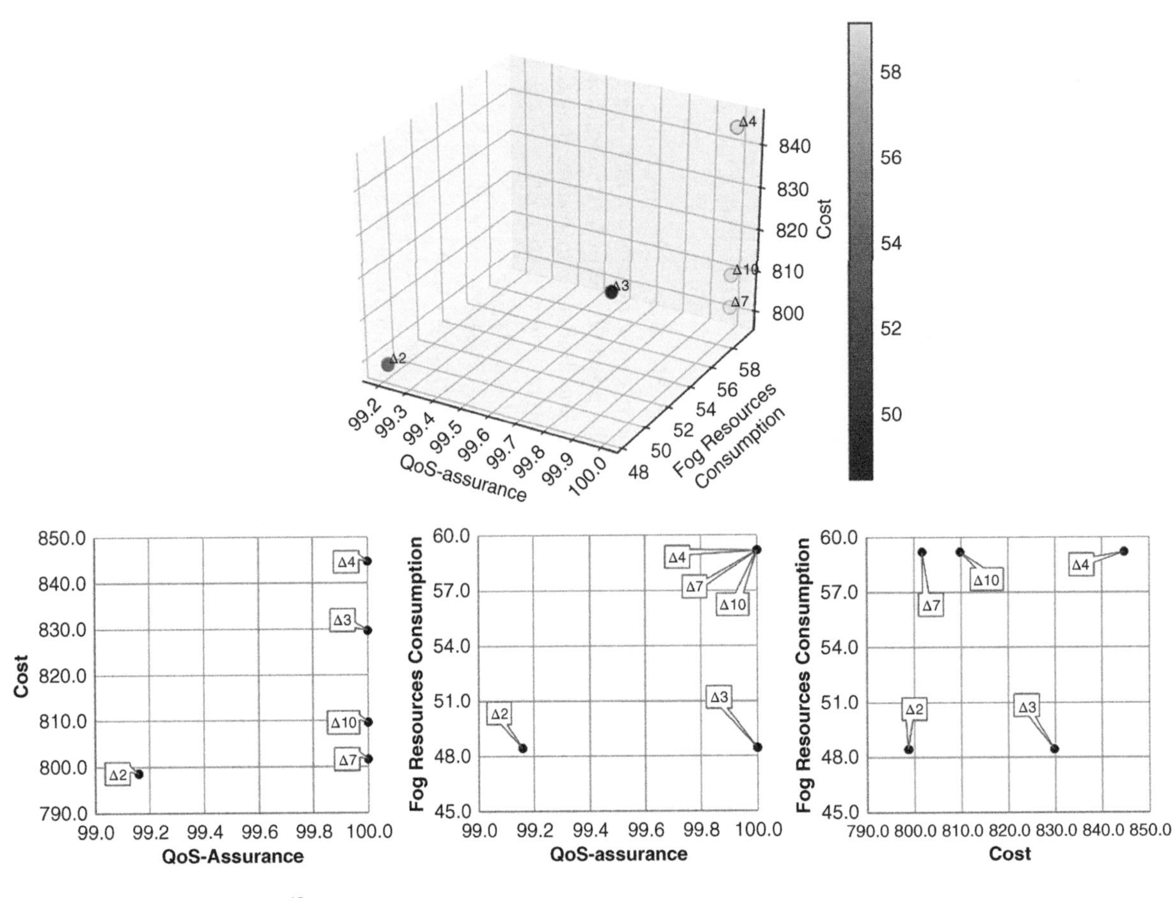

Figure 9.9 Results for Q1(a) and Q1(b).[15]

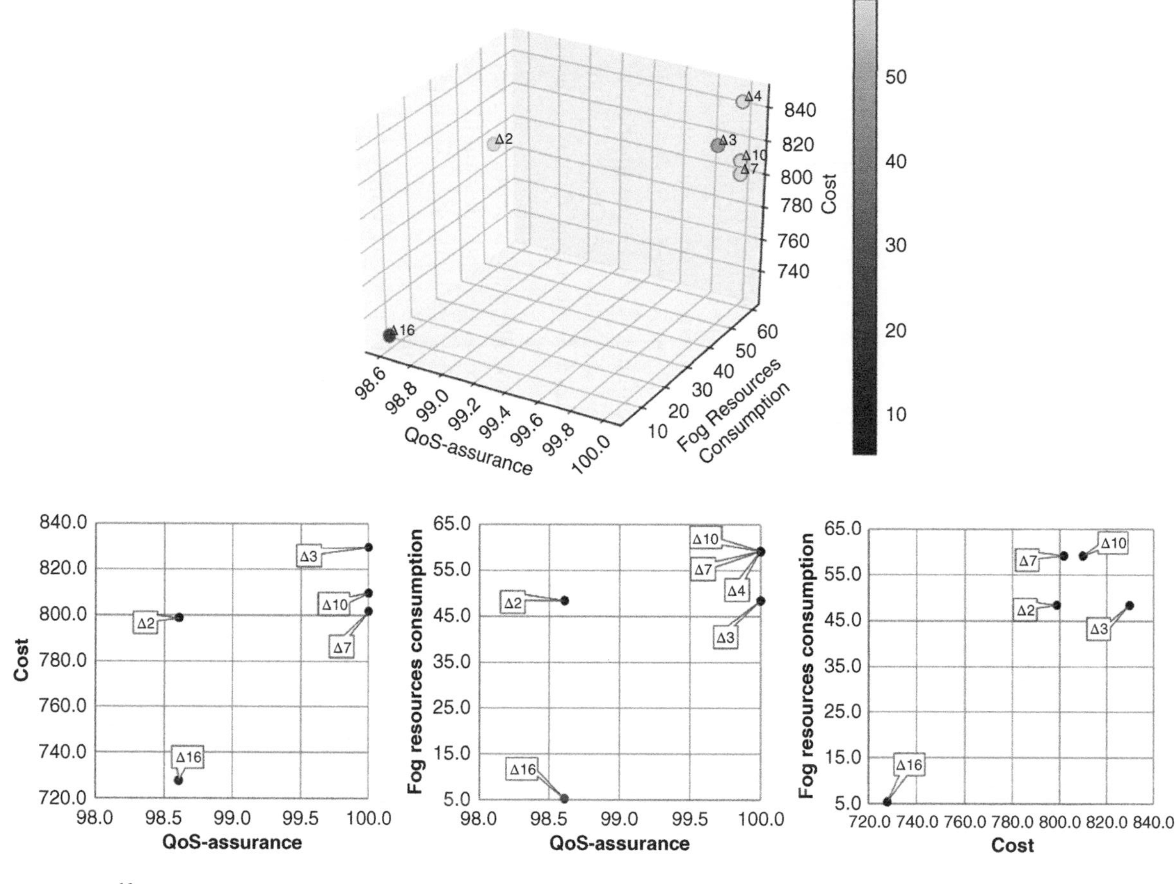

Figure 9.10 Results[16] for Q2.

Aeolus [21], or Cloud-4SOA [22], for instance, proposed model-driven optimized planning solutions to deploy software applications across multiple (IaaS or PaaS) Clouds. [23] proposed using OASIS TOSCA [24] to model IoT applications in Cloud+IoT scenarios.

With respect to the cloud paradigm, the fog introduces new problems, mainly due to its pervasive geo-distribution and heterogeneity, need for connection-awareness, dynamicity, and support to interactions with the IoT, that were not taken into account by previous work (e.g., [25–27]). Particularly, some efforts in cloud computing considered nonfunctional requirements (mainly e.g., [28, 29]) or uncertainty of execution (as in fog nodes) and security risks among interactive and interdependent components (e.g., [30]). Only recently, [31] has been among the first attempts to consider linking services and networks QoS by proposing a QoS- and connection-aware cloud service composition approach to satisfy end-to-end QoS requirements in the cloud.

Many domain-specific languages (DSLs) have been proposed in the context of cloud computing to describe applications and resources, e.g. TOSCA YAML [24] or JSON-based CloudML [32]. We aim not to bind to any particular standard for what concerns the specification of software/hardware offerings so that the proposed approach stays general and can potentially exploit suitable extensions (with respect to QoS and IoT) of such DSLs. Also, solutions to automatically provision and configure software components in cloud (or multi-cloud) scenarios are currently used by the DevOps community to automate application deployment or to lead deployment design choices (e.g., Puppet [33] and Chef [34]).

In the context of IoT deployments, formal modelling approaches have been recently exploited to achieve connectivity and coverage optimization [35, 36], to improved resource exploitation of Wireless Sensors Networks [37], and to estimate reliability and cost of service compositions [38].

Our research aims at complementing those efforts, by describing the interactions among software components and IoT devices at a higher level of abstraction to achieve informed segmentation of applications through the fog – that was not addressed by previous work.

9.5.2 Fog Application Deployment Support

To the best of our knowledge, few approaches have been proposed so far to specifically model fog infrastructures and applications, as well as to determine and compare eligible deployments for an application to an fog infrastructure under different metrics. Service latency and energy consumption were

15 The colormap in the figure shows fog resource consumption. Data displayed on the 3D axes on top are also projected to 2D in the three plots at the bottom of the figure.

16 As before, the colormap in the figure shows fog resource consumption. Data displayed on the 3D axes on top are also projected to 2D in the three plots at the bottom of the figure.

evaluated in [39] for the new fog paradigm applied to the IoT, as compared to traditional cloud scenarios. The model of [39], however, deals only with the behavior of software already deployed over fog infrastructures.

iFogSim [40] is one of the most promising prototypes to simulate resource management and scheduling policies applicable to fog environments with respect to their impact on latency, energy consumption and operational cost. The focus of iFogSim model is mainly on stream-processing applications and hierarchical tree-like infrastructures, to be mapped either cloud-only or edge-ward so as to compare results. In Section 9.5.4, we will show how iFogSim and FogTorchΠ can be used complementarily to solve deployment challenges.

Building on top of iFogSim, [41] compares different task scheduling policies, considering user mobility, optimal fog resource utilization and response time. [42] presents a distributed approach to cost-effective application placement, at varying workload conditions, with the objective of optimizing operational cost across the entire infrastructure. Apropos, [43] introduces a hierarchy-based technique to dynamically manage and migrate applications between cloud and fog nodes. They exploit message passing among local and global node managers to guarantee QoS and cost constraints are met. Similarly, [44] leverages the concept of fog colonies [45] for scheduling tasks to fog infrastructures, whilst minimizing response times. [46] provides a first methodology for probabilistic record-based resource estimation to mitigate resource underutilization, to enhance the QoS of provisioned IoT services.

All the aforementioned approaches are limited to monolithic or DAG application topologies and do not take into account QoS for the component–component and component–thing interactions, nor historical data about fog infrastructure or deployment behavior. Furthermore, the attempts to explicitly target and support with predictive methodologies the decision-making process to deploy IoT applications to the fog did not consider matching of application components to the best virtual instance (virtual machine or container), depending on expressed preferences (e.g., cost or energy targets) in this work.

9.5.3 Cost Models

While pricing models for the cloud are quite established (e.g., [18] and references therein), they do not account for costs generated by the exploitation of IoT devices. Cloud pricing models are generally divided into two types, pay-per-use and subscription-based schemes. In [18], based on given user workload requirements, a cloud broker chooses a best VM instance among several cloud providers. The total cost of deployment is calculated considering hardware requirements such as number of CPU cores, VM types, time duration, type of instance (reserved or pre-emptible), etc.

On the other hand, IoT providers normally process the sensory data coming from the IoT devices and sell the processed information as value added service to the users. [19] shows how they can also act as brokers, acquiring data from different owners and selling bundles. The authors of [19] also considered the fact that different IoT providers can federate their services and create new offers for their end users. Such end users are then empowered to estimate the total cost of using IoT services by comparing pay-per-use and subscription-based offers, depending on their data demand.

More recently, [47] proposed a cost model for IoT+Cloud scenario. Considering parameters such as the type and number of sensors, number of data request and uptime of VM, their cost model can estimate the cost of running an application over a certain period of time. In fog scenarios, however, there is a need to compute IoT costs at a finer level, also accounting for data-transfer costs (i.e., event-based).

Other recent studies tackle akin challenges from an infrastructural perspective, either focusing on scalable algorithms for QoS-aware placement of microdata centers [48], on optimal placement of data and storage nodes that ensures low latencies and maximum throughput, optimizing costs [49], or on the exploitation of genetic algorithms to place intelligent access points at the edge of the network [50].

To the best of our knowledge, our attempt to model costs in the fog scenario is the first that extends cloud pricing schemes to the fog layer and integrates them with costs that are typical of IoT deployments.

9.5.4 Comparing iFogSim and FogTorchΠ

iFogSim [40] is a simulation tool for fog computing scenarios. In this section, we discuss how both iFogSim and FogTorchΠ can be used together to solve the same input scenario (i.e., infrastructure and application). We do so by assessing whether the results of FogTorchΠ are in line with the results obtained with iFogSim. In this section, we review the VR Game case study employed for iFogSim in [40], execute FogTorchΠ over it, and then compare the results obtained by both prototypes.

The VR Game is a latency sensitive smartphone application which allows multiple players to interact with each other through EEG sensors. It is a multi-component application consisting of three components (viz., client, coordinator and concentrator) (Figure 9.11). To allow players to interact in real time, the application demands a high level of QoS (i.e., minimum latency) in between components. The infrastructure to host the application consists of a single cloud node, an ISP proxy, several gateways, and smartphones connected to EEG sensors (Figure 9.12). The number of gateways is variable and can be set (to 1, 2, 4, 8 or 16), while the number of smartphones connected to each gateway remains constant (viz., 4).

For the given input application and infrastructure, iFogSim produces and simulates a single deployment (either cloud-only or edge-ward [40]) that

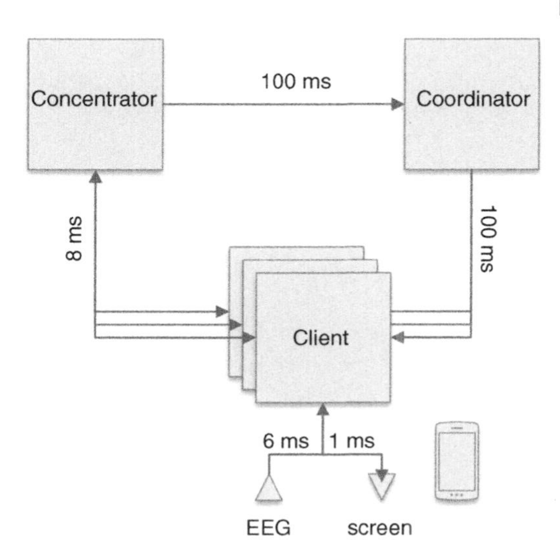

Figure 9.11 VR Game application.

satisfies all the specified hardware and software requirements. Simulation captures tuple exchange among components in the application, as in an actual deployment. This enables administrators to compare the average latency of the time-sensitive control loop ⟨EEG-client-concentrator-client-screen⟩ when adopting a cloud-only or an edge-ward deployment strategy.

On the other hand, FogTorchΠ produces various eligible deployments[17] for the same input to choose from. Indeed, FogTorchΠ outputs a set of 25 eligible deployments depending on variations in the number of gateways used in the infrastructure.

As shown in Table 9.4, FogTorchΠ output mainly includes edge-ward deployments for the VR Game application example. This is very much in line with the results obtained with iFogSim in [40], where cloud-only deployments perform much worse than edge-ward ones (especially when the number of involved devices – smartphones and gateways – increases). Also, the only output deployments that exploit the cloud as per FogTorchΠ results over this example are $\Delta 2$ and $\Delta 5$, featuring a very low QoS-assurance ($< 1\%$).

As per [40], iFogSim does not yet feature performance prediction capabilities – such as the one implemented by FogTorchΠ – in its current version. However, such functionalities can be implemented by exploiting the monitoring layer offered by the tool, also including a knowledge base that conserves historical data about the infrastructure behavior.

Summing up, iFogSim and FogTorchΠ can be seen as somewhat complementary tools designed to help end users in choosing how to deploy their fog applications by first predicting properties of a given deployment beforehand

17 https://github.com/di-unipi-socc/FogTorchPI/tree/multithreaded/results/VRGAME18

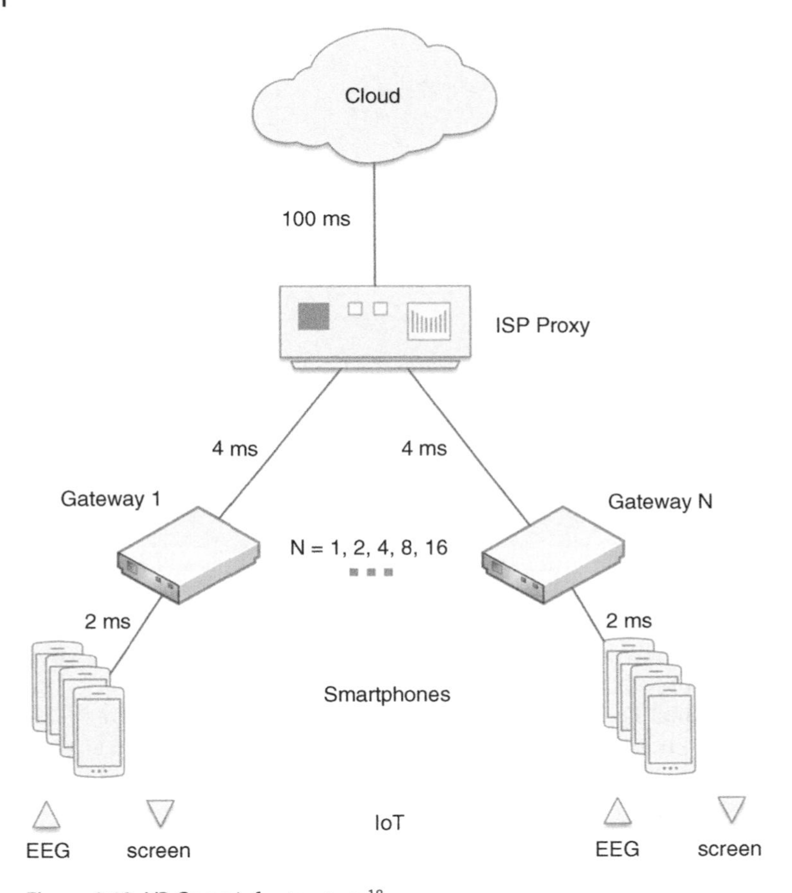

Figure 9.12 VR Game infrastructure.[18]

and by, afterward, being able to simulate most promising deployment candidates for any arbitrary time duration. The possibility of integrating predicting features of FogTorchΠ with simulation features of iFogSim is indeed in the scope of future research directions.

9.6 Future Research Directions

We see several directions for future work on FogTorchΠ. A first direction could be including other dimensions and predicted metrics to evaluate eligible

18 We assume that end-to-end communication links in the infrastructure have latency equal to the sum of latencies in the path they traverse.

Table 9.4 Result of FogTorchΠ for the VR Game.

				Number of Gateways				
Deployment ID	**Clients**	**Concentrator**	**Coordinator**	**1**	**2**	**4**	**8**	**16**
Δ1	Smartphones	Gateway 1	ISP Proxy	x	x			
Δ2		ISP Proxy	Cloud	x				
Δ3		Gateway 1	Gateway 1	x	x			
Δ4		ISP Proxy	Gateway 1	x	x	x	x	x
Δ5		Gateway 1	Cloud	x				
Δ6		ISP Proxy	ISP Proxy	x	x	x	x	x
Δ7		ISP Proxy	Gateway 2		x	x	x	x
Δ8		Gateway 2	Gateway 2		x			
Δ9		Gateway 2	Gateway 1		x			
Δ10		Gateway 2	ISP Proxy		x			
Δ11		Gateway 1	Gateway 2		x			
Δ12		ISP Proxy	Gateway 4			x	x	x
Δ13		ISP Proxy	Gateway 3			x	x	x
Δ14		ISP Proxy	Gateway 5				x	x
Δ15		ISP Proxy	Gateway 7				x	x
Δ16		ISP Proxy	Gateway 6				x	x
Δ17		ISP Proxy	Gateway 8				x	x
Δ18		ISP Proxy	Gateway 16					x
Δ19		ISP Proxy	Gateway 15					x
Δ20		ISP Proxy	Gateway 14					x
Δ21		ISP Proxy	Gateway 13					x
Δ22		ISP Proxy	Gateway 12					x
Δ23		ISP Proxy	Gateway 11					x
Δ24		ISP Proxy	Gateway 9					x
Δ25		ISP Proxy	Gateway 10					x
		Execution time (seconds)[19]		4	10	26	89	410

deployments, to refine search algorithms, and to enrich input and output expressiveness. Particularly, it would be interesting to do the following:

- Introduce estimates of *energy consumption* as a characterizing metric for eligible deployments, possibly evaluating its impact – along with financial costs – on SLAs and business models in fog scenarios,
- Account for *security constraints* on secure communication, access control to nodes and components, and trust in different providers, and

19 Run with $w = 2$ on a dual-core Intel i5-6500 @ 3.2 GHz, 8GB RAM.

- Determine *mobility* of Fog nodes and IoT devices, with a particular focus on how an eligible deployment can opportunistically exploit the (local) available capabilities or guarantee resilience to churn.

Another direction is to tame the exponential complexity of FogTorchΠ algorithms to scale better over large infrastructures, by leading the search with improved heuristics and by approximating metrics estimation.

A further direction could be to apply multiobjective optimization techniques in order to rank eligible deployments as per the estimated metrics and performance indicators, so to automate the selection of a (set of) deployment(s) that best meet end-user targets and application requirements.

The reuse of the methodologies designed for FogTorchΠ to generate deployments to be simulated in iFogSim is under study. This will permit comparisons of the predicted metrics generated by FogTorchΠ against the simulation results obtained with iFogSim, and it would provide a better validation to our prototype.

Finally, fog computing lacks medium- to large-scale test-bed deployments (i.e., infrastructure and applications) to test devised approaches. Last, but not least, it would be interesting to further engineer FogTorchΠ and to assess the validity of the prototype over an experimental lifelike testbed that is currently being studied.

9.7 Conclusions

In this chapter, after discussing some of the fundamental issues related to fog application deployment, we presented the FogTorchΠ prototype, as a first attempt to empower fog application deployers with predictive tools that permit to determine and compare eligible context-, QoS- and cost-aware deployments of composite applications to Fog infrastructures. To do so, FogTorchΠ considers processing (e.g., CPU, RAM, storage, software), QoS (e.g., latency, bandwidth), and financial constraints that are relevant for real-time fog applications.

To the best of our knowledge, FogTorchΠ is the first prototype capable of estimating the QoS-assurance of composite fog applications deployments based on probability distributions of bandwidth and latency featured by end-to-end communication links. FogTorchΠ also estimates resource consumption in the fog layer, which can be used to minimize the exploitation of certain fog nodes with respect to others, depending on the user needs. Finally, it embeds a novel cost model to estimate multi-component application deployment costs to IoT+Fog+Cloud infrastructures. The model considers various cost parameters (hardware, software, and IoT), and extends cloud computing cost models to the fog computing paradigm, while taking into account costs associated with the usage of IoT devices and services.

The potential of FogTorchΠ has been illustrated by discussing its application to a smart building fog application, performing what-if analyses at design time, including changes in QoS featured by communication links and looking for the best trade-off among QoS-assurance, resource consumption, and cost.

Needless to say, the future of predictive tools for fog computing application deployment has just started, and much remains to understand how to suitably balance different types of requirements so to make the involved stakeholders aware of the choices to be made throughout application deployment.

References

1 CISCO. Fog computing and the Internet of Things: Extend the cloud to where the things are. https://www.cisco.com/c/dam/en_us/solutions/trends/iot/docs/computing-overview.pdf, 30/03/2018.

2 CISCO. Cisco Global Cloud Index: Forecast and Methodology. 2015–2020, 2015.

3 A. V. Dastjerdi and R. Buyya. Fog Computing: Helping the Internet of Things Realize Its Potential. *Computer,* 49(8): 112–116, August 2016.

4 I. Stojmenovic, S. Wen, X. Huang, and H. Luan. An overview of fog computing and its security issues. *Concurrency and Computation: Practice and Experience,* 28(10): 2991–3005, July 2016.

5 R. Mahmud, R. Kotagiri, and R. Buyya. Fog computing: A taxonomy, survey and future directions. *Internet of Everything: Algorithms, Methodologies, Technologies and Perspectives,* Beniamino Di Martino, Kuan-Ching Li, Laurence T. Yang, Antonio Esposito (eds.), Springer, Singapore, 2018.

6 F. Bonomi, R. Milito, P. Natarajan, and J. Zhu. Fog computing: A platform for internet of things and analytics. *Big Data and Internet of Things: A Roadmap for Smart Environments,* N. Bessis, C. Dobre (eds.), Springer, Cham, 2014.

7 W. Shi and S. Dustdar. The promise of edge computing. *Computer,* 49(5): 78–81, May 2016.

8 A. Brogi and S. Forti. QoS-aware Deployment of IoT Applications Through the Fog. *IEEE Internet of Things Journal,* 4(5): 1185–1192, October 2017.

9 P. O. Östberg, J. Byrne, P. Casari, P. Eardley, A. F. Anta, J. Forsman, J. Kennedy, T.L. Duc, M.N. Mariño, R. Loomba, M.Á.L. Peña, J.L. Veiga, T. Lynn, V. Mancuso, S. Svorobej, A. Torneus, S. Wesner, P. Willis and J. Domaschka. Reliable capacity provisioning for distributed cloud/edge/fog computing applications. In *Proceedings of the* 26th *European Conference on Networks and Communications,* Oulu, Finland, June 12–15, 2017.

10 OpenFog Consortium. OpenFog Reference Architecture (2016), *http://openfogconsortium.org/ra,* 30/03/2018.

11 M. Chiang and T. Zhang. Fog and IoT: An overview of research opportunities. *IEEE Internet of Things Journal*, 3(6): 854–864, December 2016.
12 S. Rizzi, What-if analysis. *Encyclopedia of Database Systems*, Springer, US, 2009.
13 A. Brogi, S. Forti, and A. Ibrahim. How to best deploy your fog applications, probably. In *Proceedings of the 1st IEEE International Conference on Fog and Edge Computing*, Madrid, Spain, May 14, 2017.
14 A. Brogi, S. Forti, and A. Ibrahim. Deploying fog applications: How much does it cost, by the way? In *Proceedings of the 8th International Conference on Cloud Computing and Services Science*, Funchal (Madeira), Portugal, March 19–21, 2018.
15 C. Perera. *Sensing as a Service for Internet of Things: A Roadmap*. Leanpub, Canada, 2017.
16 S. Newman. *Building Microservices: Designing Fine-Grained Systems*. O'Reilly Media, USA, 2015.
17 W. L. Dunn and J. K. Shultis. *Exploring Monte Carlo Methods*. Elsevier, Netherlands, 2011.
18 J. L. Dìaz, J. Entrialgo, M. Garcìa, J. Garcìa, and D. F. Garcìa. Optimal allocation of virtual machines in multi-cloud environments with reserved and on-demand pricing, *Future Generation Computer Systems*, 71: 129–144, June 2017.
19 D. Niyato, D. T. Hoang, N. C. Luong, P. Wang, D. I. Kim and Z. Han. Smart data pricing models for the internet of things: a bundling strategy approach, *IEEE Network* 30(2): 18–25, March–April 2016.
20 A. Brogi, A. Ibrahim, J. Soldani, J. Carrasco, J. Cubo, E. Pimentel and F. D'Andria. SeaClouds: a European project on seamless management of multi-cloud applications. *Software Engineering Notes of the ACM Special Interest Group on Software Engineering*, 39(1): 1–4, January 2014.
21 R. Di Cosmo, A. Eiche, J. Mauro, G. Zavattaro, S. Zacchiroli, and J. Zwolakowski. Automatic Deployment of Software Components in the cloud with the Aeolus Blender. In *Proceedings of the 13th International Conference on Service-Oriented Computing*, Goa, India, November 16–19, 2015.
22 A. Corradi, L. Foschini, A. Pernafini, F. Bosi, V. Laudizio, and M. Seralessandri. Cloud PaaS brokering in action: The Cloud4SOA management infrastructure. In *Proceedings of the 82nd IEEE Vehicular Technology Conference*, Boston, MA, September 6–9, 2015.
23 F. Li, M. Voegler, M. Claesens, and S. Dustdar. Towards automated IoT application deployment by a cloud-based approach. In *Proceedings of the 6th IEEE International Conference on Service-Oriented Computing and Applications*, Kauai, Hawaii, December 16–18, 2013.
24 A. Brogi, J. Soldani and P. Wang. TOSCA in a Nutshell: Promises and Perspectives. In *Proceedings of the 3rd European Conference on Service-Oriented and Cloud Computing*, Manchester, UK, September 2–4, 2014.

25 P. Varshney and Y. Simmhan. Demystifying fog computing: characterizing architectures, applications and abstractions. In *Proceedings of the 1st IEEE International Conference on Fog and Edge Computing*, Madrid, Spain, May 14, 2017.

26 Z. Wen, R. Yang, P. Garraghan, T. Lin, J. Xu, and M. Rovatsos. Fog orchestration for Internet of Things services. *iEEE Internet Computing,* 21(2): 16–24, March–April 2017.

27 J.-P. Arcangeli, R. Boujbel, and S. Leriche. Automatic deployment of distributed software systems: Definitions and state of the art. *Journal of Systems and Software,* 103: 198–218, May 2015.

28 R. Nathuji, A. Kansal, and A. Ghaffarkhah. Q-Clouds: Managing Performance Interference Effects for QoS-Aware Clouds. In *Proceedings of the 5th EuroSys Conference,* Paris, France, April 13–16, 2010.

29 T. Cucinotta and G.F. Anastasi. A heuristic for optimum allocation of real-time service workflows. In *Proceedings of the 4th IEEE International Conference on Service-Oriented Computing and Applications,* Irvine, CA, USA, December 12–14, 2011.

30 Z. Wen, J. Cala, P. Watson, and A. Romanovsky. Cost effective, reliable and secure workflow deployment over federated clouds. *IEEE Transactions on Services Computing,* 10(6): 929–941, November–December 2017.

31 S. Wang, A. Zhou, F. Yang, and R. N. Chang. Towards network-aware service composition in the cloud. *IEEE Transactions on Cloud Computing,* August 2016.

32 A. Bergmayr, A. Rossini, N. Ferry, G. Horn, L. Orue-Echevarria, A. Solberg, and M. Wimmer. *The Evolution of CloudML and its Manifestations.* In *Proceedings of the 3rd International Workshop on Model-Driven Engineering on and for the Cloud*, Ottawa, Canada, September 29, 2015.

33 Puppetlabs, Puppet, https://puppet.com. Accessed March 30, 2018.

34 Opscode, Chef, https://www.chef.io. Accessed March 30, 2018.

35 J. Yu, Y. Chen, L. Ma, B. Huang, and X. Cheng. On connected Target k-Coverage in heterogeneous wireless sensor networks. *Sensors,* 16(1): 104, January 2016.

36 A.B. Altamimi and R.A. Ramadan. Towards Internet of Things modeling: a gateway approach. *Complex Adaptive Systems Modeling,* 4(25): 1–11, November 2016.

37 H. Deng, J. Yu, D. Yu, G. Li, and B. Huang. Heuristic algorithms for one-slot link scheduling in wireless sensor networks under SINR. *International Journal of Distributed Sensor Networks,* 11(3): 1–9, March 2015.

38 L. Li, Z. Jin, G. Li, L. Zheng, and Q. Wei. Modeling and analyzing the reliability and cost of service composition in the IoT: A probabilistic approach. In *Proceedings of 19th International Conference on Web Services,* Honolulu, Hawaii, June 24–29, 2012.

39 S. Sarkar and S. Misra. Theoretical modelling of fog computing: a green computing paradigm to support IoT applications. *IET Networks*, 5(2): 23–29, March 2016.
40 H. Gupta, A.V. Dastjerdi, S.K. Ghosh, and R. Buyya. iFogSim: A Toolkit for Modeling and Simulation of Resource Management Techniques in Internet of Things, Edge and Fog Computing Environments. *Software Practice Experience*, 47(9): 1275–1296, June 2017.
41 L.F. Bittencourt, J. Diaz-Montes, R. Buyya, O.F. Rana, and M. Parashar, Mobility-aware application scheduling in fog computing. *IEEE Cloud Computing*, 4(2): 26–35, April 2017.
42 W. Tarneberg, A.P. Vittorio, A. Mehta, J. Tordsson, and M. Kihl, Distributed approach to the holistic resource management of a mobile cloud network. In *Proceedings of the 1st IEEE International Conference on Fog and Edge Computing*, Madrid, Spain, May 14, 2017.
43 S. Shekhar, A. Chhokra, A. Bhattacharjee, G. Aupy and A. Gokhale, INDICES: Exploiting edge resources for performance-aware cloud-hosted services. In *Proceedings of the 1st IEEE International Conference on Fog and Edge Computing*, Madrid, Spain, May 14, 2017.
44 O. Skarlat, M. Nardelli, S. Schulte and S. Dustdar. Towards QoS-aware fog service placement. In *Proceedings of the 1st IEEE International Conference on Fog and Edge Computing*, Madrid, Spain, May 14, 2017.
45 O. Skarlat, S. Schulte, M. Borkowski, and P. Leitner. Resource Provisioning for IoT services in the fog. In *Proceedings of the 9th IEEE International Conference on Service-Oriented Computing and Applications*, Macau, China, November 4–6, 2015.
46 M. Aazam, M. St-Hilaire, C. H. Lung, and I. Lambadaris. MeFoRE: QoE-based resource estimation at Fog to enhance QoS in IoT. In *Proceedings of the 23rd International Conference on Telecommunications*, Thessaloniki, Greece, May 16–18, 2016.
47 A. Markus, A. Kertesz and G. Kecskemeti. Cost-Aware IoT Extension of DISSECT-CF, *Future Internet*, 9(3): 47, August 2017.
48 M. Selimi, L. Cerdà-Alabern, M. Sànchez-Artigas, F. Freitag and L. Veiga. Practical Service Placement Approach for Microservices Architecture. In *Proceedings of the 17th IEEE/ACM International Symposium on Cluster, Cloud and Grid Computing*, Madrid, Spain, May 14–17, 2017.
49 I. Naas, P. Raipin, J. Boukhobza, and L. Lemarchand. iFogStor: an IoT data placement strategy for fog infrastructure. In *Proceedings of the 1st IEEE International Conference on Fog and Edge Computing*, Madrid, Spain, May 14, 2017.
50 A. Majd, G. Sahebi, M. Daneshtalab, J. Plosila, and H. Tenhunen. Hierarchal placement of smart mobile access points in wireless sensor networks using

fog computing. In *Proceedings of the 25th Euromicro International Conference on Parallel, Distributed and Network-based Processing*, St. Petersburg, Russia, March 6–8, 2017.

51 K. Fatema, V.C. Emeakaroha, P.D. Healy, J.P. Morrison and T. Lynn. A survey of cloud monitoring tools: Taxonomy, capabilities and objectives. *Journal of Parallel and Distributed Computing*, 74(10): 2918–2933, October 2014.

52 Y. Breitbart, C.-Y. Chan, M. Garofalakis, R. Rastogi, and A. Silberschatz. Efficiently monitoring bandwidth and latency in IP networks. In *Proceedings of the 20th Annual Joint Conference of the IEEE Computer and Communications Societies*, Alaska, USA, April 22–26, 2001.

10

Using Machine Learning for Protecting the Security and Privacy of Internet of Things (IoT) Systems

Melody Moh and Robinson Raju

10.1 Introduction

Today, IoT devices are ubiquitous and have pervaded almost every sphere of our lives, ushering an era of *smart things:*

- Smart homes have appliances, lights, and thermostat connected to the Internet [1].
- Smart medical appliances not only monitor remotely but also administer medicines timely [2].
- Smart bridges have sensors to monitor loads [3].
- Smart power grids detect disruptions and manage distribution of power [4].
- Smart machinery in industries have embedded sensors in heavy machinery to increase worker safety and improve automation [5].

To get a better understanding of the scale of IoT, here are some numbers for review:

- In 2008, the number of devices connected to the Internet surpassed the world population of approximately 6.7 billion people.
- In 2015, approximately 1.4 billion smartphones were shipped by manufacturers.
- By 2020, the prediction is that there will be 6.1 billion smartphone users and an anticipated 50 billion things connected to the Internet [6].
- By 2027, the expectation is that there will be 27 billion machine-to-machine connections in the industrial sector.

Now, if the focus shifts to the amount of data that gets generated, one gets a glimpse of the dawn of the zettabyte era [7]. To put a zettabyte into perspective,

Fog and Edge Computing: Principles and Paradigms, First Edition.
Edited by Rajkumar Buyya and Satish Narayana Srirama.

36,000 years of high-definition television video would be the equivalent of one zettabyte.

- In 2013, devices connected to the Internet generated 3.1 zettabytes of data.
- In 2014, that number jumped to 8.6 zettabytes.
- In 2018, that number is expected to soar to 400 zettabytes [8].

10.1.1 Examples of Security and Privacy Issues in IoT

While previous chapters talked about the ubiquitousness of IoT, the amount data generated, and the technologies used, this chapter focuses on the type of data that is transmitted and the security and privacy implications of this. Ubiquitousness is a double-edged sword. The reach is higher and more widespread than human comprehension, but so is the vulnerability. Hence security and privacy implications of a system that has myriads of devices manufactured independently and communicating using different protocols and generates zettabytes of data are broad and deep. Cisco's whitepaper on Global Cloud Index [6] talks about the types of data in the cloud. A total of 7.6% of documents in file-sharing services contain confidential data. Personally identifiable information (e.g., Social Security numbers, tax ID numbers, phone numbers, addresses, and so on) follows this at 4.3% of all documents. Next, 2.3% of documents contain payment data (e.g., credit card numbers, debit card numbers, bank account numbers, and so on). Finally, 1.6% of documents contain protected health information (e.g., patient diagnoses, medical treatments, medical record IDs, and so on).

As IoT usage grows, the amount of data uploaded to the cloud by IoT systems far exceeds that done by users. Because IoT data is on the cloud and IoT devices have connectivity to the Internet, they become vulnerable to attacks of different types. In fact, more often than not, we read about breaches on a daily basis:

- Water treatment plant is hacked and chemical mix changed for tap supplies [9].
- Nuclear power plant in Ukraine is breached [10].
- Security researchers from Rapid7 security firm discover many security vulnerabilities affecting several video baby monitors [11].
- Data from wearable devices are used to plan robberies [12].
- There are reports on how hackers could target pacemakers [13].

As per Cybercrime report in 2016 [14], cybercrime damages will cost the world $6 trillion annually by 2021, up from $3 trillion in 2015.

10.1.2 Security Concerns at Different Layers in IoT

A review of the 2015 IBM Point of View on IoT security [15] shows threats at multiple points in the IoT ecosystem and protections that are applicable at every layer (see Figure 10.1).

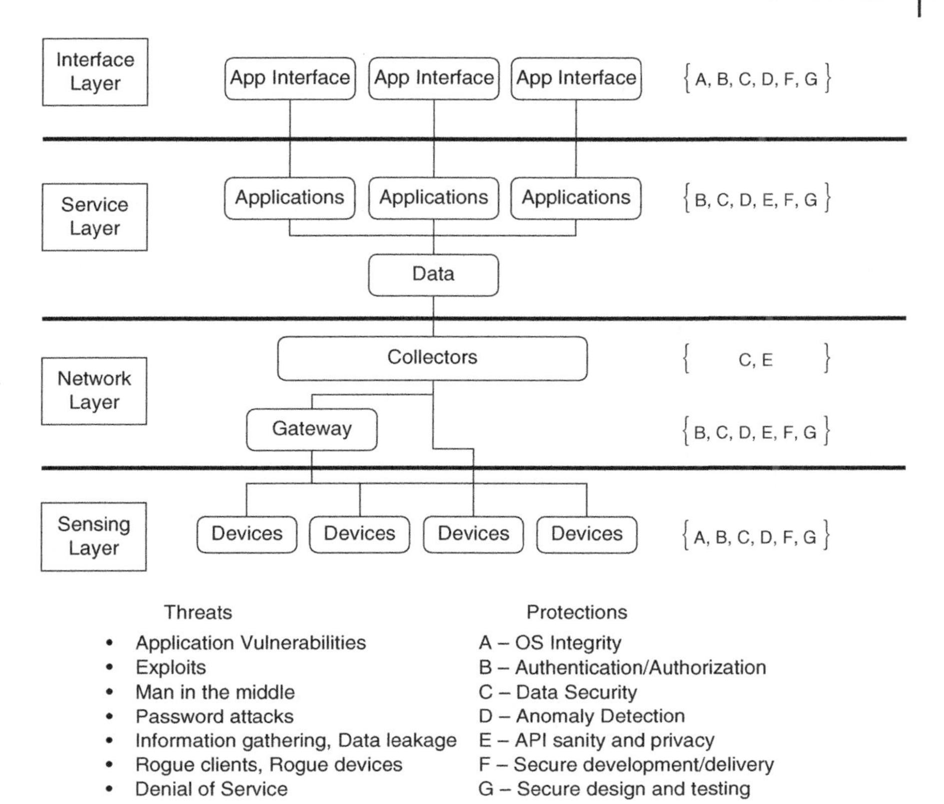

Figure 10.1 IoT system with threats and protections annotated.

10.1.2.1 Sensing Layer

In most of the scenarios just described, hackers were able to do the most damage when they gained access to sensors like baby monitors or pacemakers. So, it is critical to have sensors protected and monitored so that one can either prevent the intrusion or alert the user when there is one, in the fastest possible time. The possible threats at the sensing layer are the following:

- Unauthorized access to data
- Denial of service attack
- Malware on the device to send wrong information
- Malware on the device to send data to the wrong party
- Information gathering or data leakage leading to planned attacks

10.1.2.2 Network Layer

The availability, manageability, and scalability of the network are crucial for the operation of IoT. If the monitoring applications are not able to get data in time, IoT devices are rendered useless. Hence, hackers target networks more often to cripple the effectiveness of smart systems. Attacking the network by sending a lot of data at once to congest the network and pave the way to denial of service attacks is very common.

10.1.2.3 Service Layer

The service layer acts as a bridge between the hardware layer at the bottom and the interface layer at the top. An attack on the service layer impacts critical functions such as device management and information management, leading to the end users not being serviced. Privacy protection, access control, user authentication, communication security, data integrity, and data confidentiality are vital aspects of service layer security.

10.1.2.4 Interface Layer

In many ways, the interface layer is the most vulnerable part of the IoT ecosystem because this layer is at the top and is a gateway to all the other layers below. If there is a compromise in the authentication and authorization mechanisms of the interface, the ripple effects could permeate to the edge. The end user is a possible attack mechanism since attackers could gain sensitive information via phishing or other similar attacks. The web and the app interfaces can be subject to frequent attacks like SQL injection, cross-site scripting, known default credentials, insecure password recovery mechanism and so forth.

OWASP (Open Web Application Security Project) has a very neat summarization of the attack surface areas for IoT [16] and is a handy reference (see Table 10.1).

10.1.3 Privacy Concerns in IoT Devices

A 2015 report of Internet of Things research study [17] done by Hewlett Packard reported that 80% of devices raised privacy concerns. Many devices collect some of the other form of personal data such as name, address, date of birth, payment information, health data, light and sound information from home, activities within a home, and so forth (Figure 10.2). Most of these devices are transmitting data within the home network in an unencrypted fashion, and since data go out from home into the cloud, most people are just one misconfiguration away from exposing the data to the outside world. The report found that, on average, 25 vulnerabilities were found per device, totaling 250 vulnerabilities.

An article in *FastCompany* by Lauren Zanolli [18] talks about IoT being a "Privacy Hell." Another article in *Wall Street Journal* [19] talks about IoT opening up new privacy litigation risks. Italian retailer Benetton was boycotted for having RFID tracking in clothes [20]. There was a sense of real urgency in FTC report on IoT in Jan 2015 [21] that asked companies to adopt best practices to

Table 10.1 OWASP IoT attack surface areas.

Attack surface	*Vulnerability*
Ecosystem Access Control	• Implicit trust between components • Enrollment security • Lost access procedures
Device Memory	• Cleartext usernames and passwords • Third-party credentials • Encryption keys
Device Web Interface	• SQL injection • Cross-site scripting • Cross-site request forgery • Username enumeration • Weak passwords • Account lockout • Known default credentials
Device Firmware	• Hardcoded credentials • Sensitive information disclosure • Sensitive URL disclosure • Encryption keys • Firmware version display and/or last update date
Device Network Services	• Information disclosure • User CLI • Administrative CLI • Injection • Denial of service • Unencrypted services • Poorly implemented encryption • Vulnerable UDP services • DoS
Administrative Interface	• SQL injection • Cross-site scripting • Cross-site request forgery • Username enumeration • Weak passwords • Account lockout • Known default credentials • Logging options • Two-factor authentication • Inability to wipe device

address consumer privacy and security risks. There has been much research into security aspects of IoT, and most of them have been a continuation of security challenges with networking and routing. In comparison, the research into privacy issues has been decidedly less.

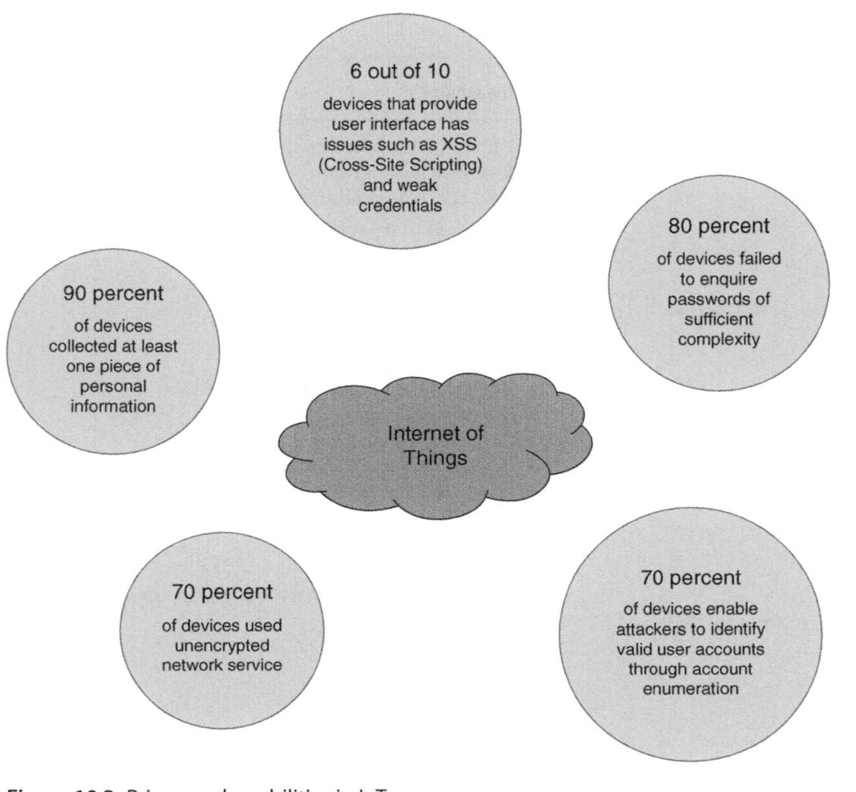

Figure 10.2 Privacy vulnerabilities in IoT

10.1.3.1 Information Privacy

Privacy is a comprehensive term, and historically it has meant media, place, communication, body privacy. Today, the term is increasingly used to mean *information privacy.* Privacy was defined by Westin in 1968 as "the claim of individuals, to determine for themselves when, how, and to what extent information about them is communicated" [22].

Ziegeldort et al. in their paper on privacy in IoT [23], concretized the definition as follows. Privacy in the Internet of Things is the threefold guarantee that addresses these subjects:

1. Awareness of privacy risks imposed by smart things and services surrounding the data subject.
2. Individual control over the collection and processing of personal information by the surrounding smart things.
3. Awareness and control of subsequent use and dissemination of personal information by those entities to any entity outside the subject's control sphere.

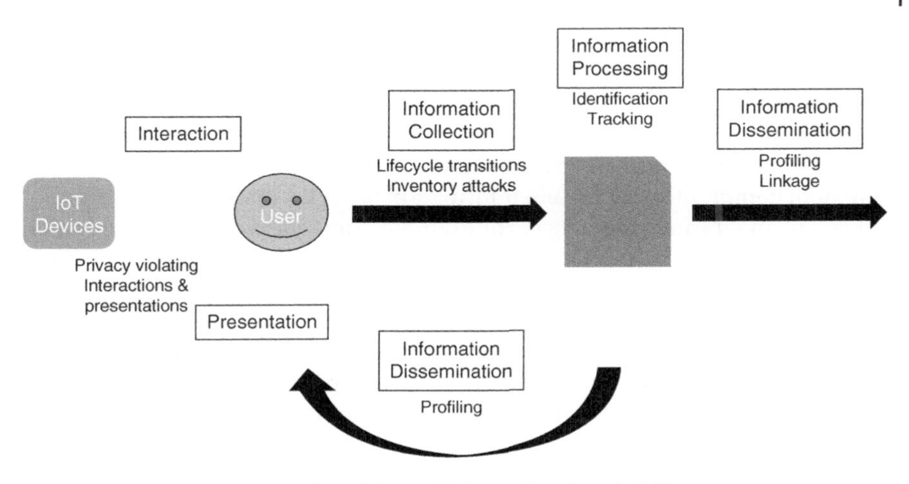

Figure 10.3 Privacy threats with entities and information flows in IoT.

Ziegeldort et al. [23] also defined a reference model to quickly understand and analyze the privacy concerns regarding anything that is interconnected anywhere via a network. The reference model contained four main types of entities: (i) smart things; (ii) subject; (iii) infrastructure; and (iv) services. It includes five types of information flows: (i) interaction; (ii) collection; (iii) processing; (iv) dissemination; and (v) presentation.

10.1.3.2 Categorization of IoT Privacy Issues

Ziegeldort et al. [23] also categorized the privacy threats (see Figure 10.3) into the following: (i) identification; (ii) localization and tracking; (iii) profiling; (iv) privacy-violating interaction and presentation; (v) lifecycle transitions; (vi) inventory attack; and (vii) linkage.

Identification. Identification is the threat of associating an identifier, e.g., a name and address, with an individual. It also enables and aggravates other threats, e.g., profiling and tracking of people.

Localization and Tracking. Localization and tracking is the threat of determining and recording a person's location through time and space. Since localization is an essential functionality in many IoT systems, the data are fetched by most applications. However, this leads to disclosure of private information such as illness, vacation plans, work schedules, and so forth.

Profiling. Profiling is the threat of categorizing individuals into groups by using data from IoT devices. Personalization in e-commerce, e.g. recommender systems, newsletters, and advertisements use profiling methods to optimize and to give targeted content. Examples, where profiling leads to a violation of privacy, are price discrimination, unsolicited advertisements, social engineering, or erroneous automatic decisions, e.g., by Facebook's automatic detection of sexual offenders. Also, several data marketplaces collect and sell profile information.

Privacy-Violating Interaction and Presentation. Privacy violating interaction is the threat of communicating private information in such a manner that it gets disclosed to an unwanted audience. For example, someone wearing a smartwatch and traveling in a public transit could inadvertently let strangers read their SMSes since the messages pop up on the watch screen as they come in.

Lifecycle Transitions. When smart things undergo upgrades, configurations and data are backed up and restored. In the process, sometimes, wrong data can end up in the wrong device, leading to a privacy violation, e.g. photos and videos on one device available on another.

Inventory Attack. Since smart things are queryable on the Internet, hackers can query devices to compile an inventory of things at a specific location, such as whether a home contains a smart meter, smart thermostat, smart lighting, and so forth.

Linkage. Linkage is a threat where one gathers insights about a subject by combining data from different sources, collected in different contexts. The revelation might be erroneous, and users may not have given permission to do this.

In summary, privacy is a critical issue in IoT devices and needs to be handled promptly from the manufacturing to deployment at every layer in the IoT ecosystem.

10.1.4 IoT Security Breach Deep-Dive: Distributed Denial of Service (DDoS) Attacks on IoT Devices

10.1.4.1 Introduction to DDoS

A denial of service (DoS) attack is a cyberattack where an attacker makes a network resource unavailable by interrupting services of a machine connected to the Internet. It is typically accomplished by flooding the target machine with fake requests in order to overload the system. A distributed denial of service (DDoS) attack is one that uses multiple network resources as the source of the attack. A DDoS is mainly intended not only as a method to multiply the capabilities of a single attacker but also to conceal the identity of the attacker and thwart mitigation efforts. Most botnets use compromised computer resources without the owner's knowledge. In the CIA (confidentiality, integrity, availability) triad of information security, DDoS attack falls in the *availability* category. Figure 10.4 depicts how an attacker could initiate one attack and transform it into a multitude of attacks on a victim [24].

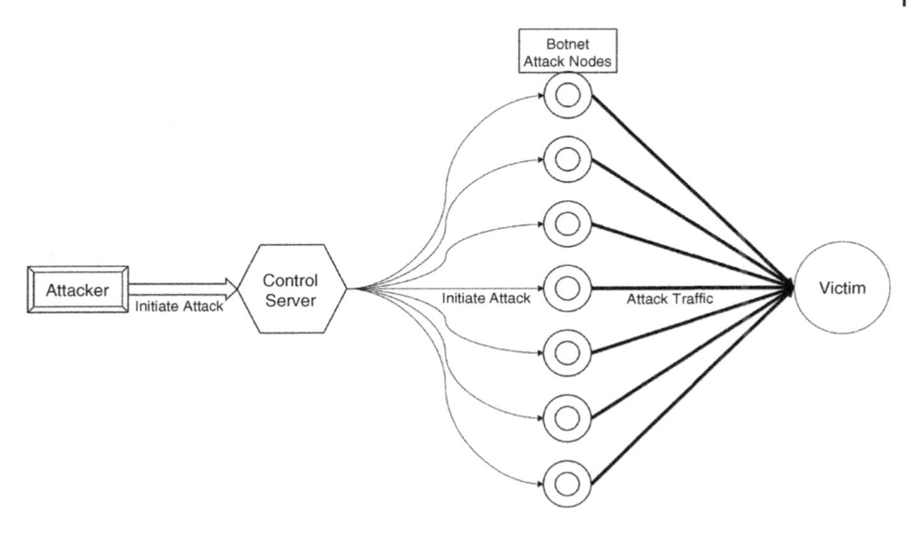

Figure 10.4 DDoS attack.

Though the motivations for DDoS can be multiple – extortion, hacktivism, cyberterrorism, personal vendetta, business rivalry, etc. – the impact is very severe in many instances. It can cause damage to reputation, huge revenue loss, and tens of thousands of hours of lost productivity. The scale of DDoS attacks has continued to rise over recent years, by 2016 exceeding a terabit per second.

10.1.4.2 Timeline of Notable DoS Events [25]

- 1988: Robert Tappan Morris launches a self-replicating worm that spreads uncontrollably throughout the Internet and causes a massive unintentional DoS.
- 1997: The "AS 7007 incident," is the first notable BGP hijacking and results in a massive DoS to significant portions of the Internet.
- 1999: Creation of trin00, TFN, and Stacheldraht botnets. The first instance of a botnet DDoS attack was a trin00 attack on the University of Minnesota.
- 2000: Michael Calce (aged 15) launched successful DoS attacks against Yahoo!, Fifa.com, Amazon.com, Dell, E*TRADE, eBay, and CNN.
- 2004: Hackers on 4chan develop the Low Orbit Ion Cannon (LOIC), a DDoS tool that would be used extensively by Anonymous and other groups to launch DDoS attacks.
- 2007: A series of DDoS attacks target various Estonian organizations. These attacks are notable as the first government-sponsored DDoS attacks, since Russian government was suspected to be behind them.
- 2008: Hacktivist collective Anonymous launches its first significant DDoS attack, successfully targeting the Church of Scientology.

- 2009: Launch of a coordinated DDoS attack targeting Facebook, Google Blogger, LiveJournal, and Twitter targeting a Georgian blogger critical of Russia.
- 2010: Hacktivist collective Anonymous launches "Operation Avenge Assange" targeting banks that froze donations to Wikileaks.
- 2013: A massive DDoS attack targeting anti-spam organization Spamhaus .org breaks records with traffic peaking at 300 Gbps.
- 2014: The hacking group Lizard Squad initiates successful DDoS attacks against the Sony Playstation Network and Microsoft Xbox Live.
- 2015: A network security hardware manufacturer reports a DDoS attack more than 500 Gbps against an unnamed customer.
- 2017: On October 21, a large-scale DDoS attack on Dyn [26], which is a primary provider of DNS services to many companies, took down many high-profile websites like Twitter, Pinterest, Reddit, GitHub, Amazon, Verizon, Comcast, and so forth.

10.1.4.3 Reason for the Recent Success of the DDoS Attacks

The most recent DDoS attack on Dyn [26] was made possible by the large number of unsecured IoT devices, such as home routers and surveillance cameras. The attackers employed thousands of such devices that had been infected with malicious code to form a botnet. The devices themselves were not powerful, but collectively they generated a massive amount of traffic to overwhelm targeted servers. The moment someone places a device on the Internet without changing the default password, it gets added to the army of vulnerable machines used for DDoS attacks. A report from welivesecurity.com [27] mentions that ESET tested more than 12,000 home routers to find 15% of them being unsecured. In the article "10 things to know about October 21 IoT DDoS attack" [28], Stephen Cobb lists default password as the leading cause. A mashable .com report in 2014 [29] mentions that 73,000 webcams were discovered in the Internet because people did not change default passwords.

To summarize, one could attribute the success of recent DDoS attacks despite decades of research and tools to mitigate, to the following:

- The proliferation of IoT devices.
- Increase in the number devices on the Internet with default passwords, and this could be due to the increase of nonsavvy technology users of smart devices.

10.1.4.4 Directions for Prevention of Specific Attacks on IoT Devices

As mentioned, in many instances above, IoT devices are growing at an alarming pace, and it is imminent that the devices be made secure. The attacks increasingly have a crippling effect on the economy and have become the new currency of global warfare. With this in mind, the US Senate introduced legislation in August 2017 [30] to improve the cybersecurity of IoT devices.

Specifically, if enacted, the Internet of Things (IoT) Cybersecurity Improvement Act of 2017 [31] would:

- Require vendors of Internet-connected devices purchased by the federal government to ensure their devices are patchable, rely on industry standard protocols, do not use hard-coded passwords, and do not contain any known security vulnerabilities.
- Direct the Office of Management and Budget (OMB) to develop alternative network-level security requirements for devices with limited data processing and software functionality.
- Direct the Department of Homeland Security's National Protection and Programs Directorate to issue guidelines regarding cybersecurity coordinated vulnerability disclosure policies to be required by contractors providing connected devices to the US government.
- Exempt cybersecurity researchers engaging in good-faith research from liability under the Computer Fraud and Abuse Act and the Digital Millennium Copyright Act when engaged in research pursuant to adopted coordinated vulnerability disclosure guidelines.
- Require each executive agency to inventory all Internet-connected devices in use by the agency.

10.1.4.5 Steps to Prevent Attacks on IoT Devices

The overarching strategy to secure IoT devices should be twofold: reduce the number of devices that can be abused and convince the would-be attackers like hacktivists on the gravity of the situation. Also, there needs to be a global strategy to punish the guilty. There have been multiple efforts to reduce the number of devices that can be abused. The Cybersecurity Improvement Act mentioned above, alerts sent out by the Department of Homeland Security, WaterISAC's 10 Basic Cybersecurity measures [32], are few initiatives from the government toward this. Here are the top four actions recommended by US-CERT [33] in the wake of the latest attacks:

1. Ensure all default passwords are changed to strong passwords. (Default usernames and passwords for most devices can easily be found on the Internet, making devices with default passwords extremely vulnerable.)
2. Update IoT devices with security patches as soon as patches become available.
3. Disable Universal Plug and Play (UPnP) on routers unless absolutely necessary.
4. Purchase IoT devices from companies with a reputation for providing secure devices.

10.2 Background

10.2.1 Brief Overview of Machine Learning

Machine learning, a term coined by Arthur Samuel, an American pioneer in the field of computer gaming and artificial intelligence [34], is the science of getting computers to learn and act without being explicitly programmed. The idea behind machine learning is to have an algorithm that can analyze data, identify patterns, and create a model that the machine could use to analyze data that it has not seen before. As systems provide more data to it, the algorithm learns continuously and will be able to produce reliable decisions repeatedly. In the past decade or so, with the increase of computing power and development of systems like Hadoop to do massive data processing at a short period, machine learning has pervaded many things that people use. From speech recognition, image recognition, fingerprint scanning, to self-driving cars, machine learning is used almost everywhere and is arguably the most impactful invention in recent times.

There are many machine-learning algorithms used in a variety of scenarios. Broadly, they could be categorized either by the nature of learning available to the system or by the desired output.

Depending on the nature of the learning, machine-learning algorithms can be categorized as follows [35]:

- **Supervised learning**. In this, one gives the computer a training set that contains data with corresponding labels. The algorithm then creates a model that maps future unknown inputs to known outputs.
- **Unsupervised learning**. In this type of learning, the training set does not contain output labels. The algorithm discovers hidden patterns in the data and then uses this to map future unknown inputs to the pattern.
- **Reinforcement learning**. The program operates in a dynamic environment where it gets inputs continuously, and the program's outputs are provided feedback whether they are right or wrong.

Depending on the desired output, machine-learning algorithms can be categorized as follows:

- **Classification**. The output is a finite number of discrete categories/classes. The algorithm should produce a model from the training data that can assign one of these classes to the new inputs. Spam filtering and credit card companies determining if a person is creditworthy or not are examples of classification problems.
- **Regression**. The output is not discrete but is one or more continuous variables. Examples include predicting output sales given the budget for TV and radio ads, predicting house prices given a set of variables, and so forth.

- **Clustering**. The objective is to group input data into clusters that contain similar data points. Examples include segmenting users based on purchase patterns, detecting activity types using motion sensors, and so forth.
- **Dimensionality reduction**. The objective is to reduce the number of dimensions with the intent of focusing on dimensions (features) that are important to the problem. It also helps in reducing complexity, space, and time to compute.

10.2.2 Frequently Used Machine-Learning Algorithms

In this section, we briefly touch on the most commonly used machine-learning (ML) algorithms [36], and this would help get a better context for the review of machine-learning algorithms utilized for IoT.

10.2.2.1 Classification

- **Logistic regression**. Predictions are mapped to be between 0 and 1 through the logistic function.
- **Classification tree**. The data are repeatedly split into separate branches to arrive at the output label.
- **Support vector machine (SVM)**. In SVM, the program views the data elements as points in an n-dimensional space. The algorithm finds a hyperplane (decision boundary) that maximizes the distance between closest points of separate classes.
- **Naïve Bayes**. In Naïve Bayes, the model is a probability table that gets created using the probability of occurrences of training data. The algorithm predicts the new output by looking up the probabilities of the input variables and using conditional probability.
- **K-nearest neighbors (KNN)**. In KNN, the algorithm predicts the class by searching through the training set for K most similar neighbors of the new input.

10.2.2.2 Regression

- **Linear regression**. In linear regression, the algorithm creates a model by fitting a straight line (or a hyperplane for n-dimensions) through the data.
- **Regression tree / decision tree**. In regression tree, the data are repeatedly split into separate branches to arrive at the output.
- **K-nearest neighbors**. In KNN, the algorithm predicts the value by searching through the training set for K most similar neighbors of the new input and summarizing the outputs.

10.2.2.3 Clustering

- **K-means**. In K-means, the algorithm creates clusters based on geometric distances between points. At the outset, the algorithm randomly assigns the data points to k clusters, computes centroids for each cluster, computes points closest to each centroid and then re-computes the centroids. The algorithm repeats the process till there are no more improvements possible. The clusters tend to be globular for K-means.

- **DBSCAN**. In DBSCAN (Density-based spatial clustering of applications with noise), clusters are created based on density. The algorithm makes an n-dimensional sphere of radius epsilon for each data point and counts the number of points inside the sphere. If the number is less than min_points, the algorithm disregards the point. If not, it computes the centroid for the sphere and continues the same process.
- **Hierarchical clustering**. The algorithm starts with n clusters for n data points. It combines two nearest clusters to create a new cluster. The algorithm repeats the process until only one cluster remains. One can view the result as a dendrogram with the height representing the distance between the clusters. If we can imagine a horizontal line that traverses the dendrogram vertically, the maximum distance it covers without intersecting another cluster gives the minimum distance between clusters. The number of vertical lines cut gives the number of clusters.

10.2.2.4 Dimensionality Reduction

- **PCA**. A principal component is a normalized linear combination of the variables in a dataset. In PCA (principal component analysis), the objective is to orthogonally project data points onto an L dimensional linear subspace that has the maximal projected variance. For PCA, the variable values need to be numerical. Hence, categorical variables are converted to numerical.
- **CCA**. Canonical correlation analysis (CCA) deals with two or more variables, and its objective is to find a corresponding pair of highly cross-correlated linear subspaces so that within one of the subspaces there is a correlation between each component and a single component from the other subspace.

10.2.2.5 Combining Models (Ensemble ML)

In many instances, a single type of algorithm may not be able to give optimal results due to the variety of the types of data or other reasons. In these cases, different algorithms are combined to give more accurate predictions than individual models.

- **CART**. In classification and regression trees (CART), The data repeatedly split into separate branches to arrive at the output label or value. Though the trees used for regression and those used for classification have some similarities, they differ in some respects, e.g., the algorithm to determine where to split.

- **Random forests**. In random forests, instead of training a single tree, a multitude of trees are trained. The algorithm outputs a class that is the mode of the training classes or the mean of the training values.
- **Bagging**. Bootstrap aggregation, also called bagging, is a general procedure that can be used to reduce the variance for an algorithm that has high variance. CART/decision tree is an algorithm that has a high variance and is sensitive to training data.

10.2.2.6 Artificial Neural Networks

Artificial neural networks (ANNs) are computing systems that model neural networks and brain in humans. ANN contains units called *neurons*. Neurons are connected to each other via synapses and communicate signals to each other. Each neuron receives inputs from other neurons connected to it and computes an output to be transmitted upstream. Each input signal has a corresponding weight, and the neuron applies a function to the weighted sum of the inputs it gets. Feed forward neural networks (FFNN), also known as multilayer perceptrons (MLP), is the most common type of neural networks in practical applications. There are other types of ANNs such as CNN (convolutional neural network), RNN (recurrent neural network), DBN (deep belief network), TDNN (time delay neural network), DSN (deep stacking network) and so forth.

10.2.3 Examples of Machine-Learning Algorithms in IoT

10.2.3.1 Overview

The main ingredient in an ML system is data. With the spread of IoT, there is a massive amount of data that gets generated on a daily basis, and this is a goldmine for machine learning. The adoption of supervised and unsupervised machine-learning techniques in IoT smart data analysis is broad. All of the smart things discussed in Section 10.1.1 – smart homes where appliances, lights, and thermostat connect to the Internet [1], smart medical appliances that not only monitor remotely but also administer medicines [2], smart bridges that have sensors to monitor load [3], smart power grids to detect disruptions and manage distribution of power [4] and smart machinery in industries that have embedded sensors in machinery to increase worker [5] – would be using or have the potential to use machine learning in some form or the other.

10.2.3.2 Examples

There are many concrete examples where machine learning saved millions of dollars for corporations:

- **Google Deepmind AI**. Google applied machine learning to 120+ variables from sensor in its data center to optimize cooling, and that cut its overall energy consumption by 15% [37].

- **Roomba 980**. This Roomba is connected to the Internet and comes with a camera that captures the images of a room and software that compares these images to gradually build up a map of the robot's surroundings to determine its location [38]. It is able to "remember" a home layout, adapt to different surfaces or new items, clean a room with the most efficient movement pattern, and dock itself to recharge its batteries.
- **NEST Thermostat**. NEST "learns" the regular temperature preferences of its users, and also adapts to the work schedule of its users by turning down energy use [39]. The input is the temperature preference of the user, time and day, presence of the user at home, etc. and the output is a discrete set of temperatures making this a classification problem.
- **Tesla cars**. Tesla enabled auto pilot service in its cars that helps in hands-free driving, including complex tasks like lane changes. Tesla cars built since 2014 have 12 sensors on the bottom of the vehicle, a front-facing camera next to the rear-view mirror, and a radar system under the nose [40]. These sensing systems are not only constantly collecting data to help the autopilot work on the road, but also to amass data that can make Teslas operate better in the future. Because all Tesla cars have an always-on wireless connection, data from driving and using autopilot is collected, sent to the cloud, and analyzed with software.

10.2.4 Machine-Learning Algorithms by IoT Domains

In this section, we summarize the machine-learning algorithms that could be used for various use cases for different domains. The data are a summarization of information from examples above and also from papers *Machine Learning for Internet of Things Data Analysis: A survey* from Mahdavinejad et. al. [41] and *Unlocking the Value of the Internet of Things (IoT) – A Platform Approach* by Misra et. al. [42].

10.2.4.1 Healthcare

Metrics to Optimize. Healthcare systems in hospitals and at home have sensors to monitor patients or surrounding. Some metrics that could use machine learning could be *remote monitoring and medication, disease management,* and *health prediction.*

Machine-Learning Algorithms

- **Classification** algorithms could be used to classify patients into different groups based on their health condition.
- **Anomaly detection** could be used to identify if someone has a problem that needs to be looked into.
- **Clustering** algorithms like *K-means* could be used to group people with similar health conditions to create profiles.

- **Feed forward neural network** could be used to make fast decisions based on a patient's continuously changing condition during illness.

10.2.4.2 Utilities – Energy/Water/Gas

Metrics to Optimize. Readings from smart meters for electricity, water, or gas could be used for *usage prediction, demand supply prediction, load balancing,* and other scenarios.

Machine-Learning Algorithms

- **Linear regression** could be used to predict usage for a particular day or time.
- **Classification algorithms** could be used to classify consumers as high-, medium-, or low-usage consumers.
- **Clustering algorithms** could be used to group consumers of similar profile together and analyze their usage patterns.
- **Artificial neural networks** could be used to dynamically balance loads if there is a surge in usage in certain areas.

10.2.4.3 Manufacturing

Metrics to Optimize. Many industries have sensors on equipment for continuous monitoring, mechanisms to track production volumes and security systems to continuously monitor. So, the metrics to optimize would be to *diagnose problems* when they occur, very *quickly,* to *predict failure* so that evasive action could be taken, *detect security breaches* into the facility or theft of goods.

Machine-Learning Algorithms

- **CART/decision tree** could be used to diagnose problems with machines.
- **Linear regression** could be used to predict failure
- **Anomaly detection** could be used to detect security breaches or anything that occurs out of the ordinary.

10.2.4.4 Insurance

Metrics to Optimize. Insurance companies would be interested in knowing what kind of cars or profiles of people are more likely to be connected with accidents. The usage pattern could be obtained by sensors in the cars. They could use that information to charge appropriate insurance premiums. Machine learning could be applied to obtain *home or car usage pattern, prediction of property damage, remote assessment of damage,* and so forth.

Machine-Learning Algorithms

- **Clustering** algorithms like *K-means* or *DBSCAN* could be used to create profiles of users who drive similarly.

- **Classification** algorithms like *Naïve Bayes* could be used to classify a customer as risky or not and also to predict whether he/she should be given insurance.
- **Decision trees** could be used to classify users or to arrive at the premium to be charged or discounts to be given.
- **Anomaly detection** could be used to determine theft or destruction of property.

10.2.4.5 Traffic

Metrics to Optimize. Traffic is a very important metric to be monitored, especially in big cities. Traffic data could be obtained via sensors in cars, data from mobiles phones, tracking devices on people, and so forth. Machine-learning algorithms could be used to *predict traffic, identify traffic bottlenecks, detect accidents* or even *predict accidents.*

Machine-Learning Algorithms

- **DBSCAN** could be used to identify roads and intersections that have high traffic.
- **Naïve Bayes** could be used to identify if a road needs maintenance or whether it is susceptible to accidents.
- **Decision trees** could be used to divert users onto a less-trafficked road.
- **Anomaly detection** could be used to determine if there is an accident on the road.

10.2.4.6 Smart City – Citizens and Public Places

Metrics to Optimize. In a smart city, it is essential to optimize facilities for citizens. Based on data from smartphones, ATMs, vending machines, traffic cameras, bus/train terminals or other tracking devices, and machine-learning algorithms can *predict the travel patterns of people, density of population at certain places, predict abnormal behaviors, forecast energy consumption, forecast needs for public infrastructure* like housing, transportation, shopping, and more.

Machine-Learning Algorithms

- **DBSCAN** could be used to identify places in the city that have high concentrations of people during different times of the day.
- **Linear regression** or **Naïve Bayes** could be used to forecast energy consumption or the need for improvement of public infrastructure.
- **CART** could be used for real-time passenger travel prediction as well as to identify travel patterns.

- **Anomaly detection** could be used to determine unusual behavior like terrorism or financial fraud.
- **PCA** could be used to reduce the number of dimensions to simplify analysis since the sheer volume of data generated by multiple devices in a city is huge.

10.2.4.7 Smart Homes

Metrics to Optimize. Smart homes are one area where IoT devices have increased multifold in the past decade. They are equipped with smart meters to monitor energy, devices like Nest and Ecobee to control temperature automatically and remotely, smart bulbs like Philips Hue that could be automated and controlled remotely, smart switches, fitness bands, smart locks, security cameras, and so forth. Multiple sensors and the amount and quality of generated data can be harnessed by machine-learning algorithms to provide valuable insights like *occupancy awareness, intrusion detection, gas leakage, energy consumption prediction, television viewing preferences and prediction*, and so forth.

Machine-Learning Algorithms

- **K-means** could be used to analyze load and consumption frequency of energy.
- **Linear regression** or **Naïve Bayes** could be used to forecast energy consumption or occupancy prediction.
- **Anomaly detection** could be used to determine intrusion detection, tampering with devices, burglary, device malfunction, and so forth.

10.2.4.8 Agriculture

Metrics to Optimize. As the demand for food increases with rise in population, large-scale farms are beginning to use sensors in the fields, drones to take pictures, and other IoT devices to be able to *optimize resource usage, detect crop diseases faster*, and *predict production.* AgTech (agriculture technology) is a growing field of active research.

Machine-Learning Algorithms

- **Naïve Bayes** could be used to determine if a crop is healthy or not.
- **Anomaly detection** could be used to determine if there is a water leakage, uneven supply of water.
- **Neural networks** could be used to analyze pictures taken by drones to identify weed growth or if patches in the field are growing slower than others.

In many ways, machine learning and IoT have a symbiotic relationship. IoT provides machine learning with large amount of data and machine learning is revolutionizing IoT by making the simple devices much smarter than they are.

In an article about machine learning revolutionizing IoT, Ahmed [43] mentions three ways in which ML is changing IoT:

1. Making IoT data useful
2. Making IoT more secure
3. Expanding the scope of IoT

In the next section, we review how machine learning is making IoT more secure.

10.3 Survey of ML Techniques for Defending IoT Devices

10.3.1 Systematic Categorization of ML Solutions for IoT Security

In the previous section, we did a review of a lot of use cases where machine-learning algorithms were used for IoT. Some of the key tasks like discovering a pattern in existing data, detecting outliers, predicting values, and feature extraction are critical to IoT security. Some of the machine-learning algorithms used for these tasks are tabulated in Table 10.2.

In most papers studied in this research, the main objective has been to detect a security breach. Hence, the second point in Table 10.2 becomes very critical from a security perspective. From the point of detecting outliers, the use cases can be further divided into the following:

- Malware detection
- Intrusion detection
- Data anomaly detection

Since *anomaly detection* is basically a classification problem, it follows that the most used machine learning techniques are the ones that are commonly

Table 10.2 Categorization of ML solutions for IOT security.

Use case	ML algorithm
Pattern discovery	• K-means [44] • DBSCAN [45]
Discovery of unusual data points	• Support vector machine [46] • Random forest [47] • PCA [48] • Naïve Bayes [48, 49] • KNN [48]
Prediction of values and categories	• Linear regression [41] • Support vector regression [41] • CART [41] • FFNN [41]
Feature extraction	• PCA [41] • CCA [41]

Table 10.3 Categorization of ML solutions for outlier detection.

Use case	ML algorithm
Malware detection	• SVM [46] • Random Forest [47]
Intrusion detection	• PCA [48] • Naïve Bayes [48, 49] • KNN [48]
Anomaly detection	• Naïve Bayes [48] • ANN [50, 51]

used in classification. These include *decision trees, Bayesian networks, Naïve Bayes, random forests,* and *support vector machines (SVM).* In many new instances, *artificial neural networks (ANNs)* have been used. ANNs are generally not used for malware detection since it takes longer time for training. Machine-learning algorithms for these use cases are tabulated in Table 10.3.

The next section reviews examples of machine-learning algorithms used for the use cases in Table 10.3 by summarizing results from research paper on each of the machine-learning algorithms.

10.3.2 Examples of ML Algorithms for IoT Security

10.3.2.1 Malware Detection Using SVM

In their paper for Android Malware detection using Linear SVM, Ham et al. [46] review various approaches for detecting malware, such as *signature based, behavior based,* and *taint analysis based* detection, and show that Linear SVM showed high performance among ML algorithms used to effectively detect malware. In a *behavior-based detection* system, in order to detect abnormal patterns, event information on the device like memory usage, data content, and energy consumption are monitored. ML techniques are used to analyze the data, and hence, the choice of features is very important.

10.3.2.2 Malware Detection Using a Random Forest

In their paper for Android malware detection using a random forest, Alam et al. [47] apply ML ensemble learning algorithm random forest on an Android feature dataset of 48919 points of 42 features each. Their goal was to measure the accuracy of random forests in classifying Android application behavior to classify applications as malicious or benign. They also analyzed the detection accuracy as the parameters of RF algorithm, such as the number of trees, depth of each tree, and number of random features were changed. The results based on fivefold cross validation showed that RF performed very well with an accuracy of over 99% in general, an optimal out-of-bag (OOB) error rate of 0.0002 for forests with 40 trees or more, and a root mean squared error of 0.0171 for 160 trees.

10.3.2.3 Intrusion Detection Using PCA, Naïve Bayes, and KNN

In their paper for anomaly-based intrusion detection, Pajouh et al. [48] present a novel model for intrusion detection based on two-layer dimension reduction and two-tier classification module, designed to detect malicious activities such as user to root (U2R) and remote to local (R2L) attacks. Their proposed model used PCA and linear discriminate analysis (LDA) to reduce the high dimensional dataset to a lower one with lesser features. They then applied a two-tier classification module utilizing Naïve Bayes and certainty factor version of K-nearest neighbor to identify suspicious behaviors.

10.3.2.4 Anomaly Detection Using Classification

In their paper for designing an IoT device for the safety of women, Jatti et al. [49] describe the design of a device that determines whether the wearer is in danger. The device transmits data related to physiology and body position of the person. The physiological signals that are transmitted are galvanic skin response (GSR) and body temperature. Body position is determined by acquiring raw accelerometer data from a triple axis accelerometer. The premise is that when a person is faced with a dangerous situation, secretion of adrenalin affects different systems in the body, resulting in increased blood pressure and heart rate and also sweating. This increases skin conductance, measured by GSR. The data are analyzed by an ML classifier that determines if the individual is in a dangerous situation, such as threat of rape.

10.3.3 Use of Artificial Neural Networks (ANN) to Forecast and Secure IoT Systems

Before the data get to the Internet and into the cloud, it could come from two kinds of IoT devices – edge devices or gateway devices. In general terms, when we refer to the billions of IoT devices that are gathering information, we talk about edge devices, which in themselves are dumb devices that are programmed to do a specific simple task, say measuring temperature. In comparison to edge devices, gateway devices have more resources and computing power. Hence, instead of focusing on security configurations at every edge device, one could focus energy on gateway devices to have a larger impact. In fact, in *Neural Network Approach to Forecast the State of the Internet of Things Elements* [50], Kotenko et al. talk about the use of artificial neural networks to predict the state of an IoT element and that this could reduce the labor costs of IoT administration. Here there is an implicit acknowledgment that security configurations at the edge are labor cost intensive. The approach in the paper combined a multi-layered perceptron network along with a probabilistic neural network. The experiments revealed that by using the multilayer perceptron network to explore similar values in the past, one could use a probabilistic neural network to determine the state of the device.

Canedo et al. [51] propose using machine learning within an IoT gateway to help secure the system. The proposal was to use an ML technique, specifically ANN, in gateway and application layers; in gateway to monitor subsystem components and in the application layer to monitor the state of the entire system. After setting up the system with training data and warming it up, the researchers manipulated the sensors to add invalid data for a 10-minute period. When the invalid data was run against the system, the neural network was able to detect the differences between the valid and invalid data. They then added a delay between transmissions as the third input to simulate man-in-the-middle attacks and they were able to predict whether the data was valid or invalid for the approximately 360 samples in the testing set and summarized that the use of ANN is very beneficial for making an IoT system more secure.

10.3.4 New Flavors of Attacks on IoT Devices

Although in the past hacking into a device to steal data, snooping to determine the information at the remote end, and so forth, were common types of attacks, the attacks in recent times have changed the landscape for IoT and put IoT devices as the leading potential cause for bringing the Internet down. In the article "Someone Is Learning How to Take Down the Internet" [52], Bruce Schneier says that based on the analysis of recent attacks, the attacker may not be the traditionally assumed types like activists, researchers, or criminals. The attack could be state-sponsored, and the world might be embarking on an era of cyber warfare. Here are some recent examples of IoT malware attacks from Perry [53].

10.3.4.1 Mirai

This DDoS attack is covered in Section 12.3.4. It took down half the Internet in the United States and Europe for hours. Mirai scans the Internet for hosts with an open telnet port and gains access if the password is weak. After it gets inside, it installs the malware and monitors the CNC (command and control) center. During the attack, the CNC instructs all the bots to create a flood of traffic and overwhelm the target. Perry [53] suggests that to protect the devices, one should take the following measures:

- Always change default password.
- Remove devices with telnet backdoors.
- Limit exposing a device directly to the Internet.
- Run port scans of all the devices.

10.3.4.2 Brickerbot

This bot makes the device under attack unusable, i.e. turns it into a *brick*. Once the malware obtains access to the device, it runs a series of commands to wipe data from the device's storage. This renders the device useless.

10.3.4.3 FLocker

FLocker (short for *Frantic Locker*) is a bot that locks the target device and prevents valid users from accessing it. Users could be asked to pay ransom or might lose access to the device and may have to hard-delete all data. Norton Security [54] has noted its use for targeting Android Smart TVs.

10.3.4.4 Summary

In summary, IoT attacks are increasing, and new variants of the attacks are created often. A report from F5 labs [55] shows that IoT attacks exploded by 280% in the first half of 2017 with a large chunk of this growth stemming from Mirai. Moreover, the report claims that 83% of attacks came from a single hosting provider in Spain called SoloGigabit that had a "bulletproof" reputation.

10.3.5 Proposal for Effective ML Techniques to Achieve IoT Security

10.3.5.1 Insights from the Research

Based on the research done on ML techniques used for IoT security, it is evident that different techniques need to be used for different scenarios. There is no one-size-fits-all solution because of the complexity of the problem statement. Also, anomalies in data can occur at different layers in the IoT ecosystem. Multiple devices could be hacked, resulting in wrong access patterns or data dispatch, or a gateway could be hacked, resulting in data routing. This would mean that the training system could get incomplete data or different types of data. In these cases, classic ML algorithms might fail to operate – SVM needs standardized numerical data, as the input to a decision tree cannot traverse through a branch in the tree when values are missing. In these cases, the best option is ensemble machine learning.

The other insight that came out of the research is that there are increasing use cases where IoT data must be analyzed as data are streamed, and decisions must be taken quickly. This means that the data cannot wait to be sent to the cloud and processed. Hence, new paradigms like fog computing and edge computing are more relevant for IoT security than others. Table 10.4 shows characteristic of data in smart city use case mentioned in Mahdavinejad et.al. [41] and it is clear that there are many use cases that need data to be processed near the device for quicker turnaround.

To summarize the insights:

1. IoT devices and data are diverse and need different machine-learning algorithms to analyze different aspects of the system.
2. IoT data need to be analyzed closer to the device than in the cloud.

Table 10.4 Where data should be processed.

Use case	Type of data	Where it is best to be processed
Smart Traffic	Stream/massive data	Edge
Smart Health	Stream/massive data	Edge/cloud
Smart Environment	Stream/massive data	Cloud
Smart Weather Prediction	Stream data	Edge
Smart Citizen	Stream data	Cloud
Smart Agriculture	Stream data	Edge/cloud
Smart Home	Massive/historical data	Cloud
Smart Air Controlling	Massive/historical data	Cloud
Smart Public Place Monitoring	Historical data	Cloud
Smart Human Activity Control	Stream/historical data	Edge/cloud

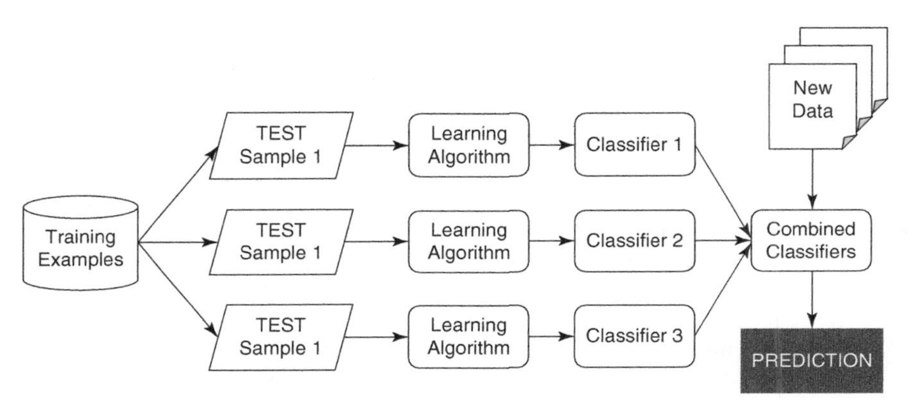

Figure 10.5 Ensemble machine learning.

10.3.5.2 Proposals

Proposal #1. Use ensemble machine learning method for IoT data analysis in the cloud. Ensemble machine learning method uses multiple machine-learning algorithms to obtain better predictive performance than what could be obtained from a single algorithm alone. It would also perform much better for different types of data and missing data. Figure 10.5 depicts the general idea behind ensemble machine learning.

Proposal #2. Use fog computing for data analysis closer to the edge. This would mean that decisions could be taken faster. Also, it would be more relevant to the device or groups of devices serviced by the fog computing node.

It is with this intent that the next two sections are entirely focused on fog computing and machine-learning algorithms used in fog computing use cases.

10.4 Machine Learning in Fog Computing

10.4.1 Introduction

As noted earlier, the amount of data generated by IoT devices is expected to soar to 400 zettabytes by 2018 and grow exponentially every year. There are multiple issues with a cloud-only architecture where data from IoT devices make it to the cloud to be processed and analyzed:

- **Network traffic congestion**. By 2020, there will be over 50 billion things connected to the Internet, and if processing of the data happens in the cloud, there would be a network congestion and data may not get to the server and back fast enough.
- **Data bottleneck**. If data storage and analysis is done only in the cloud, there could be a bottleneck if the server is slower in analyzing due to the volume of the data or for other reasons.
- **Security issues**. Since data must travel through multiple layers from sensors to gateways to services to the cloud, there are numerous points of a breach. Also, a security solution in the cloud may address issues that are common to most devices and may not be able to take care of specific sensors or nodes on the edge.
- **Data staleness**. In many instances, data loses its value when it cannot be analyzed fast enough. Security cameras, phones, cars, ATMs, and so forth, could generate data that need immediate analysis if there is a security or a privacy issue.

Fog computing solves this by selectively moving compute, storage, and decision-making closer to the network edge where data are being generated. OpenFog Reference Architecture for fog computing defines fog computing as "*A horizontal, system-level architecture that distributes computing, storage, control and networking functions closer to the users along a cloud-to-thing continuum*" [56]. Essential characteristics of fog computing platforms include low latency, location awareness, and wired or wireless access. There are numerous benefits to this:

- **Real-time analytics**. As IoT usage grows, the number of scenarios where real-time analytics is needed occurs too often (e.g. security camera capturing a potential intruder lurking in front of a home or a fraudster gaining access to someone's account). By the time data get uploaded to the cloud and get analyzed; it may be too late. These scenarios need near-instant intelligence that fog computing provides.

- **Improved security**. Since fog is nearer to the edge, it has the capability to configure security that is tailored to the devices and their functions. Also, security decisions regarding whether to block access during a breach can be taken almost instantaneously.
- **Data thinning at the edge**. Fog consumes the raw data and makes decisions or provides insights. It sends only relevant, consolidated information upward in the hierarchy. This dramatically reduces the amount of data that gets transmitted to a central data center.
- **Cost savings**. Fog may have higher setup costs due to distributed nature of deployments, but operational costs and long-term benefits of the overall system would outweigh this.

10.4.2 Machine Learning for Fog Computing and Security

One of the main advantages of fog computing is the ability to do near real-time analytics, and in many cases, this means utilizing machine learning at the fog nodes.

We could find many examples from the case studies reviewed in Section 10.4.3 where machine learning could be used. One example could be in industries where machine learning could help in fault isolation and fault detection of machines and thus improve MTTR (mean time to repair) of a failed system to achieve higher availability. Another example could be a train station in a smart city, where machine learning could be used to optimize operations by monitoring occupancy, movement, and overall system usage and over time. More examples are reviewed in the next section.

At the fog nodes, analytics can be both reactive as well as predictive. The fog nodes closer to the edge will most likely have reactive analytics, and the nodes farther from the edge will have more predictive analytics since it needs more computation power. The basic premise is that computing power is highest in the cloud and it goes down in the hierarchy referred to section 10.4.4 on *n*-tier architecture. Machine-learning algorithms can be run at fog nodes that have the processing power to compute corresponding to the task at that layer (see Table 10.5). Machine-learning models are created at the nodes near the cloud or in the cloud itself. The models could be downloaded to middle-tier nodes to help in execution.

10.4.3 Examples of Machine Learning in Fog Computing

10.4.3.1 ML in Fog Computing in Industry

Traditional cloud-based or noncloud centralized analytics infrastructures rely on training a machine learning algorithm by using data from past failures. The algorithm would create a model that could be used to predict failure. But in

Table 10.5 ML Use cases for fog computing.

Use case	ML algorithm
Fog computing in industry – Remote monitoring for oil & gas operations [57]	• Anomaly detection models • Predictive models • Optimization methods
Fog computing in retail – Retail customer behavior analysis [57]	• Statistical methods • Time series clustering
Fog computing in self-driving cars [57]	• Image processing • Anomaly detection • Reinforced learning

many instances, failure prediction is too late to prevent the breakdown and is used to minimize the effect of damage. In comparison, if near-instant analytics is done locally using fog computing, the system would be able to take steps to prevent the occurrence of the issue. That is because the analytics system is nearer to the edge and has more context.

10.4.3.2 ML in Fog Computing in Retail

Retail stores, in general, do product placement based on analytics derived from customer purchases and also seasonal preferences. So, we see product placements change during Halloween, Thanksgiving, Christmas, and so forth. If fog computing is used with analytics being done for a store or a group of stores in an area, the system would be able to analyze buying patterns of the users in the locality and help the store to target merchandise better and improve customer experience.

10.4.3.3 Fog Computing for Self-Driving Cars

With Google, Tesla, Uber, GM, and other mainstream companies testing self-driving cars, the reality of having these vehicles for mainstream use cases is very near. Self-driving automobiles are excellent examples of fog computing, since a lot of computing and decision-making happens on the edge. Nevertheless, each car transmits a lot of data for processing in the cloud. An N-tier model would make the system considerably more efficient. Machine-learning algorithms used are ANN for image processing, Naïve Bayes or similar algorithms for anomaly detection, reinforced learning, and so forth.

10.4.4 Machine Learning in Fog Computing Security

Tang et al. [58] present a hierarchical structure for fog computing architecture to support the integration of massive number of infrastructure components and services in future smart cities. The architecture laid out in the paper is a four-layer model, with the first layer being the cloud and the last being the

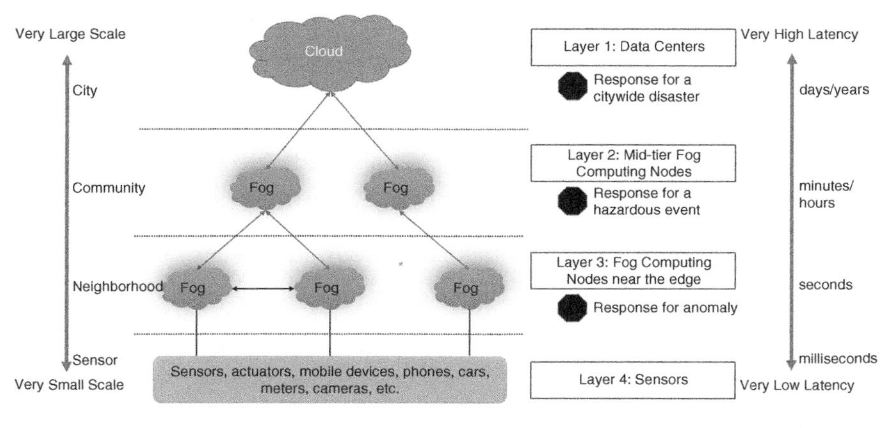

Figure 10.6 Fog computing security at multiple layers.

sensors. The layers in between are the fog layers. Figure 10.6 shows the different layers and the primary security handling at each layer.

Layer 3 contains fog nodes that get raw data from the sensors. The nodes at this layer perform two functions. One identifies potential threat patterns on the incoming data streams from sensors using machine-learning algorithms, and the other performs feature extraction for reducing the amount of data to be sent upstream. The paper [58] does not specify how anomaly detection is done. Algorithms like KNN, Naïve Bayes, random forests, or DBSCAN could be used to do anomaly detection.

Layer 2 contains fog nodes that get data from nodes below them, and the data represent information from hundreds of sensors across locations. In the paper, HMM (hidden Markov model) and MAP (maximum aposteriori) algorithms are used for classification and alert if there is a hazardous event. Table 10.6 summarizes the machine-learning algorithms at each fog layer.

Table 10.6 Machine-learning algorithms at different fog layers.

Layer	Disaster response	ML algorithm
Layer 4 - Sensors	None	None
Layer 3 – Fog nodes for the neighborhood	Response for anomaly	KNN, Naïve Bayes, random forest, DBSCAN
Layer 2 – Fog nodes for the community	Response for hazardous event	HMM, MAP [58] Regression, ANN, decision trees
Layer 1 – Cloud	Response for city-wide disaster, long-term forecasting	ANN, Deep learning, decision trees, reinforcement learning, Bayesian networks

10.4.5 Other Machine-Learning Algorithms for Fog Computing

Section 10.3.1 categorized an ML solution for IoT security into pattern discovery, anomaly detection, value/label prediction, and feature extraction. We reviewed essential ML algorithms like K-means, DBSCAN, Naïve-Bayes, random forest, CART, PCA, and so forth. We also did a deep-dive on anomaly detection use cases specifically focusing on malware and intrusion detection. All these use cases and examples apply to fog computing, such as malware detection using SVM [46], Malware detection using random forest [47], and intrusion detection [48] can be done in the fog nodes instead of on the cloud. In fact, anomaly detection using ANN by Kotenko [50] particularly talks about doing machine learning at the gateway layer, which is synonymous with doing it at a mid-tier fog node.

In conclusion, fog computing can make IoT ecosystem more secure by being more contextual, being able to detect issues faster and reacting quicker to events.

10.5 Future Research Directions

As discussed, application of machine learning is very critical to IoT security due to the volume and variety of data. AI and ML are fast-growing fields and IoT data analysis needs to be on par with the latest trends in these areas. Review of numerous machine-learning techniques and several examples in IoT point to the fact that analyzing data in near real-time at the proximity of the node is important. Hence, research on machine-learning algorithms that need lesser memory and can process large amounts of time series data quickly is needed.

We could categorize future research directions as follows:

- Usage of latest trends in AI and ML toward IoT security
- ML algorithms for fog computing security focused on techniques that use lesser memory and can process large amount of data quickly
- ML algorithms in new areas of IoT sensor development in multiple industries
- ML algorithms to analyze healthcare data – specific focus could be done on WIBSNs (wireless and implantable body sensor networks)

10.6 Conclusions

In this chapter, we covered a range of topics starting from introduction to IoT, IoT architecture, IoT security, and privacy concerns, fog computing, machine learning for IoT security and machine learning in IoT security through fog computing. In each section, we defined the concept and then proceeded to expand the topic with references and examples.

First, we introduced the concept of the Internet of Things (IoT), common IoT devices, IoT architecture with a focus on four-layer architecture, IoT applications, especially in the healthcare domain. With various examples, we showed how IoT devices have become ubiquitous and have pervaded almost every sphere of our lives ushering an era of smart things. Then we reviewed critical security and privacy issues with IoT devices and the ecosystem. With examples such as hacks of water treatment plants, nuclear power plant, baby monitor videos, wearable devices, and so forth, we showed the seriousness of the security issue. We used DDoS (distributed denial of service) as an example to show how IoT devices have been used to cripple the internet and bring down essential services to people in different parts of the world. Then we did a quick study of machine learning and commonly used machine-learning algorithms and then delved into examples of machine learning used in IoT.

We took a look at examples like smart home, smart medical appliances, smart power grids, Roomba vacuum, Tesla, and so forth. Then we further reviewed use cases per domains like manufacturing, healthcare, utilities, and so forth and gave examples of ML algorithms in each. Then we focused on machine-learning techniques for IoT security. By reviewing several papers and websites, we categorized the fundamental ML tasks used in defending IoT systems and then summarized a few papers focused on machine learning for IoT security with focus on malware detection, intrusion detection, and anomaly detection. In the end, we concluded that bringing computing closer to the edge and using ensemble learning techniques could provide reliable defense against attacks on IoT devices. We also concluded that fog computing is a critical emerging field within IoT domain and machine-learning algorithms used in fog nodes are critical to the success and scalability of IoT.

References

1 IBM Electronics. The IBM vision of a smart home enabled by cloud technology, December 2010. https://www.slideshare.net/IBMElectronics/15-6212631. Accessed September 2017.

2 M. Cousin, T. Castillo-Hi, G.H. Snyder. Devices and diseases: How the IoT is transforming MedTech. *Deloitte Insights* (2015, September). https://dupress.deloitte.com/dup-us-en/focus/internet-of-things/iot-in-medical-devices-industry.html. Accessed September 2017.

3 S. Wende and C. Smyth. The new Minnesota smart bridge. http://www.mnme.com/pdf/smartbridge.pdf. Accessed September 2017.

4 D. Cardwell. Grid sensors could ease disruption of power. *The New York Times* (2015, February). https://www.nytimes.com/2015/02/04/business/energy-environment/smart-sensors-for-power-grid-could-ease-disruptions.html. Accessed September 2017.

5 K.J. Wakefield. How the Internet of Things is transforming manufacturing. *Forbes* (2014, July). https://www.forbes.com/sites/ptc/2014/07/01/how-the-internet-of-things-is-transforming-manufacturing. Accessed September 2017.
6 Cisco. Cisco global cloud index: forecast and methodology, 2015–2020, 2016. https://www.cisco.com/c/dam/m/en_us/service-provider/ciscoknowledgenetwork/files/622_11_15-16-Cisco_GCI_CKN_2015-2020_AMER_EMEAR_NOV2016.pdf. Accessed September 2017.
7 T. Barnett Jr. The dawn of the zettabyte era [infographic], 2011. http://blogs.cisco.com/news/the-dawn-of-the-zettabyte-era-infographic. Accessed September 2017.
8 D. Worth. Internet of things to generate 400 zettabytes of data by 2018, November 2014. http://www.v3.co.uk/v3-uk/news/2379626/internet-of-things-to-generate-400-zettabytes-ofdata-by-2018. Accessed September 2017.
9 J. Leyden. Water treatment plant hacked, chemical mix changed for tap supplies. *The Register* (2016, March). http://www.theregister.co.uk/2016/03/24/water_utility_hacked. Accessed September 2017.
10 K. Zetter. Everything we know about Ukraine's power plant hack. *Wired* (2016, January). https://www.wired.com/2016/01/everything-we-know-about-ukraines-power-plant-hack. Accessed September 2017.
11 P. Paganini. Hacking baby monitors is dramatically easy, September 2015. http://securityaffairs.co/wordpress/39811/hacking/hacking-baby-monitors.html. Accessed September 2017.
12 A. Tillin. The surprising way your fitness data is really being used. *Outside* (2016, August). https://www.outsideonline.com/2101566/surprising-ways-your-fitness-data-really-being-used. Accessed September 2017.
13 L. Cox. Security experts: hackers could target pacemakers. *ABC News* (2010, April). http://abcnews.go.com/Health/HeartFailureNews/security-experts-hackers-pacemakers/story?id=10255194. Accessed September 2017.
14 S. Morgan. Hackerpocalypse: a cybercrime revelation, 2016. https://cybersecurityventures.com/hackerpocalypse-cybercrime-report-2016/. Accessed September 2017.
15 IBM Analytics. The IBM Point of View: Internet of Things security. (2015, April). https://www-01.ibm.com/common/ssi/cgi-bin/ssialias?htmlfid=RAW14382USEN. Accessed October 2017.
16 OWASP. IoT attack surface areas. (2015, November). https://www.owasp.org/index.php/IoT_Attack_Surface_Areas. Accessed November 2017.
17 Hewlett Packard. Internet of things research study, 2015. http://www8.hp.com/h20195/V2/GetPDF.aspx/4AA5-4759ENW.pdf. Accessed March 10, 2016.
18 L. Zanolli, Welcome to privacy hell, also known as the Internet of Things. *Fast Company* (2015, March 23). http://www.fastcompany.com/3044046/

tech-forecast/welcome-to-privacy-hell-otherwise-known-as-the-internet-of-things. Accessed March 24, 2016.

19 J. Schectman. Internet of Things opens new privacy litigation risks. *The Wall Street Journal* (2015, January 28). http://blogs.wsj.com/riskandcompliance/2015/01/28/internet-of-things-opens-new-privacy-litigation-risks. Accessed March 24, 2016.

20 B. Violino. Benetton to Tag 15 Million Items. *RFiD Journal* (2003, March). http://www.rfidjournal.com/articles/view?344. Accessed March 23, 2016.

21 FTC. FTC Report on Internet of Things urges companies to adopt best practices to address consumer privacy and security risks (2015, January 27). https://www.ftc.gov/news-events/press-releases/2015/01/ftc-report-internet-things-urges-companies-adopt-best-practices. Accessed March 24, 2016.

22 A. F. Westin. Privacy and freedom. *Washington and Lee Law Review,* 25(1): 166, 1968.

23 J. H. Ziegeldorf, O. G. Morchon, K. Wehrle. Privacy in the internet of things: Threats and challenges. *Security Community Network,* 7(12): 2728–2742, 2014.

24 Keycdn. DDoS Attack. (2016, July). https://www.keycdn.com/support/ddos-attack/. Accessed October 2017.

25 Ddosbootcamp. Timeline of notable DDOS events. https://www.ddosbootcamp.com/course/ddos-trends. Accessed October 2017.

26 J. Hamilton. Dyn DDOS Timeline. (2016, October). https://cloudtweaks.com/2016/10/timeline-massive-ddos-dyn-attacks. Accessed October 2017.

27 P. Stancik. *At least 15% of home routers are unsecured.* (2016, October). https://www.welivesecurity.com/2016/10/19/least-15-home-routers-unsecure/. Accessed October 2017.

28 S. Cobb. 10 things to know about the October 21 IoT DDoS attacks. (2016, October). https://www.welivesecurity.com/2016/10/24/10-things-know-october-21-iot-ddos-attacks/. Accessed October 2017.

29 L. Ulanoff. 73,000 webcams left vulnerable because people don't change default passwords. (2014, November). http://mashable.com/2014/11/10/naked-security-webcams. Accessed October 2017.

30 M. Warner. Senators Introduce Bipartisan Legislation to Improve Cybersecurity of "Internet-of-Things" (IoT) Devices. (2017, August). https://www.warner.senate.gov/public/index.cfm/2017/8/enators-introduce-bipartisan-legislation-to-improve-cybersecurity-of-internet-of-things-iot-devices. Accessed November 2017.

31 M. Warner. Internet of Things Cybersecurity Improvement Act of 2017 (2017, August). https://www.scribd.com/document/355269230/Internet-of-Things-Cybersecurity-Improvement-Act-of-2017. Accessed November 2017.

32 WaterISAC. 10 Basic Cybersecurity Measures. (2015, June). https://ics-cert.us-cert.gov/sites/default/files/documents/10_Basic_Cybersecurity_Measures-WaterISAC_June2015_S508C.pdf. Accessed November 2017.

33 US-CERT. Heightened DDoS threat posed by Mirai and other botnets. (2016, October). https://www.us-cert.gov/ncas/alerts/TA16-288A. Accessed November 2017.

34 A.L. Samuel. Some studies in machine learning using the game of checkers. *IBM Journal of Research and Development,* 44 (1–2): 206–226, 2000.

35 SAS. Machine Learning: What it is and why it matters. https://www.sas.com/en_us/insights/analytics/machine-learning.html. Accessed November 2017.

36 P.N. Tan, M. Steinbach, and V. Kumar (2013). *Introduction to Data Mining.*

37 J. Vincent. Google uses DeepMind AI to cut data center energy bills. (2016, July). Retrieved November, 2017, from https://www.theverge.com/2016/7/21/12246258/google-deepmind-ai-data-center-cooling. Accessed November 2017.

38 W. Knight. The Roomba now sees and maps a home. *MIT Technology Review* (2015, September 16). https://www.technologyreview.com/s/541326/the-roomba-now-sees-and-maps-a-home/. Accessed October 2017.

39 Nest Labs. Nest Labs introduces world's first learning thermostat. (2011, October). https://nest.com/press/nest-labs-introduces-worlds-first-learning-thermostat/. Accessed October 2017.

40 K. Fehrenbacher. How Tesla is ushering in the age of the learning car (2015, October). http://fortune.com/2015/10/16/how-tesla-autopilot-learns/. Accessed October 2017.

41 M. S. Mahdavinejad, M. Rezvan, M. Barekatain, P. Adibi, P. Barnaghi, and A.P. Sheth. Machine learning for Internet of Things data analysis: A survey. *Digital Communications and Networks*, 4(3) (August): 161–175, 2018.

42 P. Misra, A. Pal, P. Balamuralidhar, S. Saxena, and R. Sripriya. Unlocking the value of the Internet of Things (IoT) – A platform approach. *White Paper*, 2014.

43 M. Ahmed. Three ways machine learning is revolutionizing IoT. (2017, October). https://www.networkworld.com/article/3230969/internet-of-things/3-ways-machine-learning-is-revolutionizing-iot.html. Accessed November 2017.

44 A.M. Souza and J.R. Amazonas. An outlier detect algorithm using big data processing and Internet of Things architecture. *Procedia Computer Science* 52 (2015): 1010–1015.

45 M.A. Khan, A. Khan, M.N. Khan, and S. Anwar. A novel learning method to classify data streams in the Internet of Things. In *Software Engineering Conference (NSEC),* November 2014, National: 61–66.

46 H.S. Ham, H.H. Kim, M.S. Kim, and M.J. Choi. Linear SVM-based android malware detection for reliable IoT services. *Journal of Applied Mathematics* (2014).

47 M.S. Alam, and S.T. Vuong. Random forest classification for detecting android malware. In *Green Computing and Communications (GreenCom),*

2013 IEEE and Internet of Things (iThings/CPSCom), IEEE International Conference on and IEEE Cyber, Physical and Social Computing. (2013, August): 663–669.

48 H. H. Pajouh, R. Javidan, R. Khayami, D. Ali, and K.K.R. Choo. A two-layer dimension reduction and two-tier classification model for anomaly-based intrusion detection in IoT backbone networks. *IEEE Transactions on Emerging Topics in Computing,* 2016.

49 A. Jatti, M. Kannan, R.M. Alisha, P. Vijayalakshmi, and S. Sinha. Design and development of an IOT-based wearable device for the safety and security of women and girl children. In *Recent Trends in Electronics, Information & Communication Technology (RTEICT), IEEE International Conference* on (pp. 1108–1112), 2016, May. IEEE.

50 I. Kotenko, I. Saenko, F. Skorik, S. Bushuev. Neural network approach to forecast the state of the Internet of Things elements. *2015 XVIII International Conference on Soft Computing and Measurements (SCM),* 2015. doi:10.1109/scm.2015.7190434.

51 J. Canedo, and A. Skjellum. Using machine learning to secure IoT systems. *2016 14th Annual Conference on Privacy, Security and Trust (PST),* 2016. doi:10.1109/pst.2016.7906930.

52 B. Schneier. Someone is learning how to take down the Internet. (2016, September). https://www.lawfareblog.com/someone-learning-how-take-down-internet. Accessed November 2017.

53 J.S. Perry. Anatomy of an IoT malware attack. (2017, October). https://www.ibm.com/developerworks/library/iot-anatomy-iot-malware-attack/. Accessed November 2017.

54 N. Kovacs. FLocker ransomware now targeting the big screen on Android smart TVs. (2016, June). https://community.norton.com/en/blogs/security-covered-norton/flocker-ransomware-now-targeting-big-screen-android-smart-tvs. Accessed November 2017.

55 S., Boddy, K. Shattuck, The hunt for IoT: The Rise of Thingbots. (2017, August). https://f5.com/labs/articles/threat-intelligence/ddos/the-hunt-for-iot-the-rise-of-thingbots. Accessed November 2017.

56 OpenFog Consortium Architecture Working Group. OpenFog Reference Architecture for Fog Computing. *OPFRA001,* 20817 (2017, February). 162.

57 H. Vadada. Fog computing: Outcomes at the edge with machine learning. (2017, May). https://towardsdatascience.com/fog-computing-outcomes-at-the-edge-using-machine-learning-7c1380ee5a5e. Accessed November 2017.

58 B. Tang, Z. Chen, G. Hefferman, T. Wei, H. He, and Q. Yang. A hierarchical distributed fog computing architecture for big data analysis in smart cities. In *Proceedings of the ASE BigData & SocialInformatics* 2015 (p. 28). ACM.

Part III

Applications and Issues

11

Fog Computing Realization for Big Data Analytics

Farhad Mehdipour, Bahman Javadi, Aniket Mahanti, and Guillermo Ramirez-Prado

11.1 Introduction

Internets of Things (IoT) deployments generate large quantities of data that need to be processed and analyzed in real time. Current IoT systems do not enable low-latency and high-speed processing of data and require offloading data processing to the cloud (example applications include smart grid, oil facilities, supply chain logistics, and flood warning). The cloud allows access to information and computing resources from anywhere and facilitates virtual centralization of application, computing, and data. Although cloud computing optimizes resource utilization, it does not provide an effective solution for hosting big data applications [1]. There are several issues, which hinder adopting IoT-driven services, namely:

- Moving large amounts of data over the nodes of a virtualized computing platform may incur significant overhead in terms of time, throughput, energy consumption, and cost.
- The cloud may be physically located in a distant data center, so it may not be possible to service IoT with reasonable latency and throughput.
- Processing large quantities of IoT data in real time will increase as a proportion of workloads in data centers, leaving providers facing new security, capacity, and analytics challenges.
- Current cloud solutions lack the capability to accommodate analytic engines for efficiently processing big data.
- Existing IoT development platforms are vertically fragmented. Thus, IoT innovators must navigate between heterogeneous hardware and software services that do not always integrate well together.

Fog and Edge Computing: Principles and Paradigms, First Edition.
Edited by Rajkumar Buyya and Satish Narayana Srirama.

To address these challenges data analytics could be performed at the network edge (or the fog) – near where the data are generated – to reduce the amount of data and communications overhead [2–6]. Deciding what to save and what to use is as important as having the facility to capture the data. Rather than sending all data to a central computing facility such as the cloud, analytics at the edge of the physical world, where the IoT and data reside introduces an intermediate layer between the ground and the cloud. The main question is which data needs to be collected, which data needs to be cleaned and aggregated, and which data needs to be used for analytics and decision making. We proposed a solution called fog-engine (FE) that addresses the above challenges through:

- On-premise and real-time preprocessing and analytics of data near where it is generated
- Facilitating collaboration and proximity interaction between IoT devices in a distributed and dynamic manner

Using our proposed solution, IoT devices are deployed in a fog closer to the ground that can have a beneficial interplay with the cloud and with each other. Users can use their own IoT device(s) equipped with our fog-engine to easily become a part of a smart system. Depending on the scale of user groups, several fog-engines can interplay and share data with peers (e.g. via Wi-Fi) and offload data into the associated cloud (via the Internet) in an orchestrated manner.

The rest of this chapter is organized as follows. Section 11.2 provides background on big data analytics. Section 11.3 describes how our proposed fog-engine can be deployed in the traditional centralized data analytics platform and how it enhances existing system capabilities. Section 11.4 explains the system prototype and the results of the evaluation of the proposed solution. Two case studies describing how the proposed idea works for different applications are described in Section 11.5. Section 11.6 discussed related work. Section 11.7 provides future research directions and Section 11.8 presents the conclusions.

11.2 Big Data Analytics

Companies, organizations, and research institutions capture terabytes of data from a multitude of sources including social media, customer emails and survey responses, phone call records, Internet clickstream data, web server logs, and sensors. Big data refers to the large amounts of unstructured, semistructured, or structured data flowing continuously through and around organizations [7]. The concept of big data has been around for years; most organizations nowadays understand that they can apply analytics to their data to gain actionable insights. Business analytics serves to answer basic questions about business operations and performance, while big data analytics is a form of advanced analytics, which involves complex applications with elements such as predictive models, statistical algorithms, and what-if analyses powered by high-performance analytics systems. Big data analytics examines large amounts of data to uncover hidden patterns, correlations, and other insights. Big data processing can be performed either in a batch mode or streamline mode. This means for some applications data will be analyzed and the results generated on a store-and-process paradigm basis [8]. Many time-critical applications generate data continuously and expect the processed outcome on a real-time basis such as stock market data processing.

11.2.1 Benefits

Big data analytics have the following benefits:

- **Improved business.** Big data analytics helps organizations harness their data and use it to identify new opportunities, which facilitates smarter business decisions, new revenue opportunities, more effective marketing, better customer service, improved operational efficiency, and higher profits.
- **Cost reduction.** Big data analytics can provide significant cost advantages when it comes to storing large amounts of data while doing business in more efficient ways.
- **Faster and better decision making.** Businesses are able to analyze information immediately, make decisions, and stay agile.
- **New products and services.** With the ability to gauge customer needs and satisfaction through analytics comes the power to give customers what they want.

11.2.2 A Typical Big Data Analytics Infrastructure

The typical components and layers of the big data analytics infrastructure are as follows [9].

11.2.2.1 Big Data Platform

The big data platform includes capabilities to integrate, manage, and apply sophisticated computational processing to the data. Typically, big data platforms include Hadoop[1] as an underlying foundation. Hadoop was designed and built to optimize complex manipulation of large amounts of data while vastly exceeding the price/performance of traditional databases. Hadoop is a unified storage and processing environment that is highly scalable to large and complex data volumes. You can think of it as big data's execution engine.

1 http://hadoop.apache.org/

11.2.2.2 Data Management

Data needs special management and governance to be high-quality and well-governed before any analysis. With data constantly flowing in and out of an organization, it is important to establish repeatable processes to build and maintain standards for data quality. A significant amount of time might be spent on cleaning, removing anomalies, and transforming data to a desirable format. Once the information is reliable, organizations should establish a master data management program that gets the entire enterprise on the same page.

11.2.2.3 Storage

Storing large and diverse amounts of data on disk is more cost-effective, and Hadoop is a low-cost alternative for the archival and quick retrieval of large amounts of data. This open source software framework can store large amounts of data and run applications on clusters of commodity hardware. It has become a key technology to doing business due to the constant increase of data volumes and varieties, and its distributed computing model processes big data fast. An additional benefit is that Hadoop's open source framework is free and uses commodity hardware to store large quantities of data. Unstructured and semi-structured data types typically do not fit well into traditional data warehouses that are based on relational databases focused on structured data sets. Furthermore, data warehouses may not be able to handle the processing demands posed by sets of big data that need to be updated frequently – or even continually, as in the case of real-time data on stock prices, the online activities of website visitors or the performance of mobile applications.

11.2.2.4 Analytics Core and Functions

Data mining is a key technology that helps examine large amounts of data to discover patterns in the data – and this information can be used for further analysis to help answer complex business questions. Hadoop uses a processing engine called MapReduce to not only distribute data across the disks but to apply complex computational instructions to that data. In keeping with the high-performance capabilities of the platform, MapReduce instructions are processed in parallel across various nodes on the big data platform, and then quickly assembled to provide a new data structure or answer set. Just as big data varies with the business application, the code used to manipulate and process the data can vary. For instance, for identifying the customers' satisfaction level on a particular product that they have bought, a text-mining function might scrape through the users' feedback data and extract the expected information.

11.2.2.5 Adaptors

It is vital to ensure that existing tools in an organization can interact and exchange data by the big data analytics tool with the skill sets available in-house. For example, Hive[2] is a tool that enables raw data to be restructured into relational tables that can be accessed via SQL-based tools such as relational databases.

11.2.2.6 Presentation

Visualizing data using existing tools or customized tools allows the average business person to view information in an intuitive, graphical way, and extract insights for the process of decision-making.

11.2.3 Technologies

The size and variety of data can cause consistency and management issues, and data silos can result from the use of different platforms and data stored in a big data architecture. In reality, there are several types of technology that work together to realize big data analytics. Integrating existing tools such as Hadoop with other big data tools into a cohesive architecture that meets an organization's needs is a major challenge for platform engineers and analytics teams, which have to identify the right mix of technologies and then put them together [10].

11.2.4 Big Data Analytics in the Cloud

Early big data systems were mostly deployed on-premises, whereas Hadoop was originally designed to work on clusters of physical machines. With the currently available public clouds, Hadoop clusters can be set up in the cloud. An increasing number of technologies facilitate processing data in the cloud. For example, major Hadoop suppliers such as Cloudera[3] and Hortonworks[4] support their distributions of the big data framework on the Amazon Web Services (AWS)[5] and Microsoft Azure[6] clouds. The future state of big data will be a hybrid of on-premise solution and the cloud [11].

11.2.5 In-Memory Analytics

Hadoop's batch scheduling overhead and disk-based data storage have made it unsuitable for use in analyzing live, real-time data in the production environment. Hadoop relies on a file system that generates a lot of input/output files, and this limits performance of MapReduce. By avoiding Hadoop's batch scheduling, it can start up jobs in milliseconds instead of tens of seconds.

---

2 https://hive.apache.org/
3 https://www.cloudera.com/
4 https://hortonworks.com/
5 https://aws.amazon.com/
6 https://azure.microsoft.com/

In-memory data storage dramatically reduces access times by eliminating data motion from the disk or across the network. SAS and Apache Ignite provide the Hadoop distributions featuring in-memory analytics.

11.2.6 Big Data Analytics Flow

Big data analytics describes the process of performing complex analytical tasks on data that typically include grouping, aggregation, or iterative processes. Figure 11.1 shows a typical flow for big data processing [7]. The first step is to perform collection/integration of the data coming from multiple sources. Data cleaning is the next step that may consume large processing time, although it may significantly reduce the data size that leads to less time and effort needed for data analytics. The raw data are normally unstructured such that neither has a predefined data model nor is organized in a predefined manner. Thus, the data are transformed to semistructured or structured data in the next step of the flow. Data cleaning deals with detecting and removing errors and inconsistencies from data to improve its quality [12]. When multiple data sources need to be integrated (e.g., in data warehouses), the need for data cleaning significantly increases. This is because the sources often contain redundant data in different representations.

One of the most important steps in any data processing task is to verify that data values are correct or, at the very least, conform to a set of rules. Data quality problems exist due to incorrect data entry, missing information, or other invalid data. For example, a variable such as gender would be expected to have only two values (M or F), or a variable representing heart rate would be expected to be within a reasonable range. A traditional ETL (extract, load, and transform) process extracts data from multiple sources, then cleanses, formats, and loads

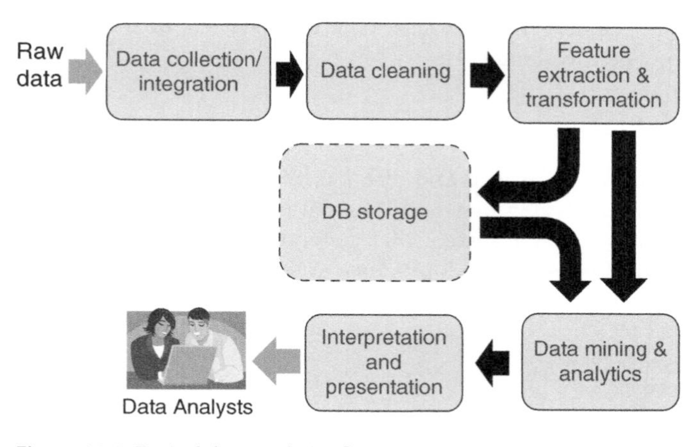

Figure 11.1 Typical data analytics flow.

it into a data warehouse for analysis [13]. A rule-based model determines how the data analytic tools handle data.

A major phase of big data processing is to perform discovery of data, which is where the complexity of processing data lies. A unique characteristic of big data is the way the value is discovered. It differs from conventional business intelligence, where the simple summing of known value generates a result. The data analytics is performed through visualizations, interactive knowledge-based queries, or machine learning algorithms that can discover knowledge [14]. Due to the heterogeneous nature of the data, there may not be a single solution for the data analytics problem and thus, the algorithm may be short-lived.

The increase in the volume of data raises the following issues for analytic tools:

1. The amount of data increases continuously at a high speed, yet data should be up-to-date for analytics.
2. The response time of a query grows with the amount of data, whereas the analysis tasks need to produce query results on large datasets in a reasonable amount of time [15].

11.3 Data Analytics in the Fog

Fog computing is a highly virtualized platform that provides compute, storage, and networking services between end devices and traditional cloud computing data centers, typically, but not exclusively located at the edge of the network. The fog is composed of the same components as in the cloud, namely, computation, storage, and networking resources. However, the fog has some distinctive characteristics that make it more appropriate for the applications requiring low latency, mobility support, real-time interactions, online analytics, and interplay with the cloud [11, 16]. While data size is growing very fast, decreasing the processing and storage costs, and increasing network bandwidth make archiving the collected data viable for organizations. Instead of sending all data to the cloud, an edge device or software solution may perform a preliminary analysis and send a summary of the data (or metadata) to the cloud. For example, Google uses cloud computing to categorize photos for its Google Photos app. For a picture taken and uploaded to Google Photos, the app automatically learns and classifies with respect to the photo's context. A dedicated chip referred to as Movidius[7] with the capability of machine learning on the mobile devices, allows processing the information in real time, instead of in the cloud [17]. It is critical to decide what should be done near the ground, in the cloud, and in-between.

---

7 https://www.movidius.com/

11.3.1 Fog Analytics

Collecting and transferring all the data generated from IoT devices and sensors into the cloud for further processing or storage poses serious challenges on the Internet infrastructure and is often prohibitively expensive, technically impractical, and mostly unnecessary. Moving data to the cloud for analytics works well for large volumes of historical data requiring low-bandwidth, but not for real-time applications. With the emergence of the IoT, which enables real-time, high data-rate applications, moving analytics to the source of the data and enabling real-time processing seems a better approach. Fog computing facilitates processing data before they even reach the cloud, shortening the communication time and cost, as well as reducing the need for huge data storage. In general, it is an appropriate solution for the applications and services that come under the umbrella of the IoT [18, 19].

With the fog providing low latency and context awareness, and the cloud providing global centralization, some applications such as big data analytics benefit from both fog localization and cloud globalization [11]. The main function of the fog is to collect data from sensors and devices, process the data, filter the data, and send the rest to the other parts for local storage, visualization, and transmission to the cloud. The local coverage is provided by the cloud, which is used as a repository for data that has a permanence of months and years, and which is the basis for business intelligence analytics.

Fog computing is still in its early stages and present new challenges such as fog architecture, frameworks and standards, analytics models, storage and networking resource provisioning and scheduling, programming abstracts and models, and security and privacy issues [20]. Fog analytics requires standardization of device and data interfaces, integration with the cloud, streaming analytics to handle continuous incoming data, and a flexible network architecture where real-time data processing functions move to the edge. Less time-sensitive data can still go to the cloud for long-term storage and historical analysis. Other capabilities such as machine learning to enable performance improvement of IoT applications over time, and data visualization functions are the important features for the future.

Tang et al. [9] proposed an architecture based on the concept of fog for big data analytics in smart cities, which is hierarchical, scalable, and distributed, and supports the integration of a massive number of things and services. The architecture consists of four layers, where layer-4 with numerous sensors is at the edge of the network, layer-3 processes the raw data with many high-performance and low-power nodes, layer-2 identifies potential hazards with intermediate computing nodes, and layer-1 represents the cloud that provides global monitoring and centralized control. In [10], FogGIS a framework based on the fog computing was introduced for mining analytics from geospatial data. FogGIS had been used for preliminary analysis including compression and overlay analysis, and the transmission to the cloud had been reduced by compression techniques. Fog computing is also becoming more popular in healthcare as organizations introduce more connected medical devices into their health IT ecosystem [21]. Cisco has introduced Fog Data Services[8] that can be used to build scalable IoT data solutions.

11.3.2 Fog-Engines

An end-to-end solution called the fog-engine (FE) [22] provides on-premise data analytics as well as the capabilities for IoT devices to communicate with each other and with the cloud. Figure 11.2 provides an overview of a typical FE deployment. The fog-engine is transparently used and managed by the end user and provides the capability of on-premise and real-time data analytics.

Fog-engine is a customizable and agile heterogeneous platform that is integrated to an IoT device. The fog-engine allows data processing in the cloud and in the distributed grid of connected IoT devices located at the network edge.

It collaborates with other fog-engines in the vicinity, thereby constructing a local peer-to-peer network beneath the cloud. It provides facilities for offloading data and interacting with the cloud as a gateway. A gateway enables devices that are not directly connected to the Internet to reach cloud services.

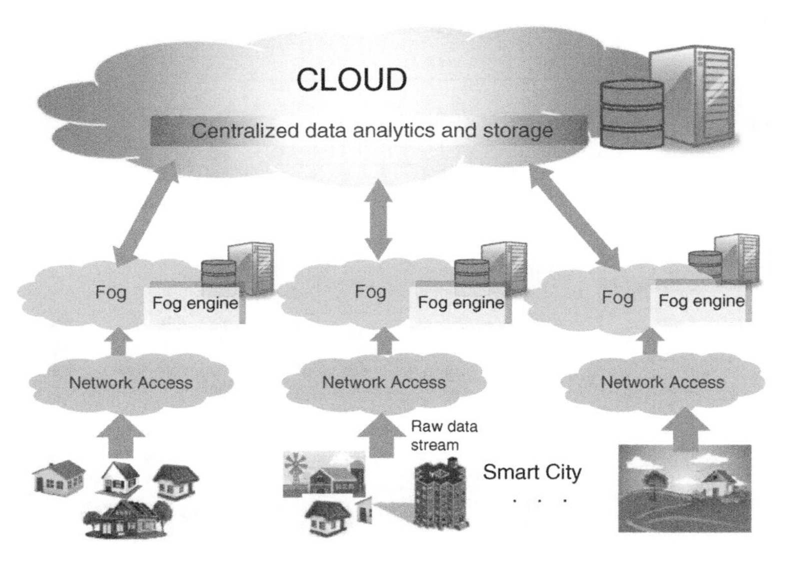

Figure 11.2 Deployment of FE in a typical cloud-based computing system.

8 https://www.cisco.com/c/en/us/products/cloud-systems-management/fog-data-services/index.html

Although the term *gateway* has a specific function in networking, it is also used to describe a class of device that processes data on behalf of a group or cluster of devices. Fog-engine consists of modular Application Programming Interfaces (APIs) for supplying the above functionalities. Software-wise, all fog-engines utilize the same API, which is also available in the cloud to ensure vertical continuity for IoT developers.

11.3.3 Data Analytics Using Fog-Engines

Figure 11.3 shows on-premise data analytics being performed near the data source using fog-engines before the data volume grows significantly. In-stream data are analyzed locally in the FE while data of the FE is collected and transmitted to the cloud for offline global data analytics. In a smart grid, for example, a fog-engine can help a user decide on the efficient use of energy. Whereas, the data of a town with thousands of electricity consumers are analyzed in the cloud of an energy supplier company to decide policies for energy use by the consumers. The analytics models employed in fog-engines are updated based on the policies decided and communicated by the cloud analytics.

As the data are preprocessed, filtered, and cleaned in the fog-engine prior to offloading to the cloud, the amount of transmitted data is lower than the data generated by IoT. Also, the analytics on fog-engine is real-time while the analytics on the cloud is offline. Fog-engine provides limited computing power

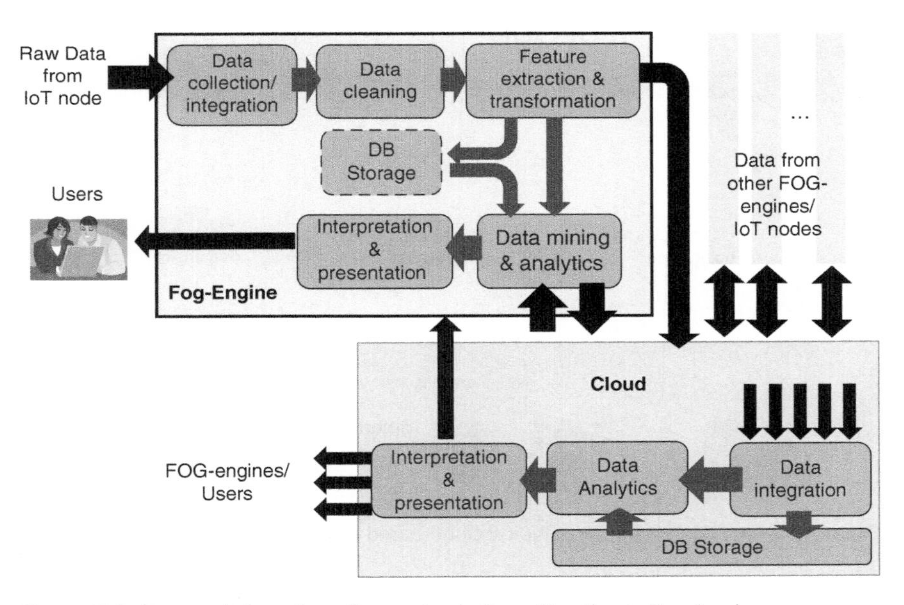

Figure 11.3 Data analytics using a fog-engine before offloading to the cloud.

and storage compared with the cloud, however, processing on the cloud incurs higher latency. The fog-engine offers a high level of fault tolerance as the tasks can be transferred to the other fog-engines in the vicinity in the event of a failure.

Fog-engine may employ various types of hardware such as multi-core processor, FPGA, or GPU with fine granularity versus a cluster of similar nodes in the cloud. Each fog-engine employs fixed hardware resources that can be configured by the user, whereas the allocated resources are intangible and out of user's control in the cloud. An advantage of fog-engine is the capability of integration to mobile IoT nodes such as cars in an intelligent transportation system (ITS) [23]. In this case, multiple fog-engines in close proximity dynamically build a fog in which fog-engines communicate and

exchange data. Cloud offers a proven model of pay-as-you-go while fog-engine is a property of the user. Depending on the IoT application, in the case of a limited access to power sources, fog-engine may be battery-powered and needs to be energy-efficient while the cloud is supplied with a constant source of power. Table 11.1 compares fog-engines with cloud computing.

Table 11.1 Data analytics using a fog-engine and the cloud.

Characteristic	Fog-engine	Cloud
Processing hierarchy	Local data analytics	Global data analytics
Processing fashion	In-stream processing	Batch processing
Computing power	GFLOPS	TFLOPS
Network Latency	Milliseconds	Seconds
Data storage	Gigabytes	Infinite
Data lifetime	Hours/Days	Infinite
Fault-tolerance	High	High
Processing resources and granularity	Heterogeneous (e.g., CPU, FPGA, GPU) and Fine-grained	Homogeneous (Data center) and coarse-grained
Versatility	Only exists on demand	Intangible servers
Provisioning	Limited by the number of fog-engines in the vicinity	Infinite, with latency
Mobility of nodes	Maybe mobile (e.g. in the car)	None
Cost Model	Pay once	Pay-as-you-go
Power model	Battery-powered/ Electricity	Electricity

11.4 Prototypes and Evaluation

We have developed and prototyped the hardware and software parts of FE architecture which we describe in the following sections. We have also conducted extensive experiments considering different deployments of the fog-engine in the system pipeline.

11.4.1 Architecture

As the fog-engine is integrated with the IoT, which mainly employs low-end devices, we need to ascertain that (i) it is agile and transparent, and (ii) adding fog-engine up to the IoT devices has no negative impact on the existing system. The fog-engine is composed of three units:

1. An analytics and storage unit for preprocessing data (i.e. cleaning, filtering, etc.), data analytics, as well as data storage.
2. A networking and communication unit consisting of the network interfaces for peer-to-peer networking and communication to the cloud and the IoT.
3. An orchestrating unit to keep fog-engines synchronized with each other and the cloud.

Figure 11.4(a) shows a general architecture of the fog-engine. Figure 11.4(b) shows the detailed FE structure. It uses several common interfaces for acquiring data through universal serial bus (USB), Wi-Fi for mid-range, and Bluetooth for small-range communication with other devices, Universal Asynchronous Receiver/Transmitter (UART), Serial Peripheral Interface (SPI) bus, and general-purpose input/output (GPIO) pins. The data may be obtained from sensor devices, other IoT devices, web, or local storage. The raw or semistructured data go through preprocessing units such as cleaning, filtering, and integration as well as extract, load, and transform (ETL). A library keeps the rules, which are used for data manipulation. For example, for the data generated by a smart meter on the energy consumption of a house, only positive values less than a few kilowatts per hour is acceptable. The preprocessed data can be transmitted or interchanged with a peer engine via peer-to-peer networking interface unit. In a cluster of fog-engines, one with higher processing capacity may act as a cluster head onto which other fog-engines offload the data. The orchestrating unit handles cluster formation and data distribution across a cluster of fog-engines. The cloud interface module is a gateway that facilitates communication between the fog-engine and cloud. The FE scheduler and task manager moderate all the above-mentioned units.

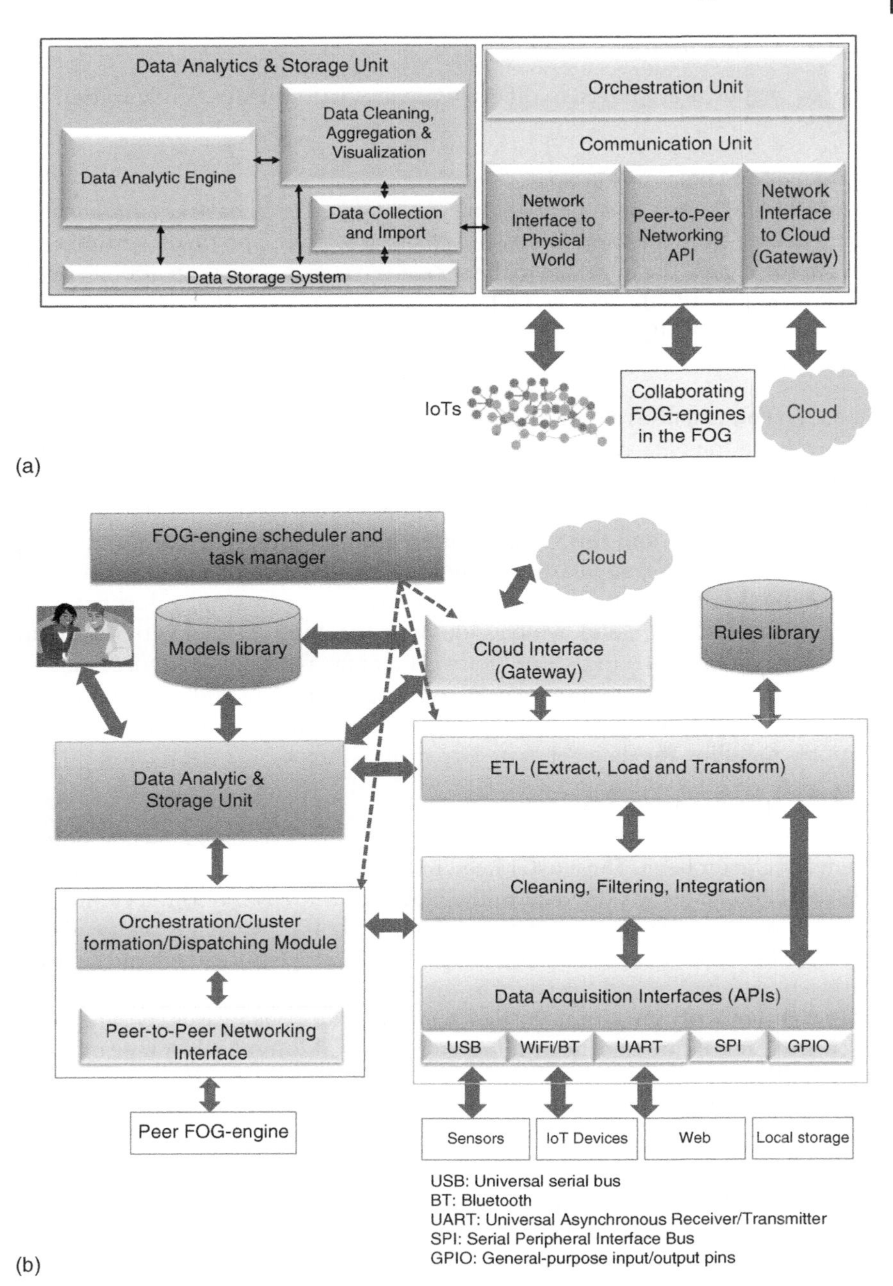

Figure 11.4 (a) General architecture of fog-engine; (b) A detailed architecture for the communication unit.

11.4.2 Configurations

The fog-engine is employed in different settings with various configurations as follows:

11.4.2.1 Fog-Engine as a Broker

Figure 11.5(a) shows how a fog-engine is configured to behave as a broker that receives data from a sensor, filters, and cleans the input, and then transmits the data to the cloud. The interface to the sensor is based on inter-integrated circuit (I2C) protocol. The data captured by the sensor is read by the I2C interface of the fog-engine. While, in a traditional scenario, the data are transmitted directly to the cloud without further processing.

11.4.2.2 Fog-Engine as a Data Analytics Engine

By involving the data analytic unit of fog-engine (Figure 11.5(b)), the data are analyzed and stored in a local storage until storage limit exceeds or any false data are detected. In this unit, a model is initially fitted for the first chunk of data (e.g., 100 samples), and this model is then used for identifying and removing outliers. The model is regularly updated (e.g., for every 100 samples) with the newer chunks of data. In this case, there is no need for streaming data between a fog-engine and the cloud, which requires a constant channel with the cloud that incurs more cost and steady network connectivity. Instead, data can be offloaded at regular time intervals. Furthermore, the data are locally analyzed in the fog-engine, which reduces the complexity of analytics on the cloud that requires handling the data generated by many sources.

11.4.2.3 Fog-Engine as a Server

In the third configuration, multiple fog-engines form a cluster while one of them is a cluster head. The cluster head receives data from all sensors, analyzes the data, and transmits the data to the cloud. In this case, all three communication units of fog-engine are engaged (Figure 11.5(c)). As an advantage of this configuration, there is no need for establishing multiple independent channels between fog-engines and the cloud as the cluster head fog-engine manages the only channel with the cloud. In this scenario, like above cases, besides having a smaller volume of data with fog-engine, an additional advantage of using a fog-engine is that the data collected from multiple sensor devices can be aggregated and transmitted to cloud in a single message by the fog-engine as a cluster head.

This configuration will save storage space and energy consumption on devices as well. In a clustered structure, it is possible to minimize the number of messages and use the maximum allowed message size, which reduces the number of message transmissions, hence causing a reduction in the cloud costs.

11.4.2.4 Communication with Fog-Engine versus the Cloud

We have conducted several experiments to examine the functionality and performance of the fog-engine. In these experiments, as depicted in Figure 11.6, we

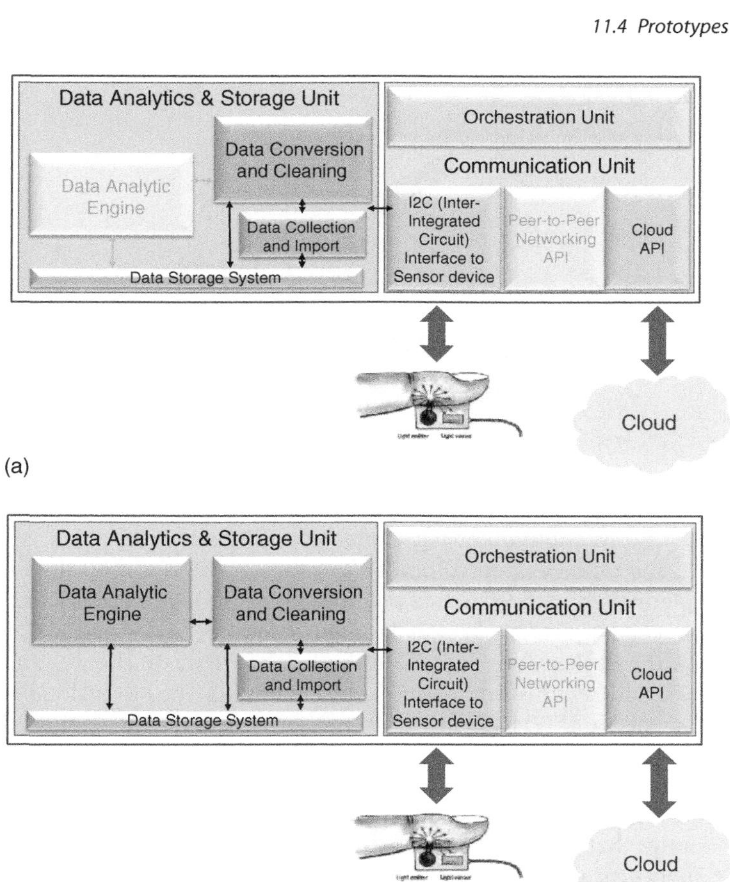

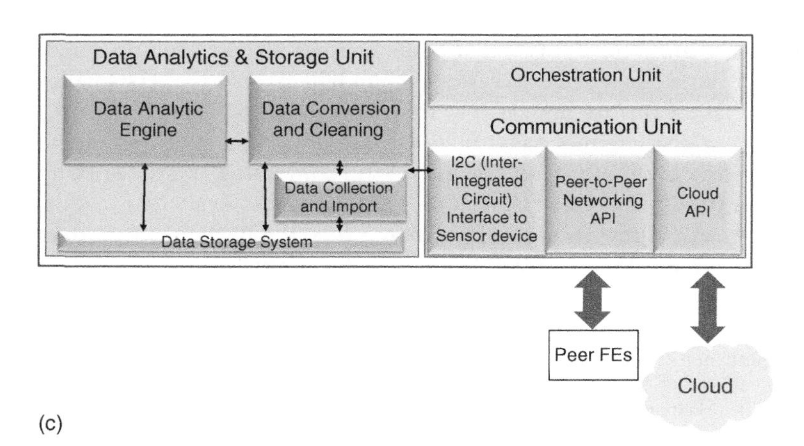

Figure 11.5 Various configurations of fog-engine: (a) as a broker; (b) as a primary data analyzer; (c) as a server.

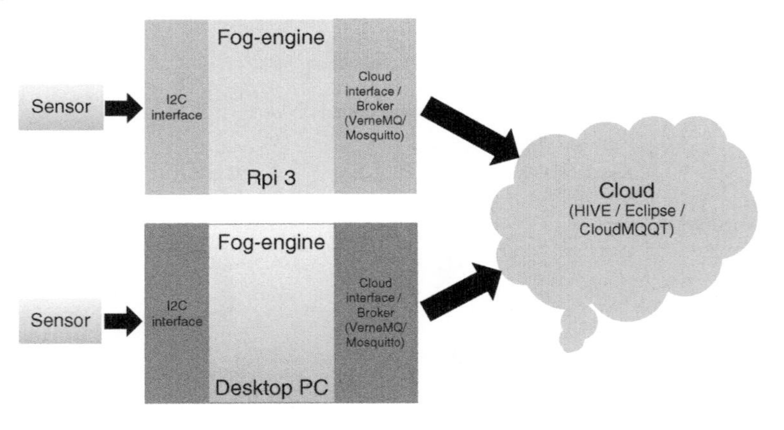

Figure 11.6 Fog-engine collects data and communicates with the cloud.

have implemented two versions of the fog-engine on a Raspberry Pi 3 board, i.e., fog-engine (Rpi), and on a desktop computer, i.e., fog-engine (PC). All the FE modules are implemented with Python. Two different MQTT brokers are used for the sending/receiving data packets including Mosquitto[9] and VerneMQ.[10] Correspondingly, the brokers are located on the IoT board or on the desktop computer. We have utilized three different clouds, including The Hive Cloud[11], Eclipse Cloud[12], and CloudMQTT.[13] The transmission time is the time required for sending a packet and receiving an acknowledgment. The packet size differs from a few bytes to more than 4 MB. The experiments have been repeated 100 times for each packet size, and the average time is measured.

Figure 11.7 shows that the transmission time from fog-engine to the cloud exponentially increases with increasing packet sizes. We observe that the transmission time for packets larger than 64 KB for Eclipse Cloud and larger than 200 KB for Hive Cloud and CloudMQTT substantially increases. Also, fog-engine to FE communication is much faster compared to fog-engine-cloud, particularly for large packet sizes (i.e. larger than 64 KB). For packets smaller than 64 KB or 200 KB, this time is less than a second while peer communication between fog-engines has still lower latency. Also, the fog-engine implemented on a desktop PC with more powerful computing and networking resources than the Rpi board performs faster. Among the evaluated clouds, since Hive does not allow packets larger than 2 MB, the packets larger than 2 MB is sent in multiple steps. We observe that exchanging packets larger than 32 MB between fog-engines is not possible, which is most likely due to

9 https://mosquitto.org/
10 https://vernemq.com/
11 http://www.thehivecloud.com
12 http://www.eclipse.org/ecd/
13 https://www.cloudmqtt.com

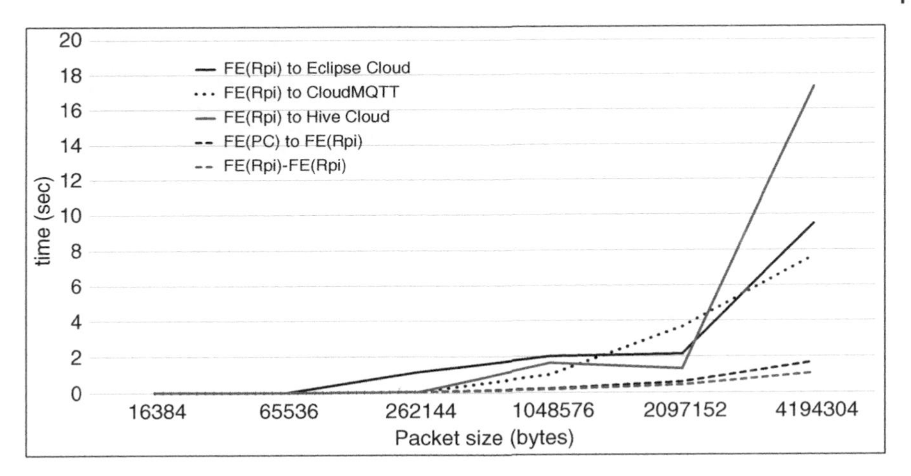

Figure 11.7 Data transmission time between FEs and cloud for various packet sizes up to 4 MB.

hardware and memory limitations of the IoT boards. Consequently, a clustered structure of devices with a fog-engine as the cluster head is a better option in terms of the reduced number of messages transmitted to the cloud resulting in reduced costs, and the increased number of devices that can be supported for a certain available bandwidth.

11.5 Case Studies

In this section, we will provide two case studies to show how the proposed fog-engine can be utilized in different applications.

11.5.1 Smart Home

In this case study, we developed a smart home application including a heart rate monitoring and activity monitoring system. Figure 11.8 shows the deployment of a fog-engine in the system while it operates as an interface between the user and the cloud. We have implemented a prototype of fog-engine on a Raspberry Pi 3 board with Python. All modules communicate over a Wi-Fi network. Fog-engine can play different roles as follows.

11.5.1.1 Fog-Engine as a Broker

In this setting, a sensor captures heart rate of a monitored patient who is residing at home. The data are read by the I2C interface of the fog-engine. It is then converted to the numeric format and goes through a filtering function. The heart rate sampling rate is 50 samples per second. The 20 msec time interval is

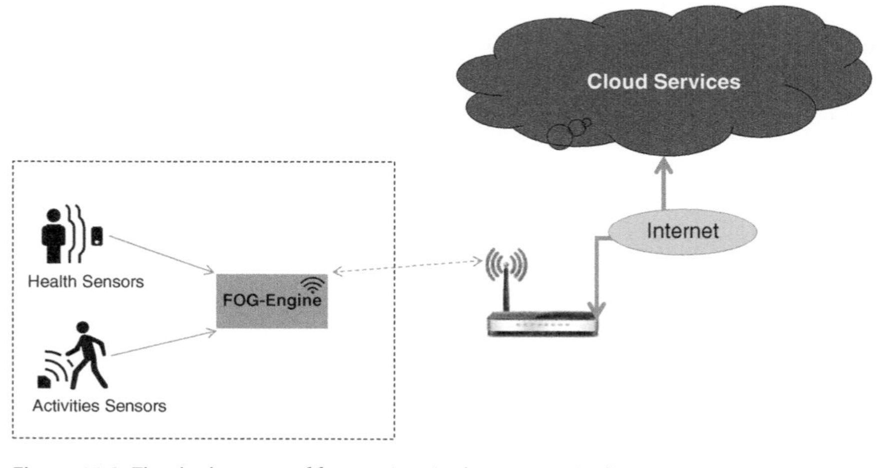

Figure 11.8 The deployment of fog-engines in the system pipeline.

sufficient for performing required processing on the collected data. According to the experiments, around 40% of heart rate data is discarded due to replication or being out of range which still leaves enough number of samples per time unit. The data are finally transmitted to Thingspeak Cloud[14] using their API. Using the fog-engine reduces the size of data; hence, less usage of network and processing resources as well as less latency.

Considering that the pricing for the cloud is typically based on the number of messages processed and stored in a period, we evaluate the efficiency of the system with and without fog-engine. We assume the maximum bandwidth provided is 1 MB/s. Each sample size (e.g. heart rate data) is 10 bytes that includes an index for the sensor, timestamp, and a value of the heart rate. In the current scenario, a separate channel is created for each sensor, so the data transfer rates are similar with and without fog-engine. However, the size of data that is transferred via fog-engine is 40% less. For the provided bandwidth of 1MB/s, the maximum number of devices (channels) that can be covered with and without fog-engine is 2000 and 3330, respectively.

11.5.1.2 Fog-Engine as a Data Analytic Engine

By involving the data analytic unit of fog-engine (Figure 11.5(b)), the data are analyzed and stored in a local storage until storage limit exceeds or any misbehavior in the heart rate is detected. In this unit, a model is initially fitted for the first chunk of data (with around 100 samples), and this model is then used for identifying and removing outliers. The model is regularly updated (every 100 samples) with the newer chunks of data. Data can be offloaded to

14 https://thingspeak.com/

the cloud at regular time intervals. Furthermore, the data are locally analyzed in the fog-engine that reduces the complexity of analytics on the hospital cloud that needs to handle data of many patients. Referring to the assumptions already given, the system with and without fog-engine performs similarly, and the size of transferred data is 40% lesser with the fog-engine.

11.5.1.3 Fog-Engine as a Server

In the third configuration, multiple fog-engines form a cluster while one of them is a cluster head. We have used Arduino Nano boards for acquiring data from the heart rate sensors and sending to the cluster head, which is implemented on a more powerful board (Raspberry Pi 3). The cluster head receives data from all sensors, analyzes, and transmits the data to the cloud. In this case, all the three communication units of fog-engine are engaged (Figure 11.5(c)). An advantage of this configuration is that there is no need for establishing multiple independent channels between fog-engines and the cloud, as the cluster head fog-engine manages the only channel with the cloud.

In this scenario, besides having a smaller volume of data with fog-engines, an additional advantage of using fog-engines is that the data collected from multiple sensor devices can be aggregated and transmitted to the cloud in a single message by the fog-engine as a cluster head. In a clustered structure, it is possible to minimize the number of messages and use the maximum allowed message size. This reduces the number of message transmissions, and consequently reducing the cloud costs. For the given 1 MB data bandwidth, if 100 sensors are clustered with a fog-engine, whole data generated in one second, which is 3 KB, can be transmitted in a single or in more number of packets. While in the case without fog-engine, the data should be directly sent through in much smaller packet sizes (e.g. 10 bytes) that leads to a larger number of messages. This requires 50 message/sec transmission rate that increases the cost of using the cloud. With a cluster of 3330 devices, the collected data can be packed into a single 1 MB packet in the fog-engine. Table 11.2 compares different FE configurations in terms of different parameters such as the ratio of data size managed, the maximum number of devices that can be supported in each case, etc.

11.5.2 Smart Nutrition Monitoring System

In the second case study, we devise a smart nutrition monitoring system that utilizes IoT sensors, fog-engines, and hierarchical data analytics to provide an accurate understanding of the dietary habits of adults, which can be used by users themselves as a motivator for change behavior and by dieticians to provide better guidance to their patients [24]. The proposed architecture is depicted in Figure 11.9. The proposed smart nutrition monitoring system is composed of a kiosk where diverse sensors are installed. This kiosk will be

Table 11.2 Comparison of various schemes with and without a fog-engine, where bandwidth, sample rate, and sample size are 10 MB/s, 50 samples/s, and 10 bytes, respectively.

	No FE	FE as broker	FE as data analytic engine	FE as server
Ratio of data size (with FE/without FE)	1	0.6	0.6	0.6
Max. no. of devices supported (with FE, without FE)	2,000	3,330	3,330	3,330
Filtering/analytic	No	Yes	Yes	Yes
Offline processing/transmission	No	No	Yes	Yes
Maximum packet size	No	No	No	Yes

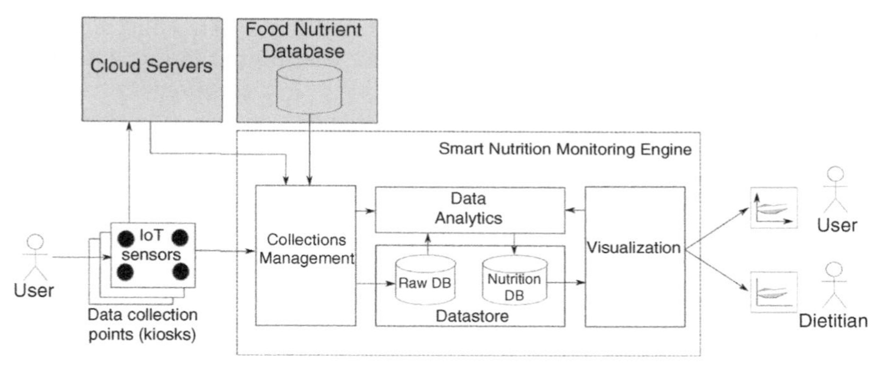

Figure 11.9 Architecture of the smart nutrition monitoring system.

equipped with various IoT sensors and fog-engine to collect weight, volume, and structure (e.g., molecular pattern) of the food. The only action required from users is to authenticate with the kiosk (via a mobile app) and deposit the food in the kiosk for a couple of seconds while relevant information is obtained by the sensors. Once data are obtained, users can cease interaction with the kiosk and can proceed with their daily activities. Therefore, the data collection will be done with a noninvasive technique where the user does not need to enter any information about the food. There are cameras located in the kiosk to capture photos from the food from different angles and transmit them to the cloud servers to generate a 3D model of the food, which is used for food volume estimation. The kiosk has a fog-engine to process and communicate the collected data with other components of the system.

The data analytics module is responsible for statistical analysis and machine learning activities in the architecture. This is used to generate reports and analyses that are relevant to users and dieticians and to identify the food that has been presented by users. The input and output for this module are datastore with two databases to store raw collected data as well as the nutritional value of the food. The visualization module displays charts showing consumption of different nutrients over time and other forms of complex data analysis that are carried out by the data analytics module.

To demonstrate the viability and feasibility of our approach, we have developed a prototype of the smart nutrition monitoring system as shown in Figure 11.10. The prototype version of the data collection points, e.g., the Kiosk, utilizes Raspberry Pi 3 Model B boards (Quad Core 1.2GHz CPU, 1GB of RAM) as fog-engine to interact with sensors and the rest of the architecture. There are five cameras with 8-megapixel resolution, each of which is augmented to one Raspberry Pi. As the sensor device, we used the SITU Smart Scale[15] that is a smart food scale that communicates with other devices via Bluetooth. The fog-engine is used to connect to the scale, receive the photos from the Raspberry Pis connected to cameras, and to interface with the architecture. To better integrate with the scale, the fog-engine in our prototype had its Operating system replaced by emteria.OS[16] (Android-compatible Operating System that is optimized to run on Raspberry Pi 3). The process of information capture is triggered by users via a mobile app we developed. In this method, previously registered users of the system deposit the food dish in the kiosk, authenticate with the app and tap a button on the app, which sends a message to a process running on the fog-engine, indicating that the

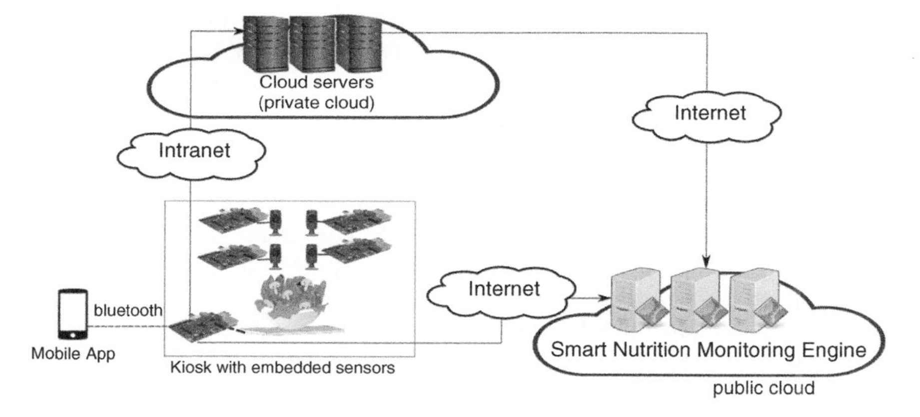

Figure 11.10 Prototype of the smart nutrition monitoring system.

15 http://situscale.com/
16 https://emteria.com/

data collection process should start. The fog-engine then collects the reading from the scale and from the other Raspberry Pis and sends all the relevant information to the smart nutrition monitoring engine via a WiFi connection. The fog-engine sends the food images taken by five cameras to a private cloud, where there is AgiSoft PhotoScan Pro[17] software to generate the 3D models. This will be used to estimate food volume in the smart nutrition monitoring engine in the public cloud.

Besides data obtained from sensors in the kiosk, another source of FE data is the external food nutrient database as depicted in Figure 11.9. Our prototype utilizes the FatSecret database[18], which is accessed via a RESTful API. The interaction with FatSecret is triggered when the data analytics module returns to the collections management module a string with a food name (which can be the result of an analysis about the likely content of the food presented by the user). This name is used for a

search in the FatSecret database (via the API) to determine nutrition facts about the food. This information is then stored in the datastore. All the nutrition data collected and stored in the database are used to generate daily, weekly, and monthly charts of intake of different nutrients and calories, for dieticians and users. Dieticians can only access data from users that are their patients (not patients from other dieticians).

11.6 Related Work

Recently, major cloud providers have introduced new services for IoT solutions with different features and characteristics. Table 11.3 shows the list of IoT solutions from five well-known cloud providers. Data collection is one of the basic aspects of these solutions, which specify the communication protocols between the components of an IoT software platform. Since IoT systems might have millions of nodes, lightweight communication protocols such as MQTT have been provided to minimize the network bandwidth. Security is another factor in these solutions where a secure communication is needed between IoT devices and the software system. As one can see in this table, link encryption is a common technique to avoid potential eavesdropping in the system.

Integration is the process of importing data to the cloud computing systems, and as mentioned in Table 11.3, REST API is a common technique to provide access to the data and information from cloud platforms. After collecting data from IoT devices, data must be analyzed to extract knowledge and meaningful insights. Data analytics can be done in several ways and each cloud provider

17 http://www.agisoft.com/
18 https://www.fatsecret.com

Table 11.3 List of IOT solutions from five major cloud providers.

	AWS	Microsoft	IBM	Google	Alibaba
Service	AWS IoT	Azure IoT Hub	IBM Watson IoT	Google IoT	AliCloud IoT
Data collection	HTTP, WebSockets, MQTT	HTTP, AMQP, MQTT and custom protocols (using protocol gateway project)	MQTT, HTTP	HTTP	HTTP
Security	Link encryption (TLS), authentication (SigV4, X.509)	Link encryption (TLS), authentication (Per-device with SAS token)	Link encryption (TLS), authentication (IBM Cloud SSO), identity management (LDAP)	Link Encryption (TLS)	Link Encryption (TLS)
Integration	REST APIs	REST APIs	REST and real-time APIs	REST APIs, gRPC	REST APIs
Data analytics	Amazon machine learning model (Amazon QuickSight)	Stream analytics, machine learning	IBM Bluemix data analytics	Cloud dataflow, BigQuery, Datalab, Dataproc	MaxCompute
Gateway architecture	Device gateway (in Cloud)	Azure IoT gateway (on-premises gateway, beta version)	General gateway	General gateway (on-premises)	Cloud gateway (in cloud)

has various packages and services including machine learning algorithms, statistical analysis, data exploration, and visualizations.

The last row in Table 11.3 is the gateway architecture, which is the main scope of this chapter. The gateway is the layer between IoT devices and cloud platform. Most providers only provide general assumptions and specifications about the gateway that will be located on the cloud platform. There are some early-stage developments for an on-premises gateway from Microsoft and Google, but none of them have implemented that completely with appropriate integration. As mentioned earlier, fog-engine can be a solution as the gateway that provides on-premise data analytics as well as capabilities for the IoT devices to communicate with each other and with the cloud.

On-premise data analytics is another type of service that has recently received lots of attention. Microsoft Azure Stack [25] is a new hybrid cloud platform product that enables organizations to deliver Azure services from their own data center while maintaining control of data center for hybrid cloud agility. CardioLog Analytics[19] offers on-premise data analytics that runs on user-side servers. Oracle [26] delivers Oracle infrastructure as a service on premises with capacity on demand that enables customers to deploy Oracle engineered systems in their data centers. IBM Digital Analytics for on-premises is the core, web analytics software component of its digital analytics accelerator solution. However, the analytic software is installed on high-performance IBM application servers. IBM PureData system for analytics is a data warehouse appliance powered by Netezza technology [27].

Cisco ParStream[20] has been engineered to enable the immediate and continuous analysis of real-time data as it is loaded. Cisco ParStream features scalable, distributed hybrid database architecture to analyze billions of records at the edge, and has patented indexing and compression capabilities that minimize performance degradation and process data in real time. ParStream can be integrated with machine learning engines to support advanced analytics. It makes use of both standard multicore CPUs and GPUs to execute queries and uses time-series analytics to combine analyzing streaming data with massive amounts of historical data. It uses alerts and actions to monitor data streams, create and qualify easy-to-invoke procedures that generate alerts, send notifications, or execute actions automatically. It derives models and hypotheses from large amounts of data by applying statistical functions and analytical models using advanced analytics.

Fog computing has received much attention in the academic community. Researchers have proposed various applications of fog computing in diverse scenarios such as health monitoring, smart cities, and vehicular networks [28].

19 http://news.intlock.com/on-premise-or-on-demand-solutions/
20 https://www.parstream.com/

As fog computing gains more traction, there have been efforts to improve the efficiency of this computing paradigm. Yousefpour et al. [29] proposed a delay-minimizing policy for fog devices. They developed an analytical model to evaluate service delay in the interplay between IoT devices, the Fog, and the cloud. The analytical model was supported by simulation studies. Alturki et al. [30] discussed analysis methods that can be distributed and executed on fog devices. Experiments conducted using Raspberry Pi boards showed that data consumption was reduced, although it led to lower accuracy in results. The authors highlighted the need to have a global view of the data to improve the accuracy of results. Jiang et al. [31] presented design adaptations in the cloud computing orchestration framework to fit into the fog computing scenario. Liu et al. [32] presented a framework for fog computing that encompasses resource allocation, latency reduction, fault tolerance, and privacy.

Liu et al. [32] highlighted the importance of security and privacy in fog computing. They suggested that biometrics-based authentication would be beneficial in fog computing. They also raised the issue of challenges in implementing intrusion detection in large-scale and mobile fog environments. They emphasized the need for running privacy-preserving algorithms such as homomorphic encryption between the fog and the cloud to safeguard privacy. Mukherjee et al. highlighted the need for new security and privacy solutions for the fog because existing solutions for the cloud cannot be directly applied to the Fog. They identified six research challenges in fog security and privacy, namely, trust, privacy preservation, authentication and key agreement, intrusion detection systems, dynamic join and leave of fog nodes, and cross-issue and fog forensic.

There have been some efforts at realizing an infrastructure to better integrate the Fog, the cloud, and IoT devices. Chang et al. [33] proposed the Indie Fog infrastructure that utilizes consumers' network devices for providing fog computing environment for IoT service providers. The Indie Fog can be deployed in various ways. A clustered Indie Fog would perform preprocessing of data collected from sensors and other devices. The infrastructure Indie Fog would be deployed in static sensor devices and provide the infrastructure for various services such as prompt data acquisition and processing. Vehicular Indie Fog would facilitate Internet of Vehicles, while Smartphone Indie Fog servers deployed on smartphones could process data on the phones. The Indie Fog system would consist of three parts, namely, the client, the server, and the registry.

Fog computing provides several interesting applications in the healthcare field. Traditional cloud-based healthcare solutions suffer from longer latency and responses times. Fog computing can potentially reduce this delay by utilizing edge devices to perform analysis, communication, and storage of the healthcare data. Cao et al. [34] employed a fall detection monitoring application for stroke patients using fog computing where the fall detection task is split between edge devices (e.g., smartphones) and the cloud. Sood and Mahajan [35] designed a fog and cloud-based system to diagnose and prevent the outbreak of Chikungunya virus, which is transmitted to humans via the bite of mosquitoes. Their system is composed of three layers: data accumulation (for collecting health, environment, and location data from the users), fog layer (for data classification into infected and other categories and alert generation), and cloud layer (for storing and processing data that cannot be managed or processed by the fog layer). Through experimental evaluation, the system was found to have high accuracy and low response times. Dubey et al. [21] presented a service-oriented fog computing architecture for in-home healthcare service. They utilized Intel Edison board as the fog computing deployment for their experiments. The first experiment involved analyzing speech motor disorders. The fog device processed the speech signals and the extracted patterns are sent to the cloud. The second experiment involved processing electrocardiogram (ECG) data. The authors concluded that their fog system reduced logistics requirements for telehealth applications, cloud storage, and transmission power of edge devices. Vora et al. [36] presented a fog-based monitoring system for monitoring patients with chronic neurological diseases using clustering and cloud-based computation. A wireless body area network collected vital health information and sent the data to a cloudlet.[21] The cloudlet cleans and segments the data and helps in decision making. Data is also sent to the cloud for classification and results are sent back to the cloudlet to detect anomalies in the processing at the fog. Performance evaluation showed that fog computing achieved higher bandwidth efficiency and lower response times. Guibert et al. [37] propose using content-centric network approach combined with fog computing for communication and storage efficiency. Their simulation results showed that delays reduced in case of fog-based content-centric networks compared to traditional content-centric networks.

There has been limited work on implementation of fog computing. For instance, a simple gateway for E-health has been implemented in a desktop PC [38]. They investigated the possibility of using the system for signal processing to decrease the latency. In [39] a gateway model for to improve the QoS in online gaming is implemented. They revealed that using the fog computing model could improve the response time by 20% for the game users. While realization of fog computing is in the early stage, fog-engine has been designed and implemented to be adapted in various applications for big data analytics. This realization has an excellent potential to be explored and developed further for other business and commercial applications.

21 https://en.wikipedia.org/wiki/Cloudlet.

11.7 Future Research Directions

There are some challenges against adopting fog-engines in big data analytics that should be considered. The benefits of the proposed solution ought to be weighed against its costs and risks, which will vary from one use case to another. Although sensors and IoT devices are normally inexpensive, the solution involving fog-engines could be expensive if many of them are involved in a wide area. Thus, further research is required with respect to the scalability and the cost of the solution.

Security is another issue, as adding fog as a new technology layer introduces another potential point of vulnerability. In addition, data management may need adjustments to address privacy concerns. Therefore, fog-engines should be part of a holistic data strategy so there are clear answers to fundamental questions such as what data can be collected, and how long the data should be retained.

Although fog-engines can be configured as redundant resources, reliability is still an important issue where we have failures in different components of the system. Given that fog-engines might be adapted for different applications, reliability mechanism should be changed based on the application requirements. As mentioned in Table 11.1, fog-engine can be battery operated so energy optimization will be a major challenge. Executing data analytics in the fog-engine is a power consuming task, so energy efficiency must be implemented, especially when there is a large number of them deployed.

Finally, resource management is a challenging task for fog-engine. A resource manager should be hierarchical and distributed where the first level of big data analytics is conducted in the fog-engine and the rest will be done in the cloud. Therefore, provisioning of resources in fog-engines for big data analytics with the requested performance and cost will be a trade-off to solve by a resource manager.

11.8 Conclusions

Data analytics can be performed near where the data are generated to reduce data communications overhead as well as data processing time. This will introduce a new type of hierarchical data analytics where the first layer will be in the fog layer and the cloud will be the last layer. Through our proposed solution, fog-engine, it is possible to enable IoT applications with the capability of on-premise processing that results in multiple advantages such as reduced size of data, reduced data transmission, and lower cost of using the cloud. fog-engine can play various roles depending on its purpose and where in the system it is deployed. We also presented two case studies where fog-engine has been adapted for the smart home as well as smart monitoring nutrition system.

There are several challenges and open issues, including resource scheduling, energy efficiency, and reliability, which we intend to investigate in the future.

References

1 A.V. Dastjerdi, H. Gupta, R. N. Calheiros, S.K. Ghosh, and R. Buyya. Fog computing: Principles, architectures, and applications. *Book Chapter in the Internet of Things: Principles and Paradigms*, Morgan Kaufmann, Burlington, Massachusetts, USA, 2016.

2 M. Satyanarayanan, P. Simoens, Y. Xiao, P. Pillai, Z. Chen, K. Ha, W. Hu, and B. Amos. Edge analytics in the Internet of Things. *IEEE Pervasive Computing*, 14(2): 24–31, 2015.

3 W. Shi, J. Cao, and Q. Zhang, Y. Li, and L. Xu. Edge computing: Vision and challenges. *IEEE Internet of Things Journal*, 3(5): 637–646, 2016.

4 L. M. Vaquero and L. Rodero-Merino. Finding your way in the fog: Towards a comprehensive definition of fog computing. *SIGCOMM Comput. Commun. Rev.*, 44(5): 27–32, 2014.

5 S. Yi, C. Li, Q. Li. A survey of fog computing: concepts, applications and issues. In *Proceedings of the 2015 Workshop on Mobile Big Data*. pp. 37–42. 2015.

6 S. Yi, C. Li, and Q. Li. A survey of fog computing: concepts, applications and issues. In *Proceedings of the Workshop on Mobile Big Data* (Mobidata '15). 2015.

7 F. Mehdipour, H. Noori, and B. Javadi. Energy-efficient big data analytics in datacenters. *Advances in Computers*, 100: 59–101, 2016.

8 B. Javadi, B. Zhang, and M. Taufer. Bandwidth modeling in large distributed systems for big data applications. *15th International Conference on Parallel and Distributed Computing, Applications and Technologies (PDCAT)*, pp. 21–27, Hong Kong, 2014.

9 B. Tang, Z. Chen, G. Hafferman, T. Wei, H. He, Q. Yang. A hierarchical distributed fog computing architecture for big data analysis in smart cities. *ASE BD&SI '15 Proceedings of the ASE BigData and Social Informatics*, Taiwan, Oct. 2015.

10 R.K. Barik, H. Dubey, A.B. Samaddar, R.D. Gupta, P.K. Ray. FogGIS: Fog Computing for geospatial big data analytics, *IEEE Uttar Pradesh Section International Conference on Electrical, Computer and Electronics Engineering (UPCON)*, pp. 613–618, 2016.

11 F. Bonomi, R. Milito, J. Zhu, and S. Addepalli. Fog computing and its role in the Internet of Things. In *Proceedings of the first edition of the MCC workshop on mobile cloud computing*, pp. 13–16, Helsinki, Finland, August 2012.

12 E. Rahm, H. Hai Do. Data cleaning: Problems and current approaches. *IEEE Data Eng. Bull.*, 23(4): 3–13, 2000.
13 Intel Big Data Analytics White Paper. *Extract, Transform and Load Big Data with Apache Hadoop*, 2013.
14 B. Di-Martino, R. Aversa, G. Cretella, and A. Esposito. Big data (lost) in the cloud, *Int. J. Big Data Intelligence*, 1(1/2): 3–17, 2014.
15 M. Saecker and V. Markl. Big data analytics on modern hardware architectures: a technology survey, business intelligence. *Lect. Notes Bus. Inf. Process*, 138: 125–149, 2013.
16 A.V. Dastjerdi and R. Buyya. Fog computing: Helping the Internet of Things realize its potential. *Computer*, 49(8) (August): 112–116, 2016.
17 D. Schatsky, *Machine learning is going mobile*, Deloitte University Press, 2016.
18 F. Bonomi, R. Milito, J. Zhu, and S. Addepalli. Fog computing and its role in the Internet of Things. *MCC, Finland*, 2012.
19 A. Manzalini. A foggy edge, beyond the clouds. *Business Ecosystems* (February 2013).
20 M. Mukherjee, R. Matam, L. Shu, L> Maglaras, M.A. Ferrag, N. Choudhury, and V. Kumar. Security and privacy in fog computing: Challenges. *IEEE Access*, 5: 19293–19304, 2017.
21 H. Dubey, J. Yang, N. Constant, A.M. Amiri, Q. Yang, and K. Makodiya. Fog data: Enhancing telehealth big data through fog computing. In *Proceedings of the ASE Big Data and Social Informatics*, 2015.
22 F. Mehdipour, B. Javadi, A. Mahanti. FOG-engine: Towards big data analytics in the fog. *In Dependable, Autonomic and Secure Computing, 14th International Conference on Pervasive Intelligence and Computing*, pp. 640–646, Auckland, New Zealand, August 2016.
23 H.J. Desirena Lopez, M. Siller, and I. Huerta. Internet of vehicles: Cloud and fog computing approaches, *IEEE International Conference on Service Operations and Logistics, and Informatics (SOLI)*, pp. 211–216, Bari, Italy, 2017.
24 B. Javadi, R.N. Calheiros, K. Matawie, A. Ginige, and A. Cook. Smart nutrition monitoring system using heterogeneous Internet of Things platform. *The 10th International Conference Internet and Distributed Computing System (IDCS 2017)*. Fiji, December 2017.
25 J. Woolsey. *Powering the Next Generation Cloud with Azure Stack*. Nano Server and Windows Server 2016, Microsoft.
26 Oracle infrastructure as a service (IaaS) private cloud with capacity on demand. *Oracle executive brief*, Oracle, 2015.
27 L. Coyne, T. Hajas, M. Hallback, M. Lindström, and C. Vollmar. IBM Private, Public, and Hybrid Cloud Storage Solutions. *Redpaper*, 2016.

28 M.H. Syed, E.B. Fernandez, and M. Ilyas. A Pattern for Fog Computing. In *Proceedings of the 10th Travelling Conference on Pattern Languages of Programs* (VikingPLoP '16). 2016.

29 A. Yousefpour, G. Ishigaki, and J.P. Jue. Fog computing: Towards minimizing delay in the Internet of Things. *2017 IEEE International Conference on Edge Computing (EDGE)*, Honolulu, USA, 2017, pp. 17–24.

30 B. Alturki, S. Reiff-Marganiec, and C. Perera. A hybrid approach for data analytics for internet of things. In *Proceedings of the Seventh International Conference on the Internet of Things* (IoT '17), 2017.

31 Y. Jiang, Z. Huang, and D.H.K. Tsang. Challenges and Solutions in Fog Computing Orchestration. *IEEE Network*, PP(99): 1–8, 2017.

32 Y. Liu, J. E. Fieldsend, and G. Min. A Framework of Fog Computing: Architecture, Challenges, and Optimization. *IEEE Access*, 5: 25445–25454, 2017.

33 C. Chang, S.N. Srirama, and R. Buyya. Indie Fog: An efficient fog-computing infrastructure for the Internet of Things. *Computer*, 50(9): 92–98, 2017.

34 Y. Cao, P. Hou, D. Brown, J. Wang, and S. Chen. Distributed analytics and edge intelligence: pervasive health monitoring at the era of fog computing. In *Proceedings of the 2015 Workshop on Mobile Big Data* (Mobidata '15). 2015.

35 S.K. Sood and I. Mahajan. A fog-based healthcare framework for chikungunya. *IEEE Internet of Things Journal*, PP(99): 1–1, 2017.

36 J. Vora, S. Tanwar, S. Tyagi, N. Kumar and J.J.P.C. Rodrigues. FAAL: Fog computing-based patient monitoring system for ambient assisted living. *IEEE 19th International Conference on e-Health Networking, Applications and Services (Healthcom)*, Dalian, China, pp. 1–6, 2017.

37 D. Guibert, J. Wu, S. He, M. Wang, and J. Li. CC-fog: Toward content-centric fog networks for E-health. *IEEE 19th International Conference on e-Health Networking, Applications and Services (Healthcom)*, Dalian, China, 2017, pp. 1–5.

38 R. Craciunescu, A. Mihovska, M. Mihaylov, S. Kyriazakos, R. Prasad, S. Halunga. Implementation of fog computing for reliable E-health applications. In *49th Asilomar Conference on Signals, Systems and Computers*, pp. 459–463. 2015.

39 B. Varghese, N. Wang, D.S. Nikolopoulos, R. Buyya. Feasibility of fog computing. arXiv preprint arXiv:1701.05451, January 2017.

12

Exploiting Fog Computing in Health Monitoring

Tuan Nguyen Gia and Mingzhe Jiang

12.1 Introduction

The number of people with cardiovascular diseases is at an alarming rate. According to the National Center for Health, more than 28.4 million people in the United States have cardiovascular diseases in 2015 [1]. Risks for heart diseases become higher for people with diabetes, obesity, and physical inactivity. Cardiovascular diseases can cause serious consequences such as kidney trauma, nerves injury, and even death [2]. For example, stroke, which is one of the cardiovascular diseases, kills about 129,000 Americans each year [2, 3]. To lessen the severe effects of cardiovascular diseases, health-monitoring systems are often used in many hospitals and healthcare centers. These systems monitor vital signals such as electrocardiography (ECG), body temperature, and blood pressure. Based on the collected biosignals, medical doctors apply suitable treatment methods.

More than 30% of people over 50 years old fall every year with severe consequences [4]. Only half of those fall cases are reported to medical doctors or caregivers [5]. Dealing with injuries from an unreported fall is difficult, time-consuming, and costly. Together with cardiovascular diseases, falling is one of the leading causes of adult disability and many other serious injuries such as brain injuries [2, 4]. Therefore, there is an urgent need for fall detection systems that can inform the incident to medical doctors or caregivers in real time. A quick response from a medical doctor to a fall case may help to reduce the severity of the injury and save a patient's life.

Conventional health-monitoring systems (e.g., ECG monitoring) often have many drawbacks, such as nonubiquitous access to data and noncontinuous monitoring. For instance, 12 leading ECG monitoring systems in many hospitals do not support mobility and their ECG measurements are applied for an instant moment or a short time period (e.g., a couple of minutes). In addition, the results of the measurements provided by these systems cannot

Fog and Edge Computing: Principles and Paradigms, First Edition.
Edited by Rajkumar Buyya and Satish Narayana Srirama.

be analyzed in real time by medical doctors or specialists. There is a need for enhanced healthcare systems that proffer continuous real-time health monitoring and other advanced services for improving quality of healthcare services. Via the systems, medical doctors can remotely access the collected data for real-time analysis. In addition, the systems can report abnormality or emergency (e.g., a fall, too low or too high heart rate) to doctors or caregivers for a quick response [6, 7].

Internet of Things (IoT), which is described as a dynamic platform where physical and virtual objects are interconnected, can be a suitable option for improving health-monitoring systems [8]. IoT-based health-monitoring systems involving wearable devices, wireless body sensor networks, and cloud computing are able to provide high-quality services (e.g., long-term history of data) with low costs while they do not interference the patient's daily activities [8]. For example, the wearable devices in IoT systems can collect different types of biosignals such as ECG, electromyography (EMG), and electroencephalography (EEG).

Some other sensors such as an accelerometer, gyroscope, and magnetometer can provide parameters related to human motions (e.g., stepping and hand moving) [9–11]. In many cases, the collected data are transmitted to gateways which primary forward to the data to cloud servers for further processing (e.g., data processing and data analysis). Correspondingly, the e-health data can be remotely monitored in real time in human-readable forms such as text or graphical forms [12]. In addition, the systems are able to detect abnormalities (e.g., a fall or a high heart rate) via algorithms running on cloud servers. The detected abnormality is informed to correspondent individuals (e.g., medical doctors) in real time [13].

However, there are challenges in these IoT systems, such as transmission bandwidth and wearable sensor nodes' energy efficiency. For instance, wearable sensor nodes in multichannel ECG or EMG monitoring IoT-based systems often collect a large amount of data with a high data rate (e.g., about 6 kbps per ECG channel) and wirelessly transmit the data over a network [6]. Gateways in these systems primarily forward the collected data to cloud servers for storing and analysis. Correspondingly, wearable sensor nodes' lifetimes cannot last for a long time period because these nodes often have to perform both computational and communicational tasks with a limited power budget. In addition, the network and cloud servers must deal with a large volume of data, which may cause the higher error rate and infringe latency requirements of real-time healthcare systems (e.g., the maximum latency of ECG signals is 500 ms [14]). Therefore, the energy consumption of the sensor nodes and the volume of the data transmitted over the network must be reduced as much as possible while maintaining a high level of quality of service (QoS).

A suitable solution for dealing with these challenges in IoT systems while maintaining the high quality of healthcare services is to exploit fog computing at smart gateways [15–17]. In detail, an extra layer called fog is added in between conventional gateways and cloud servers. Fog computing helps to reduce the burdens of wearable sensor nodes by switching computational loads from the wearable devices to smart gateways. For example, computationally heavy loads of running complex algorithms (e.g., ECG extraction algorithms based on wavelet transform) are forced to be run at a fog layer of smart gateways instead of sensor nodes [18]. Correspondingly, the sensor node's lifetime can be increased dramatically [18, 19]. Furthermore, fog computing facilitates enhanced services at the edge of the network and reduces the burdens of cloud servers [20]. Fog computing helps to bring the cloud computing paradigm to the edge of the network and provides advanced features, which are supported by cloud servers [18, 20]. For example, some of the fundamental characteristics of fog computing are location awareness, geographical distribution, interoperability, edge location, low latency, and support for online analytics [20]. To sum up, a combination of fog computing and IoT systems using smart gateways and wearable devices can be a sustainable solution for existing challenges in remote continuous health-monitoring systems.

In this chapter, we exploit fog computing in health-monitoring IoT systems for enhancing the quality of healthcare service. Fog computing and its services help to improve the energy efficiency of the sensor devices (nodes), increase the security level, and save network bandwidth. In addition, the fog-assisted system analyzes and processes data in a distributed manner at smart gateways for providing real-time analytic results. To demonstrate the benefits of fog computing in IoT systems, a complete system including wearable sensor nodes, gateways with fog computing, and end-user terminals is implemented. Two cases studies related to human fall detection and heart rate variability are presented and evaluated.

The rest of the chapter is organized as follows: Section 12.2 shows an overview of the architecture of an IoT-based system with fog computing. Section 12.3 provides Fog computing services in smart e-health gateways. Section 12.4 presents system implementation. Section 12.5 provides a case study, experimental results, and evaluation. Section 12.6 presents discussions. Section 12.7 presents the related applications in fog computing. Section 12.8 discusses future research directions. Section 12.9 concludes the work.

12.2 An Architecture of a Health Monitoring IoT-Based System with Fog Computing

Health monitoring IoT systems have to be reliable because their results indirectly or directly impact the medical doctor's analysis and decisions. An error or a delay in the results may lead to serious consequences, such as an incorrect treatment or a late response to the emergency, which may negatively affect human health. For example, a late notification from a human fall detection monitoring IoT system to a medical doctor can lead to a late response to a serious head injury, which quite possibly causes a death. In this situation, if a medical doctor was informed about the case in real time, the doctor could provide first-aid procedures (e.g., stopping bleeding) to save the patient's life. Therefore, health-monitoring IoT systems must provide high-quality data in real time. The latency requirements of e-health signals vary, depending on the characteristics of particular e-health signals. For instance, a maximum latency of EMG signals is less than 15.6 ms [14]. In addition, it is necessary for the system to provide advanced services such as push notification for reporting emergencies to respondent personnel in real time.

However, the conventional health-monitoring IoT systems built from sensor devices (nodes), gateways, and cloud server cannot fulfill the strict requirements of latency in many cases (e.g., a disconnection between the system's gateway and cloud servers). To overcome disadvantages of the conventional health-monitoring system, advance health-monitoring IoT systems with fog computing are presented. An architecture of the system with fog computing is shown in Figure 12.1. The system includes several primary components such as a sensor layer, smart gateways with a fog layer, and cloud servers with end-user terminals. The functionality of distinct layers of the architecture is described as follows.

12.2.1 Device (Sensor) Layer

A device (sensor) layer consists of sensor nodes in which each node often has three primary components, including sensors, a micro-controller, and a wireless communication chip. In some applications [21], an SD card can be integrated into a sensor node for storing temporary data. The sensors (e.g., ECG, glucose, SpO2, humidity, and temperature sensors) are used for collecting contextual data from surrounding environments and e-health data from a human body. The contextual data such as room temperature, humidity, and statuses of

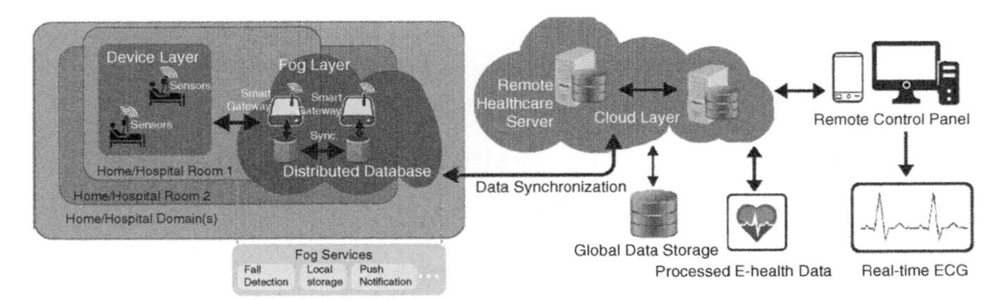

Figure 12.1 Architecture of remote real-time health-monitoring IoT system with fog computing.

the patient activities helps to improve the quality of e-health data and a doctor's decisions. For example, a heart rate of 100 beats/s of a healthy person will be normal when the person is running, whereas this rate will be high and problematic if he/she is resting on a chair. Without the activities statuses, it is difficult to achieve an accurate analysis. Collecting contextual data does not dramatically increase the burden of a sensor node in terms of weight, size, complexity, and energy consumption dramatically. For example, the activity statuses can be extracted from a single IC chip having 3-D accelerometer and 3-D gyroscope, while room temperature and humidity can be collected from another IC chip. These chips are often small, lightweight, and energy efficiency [10, 11]. The sensors often communicate with a micro-controller via one of the wire protocols such as UART, SPI, or I2C.

The micro-controller is often a low-power chip supporting sleep modes and waking up methods. The micro-controller's frequency can vary depending on applications. For example, an 8 MHz micro-controller can be used for collecting high-quality data from sensors and performing some light computation tasks (e.g., Advanced Encryption Standard – AES algorithm) while it still fulfills requirements of latency [11, 15]. The micro-controller also communicates with a wireless chip via one of the mentioned wire methods.

The wireless communication chip is various depending on the applications' requirements. In general, low-power wireless protocols (e.g., BLE and 6LoWPAN) are more preferred for low data rate applications such as fall detection or heart rate monitoring because these protocols have a maximum bandwidth of 250 kbps [10, 13]. In contrast, Wi-Fi is chosen for high-quality streaming applications (e.g., video surveillance or 24-channel EEG monitoring) where energy consumption is not the most important criteria.

12.2.2 Smart Gateways with Fog Computing

Fog computing can be described as a convergent network of interconnected smart gateways with fog services. Depending on applications' requirements, a smart gateway can be movable or fixed in a specific place. Each gateway type (i.e., mobile or fixed type) has its own advantages and disadvantages. For example, a mobile gateway provides a mobility support but it has a limited battery capacity and hardware resource constraints. In contrast, a fixed gateway is often built from a powerful device supplied from a wall socket's power. Correspondingly, the fixed gateway can perform heavy computational tasks easily and provide more advanced services with high-quality data while a mobile gateway might not be able to perform the similar tasks. In general, the fixed gateway is more preferred in many healthcare applications such as remote health-monitoring systems in the hospital and at home.

Each smart gateway in a fog layer is an embedded device that often consists of three main components such as hardware, an operating system, and software.

Depending on specific health-monitoring applications and sensor nodes, hardware can be various. For instance, a smart gateway's wireless communication chip is compatible with the wireless protocols (6LoWPAN, BLE, or Wi-Fi) used by sensor nodes. In addition, the smart gateway often is equipped with Ethernet, Wi-Fi, or 4G for connecting to cloud servers via the Internet. A smart gateway can be equipped with a hard drive or an SD card for storing data and installing an operating system. Although a storage capacity of the hard drive or an SD card is various, it is often not very large (e.g., less than 128 GB) [22].

Lightweight operating systems are often preferred in smart gateways because it does not require powerful hardware. For instance, lightweight versions of Linux kernels are used in many smart gateways [8, 15]. The operating system provides a platform for installing useful software easily and helps to manage tasks and hardware resources more efficiently and precisely.

Software in a smart gateway can consist of both basic programs and fog services. These programs and services are designed for serving particular applications' requirements such as latency, bandwidth, and interoperability. Basic programs provide fundamental features and functions of a gateway such as data transmission, gateway management, and some basic levels of security. For example, IPtable, which is a lightweight and simple software installed in Ubuntu, is used for blocking unused communication ports of the gateway. MySQL or MongoDB, which is an open-source database, can be installed in the smart gateway for flexibly, reliably, and efficiently managing database.

Fog services can consist of many advanced services for augmenting the quality of healthcare service. The services can help to reduce the burden of sensors nodes for extending their battery lifetime, saving the network bandwidth, reducing the burden of cloud, and informing emergent cases in real time. For example, push notification for informing emergent is an important service of the fog. Detail of the fog services are explained in Section 12.3.

12.2.3 Cloud Servers and End-User Terminals

In general, there is not much difference between the cloud of a remote health-monitoring IoT systems having a fog layer and the cloud of other IoT applications without a fog layer (e.g., automation, education, and entertainment). They all provide the fundamental features and basic services of a cloud (e.g., data storage and data analysis) [23]. However, the burdens of the cloud in IoT systems with a fog are less than in the IoT applications without a fog. For instance, ECG feature extraction algorithms and machine-learning algorithms can be processed at a fog while the rest of the processing can be run on a cloud. Merely, results from the processing are updated in both the fog's local storage and the cloud. Correspondingly, a large amount of transmitted data can be saved and the cloud's storage can be efficiently used. In general, a cloud of IoT systems with a fog is often customized for supporting the fog services. For example, in the IoT systems without a fog, cloud servers do not send data back to gateways. In most of the cases, cloud servers merely transmit commands and instructions to gateways, which are then forwarded to actuators. In addition to commands and instructions, cloud servers in IoT systems with a fog also transmit data to smart gateways for serving some of the fog services such as mobility support.

Similar to most of the conventional health-monitoring IoT systems, Web browsers, and mobile applications are the primary terminals of the health-monitoring IoT systems. These terminals are often ease-to-use, popular, and suitable for most of the devices including smart devices (e.g., smart phone, Ipad) and computers (e.g., laptop and desktop). End users can access real-time data in human-readable forms (e.g., text or graphical waveforms) via these terminals anytime and anywhere. In some health-monitoring IoT systems, executable programs are used together with other terminals for accessing the monitored data. For example, for reducing the risks from security attacks, end users have to use a virtual private network (VPN) and virtual platforms in order to use executable programs installed in the hospital's system for accessing patient's data.

12.3 Fog Computing Services in Smart E-Health Gateways

Fog computing services locating in a fog layer of smart gateways are diversified for serving IoT applications (e.g., healthcare, education, and autonomous industry). Fog services for healthcare are distinct for fulfilling strict requirements of latency and quality of data. In addition to the commonly used fog services such as push notification, local data storage, and data processing, fog services for healthcare can consist of security management, fault tolerance, categorization, localhost with a user interface and channel managing. These services are shown in Figure 12.2 and explained in detailed as follows.

12.3.1 Local Database (Storage)

Depending on IoT applications, a fog's local storage can be structured differently. In general, a fog's local storage can be categorized into two primary databases: an external database and an internal database [15]. The external database is used for storing data and results that are transmitted to a cloud and can be accessed by end users. The structure and the format of data stored in the external database are diversified depending on the applications. For example, the internal database can store data in a standard format of Health Level Seven (HL7). The database is always synchronized with the cloud server's database. In general, biosignals and contextual data are stored in an external database.

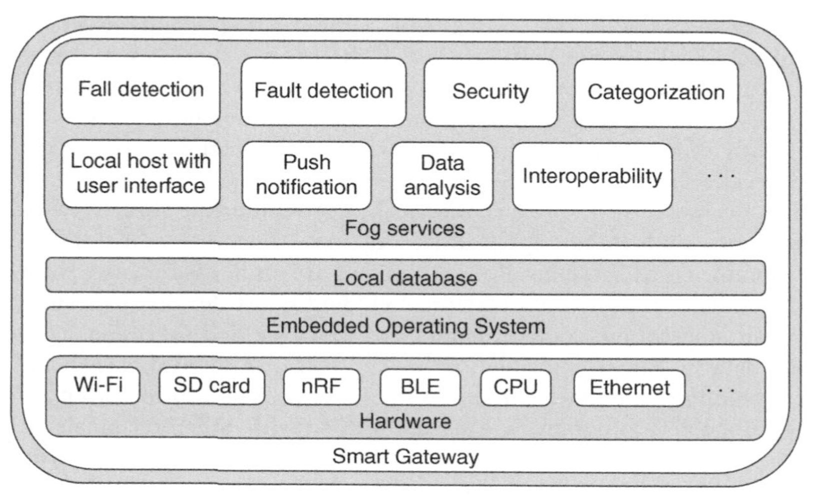

Figure 12.2 Fog services in a smart gateway.

For example, heart rates of monitored patients during a time period are stored at an external database. End users such as medical doctors or caregivers can access the heart rate data of the patients by using terminals and the local network to connect a fog's local storage when the connection between smart gateways and cloud is interrupted during a short period of time. Depending on the system's requirements and smart gateways' specifications, the storage capacity of this database varies. In general, this database has a limited storage capacity. Therefore, after a time period, old data will be replaced by incoming data. For accessing the history of data, the cloud must be used. In contrast, the internal database is used for storing configuration parameters and various parameters used for algorithms and fog services. In most cases, this database is not synchronized with the cloud servers' databases except for the back-up cases. Merely, the system and system administrators have authority to access the database.

12.3.2 Push Notification

The push notification service is one of the most important features of the fog services because it can inform abnormalities in real time. In conventional health-monitoring IoT systems, the push notification is always implemented at a cloud for informing abnormal cases. This helps to reduce the burden of the gateways; however, responsible persons may not receive the push notification messages in real time to the network traffic. For example, it may take many seconds or up to a minute to receive a notification from a Google Firebase service during the heavy traffic period in developing countries such as Vietnam, Laos, and Cambodia. To avoid this situation, the push notification service should be applied in both fog and cloud.

12.3.3 Categorization

In most healthcare IoT systems, the systems send real-time data and push messages via the cloud to responsible persons. As mentioned, the latency of the data and the push messages may be too high, as much as 30 to 60 seconds (s), in the case of heavy traffic. In the case that end users and monitored persons are in the same geographical location (hospital or home), the high latency issue can be avoided by applying the categorization service together with the fog-based push notification service. The categorization service classifies connected devices for distinguishing local and external end-users. In general, end users must use devices that are connected to the system by one of the protocols such as Ethernet, Wi-Fi, or 4G/5G. The service scans the devices periodically (about 5 s). When it detects locally connected devices, it stores information of the devices in a local database. When a device requests real-time data, the system checks the local database. If the device is currently connected to a local network, the real-time data are directly sent to the device from smart gateways. If the device requests data history, data will be retrieved from the cloud. This service helps to dramatically diminish the latency of monitoring data because the transmission path is much shorter.

12.3.4 Local Host with User Interface

Local host with an easy-to-use user interface is required for providing real-time monitoring data at smart gateways. Concisely, a local server hosts web pages that can show necessary data in both text and graphical forms in the easy-to-use interface. The web pages have a form for an end user to fill his/her username and password. When the form is submitted, the form's data are verified by comparing the credentials data stored in a local database. If they are matched, an end user is granted an access right. In the case that a password is incorrect after a few verification times, the username can be locked for a period of time (e.g., 10 minutes). For improving the security level, two-step or three-step verification (e.g., checking with an SMS message or a phone call) can be used.

12.3.5 Interoperability

In general, IoT systems are compatible with sensor nodes from different manufacturers and have different functionalities (e.g., collecting biosignals, obtaining contextual data, or controlling other electric devices). Therefore, interoperability of IoT systems primarily indicates the compatibility level of the systems toward various sensor nodes that use different wireless communication protocols. The interoperability level of an IoT system depends on the application's requirements. A health-monitoring IoT system with a high level of interoperability can be applied to different applications and helps to save healthcare costs (e.g., system deployment and maintenance). For example, the IoT systems with interoperability can support both high-quality multichannel ECG, EMG monitoring applications using Wi-Fi, and energy-efficient fall detection applications using 6LoWAN simultaneously. However, it is difficult to achieve a high level of interoperability in conventional IoT systems because of limitations on traditional gateways which merely receive and forward data. Fortunately, the target can be addressed successfully by an assistance of smart gateways and fog services. For instance, several components for supporting different wireless communication protocols such as Wi-Fi, 6LoWPAN, Bluetooth, BLE, and nRF are integrated into a smart gateway whose fog services will handle the rest of the task. Concisely, the interoperability service operates with multithreading in which each thread is used for a single wireless communication protocol. These threads can communicate with each other to exchange data if required. Incoming coming data collected from each thread are stored in a local database.

12.3.6 Security

Security is an important issue, which healthcare IoT systems have to consider attentively. A single security weakness in the systems can be exploited and hacked by cybercriminals. Correspondingly, it can cause serious consequences, such as a loss of a patient's life or a loss of sensitive data. For example, an insulin pump device can be wirelessly hacked from 300 feet away. A researcher uses his software to steal the pump's security credential and control the pump [24]. In this case, if he increases a large amount of insulin pumped into a patient's blood, the patient life can be in danger. To avoid or reduce risks from cyberattacks, the whole health-monitoring IoT system must be protected. In another word, each device, component (e.g., sensor nodes, gateways, and cloud servers) and communications between devices or components must be protected. In many health-monitoring IoT systems, end-to-end security algorithms or methods protected from sensor devices to end users are applied [25, 26]. These methods can protect the system from wireless cyberattacks, which target the communications between sensor nodes and gateways and the communications between gateways and cloud servers. In many healthcare monitoring IoT systems, the communication between sensor nodes and gateways are often more vulnerable than the communications between gateways and cloud servers. This is because it is difficult or even impossible to implement complex security algorithms in sensor nodes due to latency requirements and resource constraints. Whereas, it is feasible to perform the algorithms at gateways and cloud servers without infringing the requirements. Fortunately, sensor nodes and their communications can be still protected by applying lightweight security algorithms such as Datagram Transport Layer Security (DTLS) based algorithms [27, 28]. In some health-monitoring IoT applications, an advanced encryption standard (AES) is applied at both sensor nodes and gateways for protecting the transmitted data between them [15]. The risks of being attacked in health-monitoring IoT systems with a fog are higher because many systems often allow end users to directly connect to smart gateways with a fog for assessing data. Therefore, it is required that fog services provide a high level of security for protecting the whole health-monitoring system. In addition to end-to-end security methods, other advanced methods for protecting smart gateways are often used. For example, authentication checking and verification are used when an end user connects to a fog's local storage [15].

12.3.7 Human Fall Detection

Many algorithms (e.g., based on camera or motion) have been proposed for detecting a human fall [10, 13, 29, 30]. Algorithms based on a person's motions seem to be more popular and suitable for IoT systems because motion data can be collected easily anytime and anywhere by wearable wireless sensor nodes without interfering a monitored person's daily activities. Most motion-based algorithms use data collected from a 3-D accelerometer, a 3-D gyroscope, or both [10, 13, 30]. Gia et al. [11] show that using both a 3-D accelerometer and a 3-D gyroscope provides more accurate fall detection results than a single sensor type, although the energy consumption of a sensor node slightly increases. Fall-related parameters such as sum vector magnitude (SVM) and different SVM (DSVM) are often used as inputs for motion-based fall detection algorithms (e.g., threshold-based algorithms or a combination of threshold and hidden Markov model algorithms) [11, 30]. The fall-related parameters are calculated via the formulas presented by equations (12.1), (12.2), and (12.3) [11, 30]. Note that equation (12.2) is not applied to data from a gyroscope sensor.

$$SVM_i = \sqrt{x_i^2 + y_i^2 + z_i^2} \tag{12.1}$$

$$\theta = \arctan\left(\frac{\sqrt{y_i^2 + z_i^2}}{x_i}\right) * \frac{180}{\pi} \tag{12.2}$$

$$DSVM_i = \sqrt{(x_i - x_{i-1})^2 + (y_i - y_{i-1})^2 + (z_i - z_{i-1})^2} \tag{12.3}$$

SVM: Sum vector magnitude
DSVM: Differential sum vector magnitude
i: the sample order
x, y, z: three-dimensional values of accelerometer or gyroscope
θ: The angle between the y-axis and vertical direction

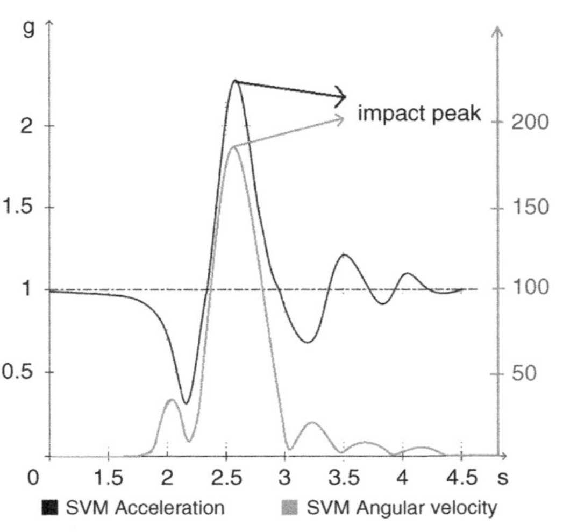

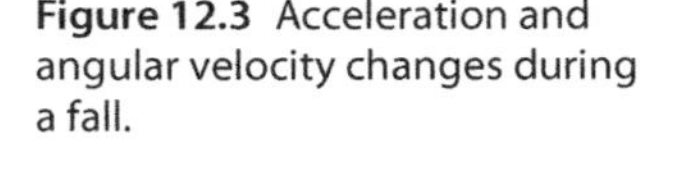
Figure 12.3 Acceleration and angular velocity changes during a fall.

The changes of an SVM acceleration and an SVM angular velocity during a fall are shown in Figure 12.3. When a person stands still or sits still, an SVM acceleration and an SVM angular velocity is 1 g and 0 deg/s, respectively. When the person falls, an SVM acceleration and an SVM angular velocity change dramatically.

In this chapter, a multilevel threshold algorithm, shown in Figure 12.4, is applied. The algorithm is simple and easy to implement, while it provides a level high of precision. First, data are filtered for removing noise and interference from surrounding environments. Then they are used for calculating the fall-related parameters (e.g., SVM of 3-D acceleration and 3-D angular velocity). The SVM values of both acceleration and gyroscope are compared

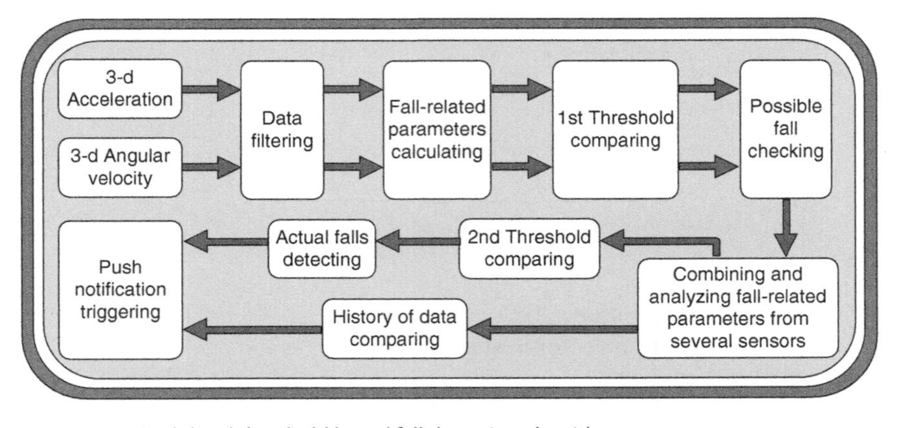

Figure 12.4 Multilevel threshold based fall detection algorithm.

with the first threshold. If both of them are higher than their first threshold, they are compared with the second threshold. If they are larger than their second threshold, a fall case is detected. The push notification service is triggered to report the case. When one of them is higher than its first threshold (e.g., 1.5 g for acceleration and 130 deg/s for angular velocity), the possible fall case is defined and the value is marked. In this case, the system compares the marked value with 20 previous values. If the result shows the pattern of a fall case illustrated in Figure 12.3, a fall case is triggered. In addition to sending a push notification message, an alarm message is sent to the system administrators to report that one of the sensors does not properly work.

12.3.8 Fault Detection

Fault detection is a very important service of a fog because it helps to avoid a long interruption of fog services. The fault detection service is responsible for detecting abnormality related to sensor nodes and smart gateways. When a smart gateway does not receive any data from a specific sensor node during a short period of time (e.g., 5 s to 10 s), the fault detection service sends predefined commands or instructions to the node. If the node does not reply to the smart gateway after several commands are sent, the fault detection service triggers a push notification service for informing the system's administrators. A similar mechanism is applied for detecting nonfunctional gateways. The gateway periodically sends predefined multicast messages to neighbor gateways and waits for the replies. If the gateway does not receive any reply from its neighbors after some periods of time and a couple of messages, it triggers the push notification service. In case only a single gateway is used in the system, the fault detection service for detecting dysfunctional gateways can be implemented at the cloud and the similar mechanism is applied.

12.3.9 Data Analysis

The raw data captured from sensors need to be processed and analyzed into information for disease diagnosis and health monitoring. However, sensor nodes usually have limited computing power to manage all the tasks such as digitalization, communicating with wireless data transmission modules, signal processing, and data analysis. The processing is even more challenging for a node integrating multiple sensors with a high data rate that requires instant data transmission or local data storage. Comparatively, fog computing in an IoT system has stronger computing power than the energy-efficient microprocessors in a sensor node, which can provide application customized data analysis and timely feedback to end users.

Data analysis methods are signal-dependent and application-dependent. However, the data analysis procedure usually contains data preprocessing and feature extraction. The extracted features are the data for statistical analysis or machine learning methods.

12.4 System Implementation

A complete remote real-time health-monitoring IoT system with fog services is built. The system consists of several wearable sensor nodes, smart gateways with fog services, cloud servers, and terminals. Detailed implementations of these components are discussed as follows.

12.4.1 Sensor Node Implementation

Two types of sensor nodes, including wearable sensor nodes and static sensor nodes, are implemented. Wearable sensor nodes are used for collecting ECG, body temperate, and body motion, while static sensor nodes are placed in a room for monitoring room temperature and humidity. Although several communication protocols such as Wi-Fi, nRF, Bluetooth, and 6LoWPAN are used in our experiments, only the implementation of sensor nodes based on nRF is described in detail in this chapter. The implementation of other sensor nodes based on Wi-Fi, Bluetooth, and 6LoWPAN are carefully discussed in our other works [6, 7, 13, 19].

As mentioned, each sensor node has three primary components consisting of a microcontroller, sensors, and a wireless communication chip. According to Gia et al. [11], an 8-bit microcontroller is more suitable than a 32-bit microcontroller for such an IoT sensor node that does not perform heavy computational tasks. In the implementation, a low-power 8 MHz Atmega328P is used because the micro-controller consumes low energy during an active mode, and it provides several sleep modes for saving energy. In our experiments, the micro-controller is in a deep sleep all the time except for when it is receiving data from sensors and transmitting the data to smart gateways. In several experiments, the micro-controller is active for performing encryption methods (e.g., AES). Although the micro-controller supports up to 20 MHz with an external oscillator, 8 MHz is one of the most suitable frequencies for the sensor node. For running at 16 MHz and 20 MHz, the micro-controller needs 5 V supply, while it requires 3 V for running at 8 MHz. According to Gia et al. [11], the micro-controller aims to communicate with sensors via 1 MHz SPI since SPI is more energy efficient than other wirer protocols such as I2C and UART [11].

For collecting ECG, an analog front-end ADS1292 component is used. ADS1292 can collect high-quality ECG with two channels. Each channel supports up to 8000 samples/s where each sample is 24 bits. ADS1292 consumes low energy and communicates with the micro-controller via SPI. In the experiments, 125 samples/s two-channel ECG are obtained via 1 MHz SPI.

MPU9250, which is a nine-axis motion sensor consisting of a 3-D accelerometer, a 3-D gyroscope and a 3-D magnetometer, is used for collecting acceleration and angular velocity. The sensor can operate with 3 V power supply, and it consumes low energy. The sensor can connect with the microcontroller via SPI. In the experiments, 100 samples from the 3-D accelerometer and 100 samples from the 3-D gyroscope are acquired every second via 1 MHz SPI.

BME280 is used in different sensor nodes for collecting a body temperature and a room temperature because BME280 consumes low energy [31]. BME280 can operate with 3V power supply, and it can connect with the micro-controller via SPI. Each temperature data sample from a monitor person is collected every 10 s because a body temperature does not change quickly. Similarly, a room temperature is obtained and transmitted in every minute for saving sensor nodes' energy consumption.

In our implementation, nRF is used because of its advantages in low energy consumption, M2M communications, and flexible bandwidth support. For example, its peak power per transmission is less than 50 mW and a time period of each transmission is about 2 ms [11]. A nRF24L01 chip is used in sensor nodes. The chip can operate with 3 V and connects to the micro-controller via SPI.

12.4.2 Smart Gateways with Fog Implementation

Smart gateways are built based on a combination of several devices and components such as Pandaboard, HC05 Bluetooth, nRF24L01, and Smart-RF06 board with TI CC2538cc25. Pandaboard is the core of a smart gateway and fog services since all fog services are installed and run on top of Pandaboard. Pandaboard has a dual-core 1.2 GHz ARM Cortex-A9, 304 MHz GPU, 1 GB of RAM. In addition, Pandaboard supports different protocols such as Wi-Fi, Ethernet, SPI, I2C, and UART. In the implementation, Ethernet is used for connecting to the Internet, while Wi-Fi is used for receiving data from sensor nodes that use Wi-Fi as the main protocol for transmitting high-quality signals. Furthermore, Pandaboard supports up to 32GB SD card for installing an embedded operating system. In the implementation, a lightweight embedded operating system based on a Linux kernel is used.

HC05 is a low-cost Bluetooth chip that supports master and slave modes. HC05 can be connected to Pandaboard via UART. A driver is not required when setting up HC05 in Pandaboard.

A nRF24L01 chip integrated into the gateway is the same nRF24L01 chip used in sensor nodes. It is also connected to Pandaboard via SPI. With nRF, Pandaboard can receive data from different sensor nodes simultaneously.

A Smart-RF06 board with TI CC2538 provides a capability of communication with 6LoWPAN. A TI CC2538 chip is placed on top of the Smart-RF06 board, which is connected to Pandaboard via Ethernet. In the implementation, a USB to Ethernet adapter is used for providing an extra Ethernet port for Pandaboard.

The local database is implemented by MongoDB, which is an open source database using document-oriented data model. For instance, in addition to biosignals and contextual data, username, password, and other important information are also stored in the local database.

AES-256 (Standard) and IPtables [32] are applied in smart gateways for providing some levels of security. AES-256 is a symmetric block cipher for protecting transmitted data. Biosignals are encrypted at sensor nodes and the encrypted data are decrypted at the smart gateways for storing and processing. IPtables is a type of firewall that contains many rules for the treatment of packets (e.g., allow or block traffic). IPtables checks all rules when trying to establish a connection to smart gateways. If the rules are not fulfilled, it performs the predefined actions. For proffering the higher security level, our advanced and complex security algorithms [25, 26, 28], which are suitable for smart gateways with fog computing, can be customized and applied in the system.

The fall detection algorithm based on the multilevel threshold discussed above is applied in the system. In the fall detection application, most of the noises come from motion artifacts and power-line noise from surrounding environments. In the experiment, 100 Hz motion data are collected from both accelerometer and gyroscope. Therefore, a second order 10–40 Hz band pass filter is used. The parameters of the filter can be different, depending on application requirements.

The filtered data, including both acceleration and angular velocity, are compared with several predefined thresholds. In the experiments, 1.6 g and 1.9 g are the first and second threshold for acceleration, while 130 deg/s and 160 deg/s are the first and the second threshold for angular velocity.

In addition, these fall-related values can be used for categorizing the activity status of a monitored person (e.g., static activity, moving activity or sleeping). In the experiments, status activities of all cases are detected successfully by a simple algorithm which counts the number of ripples whose magnitude is in between predefined ranges (e.g., 0–100 deg/s and 0.5–1.5 g). For example, the first case in Figure 12.5 is categorized into three periods, including a static (e.g., sit still or stand still) period, a falling period, and a period of standing up after a fall.

The categorization service is implemented at the fog with an assistance of the "iw, iwlist" packages, which are built for Linux kernel-based operating systems. Via the scanning methods from these packages, all necessary information (e.g., MAC address, SSID, and RSSI) of Wi-Fi devices connected to smart gateways can be easily obtained. Smart gateways scan the information regularly and update the local database. Although these packages are not fully developed, they are still suitable for the categorization service.

Data processing in the fog includes simple filtering and advanced processing algorithms. In our implementation, a 50 Hz filter is applied in Python for

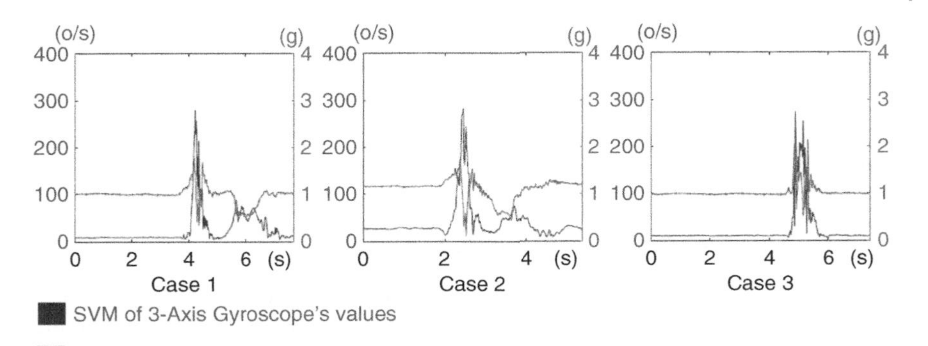

Figure 12.5 SVM of 3-D acceleration and SVM of 3-D angular velocity.

removing noise and interference from surrounding environments. Depending on particular countries, a 50 Hz or 60 Hz filter can be applied. For example, a 50 Hz notch Butterworth filter should be applied in Nordic countries, while a 60 Hz filter should be applied in American countries. Then the filtered data are applied with several algorithms (e.g., heart rate extraction algorithm) for detecting R peaks, R-R intervals, or U waves. These algorithms are implemented in Python.

12.4.3 Cloud Servers and Terminals

In the implementation, Google Cloud and its API are used. For instance, Google Cloud dataflow and Firebase are used. Cloud Dataflow is service for enriching real-time data and the history of data, while Firebase is used for push notification. The global database in the cloud is configured to have the same structure as the fog local database for achieving an easy synchronization between these databases. Different services of Google API can be used, depending on health-monitoring applications. Cloud servers host a webpage having an easy-to-use interface. The webpage is built based on up-to-date technologies such as Python, HTML5, CSS, XML, JavaScript, and JSON. Similar to the fog's webpage, the global webpage also has a form having username and password. By using an Internet browser, an end-user having valid credentials can connect to the webpage and access the monitored data in real time. The level of credentials depends on specific users. In addition, end users can use a mobile application to monitor real-time data. Similar to the webpage, the mobile app has a form consisting of a username plus a password to log in and uses the same mechanism to check the credentials. The app can show data in both text and graphical forms. Currently, the app is merely built for Android phones. In the future, another version of the mobile app will be built for IOS.

12.5 Case Studies, Experimental Results, and Evaluation

This chapter presents cases study of remote ECG monitoring together with fall detection in real time. Details of each case are discussed as follows.

12.5.1 A Case Study of Human Fall Detection

For evaluating a fall-detection feature, six volunteers, including healthy males and females whose age is about 24–32 years old, participate in the experiments organized in a lab room. Each wearable sensor node is attached to a volunteer's chest for collecting body temperature, ECG, and body motions for 4–5 hours. The collected data is wirelessly transmitted to smart gateways with a fog. In the experiments, the energy consumption of the sensor is measured in different configurations. In one configuration, data are kept intact before being sent to the gateway while data are encrypted in another configuration. At smart gateways with the fog, data are processed with advanced algorithms for detecting a human fall, analyzing ECG, and evaluating heart variability. In addition, many advanced fog services are provided for improving the quality of services. Then the data are transmitted to cloud servers, which can show analyzed and processed data in text and graphical forms.

In the experiment, each volunteer is asked to do his or her normal activities (e.g., standing still, sit still, and walking) and suddenly fall into a mattress. Each volunteer repeats his/her activities five times. The motion data acquired by sensor nodes are transmitted to a smart gateway with the fog. The real-time acceleration and angular velocity collected from three volunteers simultaneously are shown in Figure 12.5. The data is retrieved from the fog's webpage. It can be easily seen that acceleration and angular velocity are stable in the first time period (e.g., 0–4 s in the first case) and changes dramatically in the second period (e.g., 4–6 s in the first case). The data indicates that in the first case, a person falls at 4s. Similarly, the second and third persons fall at 2.3 s and 4.5 s, respectively. These collected data are processed at fog with the fall algorithm shown in Figure 12.4 for detecting fall cases. In the experiments, all fall cases are detected successfully.

In the experiments, energy consumption of sensor nodes is measured by a MonSoon professional power monitoring tool. The total energy consumption is calculated by the following formula.

$$E = \text{Average Power}_{active} * \text{Time}_{active} + \text{Power}_{idle} * \text{Time}_{idle}$$

where: $\text{AveragePower}_{active} = V_{supply} * \text{Average } I_{active}$
$\text{Power}_{idle} = V_{supply} * I_{idle}$
$\text{Time}_{idle} = \text{Total measurement time} - \text{Time}_{active}$

Table 12.1 Energy consumption of the sensor node with and without running AES.

Mode	Energy consumption (mWs)
Idle (without AES)	1.26
Active without AES	5.94
Total energy without AES	7.2
Idle (with AES)	1.044
Active with AES	8.71
Total energy with AES	9.754

The energy consumption of sensor nodes is measured during 1 s. Results are shown in Table 12.1. Results show that running AES encryption in sensor node only causes a slight increase of energy consumption about 2.2 mWs. By using the 1000 mWh battery, the sensor node can be used up to 45 hours.

12.5.2 A Case Study of Heart Rate Variability

As introduced above, ECG signals can be captured either by a wearable sensor device or by a professional monitoring machine in multiple leads. The raw ECG signal in Figure 12.6 is 1-lead ECG and was measured from a wearable device. In the preprocessing phase, moving average filter was first applied to remove the baseline wander of the signal, and then a 50 Hz notch filter was applied to remove power-line interference. In this pain assessment application, R wave is of interest in the analysis; therefore, R peaks were detected with a peak detection algorithm and R to R intervals were calculated from every two adjacent R peaks for heart rate variability analysis.

In pain assessment research, heart rate variability has been studied as a potential automatic parameter to quantify pain experience [33–35]. Heart rate variability (HRV) analysis is built on the extraction of R-R interval shown in Figure 12.6, which is usually denoted as N-N, meaning normal sinus to normal sinus. HRV can be analyzed in the time domain, frequency domain, or with nonlinear methods [36]. HRV features are extracted from time windows in a certain length, which may vary in different applications and for different purposes. In addition to pain assessment, HRV analysis is also widely applied in disease diagnoses such as cardiac arrhythmia and clinical studies such as sleep analysis. Long-term HRV analysis is usually processed.

every 24 hours; short-term HRV analysis is processed in time windows of several minutes; there is also ultra-short-term analysis in time windows that are shorter than 1 minute. Some commonly seen HRV features in the time domain analysis are:

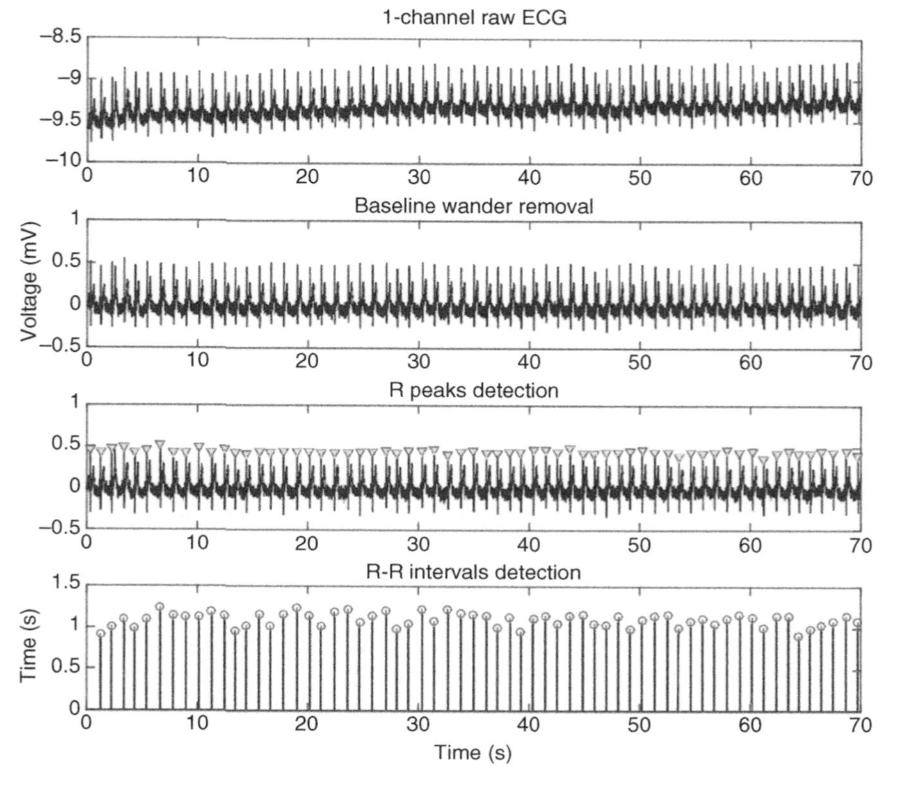

Figure 12.6 Real-time ECG monitoring and preprocess ECG data at fog.

- AVNN: an average of NN intervals
- SDNN: a standard deviation of NN intervals
- RMSSD: a root mean square of differences between adjacent NN intervals
- pNNx: a percentage of differences between adjacent NN intervals that are larger than x milliseconds

Some HRV features in the frequency domain are:

- LF: low-frequency component, the cumulative sum of the spectral power between 0.01 Hz and 0.15 Hz
- HF: high-frequency component, the cumulative sum of the spectral power between 0.15 Hz and 0.4 Hz
- LF/HF: the ratio of low-frequency component to high-frequency component

The nonlinear methods in HRV analysis include correlation dimension analysis, detrended fluctuation analysis, and entropy analysis, for example.

In pain monitoring, ultra-short-term and short-term HRV features are extracted from the real-time monitoring of ECG signals, indexing the activity of the autonomic nervous system. In a health-monitoring system, the ECG waveforms captured by a wearable sensor node (e.g., bioharness sensor) can then be analyzed in terms of HRV features and can be classified into a pain intensity in the system's fog layer. The classifier for pain intensity recognition is first trained and tested with a pain database.

The Internet of things for healthcare research group at the University of Turku built such a database from 15 healthy female and 15 healthy male volunteers under experimental pain stimulation. Each subject experienced four successive tests on the same day. In two of the tests, the pain stimulation was heat on a forearm, which was produced by a round heating element with a diameter of 3 cm. The heating element was placed on the left or right forearm when its temperature was 30 °C and first increased 1 °C every 3 seconds before 45 °C. After 45 °C, its temperature increased 1 °C every 5 seconds and the active heating stopped at 52 °C. In the rest two tests, the pain stimulation was electrical pulses with a width of 250 μs repeating at a frequency of 100 Hz. The pulses were generated by a commercial TENS device and the electrodes were placed on the left or right ring finger. The pulse intensity was controlled by a researcher that increased 1 in every 3 seconds with a maximum level of 50. The order of the four tests was randomized. In each test, each subject was able to report the time point when he or she reached his or her pain threshold and then pain tolerance. The pain stimulation was then removed from the subject either when the pain tolerance was reported or the stimulation had reached its maximum intensity. The subjective pain intensity was self-reported at the end of each test in the VAS score.

The No pain data was defined as the data from 30 s before the start of the test. The data between start and pain threshold was labeled as *Mild pain* and the data between pain threshold and pain tolerance was labeled as *Moderate/Severe pai*n because the self-reported pain score differs among subjects. The changes of some HRV features along with the test are presented as the root mean square (RMS) of the feature in each pain category, as shown in Figure 12.7. To explore the influence of the time window choice on the pattern, HRV features extracted from time windows between 10 s to 60 s are presented. To reduce the impact of different resting heart rate on the HRV analysis [37], NN intervals were normalized by the reference to AVNN. Moreover, to adjust the feature distribution into a normal distribution, some features were logarithmically transformed with natural logarithm and ln was added in their names.

The classification was then conducted between *No pain* and *Pain*, where *Pain* is a merge of the other two categories. Support vector machine classifiers were trained and tested in 10-fold cross-validation with the HRV features in Figure 12.7. The ROC curves of the classifications in each time window length are presented in Figure 12.8. The AUC values show that the classification of HRV features had better performance in a larger time window length.

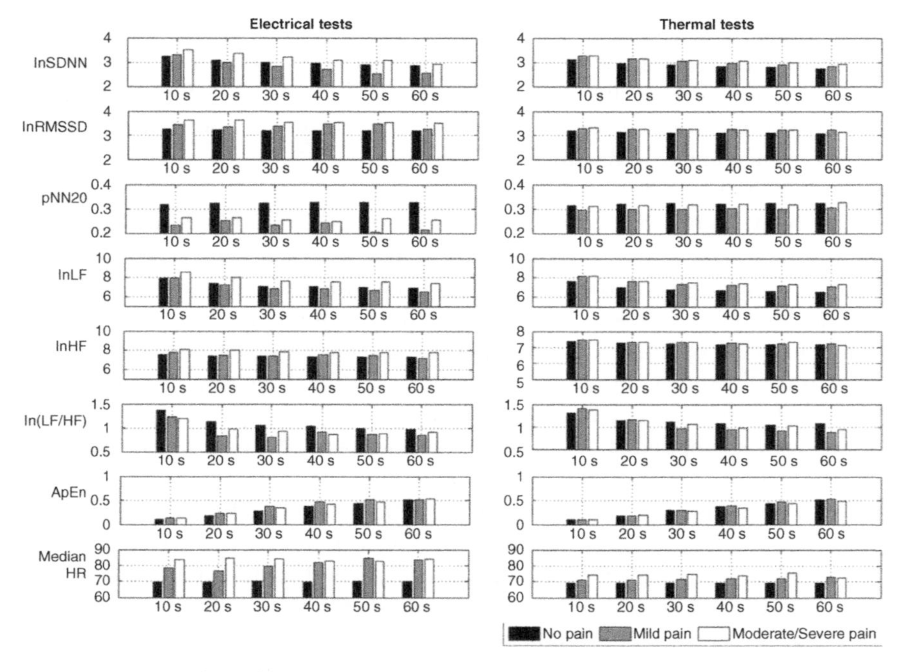

Figure 12.7 RMS of HRV features in different window lengths and different pain intensities.

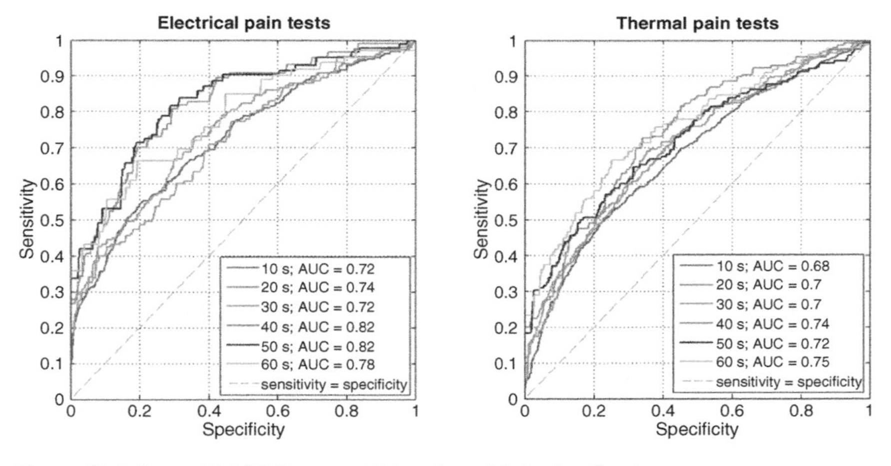

Figure 12.8 Figure 12.8 ROC curves of *No pain* and *Pain* classification.

12.6 Discussion of Connected Components

Fog computing shows its capability to reduce heavy computational loads of sensor nodes by switching the loads to fog-assisted smart gateways. Accordingly, sensor nodes can be used for a longer time per charge. Depending on the specific applications, the fog's role becomes more or less important in terms of increasing energy efficiency. For example, fog is important in real-time monitoring applications using complex algorithms (e.g., real-time multichannel ECG monitoring with ECG feature extraction) while fog does not dramatically increase the lifetime of sensor nodes in simple monitoring applications (e.g., temperature monitoring). With a fog, a high level of the energy efficiently of the entire system cannot be achieved. To address the target, all components of the system from sensor nodes, smart gateways, to enduser terminals must be carefully considered.

12.7 Related Applications in Fog Computing

Many IoT systems for healthcare have been proposing [38–40]. However, these conventional IoT systems still have several limitations, such as nonsupport interoperability, energy inefficiency, nonsupport distributed local storage, or inefficient utilization of bandwidth. Recently, several fog-based solutions have been proposed for enhancing existing healthcare IoT systems. Gia et al. [18] present a fog-based approach for ECG feature extraction. The approach extracts heart rate, P wave, T wave, from ECG signals. In addition, it helps to save about 90% bandwidth. Azimi et al. [41] present hierarchical fog-assisted computing architecture for healthcare IoT. The approach can detect arrhythmia by algorithms implemented in both fog-assisted smart gateways and cloud servers. Moosavi et al. [25] present an end-to-end security approach based on fog computing for health IoT systems. The approach provides a high level of security even though sensor nodes randomly moves from a gateway to another gateway. Rahmani et al. [22] propose a fog-based approach for enhancing healthcare monitoring IoT systems. By using the fog-based smart gateway, the proposed system provides distributed local storage, data fusion, data analysis at smart gateways. Bimschas et al. [42] propose a middleware at smart gateways with a fog layer. The approach provides some levels of interoperability supporting multi-communications between standard and non-standard protocol applications. Similarly, Shi et al. [43] present a fog-based approach for intercommunicating between different wireless protocols such as ZigBee, WiFi, 2G/3G/4G, WiMax, and 6LoWPAN. In addition, the approach provides some levels of self-decision-making for selecting a suitable format before sending both raw and processed data to cloud servers. Cao et al. [44] present the fog-based approach for fall detection systems. The approach analyzes and detects a human fall in real time by splitting the detection task between the edge devices. Similarly, Gia et al. [10] and Igor et al. [13] present the fog-based approaches for real-time fall detection. These approaches analyze motion data such as 3-D acceleration and 3-D angular velocity at smart gateways. When the collected data trespasses pre-defined threshold, push notification messages are sent to responsible persons in real time. In [45], authors present a fog computing system for monitoring mild dementia and COPD patients. Collected data are processed at the fog for reducing the communications overload and protecting a patient's privacy.

12.8 Future Research Directions

Fog computing with many advanced services has proven an important role in improving the quality of healthcare services. However, the fog services can be augmented to achieve the outstanding quality of healthcare service. For example, data analysis and data fusion can be enhanced for improving the quality of disease diagnosis (i.e. avoiding incorrect disease diagnoses). Instead of applying conventional methods using predefined thresholds, these fog services will use machine learning such as deep learning or reinforcement learning to achieve more accurate results. For instance, fixed thresholds in fall detection algorithms will become self-adaptive for suiting to particular situations such as irregular movements or dancing. In addition, machine learning helps to deploy artificial intelligence (AI) at a fog easier. The fog-based system with AI helps to improve decision-making significantly for correct and smart reactions in real time. Future health-monitoring systems may be able to provide diverse applications by applying different analysis methods to the same physiological signals, which will meet the demand of different people with more precise services. For example, in addition to pain monitoring, HRV analysis is also useful in reflecting current diseases, indicating cardiac issues, and reflecting depression or stress.

12.9 Conclusions

This chapter primarily presented fog computing in health-monitoring IoT systems. Fog provides advanced services consisting of distributed local storage, push notification, human fall detection, data analysis, security, localhost with a user interface, and fault detection. These services play important roles in improving healthcare services. With these services, a patient's health status is always continuously and remotely monitored by responsible persons while his or her daily activities are not interrupted. Cases studies consisting of human fall detection and heart rate variability demonstrated the benefits of the fog and its services. These services process and analyze data in real time. When some abnormalities occur (e.g., a fall), the push notification service is triggered to inform responsible persons such as medical doctors and caregivers in real time for in-time reactions. Furthermore, fog computing and its services address challenges in healthcare IoT. For example, fog helps to improve the energy efficiency of sensor nodes by switching heavy computational loads to smart gateways. Fog is capable of detecting system faults (e.g., hardware and software faults) and informing system administrators in real time. Fog not only facilitates advanced services but also reduces the burden of a cloud. Fog computing demonstrated that it is one of the most suitable candidates for augmenting IoT systems in healthcare and other domains.

References

1 Summary Health Statistics: National Health Interview Survey 2017, National Center for Health Statistics, 2017.

2 National diabetes statistics report: Estimates of diabetes and its burden in the United States, Centers for Disease Control and Prevention, Atlanta, 2014.

3 E.J. Benjamin, M.J. Blaha, and S.E. Chiuve, et al. Heart disease and stroke statistics, American Heart Association. *Circulation,* 135(10): e146–e603, 2017.

4 D.A. Sterling, J.A. O'Connor, J. Bonadies. Geriatric falls: injury severity is high and disproportionate to mechanism. *Journal of Trauma and Acute Care Surgery* 50(1): 116–119, 2001.

5 J.A. Stevens, P.S. Corso, E.A. Finkelstein, T. R Miller. The costs of fatal and nonfatal falls among older adults. *Injury Prevention,* 12(5): 290–295, 2006.

6 T.N. Gia, N.K. Thanigaivelan, A.M. Rahmani, T. Westerlund, P. Liljeberg, and H. Tenhunen. Customizing 6LoWPAN Networks towards Internet-of-Things Based Ubiquitous Healthcare Systems. In *Proceedings of 32nd IEEE NORCHIP,* 2014.

7 T.N. Gia, A.M. Rahmani, T. Westerlund, T. Westerlund, P. Liljeberg, and H. Tenhunen. Fault tolerant and scalable iot-based architecture for health monitoring. In *Proceedings of IEEE Sensors Applications Symposium,* 2015.

8 A.M. Rahmani, N.K. Thanigaivelan, T.N. Gia, J. Granados, B. Negash, P. Liljeberg, and H. Tenhunen. Smart e-health gateway: bringing intelligence to Internet-of-Things based ubiquitous healthcare systems. In *Proceedings of 12th Annual IEEE Consumer Communications and Networking Conference,* 2015.

9 V.K. Sarker, M. Jiang, T.N. Gia, M. Jiang, T.N. Gia, A. Anzanpour, A.M. Rahmani, P. Liljeberg. Portable multipurpose biosignal acquisition and

wireless streaming device for wearables. In *Proceedings of IEEE Sensors Applications Symposium*, 2017.

10 T.N. Gia, I. Tcarenko, V.K. Sarker, A.M. Rahmani, T. Westerlund, P. Liljeberg, and H. Tenhunen. IoT-based fall detection system with energy efficient sensor nodes. In *Proceedings of IEEE Nordic Circuits and Systems Conference*, 2016.

11 T.N. Gia, V.K. Sarker, I. Tcarenko, A.M. Rahmani, T. Westerlund, P. Liljeberg, and H. Tenhunen. Energy efficient wearable sensor node for IoT-based fall detection systems. *Microprocessors and Microsystems*, Elsevier, 2018.

12 M. Jiang, T.N. Gia, A. Anzanpour, A.M. Rahmani, T. Westerlund, S. Salanterä, P. Liljeberg, and H. Tenhunen. IoT-based remote facial expression monitoring system with sEMG signal. In *Proceedings of IEEE Sensors Applications Symposium*, 2016.

13 I. Tcarenko, T.N. Gia, A.M. Rahmani, T. Westerlund, P. Liljeberg, and H. Tenhunen. Energy-efficient IoT-enabled fall detection system with messenger-based notification. In *Proceedings of 6th International Conference on Wireless Mobile Communication and Healthcare*, Springer, 2017.

14 F. Touati and T. Rohan. U-healthcare system: State-of-the-art review and challenges, *Journal of medical systems* 37 (3) (2013).

15 T.N. Gia, M. Jiang, V.K. Sarker, A.M. Rahmani, T. Westerlund, P. Liljeberg, and H. Tenhunen. Low-cost fog-assisted health-care IoT system with energy-efficient sensor nodes. In *Proceedings of 13th IEEE International Wireless Communications & Mobile Computing Conference*, 2017.

16 B. Negash, T.N. Gia, A. Anzanpour, I. Azimi, M. Jiang, T. Westerlund, A.M. Rahmani, P. Liljeberg, and H. Tenhunen. Leveraging Fog Computing for Healthcare IoT, *Fog Computing in the Internet of Things*, A. M. Rahmani, et al. (eds), ISBN: 978-3-319-57638-1, Springer, 2018, pp. 145–169.

17 T.N. Gia, M. Ali, I.B. Dhaou, A.M. Rahmani, T. Westerlund, P. Liljeberg, and H. Tenhunen. IoT-based continuous glucose monitoring system: A feasibility study, *Procedia Computer Science*, 109: 327–334, 2017.

18 T.N. Gia, M. Jiang, A.M. Rahmani, T. Westerlund, P. Liljeberg, and H. Tenhunen. Fog computing in healthcare Internet-of-Things: A case study on ECG feature extraction. In *Proceedings of 15th IEEE International Conference on Computer and Information Technology*, 2015.

19 T.N. Gia, M. Jiang, A.M. Rahmani, T. Westerlund, K. Mankodiya, P. Liljeberg, and H. Tenhunen. Fog computing in body sensor networks: an energy efficient approach. In *Proceedings of IEEE 12th International Conference on Wearable and Implantable Body Sensor Networks*, 2015.

20 F. Bonomi, R. Milito, J. Zhu, S. Addepalli. Fog computing and its role in the Internet of Things. In *Proceedings of 1st ACM MCC Workshop on Mobile Cloud Computing*, 2012.

21 M. Peng, T. Wang, G. Hu, and H. Zhang. A wearable heart rate belt for ambulant ECG monitoring. In *Proceedings of IEEE International Conference on E-health Networking,* Application & Services, 2012.

22 A.M. Rahmani, T.N. Gia, B. Negash, A. Anzanpour, I. Azimi, M. Jiang, and P. Liljeberg. Exploiting smart e-health gateways at the edge of healthcare internet-of-things: a fog computing approach, *Future Generation Computer Systems, 78(2) (January): 641–658,* 2018.

23 M. Armbrust, A. Fox, R. Griffith, A.D. Joseph, R. Katz, A. Konwinski, G. Lee, D. Paterson, A. Rabkin, I. Stoica, and M. Zaharia. A view of cloud computing. *Communications of the ACM,* 53(4) (April): 50–58, 2010.

24 D. C. Klonoff. Cybersecurity for connected diabetes devices. *Journal of Diabetes Science and Technology,* 9(5): 1143–1147, 2015.

25 S. R. Moosavi, T.N. Gia, E. Nigussie, et al. End-to-end security scheme for mobility enabled healthcare Internet of Things. *Future Generation Computer Systems,* 64 (November): 108–124, 2016.

26 S.R. Moosavi, T.N. Gia, E. Nigussie, A.M. Rahmani, S. Virtanen, H. Tenhunen, and J. Isoaho. Session resumption-based end-to-end security for healthcare Internet-of-Things. In *Proceedings of 15th IEEE International Conference on Computer and Information Technology,* 2015.

27 T. Kothmayr, C. Schmitt, W. Hu, M. Brünig, and G. Carle. DTLS based security and two-way authentication for the Internet of Things, *Ad Hoc Networks,* 11(8): 2710–2723, 2013.

28 S. R. Moosavi, T.N. Gia, A.M. Rahmani, S. Virtanen, H. Tenhunen, and J. Isoaho. SEA: A secure and efficient authentication and authorization architecture for IoT-based healthcare using smart gateways. *Procedia Computer Science,* 52: 452–459, 2015.

29 Z.-P. Bian, J. Hou, L.P. Chau, N. Magnenat-Thalmann. Fall detection based on body part tracking using a depth camera, *IEEE Journal of Biomedical and Health Informatics,* 19(2): 430–439, 2015.

30 D. Lim, C. Park, N.H. Kim, and Y.S. Yu. Fall-detection algorithm using 3-axis acceleration: combination with simple threshold and hidden Markov model. *Journal of Applied Mathematics,* 2014.

31 M. Ali, T.N. Gia, A.E. Taha, A.M. Rahmani, T. Westerlund, P. Liljeberg, and H. Tenhunen. Autonomous patient/home health monitoring powered by energy harvesting. In *Proceedings of* IEEE Global Communications Conference, Singapore, 2017.

32 R. Russell. Linux iptables HOWTO, url: http://netfilter. samba. org, Accessed: December 2018.

33 A.J. Hautala, J. Karppinen, and T. Seppanen. Short-term assessment of autonomic nervous system as a potential tool to quantify pain experience. In *Proceedings of 38th Annual International Conference of the IEEE Engineering in Medicine and Biology Society,* 2684–2687, 2016.

34 J. Koenig, M.N. Jarczok, R.J. Ellis, T.K. Hillecke, and J.F. Thayer. Heart rate variability and experimentally induced pain in healthy adults: a systematic review *European Journal of Pain*, 18(3): 301–314, 2014.

35 M. Jiang, R. Mieronkkoski, A.M. Rahmani, N. Hagelberg, S. Salantera, and P. Liljeberg. Ultra-short-term analysis of heart rate variability for real-time acute pain monitoring with wearable electronics. In *Proceedings of IEEE International Conference on Bioinformatics and Biomedicine*, 2017.

36 U.R. Acharya, K.P. Joseph, N. Kannathal, C.M. Lim, and J.S. Suri. Heart rate variability: a review. *Medical & Biological Engineering & Computing*, 44(12): 1031–1051, 2006.

37 J. Sacha. Why should one normalize heart rate variability with respect to average heart rate. *Front. Physiol*, 4, 2013.

38 G. Yang, L. Xie, M. Mantysalo, X. Zhou, Z. Pang, L.D. Xu, S. Kao-Walter, and L.-R. Zheng. A health-iot platform based on the integration of intelligent packaging, unobtrusive bio-sensor, and intelligent medicine box. *IEEE transactions on industrial informatics*, 10(4): 2180–2191, 2014.

39 M.Y. Wu and W.Y. Huang. Health care platform with safety monitoring for long-term care institutions. In *Proceedings of 7th International Conference on Networked Computing and Advanced Information Management*, 2011.

40 H. Tsirbas, K. Giokas, and D. Koutsouris. Internet of Things, an RFID-IPv6 scenario in a healthcare environment. In *Proceedings of 12th Mediterranean Conference on Medical and Biological Engineering and Computing*, Berlin, 2010.

41 I. Azimi, A. Anzanpour, A.M. Rahmani, T. Pahikkala, M Levorato, P. Liljeberg, and N. Dutt. HiCH: Hierarchical fog-assisted computing architecture for healthcare IoT. *ACM Transactions on Embedded Computing Systems*, 16(5), 2017.

42 D. Bimschas, H. Hellbrück, R. Meitz, D. Pfisterer, K. Römer, and T. Teubler. Middleware for smart gateways. In *Proceedings of 5th International workshop on Middleware Tools, Services and Run-Time Support for Sensor Networks*, 2010.

43 Y. Shi, G. Ding, H. Wang, H.E. Roman, S. Lu. The fog computing service for healthcare. In *Proceedings of 2nd International Symposium on Future Information and Communication Technologies for Ubiquitous HealthCare*, 2015.

44 Y. Cao, S. Chen, P. Hou, and D. Brown. FAST: A fog computing assisted distributed analytics system to monitor fall for stroke mitigation. In *Proceedings of 10th International Conference on Networking, Architecture, and Storage*, 2015.

45 O. Fratu, C. Pena, R. Craciunescu, and S. Halunga. Fog computing system for monitoring mild dementia and COPD patients – Romanian case study. In *Proceedings of 12th International Conference Telecommunications in Modern Satellite, Cable and Broadcasting Service*, 2015.

13

Smart Surveillance Video Stream Processing at the Edge for Real-Time Human Objects Tracking

Seyed Yahya Nikouei, Ronghua Xu, and Yu Chen

13.1 Introduction

The past decade has witnessed worldwide urbanization because of the benefits and diverse lifestyles in bigger cities. While it brings higher living quality, it introduces new challenges to city administrators, urban planners, and policy makers. Safety and security are among the top concerns when more and more people live in an area with such a high density. Situational awareness (SAW) has been recognized as one of the key capabilities in order to timely deal with urgent issues. To serve this purpose, more and more surveillance cameras and sensors are installed in urban area to monitor the daily activities of the residents. For example, North America alone had more than 62 million cameras by 2016 [1]. The enormous surveillance data generated by these cameras requires extraordinary supervisory action to extract useful information, which implies 24/7 attention to the captured video streams. It is not realistic to rely on human operators facing the ubiquitously deployed cameras. Recent machine-learning algorithms are promising to make smarter decisions based on surveillance video in real time. However, intelligent decision-making approaches is not mature yet today.

When each frame is taken, it must be transferred from the field to the data center where further processing. Nowadays, the video data dominate the real-time traffic and creates heavy workload on the communication networks. Online video streaming accounts for 74% of the online traffic in 2017 [2], and 78% of mobile traffic will be video data by 2021 [3]. A single camera generates more than 9600 GB of data in a single day. There are a couple of important concerns the community is aware of and has been working hard to resolve. First of all, it is essential to avoid sending raw data that is not globally significant to reduce the heavy burden on the communication network. Also, the transmission time for the raw footage to reach the data center can be vital in some delay-sensitive, mission-critical applications. It is desired to reduce the communication delays as much as possible. Second, one of the important issues

Fog and Edge Computing: Principles and Paradigms, First Edition.
Edited by Rajkumar Buyya and Satish Narayana Srirama.

is data loss during transmission, or even worse, that a third-party eavesdrops on the transmission line. Considering the huge volume of video data to be stored in the data center, new challenges are introduced. While the capacity of data storage facilities is getting larger and larger, nowadays the surveillance video owners can merely keep weeks of the most recently captured footages. The limited storage capacity results in losing footage that contains important information for forensics analysis or other purposes. Thus, it is critical to be able to timely extract features from raw video such that the operators can identify and selectively store the clips of interest for longer periods.

In order to address these problems, edge and fog computing and distributed real-time data processing are attracting a lot of attention in the surveillance community [5]. The functions, including feature extraction and decision making, are migrated to the edge of the network, and a distributed environment is created instead of a single or couple points of reference. In this chapter, an edge computing based smart surveillance system is introduced [6]. Focusing on human object detection, the system takes three steps toward an intelligent decision-making. First, it recognizes and detects human objects in each given frame and each human object is tracked for feature extraction. The speed or movement direction of each human object is saved in an array along with other specific information as features for next step. The final step enables decision-making using machine-learning algorithms based on time-series features, which decides whether an alarm should be generated to the higher levels or human operators in charge. Figure 13.1 shows the network architecture and tasks allocated in each layer. In this figure, human detection and tracking are considered to be accomplished at the edge of the network and more computing intensive decision-making algorithms will take place at the fog level where tablets or notebooks are available, and then in final decision in case of an incident is sent to the person in charge or first responders.

In this chapter, the computations and algorithms used at the edge and fog levels are discussed and compared to create such automated surveillance system. The rest of this chapter is organized as the follows. Section 13.2 briefly introduces the human object identification algorithms that are potentially feasible in the edge computing environment, followed by the object tracking algorithms in the Section 13.3. Section 13.4 is focused on the design issues of a lightweight human object detection scheme and a case study using Raspberry Pi as the edge device is presented in Section 13.5. At the end, Section 13.6 summarizes this chapter with some discussions.

13.2 Human Object Detection

Although there are many publications devoted to human detection in general [4, 7], this task has not been thoroughly investigated on the devices with limited computing resources, such as the ones at the edge of the network.

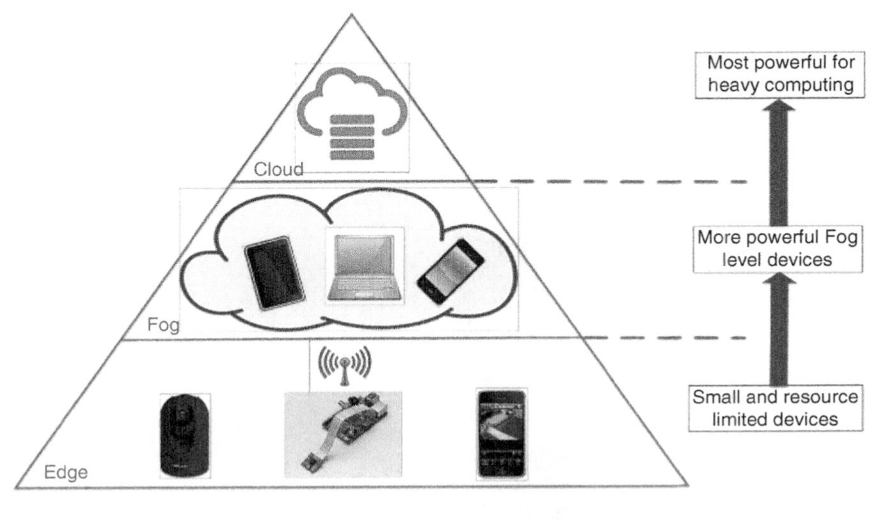

Figure 13.1 Edge-fog-cloud-based hierarchy smart surveillance architecture.

Human detection can be done using different accepts and algorithms. This chapter highlights three of them, which are potentially able to be fit into the edge computing environments.

13.2.1 Haar Cascaded-Feature Extraction

Haar cascaded-feature extraction is a well-studied method for human face or eye detection with decent performance [8]. It can also be applied for full human body detection. The algorithm subtracts pixel values from each other based on Haar-like features. There are a huge number of ways pixel values can be picked and subtracted, such that the learning process is normally conducted on a very powerful CPU. With a 24×24 image, about 160,000 features are produced. After training the algorithm is fast because only some subtractions need to take place.

Figure 13.2 shows several typical Haar-like features. There are three types of features, including two rectangular features (Figure 13.2 (a)), three rectangular features (Figure 13.2 (b)), and four rectangular features (Figure 13.2 (c)). In each feature, the pixel values in the black area are subtracted from the pixel values in the white area. In the learning phase, around 2000 positive images (containing the object of interest) and about half this size negative images are selected. These feature sets convolute over the image and a vector of values is created. Then an algorithm by the name of Adaboost will pick the best-performing features for the detection and thresholds found. Thus, the existence of the object in the image is determined by applying the selected features and a matching score above the thresholds.

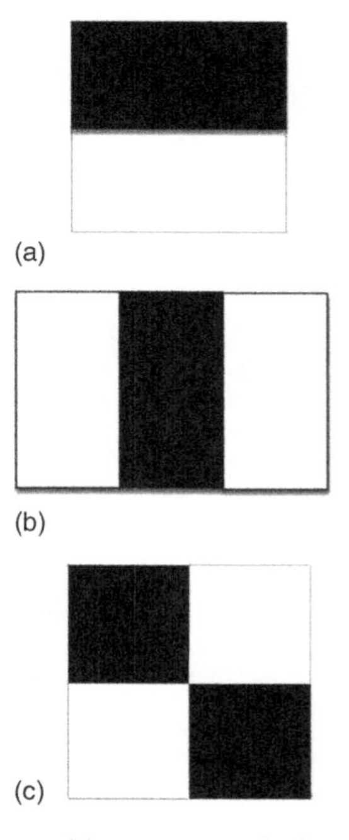

Figure 13.2 Haar-like features: (a) two rectangular features; (b) three rectangular features; (c) four rectangular features.

However, in terms of speed the performance is far from satisfactory because of the huge number of features used for higher accuracy. Therefore, a hierarchy method is introduced. It screens the input image by running the most important features first. If the result is positive, implying there might be an object of interest in the frame, more features will be tested. For instance, in first step, only one of the most dominating features is applied. A negative result means that the chance of existence of the object of interest in the frame is very low. Otherwise, a positive result leads to further tests with more features for fine position tuning and higher accuracy. In one example reported in literature, there are 28 stages in total, where the first stage has 1 feature, the second stage has 10 features, and the third stage has 25 features [9].

13.2.2 HOG+SVM

HOG+SVM is another widely used method because of its high accuracy. The name comes from a feature extraction named histogram of oriented gradients (HOG) and support vector machine (SVM) [11, 12]. These features are applied to classify or detect objects of interest. Traditionally, the high computing cost of this feature extractor makes the overall object detector not an ideal candidate for the edge computing environments. However, with more powerful devices deployed at the edge, the HOG+SVM method becomes more attractive for its accuracy. No matter how complicated or simplified a classifier is, if the features used for classification do not describe the object of interest in the best way the detection result will be inaccurate. For example, when oranges are to be separated from apples, while the orange color of the fruit is a good feature, the sphere shape does not give useful information for classification.

HOG is a well-known method for feature extraction. The difference in vertical neighboring pixels to the target pixel are considered as the vertical differential and the same method is also used to calculate the horizontal differential. It is worth mentioning that in some cases, instead of using only two immediate neighbors a vector of several pixels in each direction can be used, which contains more information for each pixel. The horizontal and vertical values are considered as an amplitude and angle instead of two derivatives. The horizontal derivative fires on vertical lines and the vertical derivative fires on horizontal lines. If there is more than one input channel such as RGB images with three channels, then the highest amplitude, along with its corresponding angle, are chosen to represent a pixel's gradient.

A histogram with nine bins is usually used with 0–20 degrees in each bin to represent the unsigned gradients and the amplitude of the respective angle is considered in the corresponding bin. If, however, the angle is closer to the border of a bin, part of the amplitude is given to the neighboring bin. The histogram is created for a window of 8×8 pixels normally. In order to resolve the effect of lightning or other temporary changes related to pixel values, normalization is needed [10]. In most object detection cases, 32×32 windows are selected to reduce processing time. This bigger window strides through the image with step of 1, which means a 4×4 of batches of windows of size 8×8 pixels is selected and then one 8×8 goes out from the 32×32 window from left and another one enters from the right. Each 32×32 window has 16 histogram bins in it, which can be represented in a 144 vector. The vector is normalized with the second norm of the vector values. Each vector is used as features for a part of image for object detection in the SVM. Figure 13.3(a) shows one of these before normalized histogram for an 8×8 super pixel and Figure 13.3(b) represents the gradient calculations in a 64×64 window (window is bigger than 8×8 pixels because of better visibility).

Another problem arises when human objects are closer or farther away from the camera in different camera placements. Taking a fixed window such as 8×8 is not useful if the person is very close or 16×16 might be too big when human objects are far away. An image pyramid is implemented in this case to change the resolution of the image and detect every possible object. Each stage reduces the pixels to create smaller image versions so that a super pixel covers bigger portions of the image. Consider that multiple stages of the image may result in multiple positive outputs for the same object. Figure 13.4 illustrates such a scenario. It is needed to change the variables in every specific usage to guarantee the best results, which means the algorithm is not good for generalization.

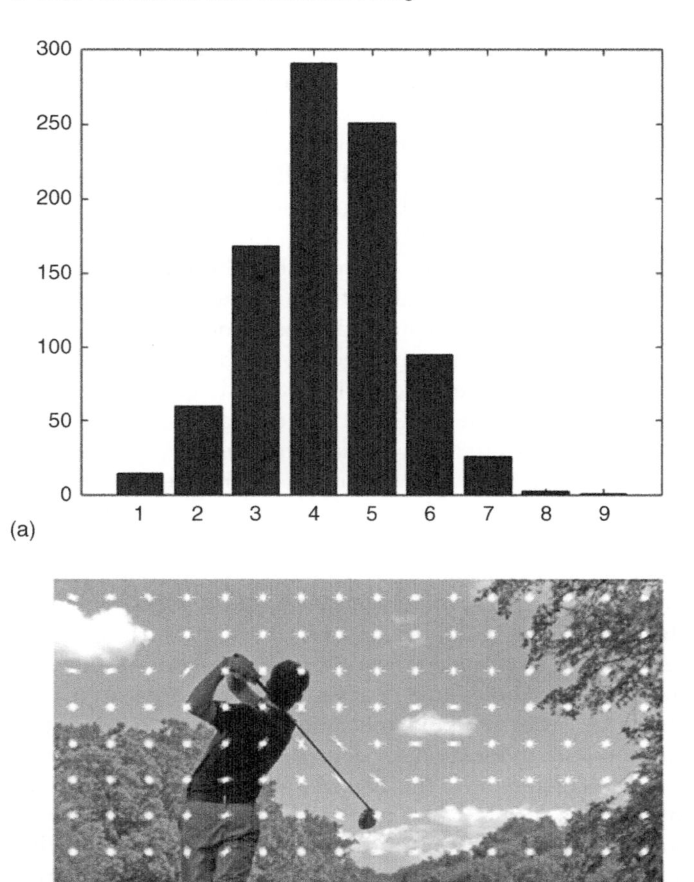

Figure 13.3 (a) Histogram of oriented gradients; (b) representation of HOG on an image.

13.2.3 Convolutional Neural Networks (CNNs)

Convolutional neural networks (CNNs) are based on multi-layer perceptron (MLP) network, one of the most famous types of neural networks, which have convolutional layers to produce feature maps.

CNNs usually have two separable parts, one is the convolutional layers and the other is a fully connected neural networks (FCNNs) or in some cases SVM classifier, which classifies objects using the feature map created by the

Figure 13.4 An example of multi-detection for a single object.

convolutional layers. In each convolutional layer, a set of filters convolute with the input and the dot product will be reconsidered as the output. After a convolutional layer, which is considered as a linear layer, a ReLU layer is added so that nonlinearity is introduced to the network. To keep the dimensionality unchanged, a padding of pixels around the input with a value of zero is added. Spatially downsizing of the feature map will be done with the pooling layers, where in a 2×2 square of values, the highest one is selected or the average value is calculated. At the final convolutional layer, only a series will be remained from the input image and that is used for classification.

Image classification generally refers to the process of giving the computer one image and the computer outputs a label about the most dominant object in the image which the image is taken from. In 2012, Alexander Krizhevsky's network showed very promising results for image classification [13]. The architecture is commonly known as AlexNet and it won the most accurate network in the ImageNet contest. In the following year, VGG [14] was introduced. This network had the same structure as the AlexNet with some little changes in the filter size and layer number. VGG was the winner of 2013. In order to compete in ImageNet, the architectures need to classify 1000 objects and so the training needs 1000–1500 images from each class, and ImageNet provides this vast labeled dataset for public use.

In 2014, Google released another architecture GoogleNet that won the best accuracy in ImageNet [15]. The architecture of GoogleNet is different from the previous models. Although it still consists of convolutional layers and fully connected network at the end, GoogleNet adopts inception modules, which are some convolutional layers executing in parallel, and then they connect to a layer's input. In 2015, ResNet [16] was released by Microsoft, and to date, many modified architectures based on the residual blocks of ResNet have been introduced. This module has outputs not only to the immediate higher level convolutions but also to layers higher than the immediate layer.

The filter size and network architecture are predefined. During the training phase, the filter values that are generated randomly are tuned for best performance. Also, the weights of the classifier at the end of the network are based on the training phase. After that, the network is relatively fast. Training is usually based on back propagation algorithm and takes about 100,000 epochs to complete, where each round through all training image sets is defined one epoch.

There are models that are specifically designed for working with CNNs. Caffe model [20] from the University of Berkley is a well-known framework. This model is considered a low-level architecture. The main advantage lies in the fast training and implementation. Meanwhile, the main disadvantage of Caffe model is that it does not have unified and complete documentation, and this may confuse a beginner.

Another widely used framework is TensorFlow from Google [21]. This model works well in a parallel GPU environment and is used as a back engine for other higher-level models such as Keras [22]. A lighter version of this model is introduced for fog level or some powerful edge devices last year. OpenCV 3.3 also has libraries necessary to load and do forward propagation on architectures created by Caffe or Tensorflow. Keras was created to make designing, training, and testing CNNs easier. It has a high-level approach accessible through Python. The model then transforms the code into TensorFlow and does the training or forward propagation. Many of the confusing details of low-level models are not accessible with Keras, but it is convenient to work with. The models in Caffe are in a simple text, and for a big architecture it is hard to handle the code. However, in Keras it is very compressed and also in Python, which is easier to manage. MxNet is another high-level model Python, which is also very good for parallelism.

In smart surveillance system, the camera needs to give the location of the detected object. Thus, image classification might not be helpful, as there might be several people in one given frame and classifying the image as containing humans does not contribute value in this category. In this context, object detection is required, where the detector gives the bounding box around the object of interest with a label of the detection. Single Shot Multi-box Detector (SSD) [17] or Regional CNN (R-CNN) [18, 19], along with other models, are introduced to create a prediction not for the whole image but for neighborhood where the object is located. Training an architecture of this kind needs images from the object that are labeled and also the object is marked within the image. In SSD structure, based on the features that are extracted from the source image, the network makes predictions of the objects that might exist in a certain region.

Although the performance of the neural networks is decent at the edge device and more recent architectures such as GoogleNet have very accurate rates, these models need a huge volume of RAM space, which might not be available in a resource-limited device. For example, when loading the VGG network on the selected edge device (Raspberry Pi), the model gave an error because the available RAM space is low and the program was interrupted. Therefore, more compact architectures to be used at the network edge is expected.

13.3 Object Tracking

Object tracking plays an important role in human behavior analysis in smart surveillance systems. The main purpose of tracking algorithms is to generate the trajectory of the object over time by calculating its position in every frame of the video stream. Compared to object detection that is responsible for isolating a specific region of frame and identifying target object, object tracking is focused on establishing correspondence between the object instances across frames [25]. In object tracking approaches, the detection and tracking can work either separately or jointly to generate trajectory of object. In the first case, object detection algorithm extracts a region of interest (ROI) from every frame. Then the tracking algorithm corresponds object instances across frames by means of marked object regions. In the latter case, object detection and tracking work jointly as one algorithm to compute trajectory of objects by iteratively updating object features obtained from previous frames. Challenges in object tracking are summarized as follows [27]:

- Loss of evidence caused by estimating the 3D realm on a 2D image
- Noise in an image
- Difficult object motion
- Imperfect and entire object occlusions
- Complex objects structures

Those challenges are mainly associated with object feature representation. The following subsections discuss feature representations and classifications of object tracking methods according to selected object features.

13.3.1 Feature Representation

Selecting the accurate feature representation is important to object tracking. Identified objects by means of object detection algorithms are represented as either shape model or appearance model. Whether it is the shape or appearance model, feature selection strictly depends on which characteristics will be used for describing object model. The object can be described using color, edge, and texture.

- **Color**. Each frame of video is an image that is represented by using certain type of color space models ranging from gray scale, RGB, YCbCr, and HSV. In each image, the data are stored as a layered matrix in which value in each cell is brightness of spectral band. For example, colorful images are denoted as a three-layered matrix that consists of red (R), green (G), and blue (B), while gray images decompose color into one channel-gray value. HSV or HLS decompose colors into their hue (H), saturation (S), and value/luminance (V) components.
- **Edge**. Edges are regions in the image with large variation in intensity in opposite directions. Edge-detection algorithms take advantage of variation in intensity to find edge regions, then draw contour of object through connecting edges. The most significant property of edges is that they are less sensitive to illumination changes compare to color features. However, demarcation boundaries between different objects are difficult, especially when multiple objects are overlapped. Edge detection is a fundamental method in image processing, especially in feature detection and feature extraction.
- **Texture**. Texture is a degree of intensity dissimilarity of a surface that enumerates properties such as smoothness and regularity. Image texture usually includes information about the spatial arrangement of color or intensities in an image or selected region, which could be useful features for object detection and tracking. Compared to color space model, texture obeys statistical properties and has similar structures. It requires an analytical processing step to calculate features. Texture analysis approaches are structural approach, statistical approach, and Fourier approach. Like edge features, the texture features are less sensitive to illumination changes than color space in image.

The model of representing object limits the type of features that can be leveraged in tracking algorithms, such as motion and deformation. For example, if an object is represented as a point, then only a translational model can be used; when a geometric shape representation such as an ellipse is used for representing the object, parametric motion models like affine or projective transformations are appropriate [26].

13.3.2 Categories of Object Tracking Technologies

Figure 13.5 shows a taxonomy of tracking approaches. In general, object tracking technologies can be categorized into three groups: point-based tracking, kernel-based tracking, and silhouette-based tracking [27].

The following subsections offer a detailed discussion in those tracking approaches by illustrating the underlying algorithms and analyzing their characteristics.

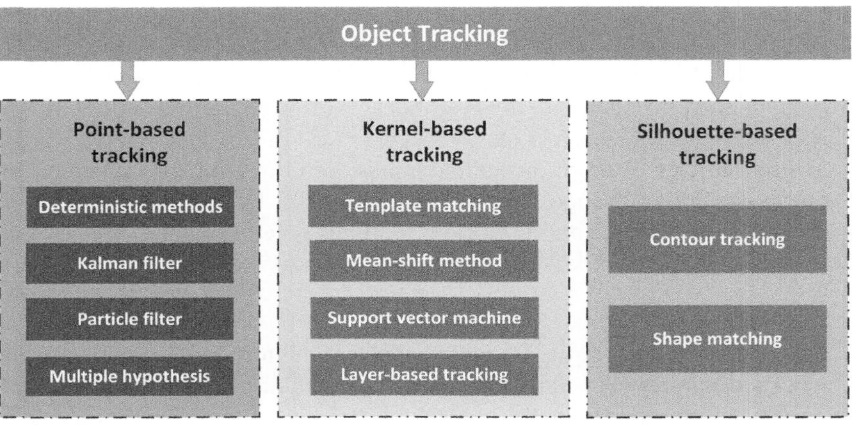

Figure 13.5 Object tracking methods.

13.3.3 Point-Based Tracking

Point-based tracking can be formulated as the correspondence of detected objects represented by points across frames [25]. In general, point-based tracking can be divided into two board categories according to the point correspondence methods: deterministic methods and statistical methods. The deterministic approaches exploit qualitative motion heuristics to solve the correspondence problem, and the statistical methods use probabilistic models to establish correspondence. Several widely applied methods such as Kalman filter, particle filter, and multiple hypotheses belong to the statistical category.

13.3.3.1 Deterministic Methods

Deterministic methods are essentially formulated as the combinatorial optimization problem that attempts to minimize the correspondence cost. The correspondence cost is usually defined by a combination of different motion constraints [27], which are shown in Figure 13.6.

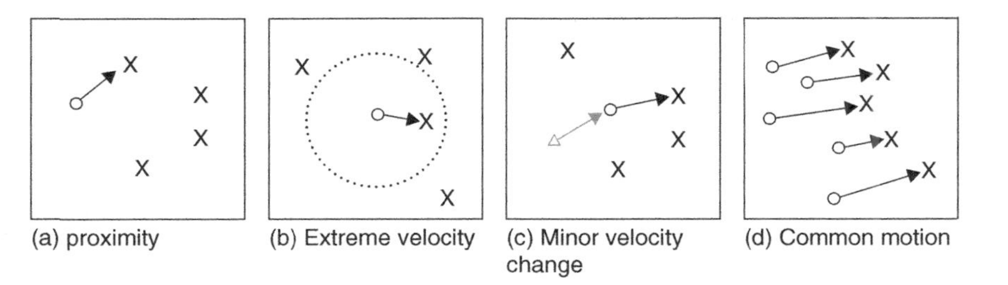

Figure 13.6 Different motion constrains.

- **Proximity** assumes the position of object would not change significantly from previous frame to current frame (Figure 13.6 (a)).
- **Extreme velocity** defines the upper bound of object's position and limits the possible correspondence to the neighborhood around the object in a circular region (Figure 13.6 (b)).
- **Minor velocity** (smooth motion) assumes both direction and speed of the object should not change notably (Figure 13.6 (c)).
- **Common velocity** (smooth motion) assumes objects in a small neighborhood should have similar direction and velocity between frames (Figure 13.6 (d)).

All of the above constraints are not specific to the deterministic methods only. They can also be used in statistical methods for point tracking. Deterministic methods are appropriate in object tracking tasks in which objects are usually very small compared with surrounding context.

13.3.3.2 Kalman Filters

A Kalman filter [28], also known as a linear quadratic estimation (LQE), is based on Optimal Recursive Data Processing Algorithm. Using a series of measurements observed over time, a Kalman filter could produce estimates of unknown variables based on recursive computational means. A Kalman filter is appropriate to estimate the optimal state of a linear system where the state and noise have a Gaussian distribution. Kalman filters work in a two-step process: prediction and correction. The prediction process predicts the new state of the variables given a current set of observations. The correction step gradually updates the predicted values and generates the optimal approximation of the next state [27].

13.3.3.3 Particle Filters

In cases where the object state is not assumed to be a Gaussian distribution, Kalman filter will give poor estimations of state variables due to the limitation of requiring the state variables to be normally distributed (Gaussian). In such situations, particle filters [29] is better to perform state estimation. A particle filter generates all models for one variable before processing the next variable. It calculates the conditional state density at time t to represent the posterior distribution of stochastic process by using a genetic mutation-selection sampling approach with a set of particles. The particle filter is actual a Bayesian sequential importance technique, which recursively approaches the later distribution using a finite set of weighted trials [27]. Contours, color, and texture are all features used in a particle filter algorithm. Like Kalman filters, particle filters also consist of two basic steps: prediction and correction.

13.3.3.4 Multiple Hypothesis Tracking (MHT)

If only two frames are used in motion correspondence processes, there is a limited chance of a correct correspondence. To get better tracking results, correspondence decisions could be performed when several frames have been evaluated. Thus, multiple hypothesis tracking (MHT) algorithms maintain multiple correspondence estimates for each object at each frame. The final object track is the trajectory, including the entire set of correspondences during the time periods of observation. MHT is an iterative algorithm. An iteration starts by feeding a set of current track hypotheses, and each hypothesis is a collection of mutual independent tracks [25]. Through establishing correspondence for each hypothesis based on the distance measurement, a new hypothesis that represents a new set of tracks is generated as the result of prediction process. MHT is good at tracking multiple objects, especially in those scenarios where objects enter and exit the field of view (FOV).

13.3.4 Kernel-Based Tracking

Kernel-based tracking methods compute motion of kernel on each frame to estimate movement of the object. In kernel-based tracking, *kernel* refers to the object representations in the form of a rectangular or ellipsoidal shape and object appearance. Kernel-based tracking algorithms are divided into four categories: template matching, mean-shift method, support vector machine (SVM), and layering-based tracking.

In template-matching tacking, a set of object template O_t is defined in the previous frame, and tracking algorithms use a brute force method to search a region that is most similar to the predefined object template. The position of possible templates in the current frame is produced after the similarity measurement. Since templates are generated by means of image intensity or color features, which is sensitive to illumination changes, template-matching algorithms are preferable to detect small pieces of a reference image. The brute force searching in measuring template similarity leads to a high computation cost. Therefore, the template-matching tracking is not suitable for multiple-object tracking scenarios in a device with limited resources.

Instead of using the brute force method, mean-shift–based algorithm takes advantage of mean-shift clustering [30] technology to detect the region of object that is most similar to a reference model. By comparing the histograms of the object and the window around the hypothesized object location, the mean-shift tracking algorithm attempts to maximize the appearance similarity iteratively. It usually takes five to six iterations until convergence is achieved; thus, mean-shift tracking requires less computational cost than template-matching tacking does. However, mean-shift tracking assumes that a portion of the object is inside the circular region in initial state. Physical initialization is necessary during initialization of the tracking task. Additionally, mean-shift algorithm is only capable of tracking one single object.

Avidan first integrated the SVM classifier into an optic-flow-based tracker [31]. Given a set of positive and negative training samples, SVM is preferable to handle binary classification problems through finding the best separating hyperplanes between two classes. In SVM-based tracking, tracked objects are labeled as positive, while untracked objects are defined as negative. The tracker could use trained SVM classifiers to estimate the position of the object by maximizing the SVM classification score over an image's region. SVM-based tracking can handle partial occlusion of the tracking object. However, it needs training process to prepare the SVM classifier before performing the tracking task.

In layering-based tracking, each frame is separated into three layers: namely, shape representation (ellipse), motion (such as translation and rotation), and layer appearance (based on intensity) [32]. In layering-based tracking, at first, layering is achieved by compensating the background motion, then object position is estimated by calculating a pixel's probability based on the object's foregoing motion and shape features. Layering-based method is appropriate in scenarios where multiple objects are tracked or fully occlusion of objects happens.

13.3.5 Silhouette-Based Tracking

For objects with complex shapes, which are difficult to be well described by simply using geometric features, for example, hands, head, and shoulders, silhouette-based algorithm provides a better solution. According to the object model, silhouette-based tracking approaches are categorized as either contour tracking or shape matching. In contour tracking, initial contour evolves to its new position in the current frame to keep track with the object. In contrast, shape matching only searches object in one frame from time to time by using density functions, silhouette boundary and object edges [32].

The above discussion provides a comprehensive summary of research in object tracking algorithms. In the next section, a detail illustration of the Kernelized Correlation Filter (KCF) tracking method is presented, which achieves good performance in terms of resource consumption.

13.3.6 Kernelized Correlation Filters (KCF)

Tracking-learning-detection (TLD) framework is widely used in modern object tracking arts of field [33]. Boosting [34] and multiple instance learning (MIL) [35] demonstrate capability in online training that makes the classifier adaptive while tracking the object. However, updating process consumes a lot of resources. The lower resource consumption with high tracking success rate makes Kernelized Correlation Filter (KCF) a preferable online tracking method in delay sensitive surveillance system.

KCF is initially inspired by successful applications of the correlation filter in tracking [36]. Compared with other complicated approaches, correlation filters have been proved to be competitive in environments with tight constraints on computational power. Object detection using KCF could be defined as a deterministic problem based on Kernel Ridge Regression [37]. The KCF algorithm is essentially a kernelized version of linear correlation filter. Exploiting powerful kernel trick allows transferring unstructured linear correlation filter to linear space, so that KCF has the same computational complexity as linear correlation filter when handles nonlinear regression problem with multiple channel features.

To determine object position in the current frame, template matching is first performed by computing a correlation with a special filter h, and subsequently searches the maximum value on the obtained correlated image c [38]:

$$(x,y)^* = \underset{(x,y)\in c}{argmax}(c), \text{ where } \; c = s \circ h \tag{13.1}$$

c: Correlated image
s: Image region for searching
h: Filter generated from the object template
°: Operator to calculate two-dimensional correlation
$(x,y)^*$: The target object position corresponding to the maximum of correlated image c

Equation (13.1) assumes that the tracking area f and the filter h have the same dimension. The correlated filter h is calculated by the Ridge regression to minimize the squired error over a template t. It is:

$$\underset{h}{min} \sum_{i}^{c} (\|f(x_i) - g\|^2 + \lambda \|h_i\|^2) \tag{13.2}$$

λ: regularization parameter, as in the SVM
$f(x_i)=t_i \circ h_i$: Correlation function between template and filter images
c: Channels of the two-dimensional images
g: Two-dimensional Gaussian distribution function, $g(u,v) = \exp[-(u^2 + v^2)/2\sigma^2]$

The purpose of the optimization problem defined in Eq. (13.2) is to find a function h that correlates with object template t to output the minimum difference from Gaussian distribution function g. It is straightforward to work in the frequency domain, where Equation (13.2) could also be directly transformed into Fourier expression:

$$H^* = \frac{G \odot T^*}{T \odot T^* + \lambda} \tag{13.3}$$

X^*: Complex-conjugation operation of X
$\odot$: Element-wise product operator
H: Filter in Fourier domain
T: Object template in Fourier domain
G: Gaussian function in Fourier domain

Given the filter H, and the search region F in the frequency domain, combining Equations (13.1) and (13.3), correlation image C can be calculated in Fourier domain as:

$$C = F \odot H^* = \frac{F \odot G \odot T^*}{T \odot T^* + \lambda} \tag{13.4}$$

Finally, given Equations (13.1) and (13.4), the object tracking algorithm is:

$$(x, y)^* = \underset{(x,y)\in \mathcal{F}^{-1}(C)}{argmax} \left(\mathcal{F}^{-1}(C)\right) \tag{13.5}$$

where $\mathcal{F}^{-1}()$ denotes the inverse DFT operation.

In KCF tracking algorithm, in order to increase the object tracking area, the template t is selected as a region with a size larger than the object size. For best results of the KCF tracking method, the template size is suggested to select as 2.5 times larger than the object size [36]. KCF takes advantage of the HOG feature to track objects given the assumption that objects have similar contour even though they have different appearance. Figure 13.7 shows the KCF object tracking process.

Detailed illustrations of the feature extraction and object tracking steps are listed as follows:

- **Gradient computing**. Normalizing the color and gamma values is the first step to calculating the feature detector, which is followed by the calculation of the magnitude and orientation of the gradient.
- **Weighted vote in orientation cell**. The image is divided based on a sliding detection window and the cell histograms are created. Each pixel within the

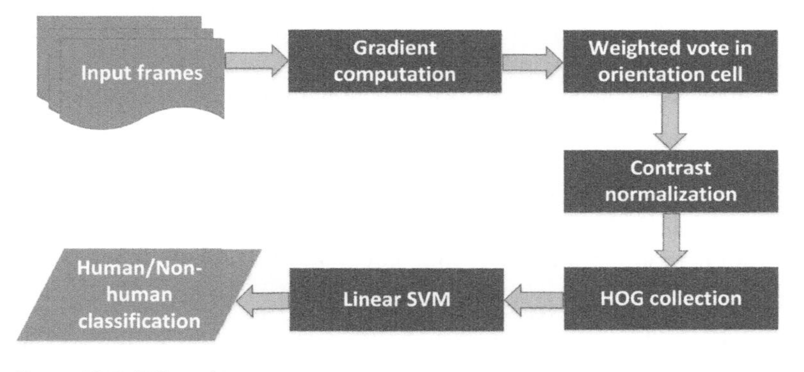

Figure 13.7 KCF tracking process.

cell is associated with a weighted vote for an orientation-based histogram channel according to the values calculated in the previous step of gradient computation.

- **Contrast normalization**. Considering the effect caused by illumination and contrast change, gradient strengths are locally normalized by grouping the overlapping cells together into larger, spatially connected blocks.
- **HOG collection**. In this step, the concatenated vector of the components of the normalized cell histograms from all block regions is calculated to create a HOG descriptor.
- **KCF tracker**. HOG descriptor that contains extracted HOG feature vectors is fed to a KCF tracker for producing hypothesis of target position.

13.4 Lightweight Human Detection

Due to the constraints on resources, lightweight algorithms are required for the edge devices. Normally there are trade-offs to be considered carefully when designing a light version of an existing algorithm, which often means sacrificing the accuracy or speed. There are two important components to building a good object detector: the feature extractor and the classifier. Considering the algorithms discussed here, classifiers are the most resource-consuming part of the algorithm, but there are not many changes that can be made to the architecture of these SVMs and FCNNs. Meanwhile, feature extraction algorithms have a lot of room for improvement. Especially in the system that is focused on human objects detection and extracting them from the surrounding environment. Hence, the extracted features are applied as the given input data to make the distinction more noticeable to the classifier.

Haar cascade algorithm is a fast algorithm and it does not map the pixel values to another space for feature extraction. In addition, this algorithm only uses simple mathematical functions such as dot product that are very fast. Therefore, this algorithm is suitable for mobile and edge devices. But its accuracy is not satisfactory.

The HOG algorithm may follow the same principle. However, the video frame can be resized before passing to algorithm. Also, there are many parameters in the algorithm that can be tuned to improve the accuracy. For example, the window that takes pixels for histograms creation can be bigger, which makes the algorithm faster but there is a bigger chance that a pedestrian is ignored. Smaller window sizes make the algorithm run very slowly but most of pedestrians are detected and the chance of having multi-bounding box for a human object increases. A set of variables for one camera positioning that can be accurate may not perform best for another video; this creates a need to fine tune the algorithm for every camera.

CNN is the focus of attention for simplifications that fit the algorithm on smaller devices. Several architectures are introduced that enable the creation of a more condenses CNN. Some of them are mathematically proven to be able to reduce computational burden. One of these architectures is the Fire module used in SqueezeNet [22]. This architecture has 50 times fewer parameters and has the same performance accuracy as AlexNet. The Fire module has two sets of filters. The first is a convolutional layer with a 1×1 convolution filter. Although it might sound like there is nothing happening in a 1×1 filter, it is reminded that number of channels can be changed. This layer is referred as squeeze layer. Another one is a convolutional layer by the name of expand that consists of a 1×1 set and a 3×3 set of convolutional filters.

MobileNet, which is introduced by Google in 2017 [23], achieves a very good performance and also has less computational burden through a separable depthwise convolutional layer [24]. Where each conventional convolutional layer is split into two parts. A conventional convolution takes F with size $D_f \times D_f \times M$ from the input and with filter K of size$D_k \times D_k \times M \times N$ maps it to G as an output of size$D_g \times D_g \times N$.

$$CB = D_k \times D_k \times M \times N \times D_f \times D_f \tag{13.6}$$

The computing complexity of this operation is as Eq. (13.6), and it can be taken into two parts. The first is a depthwise convolutional layer with size $D_k \times D_k \times 1 \times M$ that can create a $\hat{G}$ with size $D_g \times D_g \times M$. Then, a set of N pointwise convolutional filters with size of $1 \times 1 \times M$ will create the same G as before. This time the computational complexity becomes as Eq. (13.7).

$$CB = D_k \times D_k \times M \times D_f \times D_f + N \times M \times D_f \times D_f \tag{13.7}$$

This shows a reduction of computational burden by a factor of $\frac{1}{N} + \frac{1}{{D_k}^2}$ as shown in Eq. (13.8).

$$\frac{D_k \times D_k \times M \times D_f \times D_f + M \times N \times D_f \times D_f}{D_k \times D_k M \times N \times D_f \times D_f} \tag{13.8}$$

Figure 13.8 compares the network with a separable depthwise convolutional layers and the conventional layers, where the depthwise and pointwise steps are together in a single filter shape. The left-side is the separable structure and after each depthwise or pointwise convolution, a batch normalization (to normalize the data because one-time normalization in deep learning is not sufficient) and a ReLU layer are placed.

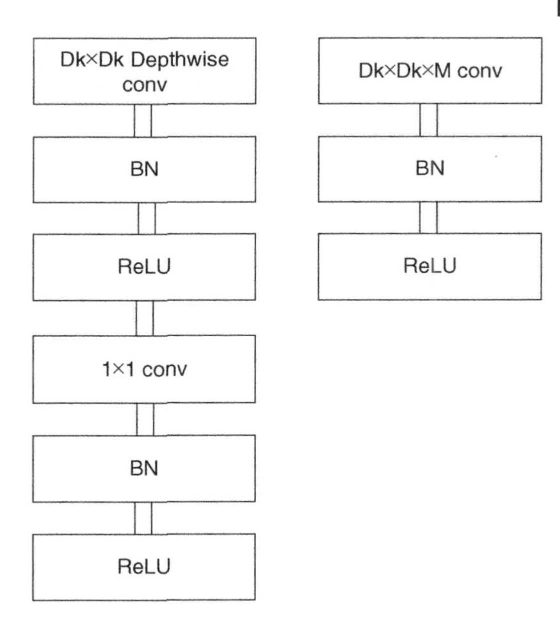

Figure 13.8 Convolutional filter vs. separable depthwise convolutional filter.

13.5 Case Study

The case study in this section provides more information about the algorithms discussed in this chapter. Implemented on physical edge devices, these algorithms are applied to process sample surveillance video streams.

The selected edge computing device is a Raspberry PI 3 Model B with a 1.2GHz 64-bit quad-core ARMv8 CPU and 1GB LPDDR2-900 SDRAM. The operating system is Raspbian based on the Linux kernel. The fog computing layer functions are implemented on a laptop with a 2.3 GHz Intel Core i7, the RAM memory is 16 GB, and the operating system is Ubuntu 16.04. The software applied for human objects detection and tracking is implemented using C++ and Python programming languages and OpenCV library (version3.3.0) [39].

13.5.1 Human Object Detection

Haar cascades algorithms are very powerful for recognizing individual objects in training data set but not very good with changes. If the positioning or angle of human object does not match the training samples, the algorithm often fails to recognize it. In real-world surveillance systems, there is no guarantee to always capture pedestrians from the same angles. Figure 13.9 shows a sample video and false positive detections the algorithm generates. In average 26.3% of detections are false in this sample surveillance video, this number may be different in different videos and initial variables. In terms of speed, the performance of this

Figure 13.9 Results of Haar cascaded human detection.

algorithm is very fast at the edge with around 1.82 frames per second (FPS). Considering the velocity of pedestrians, it is sufficient to sample twice per second. From the perspective of resource utility, in average it uses 76.9% of CPU and 111.6 MB of RAM.

In contrast, HOG+SVM algorithm on average uses 93% of the CPU, which is expensive because resources are needed by other operations and functions to achieve the goal of a smart surveillance system and 139 MB of RAM. Note that the HOG+SVM algorithm is very slow, 0.304 FPS is reached. Figure 13.10 shows different instances of the sample video in which the bounding boxes are generated by this algorithm. Some of the boxes are not exactly fitted around the human object. For example, in the bottom left screenshot parts of the vehicle is also in the bounding box, which will bring negative impact to the performance of tracking algorithms.

Based on the approach explained earlier, a lightweight version of CNNs are created. A typical CNN recognizes up to 1000 different classes of objects and the network is too big to be fit in edge devices. Even using MobileNet or SqueezeNet or other examples of such CNNs can take up to 500 megabytes of RAM. For object detection VOC07 is frequently used and it has 21 classes.

Figure 13.10 Performance of HOG+SVM algorithm.

However, the major object of smart surveillance systems is to detect human objects, which means there is only one class such that it is not necessary to keep many filters in each layer of the network. Based on this observation, a lightweight CNN network was trained to have four times fewer parameters in each convolutional layer than the MobileNet does. Figure 13.11 shows the results on a Raspberry PI 3 model B. The lightweight CNN takes less than 170 MB of RAM and is relatively accurate and can detect human objects in different angles.

13.5.2 Object Tracking

To test feasibility of tracking objects by processing video streams on edge computing devices, a concept-proof prototype of the system was built using the KCF-based object tracking algorithm. Here, the performance of the algorithm is presented on object tracking and multi-tracker lifetime handle such as phase in & out of frame, retracking after tracked object lost, etc.

13.5.2.1 Multi-Object Tracking

Figure 13.12 shows an example of the multi-object tracking results. Multi-tracker object queue is designed to manage tracker lifetime. After

Figure 13.11 Example of light version CNN for human object detection.

(a) Pedestrians (b) Vehicles

Figure 13.12 An example of multi-object tracking.

the object detection processing finished, all detected objects are fed to the tracker filter, which compares detected target region and the multitracker object queue to rule out the duplicated trackers. Only those newly detected objects are initialized as KCF trackers and appended to the multitracker object queue. During execution time, each tracker runs KCF tracking algorithm independently on target region through processing the video stream frame by frame until the object phases out or it loses the object in the scenario.

(a) Enter frame

(b) Exit frame

Figure 13.13 An example of object tracker phase in and out.

13.5.2.2 Object Tracking Phase In and Out

The boundary region is defined to address scenarios when the moving objects enter or exit the current view of frame. In object tracker phase in cases, as objects entered the boundary region, they are detected as new tracking targets with active status and appended to the multi-tracker queue. In object tracker phase out scenarios, those tracked objects that are moving out of the boundary region will be deleted and the corresponding trackers transfer to inactive status. After a frame is processed, those inactive trackers will be removed from the multi-tracker object queue such that the computing resources are relieved for future tasks. The movement history is exported to tracking history log for further analysis. Figure 13.13 presents an example of the object tracker phase in & out results.

13.5.2.3 Tracking Object Lost

Because of the occlusion resulted from variance of color appearance and illumination conditions between the background environment and the tracked objects, the tracker may fail to track the target objects. It is necessary to handle such scenarios. Trackers that lose tracking objects could be cleared from multi-tracker queue and the lost objects can be re-detected and re-tracked as new objects of interest. Figure 13.14 shows a scenario that the tracker loses the object (the car in on left, marked as object #3) when it moved across the shadow of trees. In subsequent frames, the detection algorithm identified this car as a new object and assigned it to a new active tracker to retract object (marked as object #8).

Above experimental results demonstrate that KCF method based on the HOG feature has a high reliability in object tracking. However, color appearance and illumination have significant influence on tracking accuracy. As the example illustrated in Figure 13.14, if background environment and the tracked objects have similar color appearance and illumination, only using HOG features is not sufficient to estimate region of interest so that tracker

(a) Lost object

(b) Re-tracking

Figure 13.14 An example of re-tracking after target lost.

may fail to keep tracking target object. Therefore, a more efficient and precise approach is still in need to re-track objects by establishing connection between tracker and object when occlusion happens.

13.6 Future Research Directions

There are some open challenges yet to be addressed to make the object detection and tracking a practical implementation at the edge. One of the most critical issues is the intelligent but lightweight decision making algorithms. The ideal model should be general to cover common incidents that may happen for pedestrians. Unlike classifiers, decision-making algorithms or prediction models do not need to be very accurate and in each case the algorithm can be fine-tuned. Designing a general machine-learning algorithm to actively detect any unpredicted occurrence is a challenge. However, there is a general rule. In order to predict correctly or detect accurately, it is helpful to look into the historical data. Checking several frames before the current one may give more information to make a decision. There are algorithms such as long short-term memory (LSTM) or hidden Markov model (HMM) that are designed to maintain a memory and keep the information from the previous steps. A deep investigation of these algorithms is highly expected in the future.

We made an effort to atomize surveillance environment and minimize the delay using edge-level devices with constraints and issues that need to be considered. There are still a number of questions to be answered. The first is an open question about how to achieve better performance but use less RAM and less computing power. This question exists in every corner of engineering, but it shows itself in newly developing areas such as fog computing systems. CNN architectures have been extensively investigated in recent years, which are very small in size, but keep performance accuracy. Another important question to be answered is the connection and networking side of fog systems and edge devices. New protocols are expected to be introduced for this purpose and more research is to be conducted as the field matures.

This chapter focused on the functional development side. However, a surveillance system needs a robust security measure. Due to the lack of a sophisticated operating system and limited energy source, it is more challenging for small fog/edge devices to protect themselves. Blockchain appears to be promising to make a network of small sensors and fog systems protected, but more research work is required.

13.7 Conclusions

This chapter provides an overview on a critical issue in modern surveillance system: online human object detection and tracking at the network edge, which provides benefits such as real-time tracking and video marking as important so that less footage need saving. After an introduction to several popular algorithms, including neural networks, a thorough discussion on their pros and cons is presented. Based on the insights, a lightweight CNN is introduced and a comparison experimental study is conducted by implementing these algorithms on a selected edge device and applying them on a real-world sample surveillance video stream. There are well-designed detectors and trackers that can be fit into the edge environment with some fine tuning corresponding to the requirements of the given tasks, such as with the lightweight CNN introduced in this chapter.

Moreover, several tracking algorithms were reviewed and discussed, along with their performance and accuracy in tracking object of interest and frame per second achieved in edge device of choice.

References

1 N. Jenkins. North American security camera installed base to reach 62 million in 2016, https://technology.ihs.com/583114/north-american-security-camera-installed-base-to-reach-62-million-in-2016, 2016.
2 Cisco Inc. Cisco visual networking index: Forecast and methodology, 20162021 White Paper. https://www.cisco.com/c/en/us/solutions/collateral/

service-provider/visual-networking-index-vni/mobile-white-paper-c11-520862.html, 2017.

3 L.M. Vaquero, L. Rodero-Merino, J. Caceres, and M. Lindner. A break in the clouds: towards a cloud definition. *SIGCOMM Computer Communications Review,* 39(1): 50–55, 2008.

4 Y. Pang, Y. Yuan, X. Li, and J. Pan. Efficient hog human detection. *Signal Processing,* 91(4): 773–781, April 2011.

5 O. Mendoza-Schrock, J. Patrick, and E. Blasch. Video image registration evaluation for a layered sensing environment. *Aerospace & Electronics Conference (NAECON), Proceedings of the IEEE 2009 National*, Dayton, USA, July 21–23, 2009.

6 S. Y. Nikouei, R. Xu, D. Nagothu, Y. Chen, A. Aved, E. Blasch, "Real-time index authentication for event-oriented surveillance video query using blockchain", arXiv preprint arXiv:1807.06179.

7 N. Dalal and B. Triggs. Histograms of oriented gradients for human detection. *IEEE Conference on Computer Vision and Pattern Recognition,* San Diego, USA, June 20–25, 2005.

8 P. Viola and M. Jones. Robust real-time face detection. *International Journal of Computer Vision,* 57(2): 137–154, May 2004.

9 P. Viola and M. Jones. Rapid object detection using a boosted cascade of simple features. *Proceedings of the 2001 IEEE Computer Society Conference on Computer Vision and Pattern Recognition,* Kauai, USA, December 8–14, 2001.

10 J. Guo, J. Cheng, J. Pang, Y. Gua. Real-time hand detection based on multi-stage HOG-SVM classifier. *IEEE International Conference on Image Processing*, Melbourne, Australia, September 15–18, 2013.

11 H. Bristow and S. Lucey. Why do linear SVMs trained on HOG features perform so well? *arXiv:1406.2419,* June 2014.

12 N. Cristianini and J. Shawe-Taylor. *An Introduction to Support Vector Machines and other kernel-based learning methods.* Cambridge University Press, UK, 2000.

13 A. Krizhevsky, I. Sutskever, and G.E. Hinton. ImageNet Classification with Deep Convolutional Neural Networks. *Advances in Neural Information Processing Systems,* pp. 1072–1105, 2012.

14 K. Simonyan and A. Zisserman. Very deep convolutional networks for large-scale image recognition. *arXiv:1409.1556,* April 2015.

15 C. Szegedy, W. Liu, Y. Jia, P. Sermanet, S. Reed, D. Anguelov, D. Erhan, V. Vanhoucke, A. Rabinovich. Going deeper with convolutions. *IEEE Conference on Computer Vision and Pattern Recognition*, Boston, USA, June 07–12, 2015.

16 K. He, X. Zhang, S. Ren, and J. Sun. Deep residual learning for image recognition. *IEEE Conference on Computer Vision and Pattern Recognition.* Seattle, USA, *June* 27–30, 2016.

17 G. Cao, X. Xie, W. Yang, Q. Liao, G. Shi, J. Wu. Feature-Fused SSD: Fast Detection for Small Objects. *arXiv:1709.05054*, October 2017.

18 R. Girshick. Fast R-CNN. *arXiv preprint arXiv:1504.08083*, 2015.

19 S. Ren, K. He, R. Girshick, and J. Sun. Faster R-CNN: Towards Real-Time Object Detection with Region Proposal Networks. *Advances in Neural Information Processing Systems*, 91–99, 2015.

20 Y. Jia, E. Shelhamer, J. Donahue, S. Karayev, J. Long, R. Girshick, S. Guadarrama, and T. Darrell. Caffe: Convolutional architecture for fast feature embedding, *In Proceedings of the 22nd ACM international conference on Multimedia*, Orlando, USA, November 3–7, 2014.

21 M. Abadi, A. Agarwal, P. Barham, E. Brevdo, Z. Chen, C. Citro, G. S. Corrado, A. Davis, J. Dean, M. Devin, S. Ghemawat, I. Goodfellow, A. Harp, G. Irving, M. Isard, R. Jozefowicz, Y. Jia, L. Kaiser, M. Kudlur, J. Levenberg, D. Mané, M. Schuster, R. Monga, S. Moore, D. Murray, C. Olah, J. Shlens, B. Steiner, I. Sutskever, K. Talwar, P. Tucker, V. Vanhoucke, V. Vasudevan, F. Viégas, O. Vinyals, P. Warden, M. Wattenberg, M. Wicke, Y. Yu, X. Zheng. TensorFlow: Large-scale machine learning on heterogeneous systems. *arXiv preprint arXiv:1603.04467,* March 2016.

22 F. N. Iandola, S. Han, M. W. Moskewicz, et al. SqueezeNet: AlexNet-level accuracy with 50x fewer parameters and <0.5MB model size, *arXiv:1602.07360,* November 2016.

23 A. G. Howard, M. Zhu, B. Chen, K. Ashraf, W. J. Dally, and K. Keutzer. MobileNets: Efficient Convolutional Neural Networks for Mobile Vision Applications. *arXiv:1704.04861*, April 2017.

24 L. Sifre. Rigid-motion scattering for image classification, *Diss. PhD thesis*, 2014.

25 A. Yilmaz, O. Javed, and M. Shah. Object tracking: A survey. *ACM Computing Surveys,* 38(4): 13, December 2006.

26 M. Isard and Maccormick. Bramble: A bayesian multiple-blob tracker. *IEEE International Conference on Computer Vision*, Vancouver, Canada, July 7–14, 2001.

27 S. Y. Nikouei, Y. Chen, T. R. Faughnan, "Smart Surveillance as an Edge Service for Real-Time Human Detection and Tracking", ACM/IEEE Symposium on Edge Computing, 2018.

28 R. E. Kalman. A new approach to linear filtering and prediction problems. *Journal of Basic Engineering,* 82(1): 35–45, 1960.

29 P. Del Moral. Nonlinear Filtering: Interacting Particle Solution. *Markov Processes and Related Fields,* 2(4): 555–581, 1996.

30 D. Comaniciu, P. Meer. Mean shift: A robust approach toward feature space analysis, *IEEE Transactions on Pattern Analysis and Machine Intelligence,* 24(5): 603–619, May 2002.

31 S. Avidan. Support vector tracking. *IEEE Transactions on Pattern Analysis and Machine Intelligence,* 26(8): 1064–1072, August 2004.

32 V Tsakanikas and T. Dagiuklas. Video surveillance systems-current status and future trends. *Computers & Electrical Engineering*, November 2017.

33 Z. Kalal, K. Mikolajczyk, and J. Matas. Tracking-learning-detection. *IEEE Transactions on Pattern Analysis and Machine Intelligence,* 34(7): 1409–1422, July 2012.

34 H. Grabner, M. Grabner, and H. Bischof. Real-time tracking via on-line boosting. *BMVC,* 1(5): 6, 2006.

35 B. Babenko, M.-H. Yang, and S. Belongie. Visual tracking with online multiple instance learning. *IEEE Conference on Computer Vision and Pattern Recognition,* Miami, USA, June 20–25, 2009.

36 J. F. Henriques, R. Caseiro, P. Martins, and J. Batista. High-speed tracking with kernelized correlation filters. *IEEE Transactions on Pattern Analysis and Machine Intelligence,* 37(3): 583–596, August 2014.

37 R. Rifkin, G. Yeo, and T. Poggio. Regularized least-squares classification. *Science Series Sub Series III Computer and Systems Sciences,* 190: 131–154, 2003.

38 A. Varfolomieiev and O. Lysenko. Modification of the KCF tracking method for implementation on embedded hardware platforms. *IEEE International Conference on Radio Electronics & Info Communications (UkrMiCo),* Kiev, Ukraine, September 11–16, 2016.

39 opencv.org, http://www.opencv.org/releases.html, 2017.

14

Fog Computing Model for Evolving Smart Transportation Applications

M. Muzakkir Hussain, Mohammad Saad Alam, and M.M. Sufyan Beg

14.1 Introduction

Due to the increased number of connected things in smart and industrial applications – more specifically, intelligent transportation systems (ITS), the growing volume and velocity of Internet of Things (IoT) data exchange – there is a great urgency for rigorous communication resources to address the bottlenecks in terms of data processing, data latency, and traffic overhead [1]. Fog computing emerges as an substitute for traditional cloud computing to support geographically distributed, latency sensitive, and QoS-aware IoT applications while reducing the burden of data centers in traditional cloud computing [2]. In particular, fog computing due to its peculiarities (e.g., low latency, location awareness, and capacity of processing large number of nodes with wireless access) to support heterogeneity and real-time applications is a potentially attractive solution to the delay and resource-constrained large-scale industrial applications [3].

However, with the benefits of fog computing, the research challenges arise while realizing fog computing for such applications [4]. For instance, how should we handle different protocols and data format from highly dissimilar data sources in fog layer? How do we determine which data should be processed in cloud or be processed in fog layer (task association, resource allocation/provisioning, VM migration) [5]? How can real-time responses and simultaneous data collection be achieved from large heterogeneous sources in industrial applications? This chapter makes a rigorous assessment toward the viability of fog computing approaches on emerging smart transportation architectures [6]. As a proof of concept, we perform a case study on the fog computing requirements of intelligent traffic light management (ITLM) system; see how the previous questions, and others, can be addressed [7]. Orchestrating such applications can simplify maintenance and enhance data security and system reliability [8]. For efficient management of those activities

Fog and Edge Computing: Principles and Paradigms, First Edition.
Edited by Rajkumar Buyya and Satish Narayana Srirama.

in ITS domain, we define a distributed fog orchestration framework that defines the dynamic, policy-based life-cycle management of fog services. Finally, the chapter concludes with an overview of the core issues, challenges, and future research directions in fog-enabled orchestration for IoT services in smart transportation domain.

The chapter is organized as follows. Section 14.2 introduces the needs and prospects of adopting data-drive transportation architectures and the landscape of smart applications supported over adoption of such data-driven mobility models. It discusses which computer requirements can be best fulfilled through cloud computing and which require fog rollout. Section 14.3 identifies the fog computing requirements of ITS such as mission-critical architectures. It assesses the state of cloud platforms to store and compute support for such applications and discusses the proper mix of both computational models to best meet the mission-critical computing needs of smart transportation applications. Section 14.4 presents a fog computing framework customized to support latency sensitive ITS applications. Its four advantages are captured in the acronym CEAL, for cognition, efficiency, agility, and latency. The fog orchestrating requirements in ITS domain are substantiated in Section 14.5 through an intelligent traffic lights management (ITLM) system case study. In Section 14.6 the key big data issues, challenges, and future research opportunities are outlined, while developing a viable fog orchestrator for smart transportation applications.

14.2 Data-Driven Intelligent Transportation Systems

Due to rigorous research and development in state-of-the-art information and communication technologies (ICT) and upsurge in human population, intelligent transportation systems (ITS) have become an integral part of contemporary human life [9]. The ITS architecture comprises a set of advanced applications aimed at applying ICT amenities to provide QoS and QoE guaranteed service for traffic management and transport [10, 11]. Figure 14.1 depicts the fundamental components of a typical ITS architecture [12]. The dependence on transportation systems is indispensable, as is clear from the fact that nearly 40% of the global population devotes at least one hour commuting on road every day [13, 14]. In fact, the competitiveness of a nation, its economic forte, and its productivity rely heavily on how robustly its transportation infrastructures are installed [15]. However, the current landscape of vehicle penetration into transportation architectures comes with numerous opportunities and challenges [16]. It may be in the form of traffic congestion, parking issues, carbon footprints, or accidents, for example [17]. Efficient transportation protocols and policies need to be employed to confront such issues. Thanks to odd/even policies adopted by China in the Beijing Olympics

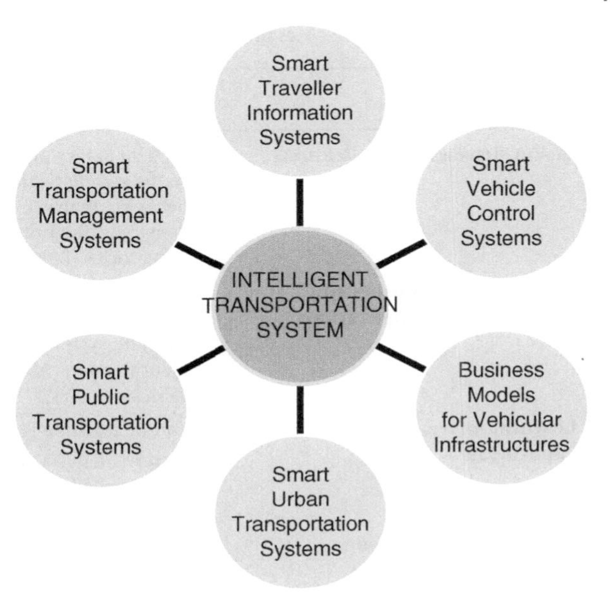

Figure 14.1 Key components of a data-driven ITS [12].

2008 [18] and the same by the Delhi government in 2016, one of the notable attempts to alleviate fleet congestion and air pollution in cities [19].

But such an approach works well only for specific events and time frames, not scalable to nationwide and every-time transportation services. Augmenting with additional infrastructures such as new road construction and road widening might have significant effect but will be trapped in cost- and space-related silos. The optimal strategy is to efficiently utilize the available transportation resources through data-driven analytics of ITS data streams. The data generated from IoT-aided transportation telematics such as cameras, inductive-loop detectors, global positioning system (GPS)-based receivers, and microwave detectors can be collected and analyzed to unlock latent knowledge, ultimately used for intelligent decision making [20].

Table 14.1 highlights the key categories of applications supported by ITS in the realm of IoT [21]. Many efforts, such as developing vehicular networking and traffic communications protocols and standards, have been being devoted by ITS utilities in order to find reliable and ubiquitous transportation solutions in contemporary smart cities [22]. For instance, the US Federal Communications Commission (FCC) has allocated 75 MHz of spectrum in the 5.850 GHz to 5.925 GHz band for the exclusive use of dedicated short-range communications (DSRC) [23]. In addition, some approved amendments have been dedicated to ITS technology such as Wireless Access in Vehicular Environments (WAVE IEEE 802.11p) and Worldwide Interoperability for Microwave Access (WiMAX IEEE 802.16) [11]. The difference between conventional technology-driven ITS and a data-driven ITS is that conventional

Table 14.1 Application use cases for data-driven intelligent transportation applications.

Applications	Usage
Vision-driven ITS applications	Vehicle detection [27], Pedestrian detection [28] traffic sign detection, lane tracking, traffic behavior analysis, vehicle density, pedestrian density estimation, construction of vehicle trajectories [28], statistical traffic data analysis
Multi-source- (sensors and IoT) driven ITS applications	Vision-driven automatic incident detection (AID) [29], DGPS [30], Cooperative collision warning system (CCWS) [31], automatic vehicle identification (AVI) [32]. Unmanned aerial vehicles (UAVs).
Learning-driven ITS application	Online learning [16], trajectory/motion pattern analysis, data fusion, rule extraction, ADP-based learning control, reinforcement learning (RL), ITS-oriented learning.
Datasets for perceived visualization	Line charts; bidirectional bar charts; rose diagrams; data images.

ITS mainly depends on historical and human experiences and places less emphasis on the utilization of real-time ITS data or information [13]. Thanks to modern ICT facilities, currently the data can not only be processed into useful information but can also be employed to generate new functions and services in varying range of ITS domains [24].

Since the major percentage of IoT endpoints in a typical ITS are primitive, i.e. the deployment of required compute and storage resources is not guaranteed every time and everywhere; an external agent should undertake the computation and analytics tasks. The storage and processing loads in an IoT-aware transportation framework will be swarmed up from billions of static as well as mobile sensor nodes spanning over a vast geographical domain [25]. An ideal ITS infrastructure is driven by mission-critical service constraints viz. low-latency, real-time decision making, strict response times, and analytical consistencies [12].

In fact, an IoT-aware ITS ecosystem is constrained by stringent service requirements such as low-power communication backbone, optimal energy trading, proper renewable penetration, and other power monitoring utilities [16]. Such heterogeneity in the data architecture of an ITS envisaged the use of advanced store and compute platforms for overcoming various technical challenges at different levels of computation and processing. Rather than relying on the master–slave computation model as in legacy systems, the current notion is to get switched to data center level analytics operating under the client–server paradigm [7].

The objective of reaching a consensus on where to install the compute and storage resources continues to be an open question for academia, industries, R&Ds, and legislative bodies. The cloud computing had emerged to be promising technology to support ITS because of its ability to provide convenient and on-demand, anytime, anywhere network access to its shared computing resources, provisioned and released with minimal management effort or service provider interaction [21]. The cloud service also frees the IoT devices from battery-draining processing tasks by availing unlimited pay-per use resources through virtualization [26]. However, the varying modalities of services facilitated by cloud computing paradigms perish to meet the mission critical requirements of a data-driven ITS. The existing cloud computing paradigm ceases to welcome its proponents because of its adequacies in building common and multipurpose platform that can provide feasible solutions to the stringent requirements of ITS in IoT space. In the next section we analyze the computing needs of mission-critical smart transportation applications and assess the states of generic cloud models. Correspondingly, we also highlight how a paradigm shift from generic cloud-based centralized computation into geo-distributed fog computing model can turn to be a near ideal solution to carry out the mission-critical smart transportation applications.

14.3 Mission-Critical Computing Requirements of Smart Transportation Applications

Consider a typical traffic lighting use case where smart traffic lights will be able to adapt themselves to the real-time traffic circumstances within a particular region. In this case, the reaction time for one or several smart traffic lights is too short that it is virtually impossible to traffic all the application execution to a distant cloud. Therefore, such traffic lights should be programmed in a way that they autonomously cooperate with each other and with all the locally available computing resources such as roadside units (RSU) to coordinate their operations. Other such examples may be vehicular search applications [9], vehicular crowd sourcing [21], smart parking, etc. [33, 34]. From such examples, it is perceived that there is a need for computing frameworks that will provide ubiquitous and real-time analytics services for varying transportation domains. Some key data collection, processing, and disseminating requirements of smart transportation infrastructures are highlighted in this section.

14.3.1 Modularity

The contemporary intelligent transportation network is a large and complex system, as it involves heterogeneous IoT and non-IoT devices with numerous data types demanding a wide set of processing algorithms. Thus, the software platform supporting the ITS applications should have characteristic modularity and flexibility support. The applications must be incrementally deployed in a way that the system should be self-evolving and fault tolerant i.e. partial failures do not affect the whole system dynamics. Modularity also ensures different data processing algorithms to be designed and plugged into the system with minimal effort. This is important due to the diverse range of data streams generated in smart transportation infrastructures. Thus, the application development process can be done in two independent stages, developing individual modules and developing module interconnection logics. The earlier stage can be done by component or module providers while the latter can be done by smart transportation developers. The cloud platforms provide enough modularity and flexibility support for deploying ITS applications but the centralized execution strategies often lead to poor quality of experience (QoE) for the stakeholders.

14.3.2 Scalability

An ideal ITS architecture should be distributed and scalable enough to efficiently serve a large vehicle population. Though cloud provides scalable resource pools, due to the huge volume of real-time data generated by the ITS environment, it might not able to sustain with the smart transportation applications' requirement with regard to the low latency requirement. Current cloud-based ITS applications often "embrace inconsistency," thus implementing consistency preserving computational structures constitute a promising investment domain for the research & development (R&D) sector. The trend envisions a more flexible infrastructure, as in fog computing models where computation resources in dynamic objects such as moving vehicles can also participate in the application.

14.3.3 Context-Awareness and Abstraction Support

As the ITS components such as vehicles and other infrastructures are mobile and sparsely distributed over large geography, fog computing will provide context-aware computing platforms for reliable transportation services. Further, the geo-distributed context information should be exposed to developers so that they can build context-aware applications. Because of the high level of heterogeneity and the large number of IoT devices in a typical ITS application, viz. smart parking requires a high degree of abstraction of how the heterogeneous computations and processing are described, coordinated or interact with one another. The centralized cloud-based ITS solutions needs to be upgraded to dedicated fog solutions such that the model allows it to work with a pool of vehicles at once. For instance, such a programming abstraction should be able to describe the command like: "get the State of Charge (SoC) of these groups of cars in this location".

14.3.4 Decentralization

Since the ITS applications usually operate over a large number of heterogeneous and dynamic transportation telematics such as mobile/autonomous vehicles or roadside units (RSU), decentralized execution or programming model is necessary. The centralized cloud-based application has to implement all sorts of conditions and exception handling to deal with such a heterogeneity and dynamic nature. The fog platform will ensure scalable execution if the application can be developed in a modular way with components being distributed to the edge devices. Instead of relying on remote cloud data centers, fog computing provides robust decentralization support to leverage the computing resources of the ITS components such vehicles and sensors to execute the application in order to fulfill the latency requirement of the ITS applications.

14.3.5 Energy Consumption of Cloud Data Centers

The energy consumption in mega data centers is likely to triple in the coming decade [35]; thus, adopting energy-aware strategies becomes an earnest need for computational folks. Offloading the whole universe of transportation applications into the cloud data centers causes untenable energy demands, a challenge that can only be alleviated by adopting sensible energy management strategies. Also, there are plenty of ITS applications without significant energy implications and instead of overloading data centers with such trivial tasks, the analytics can be made ready at and within ITS fog nodes such as vehicular platoons, parked vehicular networks, RTUs, SCADA systems, roadside units (RSU), base stations and network gateways.

Motivated by the abovementioned mission-critical computing requirements of IoT-aware smart transportation applications, the downsides of current cloud computing infrastructures to meet those needs, and having the assumption that the transportation design community is not in a position to reinvent a dedicated Internet infrastructure or to develop computing platforms and elements from scratch that fulfill all those requirements, we in this work present a fog computing framework whose principle underlie on offloading the time and resource-critical operations From cOre to edGe. The argument here is not to cannibalize the existing centralized cloud support for ITS, but to comprehend the applicability of fog computing algorithms to interplay with the core centered cloud computing support leveraged with a new breed of real-time and latency free utilities. The objective is also to develop a viable computational prototype for an ITS architecture in the realm of IoT space, through proper orchestration and assignment of compute and storage resources to the endpoints and where the cloud and fog technologies tuned to interplay and assist one other in synergistically.

14.4 Fog Computing for Smart Transportation Applications

Figure 14.2 depicts a typical fog-assisted cloud architecture customized for smart transportation applications. It is a consensus that the fog paradigm is not envisioned to cannibalize or replace the cloud computing platforms; rather, the notion is to realize fog platforms as a perfect ally, or an extension of cooperative modules having an interplay with the cloud infrastructure. In fact, according to [4], properties like elasticity, distributed computation, etc. are defined commonly for both cloud as well as fog. However, since the computation-intensive tasks from resource-constrained entities such as sensor nodes are mapped to computational resource blocks (CRBs) of dedicated fog nodes, the response time is appreciably reduced. The distinguishing geo-distributed intelligence provided by fog deployments makes it more viable for security constrained services as the critical and sensitive is selectively processed on local fog nodes and is kept within the user control instead of offloading to the vendor-regulated mega data centers. The fog service models also improve the energy efficacy by offloading the power intensive computations to battery saving modes [12]. Additional fog nodes can be dynamically plugged-in when and wherever necessary, thereby removing the scalability issues that hinders the success

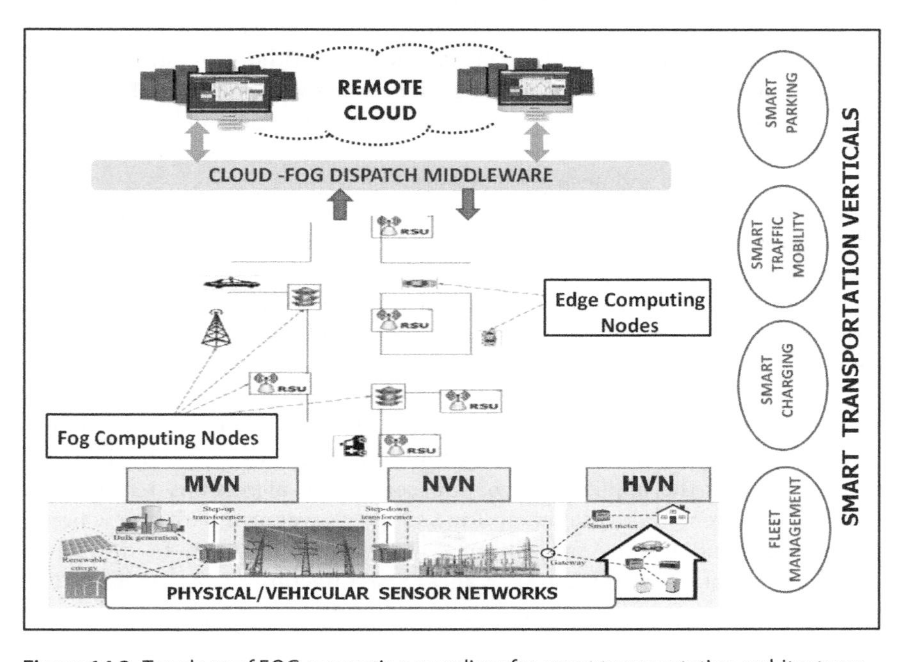

Figure 14.2 Topology of FOG computing paradigm for smart transportation architectures.

of cloud computing models. The bandwidth issues are dramatically fixed as raw application requests are filtered, processed, analyzed, and cached in local computing nodes, thus reducing the data traffic across the cloud gateways. If a robust and predictive caching algorithm is employed, the fog nodes would serve a significant portion of consumer requests from the local nodes only, thus liberating the reliance on data center connectivity. The fog nodes can be efficiently programmed to incorporate context and situational awareness about the data, thereby improving the dependability of the system.

The underlying notion of fog is the distribution of store, communicate, control, and compute resources from the edge to the remote cloud continuum. The fog architectures may be either fully distributed, mostly centralized, or somewhere in between. In addition to the virtualization facilities, specialized hardware and software modules can be employed for implementing fog applications. In the context of an IoT-aided ITS, a customized fog platform will permit specific applications to run anywhere, reducing the need for specialized applications dedicated just for the cloud, just for the endpoints, or just for the edge devices. It will enable applications from multiple vendors to run on the same physical machine without reciprocated interference. Further, a fog architecture will provide a common lifecycle management framework for all applications, offering capabilities for composing, configuring, dispatching, activating and deactivating, adding and removing, and updating applications. It will further provide a secure execution environment for fog services and applications. Among the strong list of fog specialties, we here define four key advantages of typical fog architecture acronymed as CEAL [6].

14.4.1 Cognition

The most peculiar property of a fog platform is its cognizance to client-centric objectives, also termed as geo-distributed intelligence. The framework is aware of the context of customer requirements and can best determine where to carry out the computing, storage, and control functions along the cloud-to-thing continuum. Thus, the fog applications can be populated at the vicinity ITS endpoints and are ensured to be better aware of and closely reflect customer requirements.

14.4.2 Efficiency

In fog architectures, the compute, storage, and control functions are pooled and disseminated anywhere across the cloud and the edge nodes, acquiring full advantage of the diverse resources available along the cloud-to-thing continuum. In IoT-aided ITS infrastructures, the fog model allows utilities and applications to leverage the otherwise idling computing, storage, and networking resources abundantly available both along the network edge

(HAN, NAN, MAN etc.) and at end-user devices such as smart meters, smart home appliances, connected vehicles, and network edge routers. Fog's closer proximity to the endpoints will enable it to be more closely integrated with consumer applications.

14.4.3 Agility

It is usually much faster and more affordable to experiment with client and edge devices, rather than waiting for vendors of large network and cloud boxes to initiate or adopt an innovation. Fog will make it easier to create an open marketplace for individuals and small teams to use open application programming interfaces, open software development kits (SDKs), and the proliferation of mobile devices to scale, innovate, develop, deploy, and operate new services.

14.4.4 Latency

Fog enables data analytics at the network edge and can support time-sensitive functions for ITS like cyber-physical systems. This is essential not only for developing stable control systems but also for the tactile Internet vision to enable embedded AI applications with millisecond response requirements. Such advantages, in turn, enable new services and business models, and may help broaden revenues and reduce cost, thereby accelerating IoT-aided ITS rollouts. Furthermore, Table 14.2 compares the performance of cloud and fog computing deployments in smart transportation applications.

A triple-tier fog-assisted cloud computing architecture is presented in Figure 14.2, where a substantial proportion of ITS control and computational tasks are nontrivially hybridized to geo-distributed fog computing nodes alongside the cloud computing support. The hybridization objective is to overcome the disruption caused by the penetration of IoT utilities into ITS infrastructures that calls for active proliferation of control, storage, networking, and computational resources across the heterogeneous edges or endpoints. The tier nearest to ground is termed as physical schema or data generator layer, which primarily comprises a wide range of intelligent IoT-enabled devices scattered across the ITS geography. This is the sensing network consisting of several noninvasive, highly reliable, low-cost wireless sensory nodes and smart mobile devices for capturing situational context information from ITS stakeholders.

The data capturing/generating devices are widely distributed at numerous ITS endpoints and the voluminous data streams generated from these geo-spatially distributed sensors have to be processed as a coherent whole. However, this layer may occasionally filter data streams for local consumption (edge computing) while offloading the rest to upper tiers through dedicated gateways. Such entities may be abstracted into application-specific logical

Table 14.2 Performance comparison of cloud and fog computing models in smart transportation applications.

	Characteristics and requirements	Pure cloud platform	Fog-assisted cloud platform
1	Geo-distribution	Centralized	Distributed
2	Context/location awareness	No	Yes
3	Service node distribution	Within the Internet	At core as well as edges
4	Latency	High	Low
5	Delay jitter	High	Low
6	Client-server separation	Remote/Multiple hops	Single hop
7	Security	Not defined	Defined degree of security
8	Node population	Few	Very large
9	Mobility support	Limited	Rich mobility support
10	Last-mile connectivity support	Leased line	Wired/Wireless
11	Real-time analytics	Supported	Supported
12	Enroute data attacks/DoS	High probability	Low probability

clusters, directly or indirectly influenced by the expediency of ITS operations. In connected vehicular networks, such clusters are formed from vehicular applications where the intelligent vehicles equipped with sensing units such as on-board sensors (OBS) organize themselves to form vehicular fogs. Often, the transportation telematics support such as cellular telephony, on-board sensors (OBS), roadside units (RSU), and smart wearable devices may uncover the computational as well as networking capabilities latent in the underutilized vehicular resources. The underutilized vehicular resources may occasionally be transformed into communicational and analytics use, where a collaborative multitude of end-user clients or near-user edge devices carry out communication and computation, based on better utilization of individual storage, communication, and computational resources of each vehicle [36].

Similarly, presence of clusters could also be traced in smart home networks (HAN) that have noteworthy contributions in ITS operational dynamics. The intelligent IoT-equipped home agents such as smart parking lots, CC camera, and home charging devices are potentially active data-generation entities and may also be augmented with actuators to provide storage, analysis, and computational support for satisfying the prompt and local decision-making services (edge computing).

Layer 2 constitutes the fog computing layer comprising low-power intelligent fog computing nodes (FCN) such as routers, switches, high-end proxy servers, intelligent agents, and commodity hardware, having peculiar ability of storage, computation, and packet routing. The software-defined networking (SDN) assembles the physical clusters to form virtualized intercluster private networks (ICPN) that route the generated data to the fog devices spanned across the fog computing layer The fog devices and their corresponding utilities form geographically distributed virtual computing snapshots or instances that are mapped to lower-layer devices in order to serve the processing and computing demands of ITS. Each fog node is mapped to and is responsible for a local cluster of sensors covering a neighborhood or a small community, executing data analytics in real-time. However, since the IoT devices in layer 1 are often dynamic (viz. vehicular sensors), robust mobility management techniques need to be employed to enable flexible association of those entities with the layer 2 fog nodes in order to realize a consistent and reliable data transmission policy.

Often, the FCNs in layer 2 are parallel to the nodes lying below in the hierarchy to undertake tasks. In many cases, the FCN may form further subtrees of FCNs, with each node at a higher depth in the tree managed by the ones at lower depth, in master–slave paradigm. A typical association of such hierarchies is depicted in Figure 14.3. Considering the VANET scenario, the FCNs may be assigned with spatial and temporal data to identify potential hazardous events in road transportation network such as accidents, vehicle thefts, or intruder vehicles in the network. In such circumstances, these computing nodes may interrupt the local execution for small timespans, and the data analysis results will be fed back and reported to the upper layer (from street-level to citywise traffic monitoring entities) for complex, historical, and large-scaled behavior

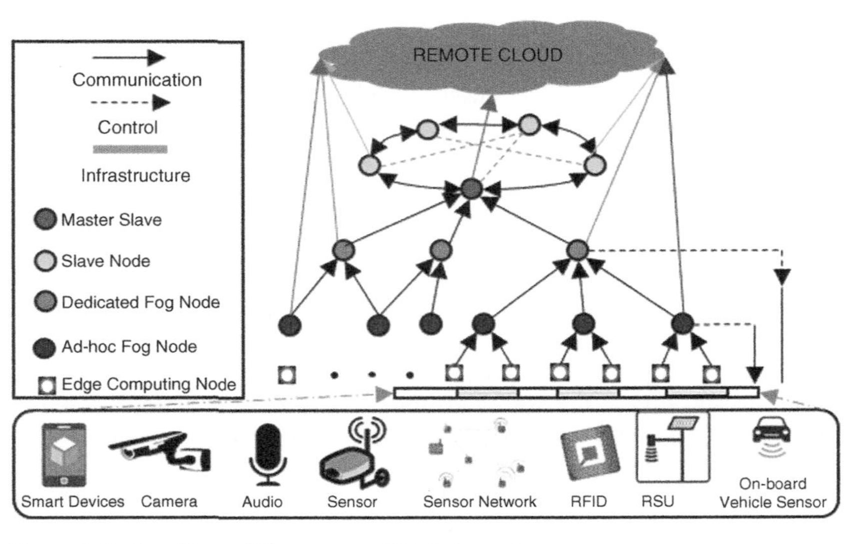

Figure 14.3 Data/Control Flow among FCNs in Layer 2

analysis and condition monitoring. In other words, the distributed analytics from multi-tier fogs (followed by aggregation analytics in many case studies) performed at proposed fog layers act as localized "reflex" decisions to avoid potential contingencies. Meanwhile, a significant fraction of generated IoT data from smart grid applications don't require that data be dispatched to the remote clouds; hence, response latency and bandwidth consumption problems could be easily solved.

The uppermost tier in customized fog architecture is the cloud computing layer consisting of mega data centers that provides citywide ITS monitoring and global centralization in contrast to localization, geo-distributed intelligence, low latency and context awareness support provided by layer 2. The computational elements at this layer are focused to produce complex, long-term, and citywide behavioral analytics such as large-scale event detection, long-term pattern recognition, and relationship modeling, to support dynamic decision making. This will ensure that ITS communities perform wide area situational awareness (WASA), wide area demand response, and resource management in the case of a natural disaster or a large-scale service interruption. The processing output of layer 2 can be categorized into two dimensions. The first one comprises analysis and status reports and the corresponding data that demand large-scaled and long-term behavior analysis and condition monitoring. Such datasets are offloaded to cloud computing mega datacenters situated in layer 3 via high-speed WAN gateways and links. The other part of analysis result is the inferences, decisions, and quick feedback control to the aligned data consumers.

14.5 Case Study: Intelligent Traffic Lights Management (ITLM) System

A smart traffic management prototype calls for the deployment of intelligent traffic lights (ITLs) equipped with sensing capabilities at each crossing. Such sensors measure the distance and speed of approaching vehicles to and from every direction. The sensors also detect and regulate the movement of pedestrian and cycle commuters intercepting every street and crossing on its way. The prime QoS attributes of ITLM architecture can be summarized as follows:

1. **Accident prevention**. The ITLs may need to trigger stop or slow-down signals to candidate vehicles or to modify their execute cycle(s) to avoid collisions in real time.
2. **Ensuring vehicles mobility**. The ITLs need efficient software programming interfaces that can learn the fleet dynamics. Accordingly, they maintain the green pulses to guarantee steady flow of traffic in near real time.

3. **Reliability**. The historical datasets generated by ITLM systems are collected, stored in back-end large databases, and then analyzed using big data analytics (BDA) tools to evaluate and enhance the architectural reliability. Thus, such activities relate to the storage and analysis of global data ranging over long time spans.

In order to illustrate the key computational requirements of such ITLMs, let us consider a green pulse signaling the movement of a vehicle at 40 mph – i.e. it travels 1.7 meters per 100 microseconds. If a probable collision with a pedestrian is anticipated, the associated ITL(s) must issue an urgent alarm to the approaching vehicles. Here fog computing comes into play, as the control loop sub-system needs to react within some 100 microseconds to few milliseconds. The aggregated local subsystem response latency for such mission-critical tasks is on the order of <10 ms. Now, triggering any action to prevent accidents may successively trump other operations. Thus, the local ITL network might also alter its execution cycle, an action that may introduce perturbation in the green lights, affecting the whole system dynamics. To dampen the effect of such perturbation, a resynchronization signal needs to be sent along all the ITLs in the global system, a task that will be accomplished on a time scale of hundreds of milliseconds to a few seconds. An interplay between the fog and the cloud is accentuated here. The research thrust is to develop a viable computational prototype for an ITLM system in the realm of IoT space, through proper orchestration and assignment of compute and storage resources to the endpoints and where the cloud and fog technologies are tuned to interplay and assist each other in a synergistic manner. Some of the critical computing requirements of a customized ITLM are identified in Table 14.3.

Table 14.3 Computing requirements of intelligent traffic light management (ITLM) systems.

Attributes	Description
Mobility	Tight mobility constraints for the commuters as well as ITLs (ideally regular red-green pulses)
Geo-distribution	Wide (across region) and dense (intersections and ramp accesses)
Low/predictable latency	Tight within the scope of the intersection
Fog-cloud interplay	Data at different time scales (sensors/vehicles at intersection, traffic info at diverse collection points)
Multi-agencies orchestration	Agencies that run the system must coordinate control law policies in real time
Consistency	Getting the traffic landscape demands a degree of consistency between collection points

The fog model leveraged with modular compute and storage devices offers common interfaces and programming environments for the ITL networking infrastructures, though having varying form factors and encasings. Since the ITLM is a highly distributed system that collects data over an extended geography, ensuring an acceptable degree of consistency between the different aggregator points is crucial for the implementation of efficient traffic policies.

The fog vision anticipates an integrated hardware infrastructure and software platform with the purpose of streamlining and making more efficient the deployment of new services and applications. The ITL fog nodes are multitenant and also provide strict service guarantees for mission-critical systems such as the ITLM, in contrast with softer guarantees (e.g., infotainment), even when run for the same provider. The network of ITLs may extend beyond the domains of a single controlling authority. Thus, the orchestration of consistent policies involving multiple agencies is a challenge unique to fog computing. A typical orchestration scenario for ITLM sub-system is presented in Figure 14.4.

The cloud–fog dispatch middleware (CFDM) defines an orchestration platform to handle a number of critical software components across the whole system, which is deployed across a wide geographical area. The CFDM employed in ITLMs have decision-making modules (DMM), which create the control policies and push them to the individual ITLs. The DMM can be implemented in a centralized, distributed, or hierarchical way. In the latter, the most likely

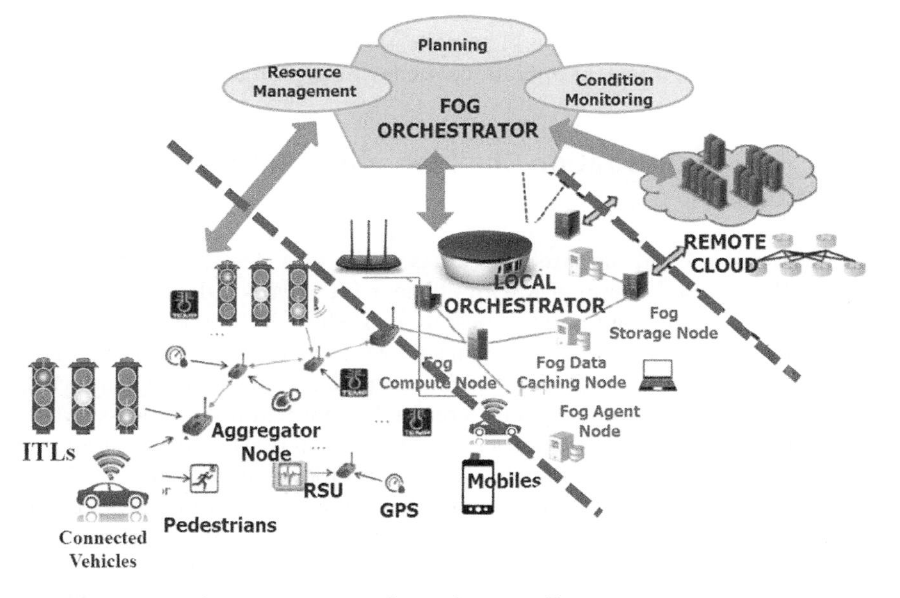

Figure 14.4 An orchestration scenario for intelligent traffic management service.

implementation nodes with DMM functionality of regional scope must coordinate their policies across the whole system. Whatever the implementation, the system should behave as if orchestrated by a single, all-knowledgeable DM. The CFDM defines a set of protocols for the federated message bus, which passes data from the traffic lights to the DMM nodes, pushes policies from the DMM nodes to the ITLs, and exchanges information between those ITLs.

In addition to the actionable real-time (RT) information generated by the sensors, and the near-RT data passed to the DMM and exchanged among the set of ITLs, there are volumes of valuable data collected by the ITLM system. This data must be ingested in a data center (DC)/cloud for deep big data analytics that extends over time (days, months, even years) and over the covered territory. The results of such historical batch analytics may be further used to improve the reliability and QoS of future executions. The outputs of such bulk analytics can be a used as solutions for:

- Evaluation of the impact on traffic (and its consequences for the economy and the environment) of different policies
- Monitoring of city pollutants
- Trends and patterns in traffic

The ITLM use-case just discussed reflects the need for robust orchestration frameworks that can simplify, maintain, and improve the ITS data security and system reliability. The data-driven ITS is an ideal example of cyber-physical systems (CPS) encompassing physical and virtual components capable of interfacing and interacting with existing network infrastructure. Thus, addressing how to efficiently deal with the ITS applications in IoT space, their dynamic variations, and the transient operational behavior is a tedious challenge.

14.6 Fog Orchestration Challenges and Future Directions

High-paced R&D and investments efforts in the past decade have led more mature cloud-based techniques with efficient frameworks, deployment platforms, simulation toolkits, and business models. However, in the context of fog deployments, such efforts, though on pace, are still in their infancy [17]. There may be plenty of studies hypothesizing the execution scenario of fog platforms, but these are still in the concept and simulation phase. Roll-out of fog services must inherit many of the properties of cloud counterparts, and the requirement of deploying computational workloads on fog computing nodes (FCN) must be properly demystified. In addition, fog comes with its inherent silos and raises many questions that seek a consensus regarding the right answers. Some of them may be where to place a workload, what are the connection policies, protocols, and standards, how to model/interpret the interaction of/among fog nodes, and how to route the workload, for example. In the next section, we highlight the key orchestration challenges in fog-enabled orchestration for ITS applications. Following this, the nascent research avenues envisioned by such issues and challenges are also explored.

14.6.1 Fog Orchestration Challenges for Intelligent Transportation Applications in IoT Space

14.6.1.1 Scalability

Since the heterogeneous sensors and smart devices employed in ITS are designed from multiple IoT manufacturers and vendors, selecting an optimal device becomes increasingly intricate while considering customized hardware configurations and personalized ITS requirements. Moreover, there may be applications that can only operate with specific hardware architectures viz. ARM or Intel etc., and through a wide range of operating systems. Additionally, the ITS applications with stringent security requirements might require specific hardware and protocols to function. An orchestration framework need not only cater to such functional requirements, it must scale efficiently in the face of increasingly larger workflows that change dynamically. The orchestrator must assess whether the assembled systems, comprised of cloud resources, sensors, and fog computing nodes (FCN), coupled with geographic distributions and constraints, are capable of provisioning complex services correctly and efficiently. In particular, the orchestrator must be able to automatically predict, detect, and resolve issues pertaining to scalability bottlenecks that could arise from an increased application scale in a customized ITS architecture.

14.6.1.2 Privacy and Security

In IoT-aided ITS case studies such as ITLMs or smart parking, a specific application is composed of multiple sensors, computer chips, and devices. Their deployment in varying different geographic locations thus results in increased attack vectors of involved objects. Examples of attack vectors may be human-caused sabotage of network infrastructure, malicious programs provoking data leakage, or even physical access to devices [37]. Holistic security and risk assessment procedures are needed to effectively and dynamically evaluate the security and measure risks, as evaluating the security of dynamic IoT-based application orchestration becomes increasingly critical for secure data placement and processing. The IoT-integrated devices for fog support such as switches, routers, and base stations, if they are brought to be used as publicly accessible computing edge nodes, need greater articulation regarding the risk associated by public and private vendors that own these devices as well as those that will employ these devices. Also, the intended objective of such devices, e.g. an Internet router for handling network traffic, cannot be compromised just because it is being used as a fog node. The fog can be made multitenant only when stringent security protocols are enforced.

14.6.1.3 Dynamic Workflows

Another significant characteristic and challenge for IoT-enabled ITS applications is their ability to evolve and dynamically change their workflow composition. This problem, in the context of software upgrades through FCNs or the frequent join-leave behavior of network objects, will change the internal properties and performance, potentially altering the overall workflow execution pattern. Moreover, handheld devices used by ITS stakeholders inevitably suffer from software and hardware aging, which will invariably result in changing workflow behavior and its device properties (e.g., low-battery devices will degrade the data transmission rate). Furthermore, performance of transportation applications will change owing to their transient and/or short-lived behavior within the ITS subsystem, including spikes in resource consumption or big data generation. This leads to a strong requirement for automatic and intelligent reconfiguration of the topological structure and assigned resources within the workflow, and importantly, that of FCNs.

14.6.1.4 Tolerance

Scaling a fog computing framework in proportion to ITS application demands increases the probability of failure. Some rare software bugs or hardware faults that don't manifest at small scale or in testing environments, such as stragglers, can have a debilitating effect on system performance and reliability. At the scale, heterogeneity, and complexity we're anticipating, different fault combinations will likely occur. To address these system failures, developers should incorporate redundant replications and user-transparent, fault-tolerant deployment, and execution techniques in orchestration design.

14.7 Future Research Directions

The challenges outlined in the previous sub-section unlock several key research directions for successful deployment of fog-supported ITS architectures. The research prospects defined for fog life cycle management can be executed in three broad phases. In the deployment phase, research opportunities include optimal node selection and routing as well as parallel algorithms to handle scalability issues. In the runtime phase, incremental design and analytics, re-engineering, dynamic orchestration, etc., are potential research thrusts for supporting dynamic QoS monitoring and providing guaranteed QoE. In the evaluation phase, big-data-driven analytics (BD^2A) and optimization algorithms are prime avenues that need to be explored to improve orchestration quality and accelerate optimization for problem solving.

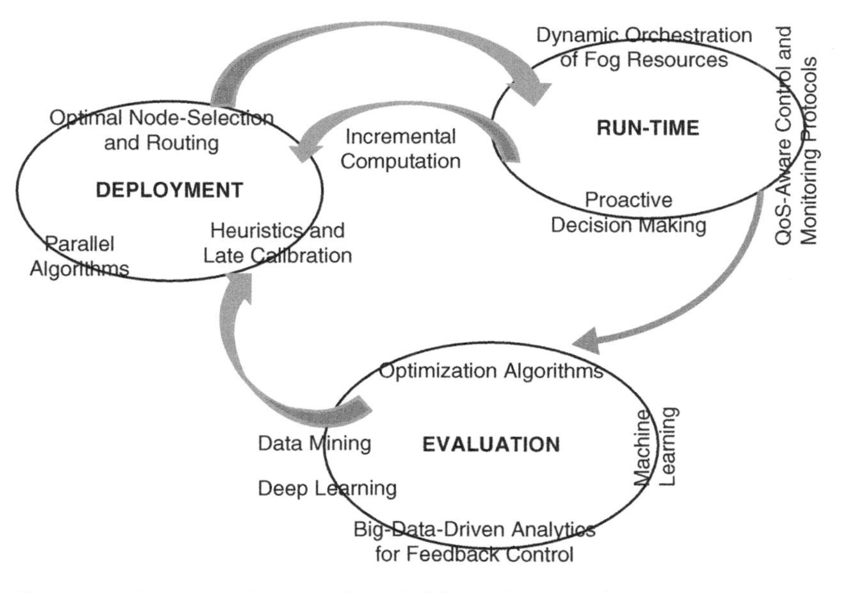

Figure 14.5 Functional elements of a typical fog orchestrator showing the key requirements and challenges at each phase.

Figure 14.5 shows the functional elements of a typical fog orchestrator, along with the key requirements and challenges at each phase.

14.7.1 Opportunities in the Deployment Phase

Fog computing provides research opportunities in node selection, routing, parallelization, and heuristics.

14.7.1.1 Optimal Node Selection and Routing

Determining resources and services in cloud paradigms is a well-explored area and easily understood, but exploiting network edges in decentralized fog settings calls for discovery mechanisms to associate optimal nodes [38]. Resource discovery in fog computing is not as easy as in both tightly and loosely coupled distributed environments, and manual mechanisms are not feasible because of the sheer volume of FCNs available at the fog layer [39]. If the ITS utility needs to execute machine learning or big-data tasks, resource allocation strategies also need to cater to datastream of heterogeneous devices from multiple generations as well as online workloads.

Benchmark algorithms must be developed for efficient estimation of FCNs' availability and capability. These algorithms must allow for seamless augmentation (and release) of FCNs in the computational workflow at varying hierarchical levels without added latencies or compromised QoE.

Autonomic node recovery mechanisms need to be devised to ensure consistency and reliability in fault detection in FCN networked architectures, as existing cloud-based solutions don't fit. Besides, the most potential research aspect to ponder is workflow partitioning in fog computing environments. Though numerous task partitioning techniques, languages, and tools have been successfully implemented for cloud data centers, research regarding work apportioning among FCNs is still in concept phase.

Without specifying the capabilities and geo-distribution of candidate FCNs, automated mechanism for realizing computation offloading among those nodes is challenging. Maintaining a ranked list of associated host nodes through priority aware resource management policies, making hierarchies or pipelines for sequential offloading of workloads, developing schedulers for dynamically deploying segregated tasks to a multiple nodes, algorithms for parallelization and multitasking of only FCNs, FCNs and data centers, or only data enters, etc., are rigorous research topics in academia as well as the R&D community.

14.7.1.2 Parallelization Approaches to Manage Scale and Complexity

Optimization algorithms or graph-based approaches are typically time- and resource-consuming when applied on a large scale, and necessitate parallel approaches to accelerate the optimization process. Recent work provides possible solutions to leverage an in-memory computing framework to execute tasks in a cloud infrastructure in parallel. However, realizing dynamic graph generation and partitioning at runtime to adapt to the shifting space of possible solutions stemming from the scale and dynamicity of IoT components remains an unsolved problem.

14.7.1.3 Heuristics and Late Calibration

To ensure near-real-time intervention during IoT application development, one approach is to use correction mechanisms that could be applied even when suboptimal solutions are deployed initially. For example, in some cases, if the orchestrator finds a candidate solution that approximately satisfies the reliability and data transmission requirements, it can temporarily suspend the search for further optimal solutions. At runtime, the orchestrator can then continue to improve decision results with new information and a reevaluation of constraints, and use task- and data-migration approaches to realize workflow redeployment.

14.7.2 Opportunities in Runtime Phase

In the runtime phase, research opportunities for fog computing include dynamic orchestration of resources, incremental strategies, QoS, and proactive decision-making.

14.7.2.1 Dynamic Orchestration of Fog Resources

Apart from the initial placement, all workflow components dynamically change in response to internal transformations or abnormal system behavior. IoT applications are exposed to uncertain environments where execution variations are commonplace. Because of the degradation of consumable devices and sensors, capabilities such as security and reliability that initially were guaranteed will vary, resulting in the initial workflow being no longer optimal or even totally invalid.

Furthermore, the structural topology might change according to the task execution progress (i.e., a computation task is finished or evicted) or will be affected by the execution environment's evolution. Abnormalities might occur, owing to the variability of combinations of hardware and software crashes, or data skew across different management domains of devices due to abnormal data and request bursting. This will result in unbalanced data communication and subsequent reduction of application reliability. Therefore, dynamically orchestrating task execution and resource reallocation is essential.

14.7.2.2 Incremental Computation Strategies

The ITS applications may often be choreographed through workflow or task graphs to assemble different IoT applications. In some domains, the orchestration is supplied with a plethora of candidate devices with different geographical locations and attributes. In some cases, orchestration would typically be considered too computationally intensive, as it is extremely time-consuming to perform operations, including prefiltering, candidate selection, and combination calculation, while considering all specified constraints and objectives. Static models and methods become viable when the application workload and parallel tasks are known at design time. In contrast, in the presence of variations and disturbances, orchestration methods typically rely on incremental scheduling at runtime (rather than straightforward complete recalculation by rerunning static methods) to decrease unnecessary computation and minimize schedule makespan.

14.7.2.3 QoS-Aware Control and Monitoring Protocols

To capture the dynamic evolution and variables (such as dynamic evolution, state transition, and new IoT operations), we should predefine the quantitative criteria and measuring approach of dynamic QoS thresholds in terms of latency, availability, throughput, and so on. These thresholds usually dictate upper and lower bounds on the metrics as desired at runtime. In a normal setting, complex QoS information-processing methods such as hyper-scale matrix update and calculation would lead to many scalability issues.

14.7.2.4 Proactive Decision-Making

Localized regions of self-updates become ubiquitous within fog environments. The orchestrator should record staged states and data produced by fog components periodically or in an event-based manner. This information will form a set of time series of graphs and facilitate the analysis and proactive recognition of anomalous events to dynamically determine such hotspots [40]. The data and event streams should be efficiently transmitted among fog components, so system outage, appliance failure, or load spikes will rapidly feed back to the central orchestrator for decision making.

14.7.3 Opportunities in Evaluation Phase: Big-Data-Driven Analytics (BD^{2}A) and Optimization

A typical ITS framework congregates the diverse transportation entities into a clique-like structure in the IoT realm and enables a bidirectional flow of energy and data among the stakeholders in order to facilitate the assets optimization. The major data sources for a data-driven ITS include ITS-sensing objects such as connected vehicles, on-board sensors (OBS), road-side units (RSU), traffic sensors and actuators, GPS devices, ITLs, and web data from recommender systems, crowdsourcing, and feedback modules.

Furthermore, the domain of IoT in ITS applications is extended to numerous geographically distributed devices that produce multidimensional, high-volume dynamic data streams requiring a noble mix of real-time analytics and data aggregation [41]. Figure 14.6 depicts the conceptual framework for BD^{2}A and optimization of an intelligent traffic management use case based on cloud and fog platforms. The fog orchestration module should employ efficient data-driven optimization and planning algorithms for reliable data management across complex IoT-aided ITS endpoints.

While developing ITS applications adhered to fog computing and making proper trade of such applications across different layers in the fog environment, the developers should employ robust optimization procedures that stabilize the schema definitions, mappings, all overlapping, and interconnection between layers (if any). In order to reduce data transmission latencies, data-processing activities and the database services may be pipelined. Rather than frequent triggering of move-data actions, use of multiple data-locality principles (e.g. temporal, spatial etc.) and efficient caching techniques can distribute or reschedule the computation tasks of FCNs near the sensors, thereby improving the delays. The data-relevant attributes related to QoS parameters such as the data-generation rate or data-compression ratio can be customized to adapt to the desired degree of performance and assigned resources to strike a balance between data quality and specified response-time targets.

A major challenge is that decision operators are still computationally time consuming. To tackle this problem, online machine learning can provision several online training (such as classification and clustering) and prediction models to capture the constant evolutionary behavior of each system element, producing time series of trends to intelligently predict the required system resource usage, failure occurrence, and straggler compute tasks, all of which

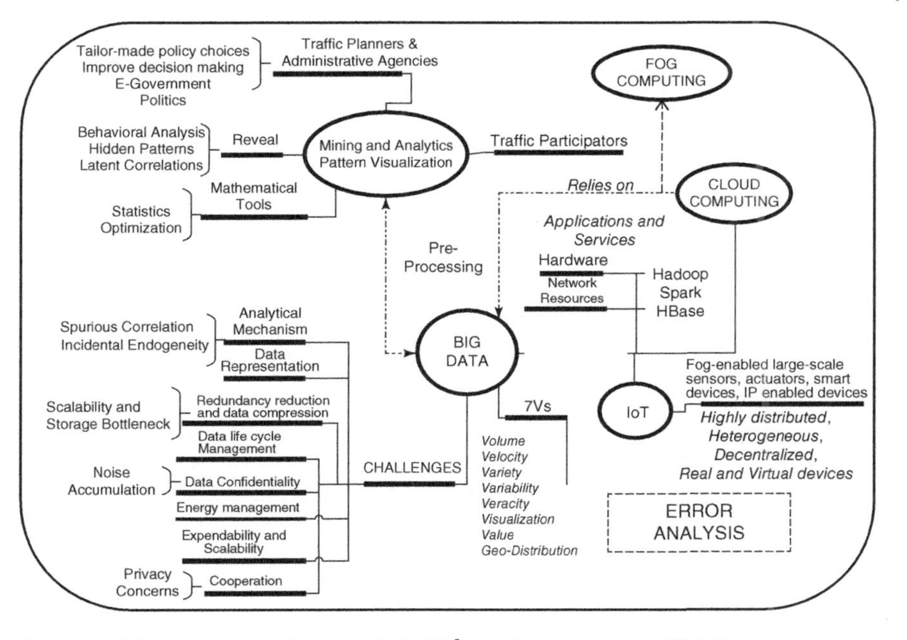

Figure 14.6 The conceptual framework for BD^2A and optimization of ITLM based on cloud and fog platforms.

can be learned from historical data and a history-based optimization (HBO) procedure. Researchers or developers should investigate these smart techniques, with corresponding heuristics applied in an existing decision-making framework to create a continuous feedback loop. Cloud machine learning offers analysts a set of data exploration tools and a variety of choices for using machine learning models and algorithms.

14.8 Conclusions

In this chapter, we revisited the need for data-driven transportation architecture, discussing the functionality of its key components and certain deployment issues associated with it. Then we identified the service-critical store and compute requirements of application supported over such data-driven transportation architectures, analyzed the current state of cloud deployments, and outlined the need for going through geo-distributed fog methodologies for fulfilling those needs. We also presented a fog computing framework customized to smart transportation applications and highlighted the requirements for fog models through an intelligent traffic management system (ITLM) use case. The successful deployment of fog models requires an orchestration framework that can simplify maintenance and enhance data security and system reliability. The chapter finally provided an overview of the core issues, challenges, and future research directions in fog-enabled orchestration for smart transportation services in the realm of IoT.

References

1 Intel Corporation. Designing Next-Generation Telematics Solutions. *White Paper*, 2018.
2 B. Varghese, N. Wang, S. Barbhuiya, P. Kilpatrick, and D. S. Nikolopoulos. Challenges and Opportunities in Edge Computing. In *Proceedings of the 2016 IEEE Int. Conf. Smart Cloud, SmartCloud 2016*, pp. 20–26, 2016.
3 O. Skarlat, S. Schulte, and M. Borkowski. Resource Provisioning for IoT Services in the Fog. *9th IEEE International Conference on Service Oriented Computing and Applications*, November 4–6, 2016, Macau, China.
4 S. Park, O. Simeone, and S.S. Shitz. Joint Optimization of Cloud and Edge Processing for Fog Radio Access Networks. *IEEE Trans. Wireless Communications*, 15(11): 7621–7632, 2016).
5 C. Perera, Y. Qin, J. C. Estrella, S. Reiff-marganiec, and A.V. Vasilakos. Fog computing for sustainable smart cities: A survey. *ACM Computing Surveys*, 50(3): 1–43, 2017.
6 M. Chiang and T. Zhang. Fog and IoT: An overview of research opportunities. *IEEE Internet Things Journal*, 3(6): 854–864, 2016.
7 M.M. Hussain, M.S. Alam, and M.M.S. Beg. Computational viability of fog methodologies in IoT-enabled smart city architectures – a smart grid case study. *EAI Endorsed Transactions*, 2(7): 1–12, 2018.
8 C. Byers and P. Wetterwald. Fog computing: distributing data and intelligence for resiliency and scale necessary for IoT. *ACM Ubiquity Symposium*, November, 2015.
9 Z. Wen, R. Yang, P. Garraghan, T. Lin, J. Xu, and M. Rovatsos. Fog orchestration for Internet of Things services. *IEEE Internet Computing*, 21(2): 16–24, 2017.
10 N.K. Giang, V.C.M. Leung, and R. Lea. On developing smart transportation applications in fog computing paradigm. *ACM DIVANet'16, November 13–17, Malta*, pp. 91–98, 2016.
11 W. He, G. Yan, L. Da Xu, and S. Member. Developing vehicular data cloud services in the IoT environment. *IEEE Trans. Industrial Informatics*, 10(2): 1587–1595, 2014.
12 S. Bitam. ITS-Cloud: Cloud Computing for Intelligent Transportation System. *IEEE Globecom 2012 – Communications Software, Services and Multimedia Symposium*, California, USA, 2054–2059.
13 J.M. Sussman. *Perspectives on Intelligent Transportation Systems (ITS)*. New York: Springer-Verlag, 2005.

14 T. Gandhi and M. Trivedi. Vehicle surround capture: Survey of techniques and a novel vehicle blind spots. *IEEE Trans. Intelligent. Transp. Syst.,* 7(3): 293–308, September 2006.

15 M.M. Hussain, M.S. Alam, and M.M.S. Beg. Federated cloud analytics frameworks in next generation transport oriented smart cities (TOSCs) – Applications, challenges and future directions. *EAI Endorsed Transactions. Smart Cities,* 2(7), 2018.

16 J. Zhang, F. Wang, K. Wang, W. Lin, X. Xu, and C. Chen. Data-driven intelligent transportation systems : a survey. *IEEE Trans. Intelligent. Transp. Systems,* 12(4): 1624–1639, 2011.

17 X. Hou, Y. Li, M. Chen, et al. Vehicular Fog Computing : A Viewpoint of Vehicles as the Infrastructures. *IEEE Trans Vehicular Tech.,* 65(6): 3860–3873, 2016.

18 A. O. Kotb, Y. C. Shen, X. Zhu, and Y. Huang. IParker – A new smart car-parking system based on dynamic resource allocation and pricing. *IEEE Trans. Intell. Transp. Systems,* 17(9): 2637–2647, 2016.

19 O. Scheme. Central Pollution Control Board. Delhi Central Pollution Control Board, Delhi, pp. 1–6, 2016.

20 X. Wang, X. Zheng, Q. Zhang, T. Wang, and D. Shen. Crowdsourcing in ITS : The state of the work and the networking. *IEEE Trans. Intell. Transp. Systems,* 17(6): 1596–1605, 2016.

21 Z. Liu, H. Wang, W. Chen, et al. An incidental delivery based method for resolving multirobot pairwised transportation problems. *IEEE Trans. Intell. Transp. System,* 17(7), 1852–1866, 2016.

22 D. Wu, Y. Zhang, L. Bao, and A. C. Regan. Location-based crowdsourcing for vehicular communication in hybrid networks. *IEEE Trans. Intell. Transp. System,* 14(2), 837–846, 2013.

23 M. Tubaishat, P. Zhuang, Q. Qi, and Y. Shang. Wireless sensor networks in intelligent transportation systems. *Wirel. Commun. Mobile. Computing. Wiley InterScience,* 2009, no. 9, pp. 87–302.

24 White Paper. Freeway Incident Management Handbook, Federal Highway Administration, *Available:* http://ntl.bts.gov/lib/jpodocs/rept_mis/7243.pdf.

25 M.M. Hussain, M.S. Alam, M.M.S. Beg, and H. Malik. A Risk averse business model for smart charging of electric vehicles. In *Proceedings of First International Conference on Smart System, Innovations and Computing, Smart Innovation, Systems and Technologies,* 79: 749-759, 2018.

26 M. Saqib, M.M. Hussain, M.S. Alam, and M.M.S. Beg. Smart electric vehicle charging through cloud monitoring and management. *Technology Economics Smart Grids Sustain Energy,* 2(18): 1–10, 2017.

27 C.-C. R. Wang and J.-J. J. Lien. Automatic vehicle detection using local features – A statistical approach. *IEEE Trans. Intell. Transp. System,* 9(1): 83–96, 2008.

28 L. Bi, O. Tsimhoni, and Y. Liu. Using image-based metrics to model pedestrian detection performance with night-vision systems. *IEEE Trans. Intell. Transp. System,* 10(1): 155–164, 2009.

29 S. Atev, G. Miller, and N.P. Papanikolopoulos. Clustering of vehicle trajectories. *IEEE Trans. Intell. Transp. System,* 11(3): 647–657, September 2010.

30 Z. Sun, G. Bebis, and R. Miller. On-road vehicle detection: A review. *IEEE Trans. Pattern Anal. Mach. Intell.,* 28(5): 694–711, 2006.

31 J. Huang and H.-S. Tan. DGPS-based vehicle-to-vehicle cooperative collision warning: Engineering feasibility viewpoints. *IEEE Trans. Intell. Transp. System,* 7(4): 415–428, 2006.

32 J.M. Clanton, D.M. Bevly, and A.S. Hodel. A low-cost solution for an integrated multisensor lane departure warning system. *IEEE Trans. Intell. Transp. System,* 10(1): 47–59, 2009.

33 K. Sohn and K. Hwang. Space-based passing time estimation on a freeway using cell phones as traffic probes. *IEEE Trans. Intell. Transp. System,* 9(3): 559–568, 2008.

34 M.M. Hussain, F. Khan, M.S. Alam, and M.M.S. Beg. Fog computing for ubiquitous transportation applications – a smart parking case study. *Lect. Notes Electrical. Engineering,* 2018 *(In Press)*.

35 T. N. Pham, M.-F. Tsai, D. B. Nguyen, C.-R. Dow, and D.-J. Deng. A cloud-based smart-parking system based on Internet-of-Things technologies. *IEEE Access,* 3: 1581–1591, 2015.

36 B.X. Yu, F. Ieee, Y. Xue, and M. Ieee. Smart grids: A cyber – physical systems perspective. In *Proceedings of the IEEE,* 24(5): 1–13, 2016.

37 E. Baccarelli, P.G. Vinueza Naranjo, M. Scarpiniti, M. Shojafar, and J.H. Abawajy. Fog of everything: energy-efficient networked computing architectures, research challenges, and a case study. *IEEE Access,* 5: 1–37, 2017.

38 A. Beloglazov and R. Buyya. Optimal online deterministic algorithms and adaptive heuristics for energy and performance efficient dynamic consolidation of virtual machines in cloud data centers. *Concurrency Comput., Practice. Experience,* 24(13): 1397–1420, September 2012.

39 H. Zhang, Y. Xiao, S. Bu, D. Niyato, R. Yu, and Z. Han. Computing resource allocation in three-tier IoT fog networks: A joint optimization approach combining stackelberg game and matching. *IEEE Internet of Things Journal,* 1–10, 2017.

40 K.C. Okafor, I.E. Achumba, G.A. Chukwudebe, and G.C. Ononiwu. Leveraging fog computing for scalable IoT datacenter using spine-leaf network topology. *Journal of Electrical and Computer Engineering, Hindawi,* 1–11, 2017.

41 J. Gubbi, R. Buyya, S. Marusic, and M. Palaniswami. Internet of Things (IoT): A vision, architectural elements, and future directions. *Future Generation Computer System,* 29(7): 1645–1660, 2013.

15

Testing Perspectives of Fog-Based IoT Applications

Priyanka Chawla and Rohit Chawla

15.1 Introduction

Fog computing facilitates the benefits of cloud computing by providing computing intelligence (in the form of virtualized resources), storage, and networking services to the edge of the network. This helps in decreasing latency (by reducing the need to communicate via cloud), uninterrupted services with intermittent connectivity, enhanced security, and support of massive machine communications. Thus, fog computing paradigm is a viable option for the development of IoT applications.

IoT is referred as an ubiquitous network of real-life physical devices (such as home appliances, medical equipment, vehicles, buildings, etc.) embedded with sensors, microchips, and software to gather and exchange information through an existing Internet connection. It is a way by which computing intelligence is directly integrated to the physical entities with a motive to enhance performance, efficiency, and financial benefits. A boom in the field of Internet of Things (IoT) in almost all vertices of the industry has motivated organizations to build IoT products to meet the market demands. As per IDC reports, global expenditure on IoT will be around $1.29 trillion by 2020 [1]. Technical report by Gartner on emerging technologies states that there will be 20.4 billion connected devices by 2020 [2]. As we expand the connectivity of the IoT, scope and capabilities of IoT systems are also increasing day by day that directly affect public safety and personal lives, like medical devices and systems and automotive safety; therefore, the consequences of a system breach or network failure are higher than ever before. However, high-velocity growth associated with rapid innovation anticipates the need of strong unique IoT testing (quality assurance) strategy to ensure the reliability of the IoT systems well before their release to the market.

Quality assurance is one of the most important phases of development to ensure the correctness and quality of developed software. Similarly, it is

Fog and Edge Computing: Principles and Paradigms, First Edition.
Edited by Rajkumar Buyya and Satish Narayana Srirama.

also crucial for IoT system, as poor design may hamper the working of the application and affects the end-user experience. The architecture of IoT is very complex, composed of heterogeneous hardware, communication module, huge volume, and variety of data, which plays a vital role in analyzing the performance and behavior of the IoT system. Functional and nonfunctional requirements (such as robustness, reliability, security, performance, etc.) of IoT systems can only be ensured if a variety of devices are tested for different kinds of operating systems (OSs), software, and hardware combinations.

The QA process for the IoT is required to perform verification and validation of the associated new technologies such as machine learning and data-mining with the aim of regularly improving existing and future systems. Moreover, the huge volume of data getting captured and sent through IoT devices to the backend makes the system prone to performance bottlenecks. This poses fresh challenges to development teams; thus, there is a dire need for comprehensive and advanced testing strategy to cover the breadth and depth of IoT systems.

This chapter starts with an explanation of the fundamental concepts of fog computing paradigm and associated benefits if adopted for the implementation of IoT applications.

Section 15.3 deliberates testing perspectives of the smart applications in the area of home, health, and transport. Testing approaches and solutions applied so far have been illustrated and compared based on their outcomes. Further, evaluation criteria relevant to the three smart technologies viz. smart home, smart health, and smart transport have been proposed to assess the existing work. Finally, Section 15.4 presents open issues and future research directions.

15.2 Background

With the emergence of IoT applications for which low latency and location awareness are of prime concern, fog computing comes into the picture. Fog computing is a conceptual model that extends compute, network, and storage services of cloud computing to the edge of the network. The paradigm of fog computing provides a decentralized architecture and extends the methodologies and characteristics of cloud computing (such as virtualization, multi-tenancy etc.) to the edge of the network. Applications such as gaming, video conferencing, geo-distributed applications (for, e.g., pipeline monitoring, sensor networks to monitor the environment), fast mobile applications (for, e.g. smart connected vehicle, connected rail), large-scale distributed control systems (for, e.g. smart grid, connected rail, smart traffic light systems), entertainment and advertising industry benefit on a large scale with fog computing paradigm due to improvement in quality of service (QoS) and reduction in latency. In addition, the fog model is well suited for data analytics and distributed data collection points by setting up end services such as setup boxes and access points. Thus, adoption of the fog computing model for the development of IoT application is very beneficial. Some of the benefits are listed below:

- **Freedom from cloud-based subscription services.** Fog computing model facilitates the developers to control, manage, and administer the IoT applications at the edge of the network without depending highly on Internet connectivity. Further, the decentralized architecture of fog computing enables the edge nodes to store the data locally for further analysis to make decision locally for IoT applications. Thus, in this way it reduces the dependency on cloud services and storage of data locally.
- **Reduction in congestion, cost, and latency.** Fog nodes process and analyze the data at a very fast rate as compared to the analytics done by a remote data center. Fog computing model prioritizes the data analytics tasks based on the time deadline requirements. The data of IoT application with real-time requirements is processed and analyzed, which results in lowering the latency and congestion of the networks. The processed data may be sent periodically to the main data center for further analytics, if required. In this way, it helps in the optimal utilization of the resources as well as bandwidth and therefore results in reduction of cost.
- **Enhanced security.** Fog computing paradigm helps in reducing the data that would be transferred over WAN by encouraging local processing of sensitive data of mission critical applications, which reduces the risks associated with data security while data are on move.
- **Fault-tolerance, reliability, and scalability.** A fog layer augments the redundancy of data processing capability in addition to the cloud nodes and thus helps in providing a high level of reliability. The large number of local nodes can also be utilized in the form of virtualized systems, which results in a significant rise of scalability. It also abolishes the core computing environment, thereby reducing a major block and a point of failure.

In view of the above benefits of the fog computing model, an IoT application that produces high volume and velocity of data requires an extensive and dense network of devices that can take advantage of the fog computing paradigm. Examples of such applications are listed below:

- Smart cities
- Smart buildings
- Smart transportation
- Smart energy
- Smart agriculture
- Smart lighting
- Smart health
- Smart power grids

- Oil refineries
- Meteorological systems

This chapter discusses testing perspectives of three case studies viz. smart home, smart health, and smart transport, along with their limitations and future research directions. The reason behind this selection is that these three applications can be considered as the main founding needs of society. Agriculture is also one of the most important fundamental needs of society, and making it smart with the adoption of high-end technology would greatly contribute to worldwide growth and prosperity. But due to time and space constraints, we will not describe this use of smart technology; it will be taken up in a future study.

15.3 Testing Perspectives

In the era of a smart technology enabled environment, devices must interact with other devices or even human beings with the purpose to share system configuration. This may hamper the working of the application and may affect the end-user experience. Hence, the software, being the soul of the smart system, must be reliable and robust, which can only be ensured by effectively testing the software. The testing perspectives and the approaches adopted by the industry and academia for various smart systems are presented in this section.

15.3.1 Smart Homes

NTS is one of the testing service providers that provides validation of home area network (HAN) devices such as smart meters, smart door locks, light controls, thermostats, and smoke sensors. It tests the interoperability of appliances and reflects the energy consumption by various devices and thus helps in an effective energy management [3, 4]. The testing tool also supports clients with self-testing by simulating the functionality of the appliances. NTS has been designated by ZigBee Alliance to test wireless products for smart energy, and ZigBee Smart Energy is nominated by the US Department of Energy and the National Institute of Standards and Technology (NIST) as an initial interoperable standard for HAN. NTS also works for iControl Platforms to test its security and home automation commodities such as smart door locks, light controls, thermostats, and smoke sensors.

Corporate major players in the field of mobile phones manufacturing (such as Apple and MI) also provide smart home applications that help in the attainment of security, effective energy management, and automated detection of smoke or gas through mobile phone applications. Security of smart homes is ensured by setting up security sensor systems for doors and windows.

Smoke or gas detectors can be turned on through mobile applications. In a similar way, smart light schedule and brightness can also be remotely controlled. Allion Smart Home provides testing and validation services, which support clients in the development, testing, and debugging of products for the three most important smart home environments – named as Cloud Service/Data Exchange," "UI/APP," and "End User Device" [5]. The lab established at Allion simulates a real home environment with three bedrooms, two living rooms, and two bathrooms including home items such as a sofa, TV cabinets, beds, desks, wardrobes, and so forth. Common appliances and electronics, such as a television, wireless speakers, computers (desktop and laptop), wireless LED lights, and so on have been installed in compartments with powerline wireless extenders, one-in-three wireless phones, and a microwave in the kitchen to introduce interference from other electrical products in the 2.4 GHz band. This is done to simulate behavior patterns and user habits in the real world [5].

eInfochips carries out performance testing for iOS and Android apps and redesigned the UI for Android and iOS platforms to improve the performance of home devices and to avoid inconsistency between iOS and Android platforms. Application response time is measured by using 24×7 performance evaluation tools and carries out bottleneck analysis to identify performance inefficiencies with the help of data flows and log files. Performance optimization techniques are implemented using cost–benefit analysis. Crash issues are resolved by utilizing detailed analysis and creating a crash log review. Code analysis is done with the help of SonarQube and XClarify tools [6].

UL has established the UL living lab in a 2500-square-foot fully furnished home situated near Silicon Valley campus and it thus enables testing of smart home devices in real-world user scenarios and provides various benefits such as ecosystem integration, large-scale interoperability, RF performance, and audio quality [7].

TUV is the third-party testing provider that tests smart home products to ensure privacy of the data as per the guidelines of data protection regulations. Various types of tests such as device default settings, local communication testing for encrypted data, interoperability testing etc. are carried out to test the effectiveness of privacy of user data. Smart home devices are tested to certify their functionality and mechanical and electrical safety by testing the products such as motion sensors and smoke alarms. In addition, usability tests are also carried out for the smart home devices [8].

Smart Home Test platform established at VDE Institute conducts tests to evaluate and certify smart home network devices for compliance, faultless functionality, user data protection and interoperability [9].

The National Renewable Energy Laboratory (NREL) has devised a smart home test bed to simulate power distribution grid for industry, manufacturers, universities, and other government organizations. The NREL test bed includes the combination of powered hardware and software simulations.

The smart home hardware comprises electric vehicle supply equipment (EVSE), home loads, a water heater, a thermostat, and an air conditioner, all powered (via red lines) by a photovoltaic inverter and an alternating current (AC) power amplifier, which emulates grid power. A high-performance computer (HPC), Peregrine, has been utilized to execute advanced home energy management system (HEMS) optimization algorithms that simulates power distribution feeder, also uses weather and price data to determine control signals sent to simulated homes and to the smart home's hardware via the HEMS. The key component of a smart home test bed is a co-simulation tool, integrated energy system model (IESM) that is responsible for managing the power system and home simulations, the HEMS algorithms, communications with the HEMS hardware, and a simulation of the smart home (using EnergyPlus) that runs on the hardware-in-the-loop (HIL) control computer in the laboratory. The IESM also provides price signals as inputs to the HEMS, allowing users to evaluate how smart home technologies respond to different retail price structures [10]. Zipperer et al. [11] have also worked in this direction and developed a mechanism for electric energy management in the smart home. Cordopatri et al. [12] established test lab in the campus of the University of Calabria to experiment with various management systems for smart homes such as energy flow and comfort management systems. The main objective of the energy and comfort management system (ECMS) developed at the University of Calabria is to attain reduction in the cost and usage of energy along with improved comfort and safety of the smart home systems. Several authors have proposed similar kind of frameworks based on fuzzy logic, neural networks, and genetic algorithms [13–16]. Hu et al. [17] developed an open and smart home test bed named as SHEMS that can be used for educational purposes. The summary of these products is depicted in Table 15.1.

15.3.2 Smart Health

The main objective of the healthcare industry is to provide patients with quality healing services round the clock in a cost-effective manner. The software industry enables the smooth functioning of the healthcare industry by providing software applications that assist in the functioning of various hospital operations and at the same also maintains the privacy of patients. Hence, crashing of an application would severely impact healthcare process and may also adversely affect the health of the patient. Therefore, testing of healthcare software is very essential as it ensures the quality and productivity of a healthcare service. The healthcare industry needs to follow strict regulatory and compliance norms, and it is bound to identify novel revenue generation strategies and to effectively utilize R&D budgets. This raises the need of software professionals to have thorough understanding of domain and industry regulations and

Table 15.1 Outline of the work done to test smart homes.

Authors/Company	Objective	Approach	Outcome
National Technical Systems (NTS)[2, 3]	ZigBee Smart Energy Certification Testing for SimpleHomeNet Appliance	• Test tools are designed to simulate the functions of the appliances to facilitate clients to carry out self-testing. • The NTS testing validates that various appliances of home network such as thermostats, meters, load controllers, pool pumps, water heaters, and display units etc. work together properly and can precisely demonstrate amount of energy is being used which helps customer to manage energy proficiently.	Smart energy device testing; increase reliability and cut costs for consumers
Allion Smart Home Testing Services [5]	Hardware development support, software apps validation and user experience optimization, cloud service validation, RF signal and interference validation and interoperability testing.	Allion carries functional testing and ensures that the products meet the specification and verification standards of the certification process; The lab established at Allion simulates a real home environment that includes simulation of users' habits and behavior patterns; Carries out testing for different products and test scenarios.	Certifies all 18 Wi-Fi certification services
eInfochops [6]	Performance testing; reliability and usability testing	SonarQube and XClarify tools are used for code analysis; Performance of the app is determined by using gap analysis between technical requirements and actual expected performance of the mobile app. Performance inefficiencies are resolved using bottleneck analysis. Mobile performance optimization techniques are realized using cost–benefit analysis. Resolution of crashes is done using detailed analysis.	Mobile app performance optimization Mobile UI redesign Code review and performance testing expertise Better app reliability

(Continued)

Table 15.1 (Continued)

Authors/Company	Objective	Approach	Outcome
TUV Smart Home Testing and Certification [8]	Security, protected privacy and testing for user friendliness	• Mechanical and electrical safety of products such as motion sensors and smoke alarms is tested thoroughly to ensure for their functionality. • Protected privacy test includes verification and validation of devices, encryption of data and IP protocol as well as local and online communication, privacy settings of mobile apps, legal requirements and expectations of the associated documents, terms and conditions of data usage; product testing; interoperability testing.	Certification named as Certipedia and Greater Transparency
UL Living Lab[7]	Interoperability testing	2500-square-foot fully furnished home to test products in a real home and in a real neighborhood	Testing real-world user scenarios: out of the box experience; Physical installation; ecosystem integration; large-scale interoperability; audio quality and RF performance
VDE Smart Home Test Platform [9]	Interoperability, information security, functional safety, and data protection	• Testing of devices such as communications devices and gateways • Back-end and cloud systems, and apps for smart phones and tablets • User documentation testing • Data protection	Conformity assessment; Certification Program" funded by Federal Ministry of Economics and Technology (BMWi)
NREL Smart Home TestBed [10]	Energy-efficiency testing	• Home energy management system (HEMS) optimization algorithms • Integrated energy system model (IESM) • Hardware-in-the loop (HIL) technology • GridLAB-D software	Controllable, flexible, and fully integrated smart home test bed

Zipperer et al. [11]	Electric energy management	• Utility-side enabling technologies • Customer-side enabling technologies	• Increase in energy efficiency • Decrease in cost of energy use • Decrease in the carbon footprint
A. Cordopatri et al.[12]	Energy and comfort management system (ECMS)	• Communication management with the peripheral devices of the system (switching box, smart plugs, etc.) through power line and/or wireless technologies and with the users through dedicated web-based and mobile graphical interface apps • Collection, interpretation, storage and elaboration in real time of all data concerning machine-to-machine and machine to-human interactions (e.g., monitoring data, users' requests, etc.) for statistical and training purposes • Prediction of home energy consumption on the basis of the stored historical data • Sending of control signals to peripheral devices in order to execute energy-control actions on the basis of a defined set of decision algorithms and interoperability rules, also by taking into account the performed predictions, as well as to specific users' requests	• Reduction in the cost and usage of energy • improvement in comfort and safety of the smart home systems
I. Dounis et al.[13]	Multi-agent control system (MACS)	TRNSYS/MATLAB	• Manage the user's preferences for thermal. • illuminance comfort, indoor air quality • energy conservation

(Continued)

Table 15.1 (Continued)

Authors/Company	Objective	Approach	Outcome
R. Baos et al. [14]	Review of the current state of the art in computational optimization methods applied to renewable and sustainable energy		Well-defined visualization of the modern research advancements
J-J. Wang et al. [15]	Review of multi-criteria decision analysis (MCDA) methods	Energy decision-making computed by the combination of weighted sum, priority setting, outranking, and fuzzy set methodology	Identification of MCDA method, and the aggregation methods for sustainable energy decision-making
T. Teich et al. [16]	Energy-efficient smart home	Neural networks	Energy saving
Q. Hu et al.[17]	Open and extensible model for energy conservation based on smart grids	Machine learning and pattern recognition algorithms	Smart home test bed named as SHEMS developed that can be used for educational purposes

standards. Significant work done in this direction is explained below and is portrayed in Table 15.2.

Virtusa has established dedicated center of excellence that provides healthcare domain testing, user acceptance testing (UAT) optimization, ICD-10 testing, and enterprise end-to-end testing [18]. Mindfiresolutions provides a manual as well as automated healthcare application testing services by using various tools such as QTP, Selenium, Appium, and Robotium over several platforms. The testing services offered are: conformance testing, interoperability testing, functional testing, security testing, platform testing, load and performance testing, system integration and interface testing, and enterprise workflow testing [19]. The healthcare testing services provided by QAInfotech include functional testing, database testing, performance testing, content QA testing and development and implementation of QA and test strategies. In addition, testing professionals also take care of HIPAA guidelines and carries out performance and security tests [20]. Cloud lab established by ALTEN Calsoft Labs' provides healthcare domain testing in the area of clinical systems, nonclinical systems, and specialized testing services. Clinical systems include EHR/EMR, hospital ERP, radiology information systems, imaging systems, and compliance-related standards and guidelines such as HIPAA. Nonclinical system contains the modules of pharmacy, billing, and revenue cycle management. Specialized testing services comprise compatibility and localization, security testing, performance testing, legacy modernization and testing, mobile healthcare, BI/analytics, and cloud migration and testing [21]. Precise Testing Solution delivers healthcare application testing in the domain of electronic medical records, patient survey solutions, quality and compliance solutions, enterprise content management, medical equipment software solution and compliance testing services [22].

ZenQ helps healthcare organizations in attaining quality, efficiency, and cost-effectiveness by providing specialized healthcare testing solutions in the area of electronic health records (EHRs) electronic medical records (EMRs), hospital management systems, healthcare data interoperability and messaging standards conformation, and mobile health. Testing services include functional/regression testing, usability testing, interoperability testing, mobile apps testing, conformance/certification testing, performance testing and security testing [23]. Testree offers a complete package of quality assurance and healthcare application testing that includes certification for automatic compliance of various standards, appropriate administration, and control of policy claims and benefits, patient and disease management, billing and reporting, etc. [24]. The healthcare testing services offered by KiwiQA encompass compliance conformance testing, product consistency testing, platform testing, and security testing [25].

Table 15.2 Outline of the work done to test smart health.

Authors/Company	Objective	Approach	Outcome
Virtusa COE [18]	Healthcare domain testing, user acceptance testing (UAT) optimization, ICD-10 testing, and enterprise end-to-end testing	Business process management, customer experience management, enterprise information management, cloud, mobility, SAP	• Transformation of business by optimizing operations • Efficiency • Expansion of target audiences • Distinctive millennial and consumer engaging experience.
MindfireSolution [19]	Conformance testing, interoperability testing, functional testing, security testing, platform testing, load and performance testing, system integration and interface testing and enterprise workflow testing	QTP, Selenium, Appium, and Robotium over several platforms	• Effective automation strategies to reduce manual effort • Production time cost • Out-of-box QA frameworks to ensure high-quality timely delivery
QA Infotech [20]	Functional testing, database testing, performance testing, content QA testing and evelopment and implementation of QA and test strategies, performance and security tests	• HIPAA guidelines followed religiously • Close interaction with functional managers to identify critical workflows for nonfunctional testing types such as performance and security tests • Trained tester	Assurance of Security, privacy and mandated compliances in healthcare application tested by QAInfotech
ALTEN Calsoft Labs' [21]	• Healthcare domain testing in the area of clinical systems, non-clinical systems and specialized testing services • Compatibility and localization, security testing, performance testing, legacy modernization and testing, mobile healthcare, BI/analytics and cloud migration and testing	• Test consulting • Testing COE • Specialized testing • Compliance	• Rapid testing framework • Improved test coverage • Reduced cycle time • Zero bugs in production

Precise Testing Solution [22]	Healthcare application testing in the domain of electronic medical records, patient survey solutions, quality and compliance solutions, enterprise content management, medical equipment software solution and compliance testing services	JMeter for load testing, ZAP proxy	Bugfree software
ZenQ [23]	Functional/regression testing, usability testing, interoperability testing, mobile apps testing, conformance/certification testing, performance testing and security testing	• Adherence to healthcare data privacy laws/regulations such as HIPAA • Dedicated in-house healthcare domain knowledge specialists	Assurance of quality, patient-centric care, high efficiency and cost-effectiveness • Minimizing errors and redundancy • Smooth transition toward preventive care
Testree [24]	Functional testing, integration testing, interoperability testing, security testing, device compatibility testing, selection of manual or automation testing methods, performance testing like load testing and scalability and compliance testing	• Health information managements systems (HIMS) • Practice and patient care • Clinical decision support system (CDSS) • Compliance solutions • Clinical IVRs systems • Personal health record and e-prescribing • Policy management • Claims management • Benefits management • Business intelligence	Comprehensive quality assurance • Effective management of policies, payments, claims and benefits • Assurance of procedural efficiencies against fraudulent claims • Seamless integration of component systems • proper automation of updates and standards compliance.
KiwiQA [25]	Compliance conformance testing, product consistency testing, platform testing and security testing	Test approach is • Analytical • Model based • Dynamic • Methodical • Directed • Regression-averse • Standard compliant	• Removes the potential threat of the software • Assured freedom from all kinds of vulnerability issues.

(Continued)

Table 15.2 (Continued)

Authors/Company	Objective	Approach	Outcome
XBOSoft [26]	• Compliant working of electronic health records (EHR), • Automated drug dispensing machines • Pharmacy management • EMAR • EPCS with mobile apps	• Careful design of test cases that ensures test coverage • Cross-platform • Multidevice • Multibrowser compatibility	• Increased efficiency and productivity • Accuracy and security of information • Improved patient relationships through business knowledge and enhanced patient experience • Accurate implementation of business rules requiring ZERO tolerance for error
Infoicon Technologies [27]	Interoperability testing, functional testing, security testing, load and performance testing, system integration testing and acceptance testing	• Multiple platforms testing • Manual as well as automation approach for testing.	• Cost-effective services. • Sustains high-quality standards • ensure compliance with healthcare industry standards and regulatory frameworks
W3Softech [28]	Testing and QA services for healthcare and pharmaceuticals industry such as claims management testing, clinical decision support system (CDSS), healthcare billing software testing Personal health record and e-prescribing, implanted application testing QA in clinical data management systems CRO workflow management system, testing support for regulatory requirements	• Agile-based healthcare and pharmaceutical testing services • Lifecycle phase-independent testing activities	• Assured excellence • Robust QA services • Amplify efficiency • Boost business efficiency

Prova [29]	Manual testing, PLM testing, and automation testing	Automation testing • Selenium Webdriver, Performance Testing • PHP and JMeter Mobile Testing • Silk Mobile	Better quality products and services • improved test coverage • Error-free software applications.
Calpion [30]	• Requirement analysis • Functional testing of healthcare workflows • Compliance testing • Interoperability testing • Mobile platform testing • Load and performance testing	HP quality center (QC), Quick Test Professional (QTP) and HP ALM • Testing solution to provide true hybrid framework and data-driven testing – Accelerated manual test executions and defect reporting using HP QC. – Batch mode of execution of test suites during different test phase – Pre-built test cases from our healthcare test case repository shorten testing cycles – Automate new processes or update existing test cases faster	• Improved quality • lower cost leveraging • re-usability and automation • Global delivery model
Abstracta [31]	Automated functional testing, security testing and performance testing services	– Continuous testing – Automation framework – Selenium or Appium • Performance testing • JMeter • Mobile test automation – Monkop	– Comply with regulations and adhere to standards (ex: Sarbanes Oxley, HIPAA, etc) • Minimize risks related to security, data accuracy, patient safety, etc. • Save time and money by nearshoring

(Continued)

Table 15.2 (Continued)

Authors/Company	Objective	Approach	Outcome
360logica labs [32]	• Healthcare billing software testing • R&D software testing • Embedded application testing • Testing and QA services for pharmaceutical and healthcare industry	• Use of open source tools assure better scalability, resource optimization, and interoperability • Testing team comprising of skilled and in-house experts	• Minimum resource wastage and maximum business optimization guaranteed • Healthcare software industry testing with focus on compatibility, reliability, security, and completeness • Ready-to-use and reusable • Reduced software testing cost • Assured on-time delivery and high quality
Renate Löffler et al. [35]	Model-based test-case generation strategy	UML 2.0	Developed model-based approach for the specification of requirements followed by integration testing for healthcare applications
Bastien et al. [36]	User-based evaluation	KALDI, Morae, Noldus	Identification of open issues in usability testing
R. Snelick [33]	Conformance testing	NIST HL7 v2 conformance test tools	Certification of EHR technologies
P. Scott et al. [34]	Conformance testing	Schematron, mind-mapping	Developed an openEHR archetype model for creating HL7 and IHE implementation artifacts

XBOSoft makes the provision of testing services in the domain of healthcare and ensures the compliant working of electronic health records (EHR), automated drug dispensing machines, pharmacy management, EMAR, and EPCS with mobile apps. This is done by careful design of test cases that ensures test coverage, cross-platform, multidevice, and multibrowser compatibility [26]. The lab setup at Infoicon Technologies Pvt. Ltd. dedicatedly provides the cost-effective healthcare testing services covering the domain of pharmaceutical industry, clinical systems, healthcare startups, body fitness, dental care, physiotherapy, doctor consultation, and homeopathy. It provides multiple platforms for manual as well as automated testing services that include interoperability testing, functional testing, security testing, load and performance testing, system integration testing, and acceptance testing [27]. W3Softech offers agile-based healthcare and pharmaceutical testing services [28].

In the similar way, Prova also provides cost-effective software testing and QA services for the healthcare industry [29]. Calpion's offers convenient and fast-testing framework that works for both web and mobile healthcare application by utilizing HP quality center (QC), quick test professional (QTP) and HP ALM [30]. Abstracta provides healthcare testing system for patient portals, medical imaging, and electronic health records (EHR) while adhering to the standards and regulations. It provides automated functional testing, security testing, and performance testing services [31]. The 360logica labs offers cost-effective, reliable, and standard compliant healthcare software testing services. The testing services are in the area of hospitals, pharmaceutical and clinical labs, which include healthcare billing software testing, R&D software testing, and embedded application testing [32].

Löffler et al. [35] devised a model-based test-case generation strategy from use case scenarios described with their newly introduced formal specification language by extending UML2.0 sequence diagrams. Test models have been derived from specifications, which are then used to generate test cases corresponding to each and every flow in the test model. J.M.C. Bastien et al. [36] carried out user-based evaluation for healthcare applications to assess the usability of the application by employing single user and paired-user testing. In this approach, users are asked to carry out certain tasks, and performance of the users is noted such as task completion rate, types of error accorded, etc. to recognize certain design flaws that causes user errors. Based on these observations, design changes can be suggested to front-end designers. Snelick [33] investigated conformance testing and the tools that are used to perform HL7 (Health Level Seven) v2-based conformance testing for certification of EHR technologies. Scott et al. [34] demonstrated the development of conformance methods based on the professional standards. Table 15.2 summarizes work done in smart health.

15.3.3 Smart Transport

The researchers at UMTRI carry out development and testing of intelligent transportation systems off-the-road to prevent collisions in passenger vehicles. Exhaustive study is carried out in the direction of automotive collision avoidance, in-vehicle driver-assistance and safety systems, and integrated technologies between the vehicle and the infrastructure [38]. Connected Vehicle Test Bed has been established in Michigan, Virginia, Florida, California, New York, and Arizona to facilitate a real environment where intersections, roadways, and vehicles are able to communicate through wireless connectivity by the US Department of Transportation (USDOT), and it comprises of a network of 50 roadside equipment (RSE) units installed along various segments of live interstate roadways, arterials, and signalized and unsignalized intersections, in Novi, Michigan. These RSEs communicate messages over 5.9 Ghz dedicated short-range communication (DSRC). This test bed provisions testing of new hardware and software for the evolution in connected vehicle technology. Various types of tests (such as signal phase and timing (SPaT) communications; security system operations; and other connected vehicle applications, concepts, and equipment) can be successfully carried out for free. In addition, there is a provision of experts to carry out complex scenario tests. Also, there is no need to make any testing arrangements because of prior contracts between the local agencies and roadway operators. Test beds frequently undergo upgrades and enrichments to provision the changing requirements of users. Clients of Connected Vehicle Test Bed include Denso, Delphi, Hirschmann, Eaton, Argenia, Wayne State University, MET Labs, Ricardo, and University of North Texas [39].

The test lab instituted at IBS provides end-to-end software testing services to travel, transportation and logistics enterprises. It provides four types of testing services which includes Enterprise QA Automation Services, Product Acceptance Test Services, Managed Testing Services and NFR Testing services. Enterprise QA Automation Services provides automation to support DevOps environment, process automation to validate build to release quality, reusable frameworks for TTL customers and transformational models to support guaranteed outcome, Product Acceptance Test Services involves system Integration, final acceptance and UAT support, domain experts to validate business requirements, reusable assets for TDM (Test Data Management), automation, and performance and multivendor management for airlines' IT solutions testing. Managed Testing Services comprises consulting services for outsourcing, transition from incumbent vendors/captive organization, end-to-end testing from functional to acceptance test, and assured output/outcome model for delivery. NFR Testing services consist of performance benchmarking and capacity planning, SMAC, usability, security, performance covered, projects supported with dedicated lab facility and compliance, industry standards and frameworks in mobility and multitenancy/cloud [40].

ETSI worked in collaboration with Telecom Italia, ERTICO, the regional government, local highway authorities and port authority to launch the ITS test bed in Livorno. The test bed contains traffic lights, IoT sensors, cameras, variable message signs, and connectivity with a highway control center. RSUs and on-board units within vehicles can be tested effectively by deploying it in the road sideways. Other ITS testing activities such as traffic sign violation, road hazards, intersections and collision warnings, and loading zones can also be carried out successfully [41].

Woo et al. [42] have designed a test bed to handle testing on various ITS and advanced driver assistance system (ADAS) technologies, such as adaptive cruise control (ACC), lane departure warning system (LDWS), cooperative intersection warning system, as well as rollover stability control (RSC) and electronic stability control (ESC). The test bed has been devised to meet the requirements of ISO/TC204 standards. The test bed for ITS encompass three tracks named as ITS high-speed track, Cooperative vehicle-infra test intersections, and Special test track. The main purpose of ITS high-speed track is to test performance of ACC, LDWS, LKAS, etc. It has three lanes of high-speed track of length equal to 1,360 m with maximum allowable speed of 204 km/h. The total length of Cooperative vehicle-infra test intersections is 1,200 m and there are three intersections. The main objective is to test pedestrian protection and intersection safety. Special test track comprises of four lanes of test road with the total area of 490×35 m. It includes Belgian road, washboard road, cobblestone road, water splash shower tunnel, for example. Durability and reliability test are carried out it these tracks.

The government of Estonia plans to restructure its public transportation system by adopting autonomous vehicles and thus legalized testing of autonomous vehicles on national and local roads of the country. Rigorous efforts are put into developing a cyber-risk management framework for autonomous vehicles in regular road and traffic conditions. The government has planned to create a fleet management system and integration of vehicles into the public transport system and the implementation of call-to-order bus stops [44].

Transit Windsor provides the development as well as testing services for intelligent transportation systems. The company has produced 10 buses furnished with an efficient, safer, and more user-friendly system. It provides onboard voice and visual announcements on the display boards for the upcoming bus stop messages. It also stipulates real-time Transit Windsor bus arrival information as well as route for the bus progress via the Internet [45].

Siphen has achieved an intelligent transportation system (ITS) product compliance with UBS II and ARAI testing. It is known for its rigorous testing procedure. It is working with the government of India to equip the country with ITS by providing 24×7 bus operation service with the features such as automatic vehicle location, vehicle health monitoring, and diagnostics. In addition, it is carrying out end-to-end testing as well as a certification process according to the timelines given by government authorities [46].

Anritsu provides ITS solutions for V2X, testing, and manufacturing in a very efficient manner with reduced test time and test cycles. Testing solutions are provided with the help of four components: MD8475A Signalling Tester, MS2830A Spectrum Analyzer, MS269xA series, and V2X 802.11p Message Evaluation Software. MD8475A Signalling Tester is similar in that it supports cellular as well as M2M standards. The services supported are eCall, IMS, VoLTE, WLAN off-load tests, and call-processing tests for vehicles. The testing tasks are easy, fast, and reliable due to GUI-based SmartStudio software and supplied test sequences for automatic remote control of the GUI. Multimode terminals and all cellular standards, such as LTE (2×2 MIMO) and LTE-Advanced (Carrier Aggregation) are well supported. SmartStudio GUI provisions easy setup of test environments and functional tests. It also carries out automated mobile terminal verification testing with the available test sequences. MS2830A Spectrum Analyzer is used for testing of 2G, 3G, LTE, and LTE-Advanced signals on a vehicle-to-vehicle or vehicle-to-x test environment. To improve the product quality, capture and replay functions are compared with the real-world effects with simulated designs and performance. The supported frequency range is 9 kHz to 26.5 GHz/43. MS269xA series units contain swept spectrum analysis, FFT signal analysis, and a precision digitizer function and are the latest high-performance signal analyzers for next-generation communication applications. It has One-Box Tester with the addition of the signal generator option. Due to the support of batch capture measurements, analysis time gets faster [47].

Penta Security Systems has launched secured smart transportation with the secure data solution AutoCrypt, implemented on the connected vehicles in the three cities of South Korea. It has also established the second-largest test bed named as K-City to test and certify autonomous cars. Public key infrastructure and V2X security system has been implemented to ensure secure and encrypted communication between vehicle-to-vehicle and vehicle-to-infrastructure, as well as the security and encryption of roadside units [43].

Simulation-based test bed has been developed at Georgia Institute of Technology by the School of Civil and Environmental Engineering, which can be used for fast assessment and incorporation of sensor and actuator systems in ITS. The test can also be used to study and examine various data networks architectural possibilities to support ITS applications. The test bed supports integrated parallel simulation ability and also involves interoperable simulations of transportation infrastructures, wired and wireless communication networks, and distributed computing applications. In addition, it possesses emulation ability that allows conducting live experiments with prototype hardware and software embedded into virtual transportation systems. The test bed incorporates data generated from sensors embedded in the vehicles (such as location, velocity, and acceleration etc.) functioning in the Atlanta metropolitan area. This data are also used for modeling and scenario development as well as validation of simulations [37]. The above stated work is summarized in Table 15.3.

15.4 Future Research Directions

This section discusses the open issues and research directions from the perspectives of testing and future enhancements for smart technologies viz. smart home, smart health, and smart transport. Certain evaluation criteria have also been presented for the assessment of existing work and to ascertain limitations and research directions.

15.4.1 Smart Homes

We have proposed the following set of criteria to evaluate the existing work in smart home test beds. The pertinence of these criteria is described in this section:

- **Energy efficiency testing.** The testing is used to verify reduction in energy consumption in the smart homes.
- **Reliability testing.** It ensures the stability of the system under various specific tests, which include stress testing, network testing, along with functional testing.
- **Functionality testing.** It is required for verification of each function of the software application in conformance with the requirement specification. It encompasses all the scenarios related to failure paths and boundary cases.
- **Interoperability testing.** Interoperability determines how devices communicate with each other and, upon receiving information, how processing is done and corresponding actions are generated. If a device can't receive information, process it, and act upon that information, it won't function as consumers hope. Without full functionality, the product may not provide value. Real-world test labs are the best way to solve interoperability issues, as they depict actual scenarios of the problem.
- **Performance testing.** It is required to ensure software applications will perform well under their expected workload. It determines responsiveness and stability of a system under various workloads and measures the quality attributes of the system, such as scalability, reliability, and resource usage.
- **Usability testing.** Usability testing measures the convenient level for learning of the system by end users, which include parameters such as level of skill required to understand the system, time requirement to attain familiarity, and user's productivity.

Table 15.3 Outline of the work done to test smart transport.

Authors/Company	Objective	Approach	Outcome
UMTRI[38]	• Development of vehicle-based technologies to avoid road accidents • In-vehicle driver-assistance • *Safety systems*	• Collision avoidance algorithm • Integrated technologies between the vehicle and the infrastructure	Vehicle safety
US DOT Connected Vehicle Test Bed [39]	• To test devices such as vehicle awareness devices (VADs), aftermarket safety devices (ASDs), in-vehicle safety devices (ISDs), radios and roadside equipment (RSEs) • Development and testing of DSRC standards • Establishment of connected vehicle security certificate credential management • Development and testing of applications using SPaT and Geometric Intersection Description (GID) data	The Test Bed operates as per the guidelines of latest IEEE 1609/802 and SAE J2735 standards – Support regular updates. – Implements latest new security features as well as latest hardware and software applications	• Systems can be tested for ability to receive and process SPaT data in a real-world environment • Increase in confidence for system before launching in real roads due to Security Certificate Management System (SCMS) or by using the SCMS emulator • Reduction in cost for testing and validation of the system due to infrastructure provided by test bed • More decentralized, simplified, and open structure • Dynamic and evolving environment
IBS Lab [40]	– End-to-end software testing • Provides four types of testing services, which includes Enterprise QA Automation Services, Product Acceptance Test Services, Managed Testing Services, and NFR Testing Services	• Requirements development • Test planning and Execution • Project coordination • Discrepancy resolution results reporting	Emphasis on the quality deliverables • Continuous improvement in the efficiency and efficacy of testing mechanism • Incorporation of new and innovative methodologies and practices

ETSI Test Bed [41]	Testing activities such as traffic sign violation, road hazards, intersections and collision warnings, and loading zones	• The infrastructure of test bed comprises of traffic lights, IoT sensors, cameras, variable message signs and connectivity with a highway control center • IoT test bed for large-scale distributed sensing and actuation.	Compliance with ETSI's ITS Release 1 standard and interoperability with radio equipment
JW Woo et al. [42]	• Performance testing of adaptive cruise control (ACC), lane departure warning system (LDWS), rollover stability control (RSC), and electronic stability control (ESC) • Testing of pedestrian protection and intersection safety • Durability and reliability testing	• The test bed for ITS encompass three tracks named as ITS high-speed track, Cooperative vehicle-infra test intersections, and Special test track • Simulators: KATECH Advanced Automotive Simulator, CarSim on dSPACE system, 3D virtual test track	As per the requirements of ISO/TC204 standards
E-Estonia [44]	To restructure the public transportation system using autonomous vehicles	• Testing of autonomous vehicles on national and local roads of the country • Cyber-risk management framework for autonomous vehicles	Provision of legal and cyber-risk management framework for testing fully autonomous vehicles in regular road and traffic conditions
Transit Windsor Testing Solutions [45]	• To improve the functionality of transportation services	Vocal announcements are in synchronization with the messages displayed on the display signs inside the bus.	• Cost-effective, secure and user-friendly system. • Launched 10 buses equipped with a system that provides automated stop announcements as well as preboarding external audible announcements to commuters waiting at bus stops

(Continued)

Table 15.3 (Continued)

Authors/Company	Objective	Approach	Outcome
Siphen [46]	• Testing as per the rigorous compliances of UBS II and ARAI testing • To provide end-to-end testing as well as to certify the system's process	• Integration of updated circuit board in accordance with the higher technical specifications • New devices manufactured that are compatible with the Indian transportation infrastructure and operating conditions	• Accomplishment of testing and certification process well in time as per the deadlines fixed the government authorities • Customized solution for 24×7 functioning in Indian operating conditions. • Reduction in response time to emergencies • Automatic vehicle location • Automatic vehicle health monitoring and diagnostics • Assurance of high-quality standards and implementation of the latest technologies
Anritsu Test Bed [47]	Functional testing; mobile terminal verification testing; testing of 2G, 3G, LTE, and LTE-advanced signals on a vehicle-to-vehicle or vehicle-to-x test environment	Four components:MD8475A Signalling Tester, MS2830A Spectrum Analyzer, MS269xA series and V2X 802.11p Message Evaluation Software; GUI-based SmartStudio software	Helped in making testing of ITS systems convenient, reliable, and efficient

Penta Security Systems K-City Testbed [43]	To carry out testing and certification of autonomous cars	AutoCrypt; Public key infrastructure and V2X security system	Reliable and secure system of ITS
Georgia Institute of Technology [37]	Prompt assessment and assimilation of sensor and actuator systems in ITS	• Transportation infrastructures, wired and wireless communication networks, and distributed computing applications are interoperable simulated • Virtual transportation systems are embedded with prototype hardware and software in order to carry out live experiments • Model, scenario developments, and validation of simulations are facilitated by using the live data received from road sensors in the Atlanta area	Framework can be used for investigation and assessment of new mechanisms under virtual operating conditions before actual deployment in real environment of intelligent transportation systems (ITS)

- **Security testing.** Security testing is a testing technique to determine whether the application or the product is secured. It aims at verifying basic principles such as confidentiality, integrity, authentication, authorization, availability and nonrepudiation.

Based on the evaluation criteria already described, limitations and research directions as well as suggestions for the future work have been elucidated here and portrayed in Table 15.4. The first hindrance in acceptance of the smart home technology is that smart homes are vulnerable to hacking. Hence, a test bed should be established that takes into account cyber-security measures to protect the smart home. The second hindrance is the high cost; measures should be taken to develop a technology that can be made available to users at a lower cost. Combinatorial testing strategy can be used to ensure the low price suggested by the pricing model that supports the pooling of distributed, dispersed resources in fog computing and the IoT. The third hindrance is the learning curve for non-tech-savvy people with smart home. Hence, usability testing should also be given topmost priority. Another most important factor that hinders the acceptance is lack of industry standardization, as use of proprietary technology can cause problems smart home users. Hence, conformance testing should also be given priority. Dependency on Internet connection should also be tackled, and reliability testing methodology specifically designed to suit the environment of the smart system need to be addressed.

15.4.2 Smart Health

To assess the existing smart health test beds, the following set of criteria have been suggested:

- **Conformance testing.** Conformance testing is performed to ensure adherence to the standards such as Sarbanes-Oxley, HIPAA, FDA etc.
- **Platform testing.** It ensures applications executes well across different platforms, which includes operating systems, different browsers, and multiple devices.
- **Interoperability testing.** Interoperability testing assesses whether connected devices and EHR systems communicate with one another effectively and correctly. It also ensures seamless operations between HL7 and DICOM transactions.
- **Functionality testing.** This is required for verification of each function of the software application in conformance with the requirement specification. It encompasses all the scenarios related to failure paths and boundary cases.
- **Enterprise workflow testing.** It checks whether the expected activities are executed and workflow data properties have correct values.
- **Performance testing.** It is required to ensure software applications will perform well under their expected workload. It determines responsiveness

Table 15.4 Summary of limitations and research directions for smart home.

Criteria	Research direction	Work	Limitations	Suggestions
Energy efficiency testing	Verify reduction in energy consumption in the smart homes.	[3, 10, 11, 12, 13, 16, 17]	1. High cost 2. Reliability 3. Security and privacy 4. User-friendliness 5. Lack of standardization 6. Dependency on Internet connection 7. Vulnerable to hacking 8. Learning curve	Test beds are required to be established for the following objectives: 1. Explore cyber-security mechanisms to protect smart home. 2. Availability of smart home technology at lower price. 3. Usability testing should be practiced. 4. Conformance testing. 5. Reliability testing strategies need to develop specifically for smart home.
Reliability testing	Ensure the stability of the system under various specific tests.	[3]		
Functionality testing	Verify each function of the software application in conformance with the requirement specification.	[3, 5, 6, 8–10]		
Interoperability testing	Ensure interoperability among devices.	[3, 5–10]		
Performance testing	Ensure software applications will perform well under their expected workload.	[5, 6]		
Usability testing	Evaluate a product or service by testing it with representative users.	[6, 8]		
Security testing	Check whether the application or the product is secured.	[8, 9]		

and stability of a system under various workloads and measures the quality attributes of the system, such as scalability, reliability, and resource usage.

- **Usability testing.** Usability testing measures the convenient level for learning of the system by end users, which include parameters such as level of skill required to understand the system, time requirement to attain familiarity, and user's productivity.
- **Security testing.** Security testing is a testing technique to determine whether the application or the product is secured. It aims at verifying basic principles such as confidentiality, integrity, authentication, authorization, availability and nonrepudiation.
- **Mobile app testing.** Mobile application testing is a procedure by which application software developed for handheld mobile devices is tested for its functionality, usability, and consistency.

Evaluation criteria described above helped in deducing the limitations and research directions in the existing smart health testing solutions. The same have been illustrated in this section and depicted in Table 15.5, along with research suggestions for the future work.

It has been implied that there is a lack of effective methodology that provides a systematic way to manage the data collected from various wearable devices. To combat this challenge, big-data, machine learning, and AI can be used. To ensure the attainment of the mentioned functionality, a test bed is required that executes blockchain-based repeatable tests with massive data received from wearable devices such as smart watches, eyeglass displays, and electroluminescent clothing, for example.

Further, it has been found that despite the various benefits of smart healthcare, it is not well adopted and market growth is restrained. It may be due to the high cost of IoT infrastructure and data privacy and security apprehensions. This can be resolved by building confidence among various stakeholders, which can be brought into practice by executing security tests specifically designed to examine the cybersecurity measures taken up to address the above-mentioned issue.

Another challenge is the management of connected devices and a lack of interoperability with EHR systems. This can be ensured by executing context-aware testing techniques. Thus, context-aware test case generation methodologies need to be worked out for smart health systems. To address the limitations associated with smart glasses (i.e. short battery life and inability to understand the medical terms of doctors by voice-control system), context-aware test data generation must be applied to ensure that the system will work. Blockchain technology should be reconnoitered to solve the problems of large-scale data sharing, ensuring data privacy and security and transparency between patient and doctors and between various healthcare providers. In this case, blockchain-based repeatable regression tests can be

Table 15.5 Summary of limitations and research directions for smart health.

Criteria	Research directions	Work	Limitations	Suggestions
Conformance testing	Ensure adherence to the standards.	[18, 19, 20, 21, 22, 23, 24, 25, 26, 27, 30, 31, 32, 34, 35]	1. No systematic way to manage the data collected from various wearable devices. 2. Smart healthcare is not well adopted and market growth is restrained. 3. Lack of interoperability of connected devices with EHR systems.	Test beds need to be developed to carry out research in the following areas: 1. Exploration of big-data, machine learning, and AI to manage and utilize huge amount of data received from wearable devices 2. To address the limitations associated with smart glasses 3. To build strong and reliable data privacy and security mechanisms 4. Cost reduction of the associated IoT infrastructure 5. Exploration of 5G applications 6. Test beds for genomics to recover from diseases like central nervous system and infectious diseases 7. Blockchai-based test beds to solve the problems of large-scale data sharing, data privacy, and security and transparency between patient and doctors and between various healthcare providers 8. Virtual reality for rehabilitation in orthopedics 9. To explore augmented reality for its use as a visualization tool during surgeries
Platform Testing	Ensure application runs across all platforms.	[19, 25, 26]		
Interoperability testing	Assess whether applications (or software systems) can communicate with one another effectively and correctly.	[19, 21, 23, 24, 27, 30]		
Functionality testing	Verify each function of the software application in conformance with the requirement specification.	[18, 19, 20, 21, 22, 23, 24, 25, 26, 27, 28, 29, 30, 31, 32]		
Enterprise workflow testing	Check whether the expected activities are executed and workflow data properties have correct value.	[18, 19, 21, 22, 23, 26, 27, 28, 30, 31, 32]		
Performance testing	Ensure software applications will perform well under their expected workload.	[18, 19, 20, 21, 23, 24, 27, 28, 29, 30, 31]		
Usability testing	Evaluate a product or service by testing it with representative users.	[18, 19, 23, 26, 27, 36]		
Security testing	Check whether the application or the product is secured.	[19, 20, 21, 23, 24, 25, 26, 27, 31]		
Mobile app testing	Ensure applications worked well for handheld devices.	[21, 23, 27, 28, 29, 30, 31]		

employed for assurance of the privacy and security of data shared between doctors and patients.

Genomics is a field that deals with genes editing and genomic sequencing in which robotics plays a major role. Such test beds that ensure the proper functioning of genomics would help patients to recover from diseases like central nervous system and infectious diseases. Thus, for this purpose an efficient testing strategy must be identified. Prospects of the utilization of virtual reality for rehabilitation in orthopedics need more exploration. Context-aware test case design would strengthen confidence in the system.

Augmented reality should also be surveyed broadly so that it can be used effectively for gathering 3D data sets of a patient in real time using sensors like magnetic resonance imaging (MRI), ultrasound imaging, or CT scans. It should also be investigated for its use as a visualization tool during surgeries. Appropriate testing mechanisms need to be identified that work well in this direction. In addition, exploration of 5G applications for its use in the smart devices (such as wearable sensors) to monitor the health condition of patients is the need of the hour. To ensure the attainment of desired functionality, a comprehensive and customized testing strategy need to be devised. Also, a transparent pricing model is required to be implemented that ensures cost reduction of the associated IoT infrastructure by promoting the pooling of distributed, dispersed resources in fog computing and the Internet of Things. This also demands the establishment of test beds that make use of customized testing strategies to ensure the attainment of desired functionality of the ubiquitous system (such as smart home, smart health, and smart transport). Such test beds should be freely available to the research community to carry out extensive studies in this area.

15.4.3 Smart Transport

The existing work toward the implementation of test beds for smart transport system has been evaluated as per the following set of verification criteria and the associated research directions along with limitation are provided in Table 15.6:

- **Privacy testing.** There is the need to ensure the privacy and security of transport devices, encryption of the data communicated between vehicles, and the roadside infrastructures for privacy and security. This can be accomplished by the establishment of dedicated transportation cybersecurity test labs for intrusion detection/prevention systems, sensor spoofing/manipulation, secure controller area network, secure software updates, resiliency and recovery, sensor spoofing/manipulation, etc.
- **Energy efficiency testing.** ITS systems need to be tested for fuel consumption. The lesser the fuel consumed by the transport vehicle to travel per unit distance, the more will be the efficiency and lower will be cost. This can be

Table 15.6 Summary of limitations and research directions for smart transport.

Criteria	Research Direction	Work	Limitations	Suggestions
Privacy testing	Ensure privacy and security of transport devices and the associated data.	[37, 39, 43, 44]	1. Inadequate work has been done in academia. 2. No work has been found describing the testing methodology that verifies security of transport vehicle. 3. No test bed has been found that quantitatively measures the extent to what level air pollution reduces and travel experience gets enriched. 4. No case study has been discussed that empirically proves the benefits of the technology. 5. Very few test beds have been developed that actively carry out testing of autonomous vehicles. 6. No work has been found that tests user friendliness of transportation systems. 7. No test bed has been suggested for pollution monitoring devices. 8. Not sufficient work has been done on reliability testing.	1. Test bed should be designed to be portable so that it can be freely used by research community. 2. The test bed should also be made available to the students. 3. Novel testing methodologies should be proposed for the comprehensive testing of collision avoidance algorithms. 4. Test bed based on the blockchain technology to carry out regression testing of smart transportation system. 5. Context-aware testing methodology may be used. 6. Development of new efficient simulator for preliminary evaluation of proposed smart transport systems is required, as real-world test beds are prone to life risks.
Energy efficiency testing	Maintain fuel efficiency of vehicles.	[37, 38, 47]		
Collision avoidance testing	Ensure the effectiveness of collision avoidance algorithm.	[38, 41]		
Autonomous vehicle testing	Validate self-steering vehicle in real environment.	[38, 43, 44]		
Traffic congestion management	Test bed to assess traffic congestion management strategy.	[38, 41, 42, 44, 47]		
Connected vehicle technology	Validate connected vehicle technology.	[38, 39, 37, 44, 47]		
Compliance with standards	Comply with standards.	[39, 41, 42, 46]		
Reliability testing	Ensure robustness and resiliency of transport devices.	[38]		
Performance testing	Ensure performance of the devices.	[37, 42]		
Usability testing	Ensure user-friendliness of mobile apps.	[37]		
Pollution Control testing	Verify functionality of roadside pollution monitoring equipment.	•		
Interopera-bility testing	Ensure interoperability among devices and roadside infrastructure.	[41, 44, 43]		

achieved by avoiding vehicles standing idle in traffic jams or circling around looking for parking spaces. The better alternative would be to design vehicles based on sustainable resources such as electric and solar vehicles. Such engines must be tested comprehensively before making them fully functional in a real-world environment.

- **Collision avoidance testing.** Collision avoidance algorithms employed to prevent road accidents need to be tested to validate the ITS effectiveness.
- **Autonomous vehicle testing.** The autonomous vehicle operates without hands on the wheel, and the safety of such vehicle is of utmost importance, as failure causes life hazards and hence must be verified comprehensively.
- **Traffic congestion management.** Traffic congestion is one of the biggest challenges faced by commuters, as it leads to the wastage of time, fuel, and money. Unnecessary burning of fuel also increases the level of carbon emissions, which causes air pollution. This issue can be addressed by installing fog nodes in vehicles and roadside to send and receive information related to traffic jams and accidents. The generated information can thus be used to trigger certain actions such as activation of automatic brakes or issue of warning messages to slow down speed or to avoid specific lanes and intersection points. The test bed should be equipped to monitor and evaluate this mechanism.
- **Connected vehicle technology.** Connected vehicle technology utilizes wireless communication to transfer information regarding road accidents, jams etc. by one vehicle to another vehicle and to the road-side infrastructures. This helps in preventing road accidents and avoids unnecessarily getting stuck in traffic jams. The test bed should incorporate such facilities where new hardware and software could be tested before putting vehicles into real operating conditions.
- **Compliance with standards.** Test beds should comply with the standards of transport so that they provide the real picture of the working system before actual launch on the road and streets among public.
- **Reliability testing.** Transport system should be robust and resilient in case of any device failure or lost Internet connectivity. There should be strong mechanism to verify the reliability of the transportation system.
- **Performance and usability testing.** The testing must ensure that the devices perform well and the mobile apps are user-friendly.
- **Pollution control testing.** Carbon emissions should be reduced and proper checks need to be maintained to control air pollution. Roadside monitoring units installed to monitor and measure emission gases (such as carbon dioxide (CO_2) or nitrogen oxide (NO)) are required to be testing for its effectiveness. It can be controlled by using sustainable vehicles such electric cars.
- **Interoperability testing.** Poor interoperability is one of the largest barriers to smart transport implementation. Interoperability determines how devices

communicate with each other and, upon receiving information, how processing is done and corresponding actions are generated. If a device can't receive information, process it, and act upon that information, it won't function as consumers hope. Without full functionality, the product may not provide value. Real-world test labs are the best way to solve interoperability issues, as they depict actual scenarios of the problem.

Although many corporate provides the testing solution for smart transport, inadequate work has been done in academia, and this requires special attention of researchers. In addition, several works discusses the importance of cyber-physical systems in transportation, but no work has been found describing the novel testing methodology that verifies the security of smart transport vehicle. Similarly, numerous works have been found that discuss the importance of connected vehicle technology in reducing air pollution and improve efficiency but no test bed has been found that quantitatively measures the percentage level of air pollution reduction and up to what percentage travel experience gets enriched. Further, no case study has been discussed that empirically proves the benefits of the technology. Also, very few test beds have been developed that actively carry out testing of autonomous vehicles, and no work has been found that tests user friendliness of transportation systems. Test-bed executing of the repeatable regression tests based on blockchain technology must be studied to address quality assurance issues of the smart transportation system.

Pollution-monitoring devices are also required to be verified for effectiveness. No test bed has been suggested that works in this direction. Reliability is one of the most crucial feature that should be possessed by transport devices and related infrastructure; hence, there should be appropriate methodology that verifies the cybersecurity measures to ensure the resilience and robustness of the system. Only one research work has been found that works in this direction. Novel testing methodologies should be proposed for the comprehensive testing of collision avoidance algorithms, and the test bed should be designed to be portable so that it can be freely used by the research community.

15.5 Conclusions

Fog computing is a paradigm that can be successfully utilized to implement smart applications, as it overcomes the disadvantages associated with edge and cloud computing. The assurance of quality and reliability of fog-based IOT application is very important before their release to the market as poor design may hamper the working of the application and affects the end-user experience.

This chapter has surveyed testing perspectives of three cases studies (viz. smart home, smart health and smart transport), along with the elucidation of their objectives, approaches, and the achieved outcomes.

Software testing in the area of fog-based IoT applications has great potential in future research toward verification and validation of reliability, better security from hacking, Internet connection independency, user-friendliness, cost cutting, and industry standardization. Practitioners can create prototype ubiquitous testing environment for fog-based smart applications using advanced testing strategies such as context-aware test case generation, combinatorial testing and blockchain-based regression testing to address the issues of quality assurance.

This area commemorates a great deal of success and recognition in the seeable future. However, as we have explained in this chapter, industry and academia need to jump on and grab the compelling challenges and risks associated with it. It will ensure favorable outcome for fog computing in smart technology in distant future. The apparent trends in this sphere include the materialization of standards, the inception of enhanced testing services by boosting and merging current compute, storage and network services, utilization of fog computing along with cloud to provide acceptable QoS and governance; the possibility of exponential growth in smart technology developers and operators, thus widening the horse race and innovation. The researchers and practitioners would find endless opportunities to invent solutions to address hindrances in smart technologies using fog computing.

References

1 Internet of Things spending forecast. https://www.businesswire.com/news/home/20170104005270/en/Internet-Spending-Forecast-Grow-17.9-2016-Led. Accessed January 4, 2018.

2 Gartner says 8.4 billion connected Things. https://www.gartner.com/newsroom/id/3598917. Accessed January 4, 2018.

3 National Technical Systems (NTS). Completes ZigBee Smart Energy Certification Testing for SimpleHomeNet Appliance, https://www.nts.com/ntsblog/national-technical-systems-nts-completes-zigbee-smart-energy-certification-testing-for-simplehomenet-appliance/. Accessed January 3, 2018.

4 NTS Selected by PG&E as first provider of ZigBee HAN device validation testing. https://www.nts.com/ntsblog/nts_pge_selection/. Accessed January 3, 2018.

5 Smart home testing: Allion creates a new smart home test environment that simulates real life to provide innovative test services. http://www.technical-direct.com/en/smart-home-testing-allion-creates-a-new-smart-home-test environment-to-simulate-real-life-to-provide-innovative-test-services/. Accessed January 3, 2018.

6 *Performance testing for smart home app.* https://www.einfochips.com/resources/success-stories/performance-testing-for-smart-home-app/#wpcf7-f4285-p12635-o1. Accessed January 3, 2018.

7 Living Lab. https://www.ul.com/media-day/living-lab/. Accessed 4 January 2018.

8 Smart home testing and certification. https://www.tuv.com/world/en/smart-home-testing-and-certification.html. Accessed January 3, 2018.

9 VDE testing and certification. https://www.vde.com/tic-en/industries/smart-home. Accessed January 3, 2018.

10 Energy System Integration. https://www.nrel.gov/docs/fy17osti/66513.pdf. Accessed January 3, 2018.

11 A. Zipperer, P. Aloise-Young, S. Suryanarayanan, R. Roche, L. Earle, and D. Christensen. Electric energy management in the smart home: perspectives on enabling technologies and consumer behavior. In *Proceedings IEEE 2013,* 101(11): 2397–2408.

12 A. Cordopatri, R. De Rose, C. Felicetti, M. Lanuzza, and G. Cocorullo. Hardware implementation of a test lab for smart home environments. *AEIT International Annual Conference (AEIT),* Naples, 2015, pp. 1–6.

13 I. Dounis, C. Caraiscos. Advanced control systems engineering for energy and comfort management in a building environment a review. *Renewable and Sustainable Energy Reviews,* 13: 1246–1261, 2009.

14 R. Baos, F. Manzano-Agugliaro, F. Montoya, and C. Gil, A. Alcayde, J. Gomez. Optimization methods applied to renewable and sustainable energy a review. *Renewable and Sustainable Energy Reviews,* 15(4): 1753–1766, 2011.

15 J-J. Wang, Y-Y. Jing, C-F. Zhang, and J-H. Zhao. Review on multi-criteria decision analysis aid in sustainable energy decision-making. *Renewable and Sustainable Energy Reviews,* 13(9): 2263–2278, 2009.

16 T. Teich, F. Roessler, D. Kretz, and S. Franke. Design of a prototype neural network for smart homes and energy efficiency. *Procedia Engineering 24th {DAAAM} International Symposium on Intelligent Manufacturing and Automation,* 69(0): 603–608, 2014.

17 Q. Hu, F. Li, and C. Chen. A smart home test bed for undergraduate education to bridge the curriculum gap from traditional power systems to modernized smart grids. *IEEE Transactions on Education,* 58(1): 32–38, February 2015.

18 Insight driven healthcare services. http://www.virtusa.com/industries/healthcare/perspective/. Accessed January 3, 2018.

19 Healthcare QA and Testing Services. http://www.mindfiresolutions.com/HealthCare-QA-and-Testing-Services.htm. Accessed January 3, 2018.

20 Healthcare. https://qainfotech.com/healthcare.html. Accessed January 3, 2018.

21 Product testing. http://healthcare.calsoftlabs.com/services/product-testing.html. Accessed January 3, 2018.

22 Healthcare and fitness. http://www.precisetestingsolution.com/healthcare-software-testing.php. Accessed January 3, 2018.

23 Healthcare. http://zenq.com/Verticals?u=healthcare. Accessed January 3, 2018.

24 Healthcare testing services. https://www.testree.com/industries/healthcare-life-sciences/healthcare. Accessed January 3, 2018.

25 Health care. http://www.kiwiqa.com/health_care/. Accessed January 3, 2018.

26 Healthcare software testing. https://xbosoft.com/industries/healthcare-software-testing/. Accessed January 3, 2018.

27 Healthcare testing. http://www.infoicontechnologies.com/healthcare-testing. Accessed January 3, 2018.

28 Healthcare and pharma. https://www.w3softech.com/healthcare.html. Accessed January 3, 2018.

29 Healthcare. http://www.provasolutions.com/industries/software-qa-application-testing-services-for-healthcare-europe/. Accessed January 3, 2018.

30 Healthcare testing as a service. http://www.calpion.com/healthcareit/?page_id=1451. Accessed January 3, 2018.

31 Healthcare testing services. https://abstracta.us/industries/healthcare-software-testing-services. Accessed January 3, 2018.

32 Healthcare software testing services. https://www.360logica.com/verticals/healthcare-testing-services/. Accessed January 3, 2018.

33 R. Snelick. Conformance testing of healthcare data exchange standards for EHR certification. *International Conference Health Informatics and Medical Systems*. Las Vegas, USA, 2015.

34 P.J. Scott, S. Bentley, I. Carpenter, D. Harvey, J. Hoogewerf, and M. Jokhani. Developing a conformance methodology for clinically-defined medical record headings: A preliminary report. *European Journal of Biomedical Informatics,* 11(2): 23–30, 2015.

35 R. Löffler, M. Meyer, and M. Gottschalk. Formal scenario-based requirements specification and test case generation in healthcare applications. In *Proceedings of the 2010 ICSE Workshop on Software Engineering in Health Care* (SEHC '10). ACM, New York, USA, 57–67, 2010.

36 J.M.C. Bastien. Usability testing: a review of some methodological and technical aspects of the method. *International Journal of Medical Informatics,* 79: e18–e23, 2010.

37 A simulation-based test bed for networked sensors in surface transportation systems. https://www.cc.gatech.edu/computing/pads/transportation/testbed/description.html. Accessed January 3, 2018.

38 Intelligent transportation systems. http://www.umtri.umich.edu/our-focus/intelligent-transportation-systems. Accessed January 3, 2018.

39 Intelligent transportation systems. https://www.its.dot.gov/research_archives/connected_vehicle/dot_cvbrochure.htm . *Accessed* January 3, 2018.

40 Independent verification and validation. https://www.ibsplc.com/services/independent-verification-and-validation. Accessed January 3, 2018.

41 K. Hill. ETSI plugfest to test smart transportation in November. *RCR Wireless News*. https://www.rcrwireless.com/20160920/wireless/etsi-test-smart-transportation-latest-plugfest-tag6. Accessed January 3, 2018.

42 J.W. Woo, S. B. Yu, S. B. Lee, et al. Design and simulation of a vehicle test bed based on intelligent transport systems. *International Journal of Automotive Technology*, 17(2) : 353–359, 2016.

43 Intelligent transportation system leads to first test bed 'K-City' for connected cars in South Korea. http://markets.businessinsider.com/news/stocks/Intelligent-Transportation-System-Leads-to-First-Test-Bed-K-City-for-Connected-Cars-in-South-Korea-1008355700. January 3, 2018.

44 Intelligent transportation systems. https://e-estonia.com/solutions/location-based- services/intelligent-transportation-system/. Accessed January 3, 2018.

45 Transit Windsor begins testing intelligent transportation system. *CTV News*, https://windsor.ctvnews.ca/transit-windsor-begins-testing-intelligent-transportation-system-1.3289934. Accessed January 3, 2018.

46 Achieved Intelligent Transportation System product compliance with UBS II and ARAI testing standards, http://www.siphen.com/case-studies/product-compliance/,Retrieved January 3, 2018.

47 Automotive, intelligent transport systems. https://www.anritsu.com/en-AU/test-measurement/industries/automotive/automotive-intelligent-transport-systems. Accessed January 3, 2018.

16

Legal Aspects of Operating IoT Applications in the Fog

G. Gultekin Varkonyi, Sz. Varadi, and Attila Kertesz

16.1 Introduction

As a growing number of communicating devices join the Internet, we will soon face a foggy and cloudy world of interconnected smart devices. Cloud systems [1] have already started to dominate the Internet; with the appearance of the Internet of Things (IoT) area [2] IoT cloud systems are formed that still needs a significant amount of research. IoT is a rapidly emerging concept where sensors, actuators, and smart devices are often connected to and managed by cloud systems. IoT environments may generate a huge amount of data to be processed in the cloud. To reduce service latency and to improve service quality, the paradigm of fog computing [5] has been introduced, where the data can be kept and processed closed to the user.

The European Commission recently implemented comprehensive European data protection rules, where the main objectives are: (i) to modernize the legal system of the European Union (EU) for the protection of personal data to respond to the use of new technologies; (ii) to strengthen users' influence on their personal data and to reduce administrative formalities; (iii) and to improve the clarity and coherence of the EU rules for personal data protection. To achieve these goals, the Commission created the General Data Protection Regulation (GDPR) [3], a regulation that sets out a general EU framework for data protection and replaced the Data Protection Directive (DPD) [4]. In IoT cloud systems, personal data are increasingly being transferred, possibly across borders, and stored on servers in multiple countries both within and outside the EU. The globalized nature of dataflow calls for strengthening the individuals' data-protection rights internationally. This requires strong principles for protecting individuals' data, aimed at easing the flow of personal data across borders while still ensuring a high and consistent level of protection without loopholes or unnecessary complexity. In these legal documents the Commission aims to introduce a single set of rules on data protection.

Fog and Edge Computing: Principles and Paradigms, First Edition.
Edited by Rajkumar Buyya and Satish Narayana Srirama.

The GDPR, unlike the former DPD, expands its jurisdiction outside of the EU and requires that all the actors that offer services to the EU citizens abide by its rules, regardless of their residence. The GDPR also introduces some of the new rights that were a natural result of the technological developments, such as data protection by design and right to be forgotten. However, the technical structure and complexity of the IoT and the fog make it hard to be implemented and, as a result, make it hard to comply with the law. For this reason, the importance of "thinking the data protection rights of the people from the early phase of the system development," called data protection by design, is also in the Regulation [3]. Data protection by design aims to reduce possible privacy harms that fog applications may cause by combining it with the data protection impact assessment (DPIA) and the data protection enhancing technologies.

In this chapter we classify fog/edge/IoT applications, analyze the latest restrictions introduced by the GDPR, and discuss how these legal constraints affect the design and operation of IoT applications in fog and cloud environments.

16.2 Related Work

Security concerns for IoT have already been investigated by Escribano [6], who presented the first opinion [7] of the Article 29 Data Protection Working Party (WP29) in this regard. They stated in this report that it is crucial to identify and realize which stakeholder is responsible for data protection. WP29 named the following challenges concerning privacy and data protection: lack of user control, low quality of user consent, secondary uses of data, intrusive user profiling, limitations for anonymous service usage, and communication- and infrastructure-related security risks.

Yi et al. [8] further extended these concerns with respect to fog computing. They argue that secure and private data computation methods are needed, and privacy must be addressed in three dimensions: data, usage, and location privacy. As fog nodes can be geographically distributed, it is even more difficult to track and monitor data and its location in real time. Furthermore, when distributed and processed data are merged, the integrity of the data should be guaranteed. Fog node can also track end user devices to support mobility (location awareness) that may be a game changing factor for location-based services and applications. This puts location privacy of the user at risk, and therefore appropriate location-preserving privacy mechanisms must be employed. From a security perspective, a man-in-the-middle attack has a high potential to become a typical attack in fog computing. In this attack, nodes serving as fog devices may be compromised or replaced by fake ones. Traditional anomaly detection methods can hardly expose man-in-the-middle attack without noticeable features of this attack collected from the fog [9].

Mukherjee et al. further detailed these challenges [10]. They envisaged a three-tier fog architecture, where communication is performed through three interfaces: fog-cloud, fog-fog and fog-things. They stated that secure communication is a key issue, and privacy-preserving data management schemes are needed. They mentioned, but did not detail, legislation challenges, which is the aim of this chapter.

16.3 Classification of Fog/Edge/IoT Applications

In the past decade, we experienced an evolution in cloud computing: the first clouds appeared in the form of a single virtualized datacenter, then broadened into a larger system of interconnected, multiple datacenters. As the next step, cloud bursting techniques were developed to share resources of different clouds, then cloud federations [11] were realized by interoperating formerly separate cloud systems. There were various reasons to optimize resource management in such federations: to serve more users simultaneously, to increase quality of service, to gain higher profit from resource renting, or to reduce energy consumption or CO_2 emissions. Once these optimization issues were addressed and mostly solved, further research directions started to focus on clouds to support newly emerging domains, such as the Internet of Things. In the case of IoT systems, data management operations are better placed close to their origins, and thus close to the users, which resulted in better exploiting the edge devices of the network.

Finally, as the latest step of this evolution, the group of such edge nodes formed the fog. Dastjerdi and Buyya defined fog computing as a distributed paradigm [5], where cloud storage and computational services are performed at the network edge. This new paradigm enables the execution of data processing and analytics application in a distributed way, possibly utilizing both cloud and nearby resources. The main goal is to achieve low latency, but it also brings novel challenges in real-time analytics, stream processing, power consumption, and security.

Concerning IoT application areas, Want et al. [12] set up three categories to classify them: (i) composable systems, built from a variety of nearby interconnected things; (ii) smart cities, including utilities of modern cities such as a traffic-light systems capable of sensing the location and density of cars in the area; and (iii) resource conservation applications, used for monitoring and optimization of resources such as electricity and water. Atzori et al. [13] proposed a survey and identified five domains: transportation and logistics, healthcare, smart environments (home, office, plant), personal and social, finally futuristic domains. In this chapter, we do not aim to classify all application fields, but to define certain architectures that fit most application cases involving cloud, IoT, and fog utilization, to enable further investigations concerning security and privacy.

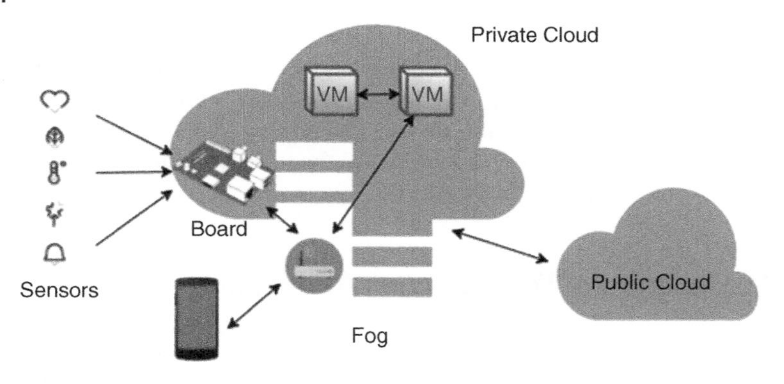

Figure 16.1 Data management in fog environments.

From this discussion, we can see that the collection, aggregation, and processing of user data can be done in various ways. Figure 16.1 presents an architecture where certain data flows can be examined.

In the next section, we summarize legislation affecting these tasks, and later we give guidelines on how to comply with such regulations in the identified cases.

16.4 Restrictions of the GDPR Affecting Cloud, Fog, and IoT Applications

The European Union is currently in the last phase of reforming the European data protection rules, where the main objectives are: (i) to modernize the EU legal system for the protection of personal data to respond to the use of new technologies; (ii) to strengthen users' influence on their personal data and to reduce administrative formalities; and (iii) to improve the clarity and coherence of the EU rules for personal data protection. To achieve these goals, the Commission created a new legislative proposal, called General Data Protection Regulation (GDPR) [3] that sets out a general EU framework for data protection to replace the currently effective DPD. Personal data are increasingly being transferred across borders and stored on servers in multiple countries both within and outside the EU. The globalized nature of data flows calls for strengthening the individuals' data protection rights internationally. This requires strong principles for protecting individuals' data, aimed at easing the flow of personal data across borders while still ensuring a high and consistent level of protection without loopholes or unnecessary complexity. According to the Article 8(1) of the Charter of Fundamental Rights of the European Union (the Charter) and Article 16(1) of the Treaty on the Functioning of the European Union (TFEU), the protection of natural persons in relation to the processing of personal data is a fundamental right. But the GDPR states that this is not an absolute right, it must be considered in relation to its function in society and be balanced against other fundamental rights.

Because of new challenges for the protection of personal data like rapid technological developments and globalization, the scale of the collection and sharing of personal data increased significantly. Both private companies and public authorities can use of personal data on an unprecedented scale in order to pursue their activities and beside that, natural persons increasingly make personal information available publicly and globally. Therefore, the European Union makes an emphasis on development of digital economy inside of the internal market with the free flow of the personal data without any barriers, but in frame of a coherent and strong data protection. The protection of individuals should be technologically neutral so it does not depend on the techniques used; otherwise, this would create a serious risk of circumvention.

16.4.1 Definitions and Terms in the GDPR

The new data protection framework of the EU, the GDPR [3], contains new rules and tools to fulfil these goals. It entered into force on May 2018, making the level of protection of the rights and freedoms of individuals with regard to the processing of such data equivalent in all member states. In the following we gather the newly introduced, relevant terms and rules of the GDPR, and later we analyze them with the operational aspects of fog computing.

16.4.1.1 Personal Data

It could be any information relating to an identified or identifiable natural person such as name, identification number, location data, and online identifier or to one or more indicators specific to the physical, physiological, genetic, mental, economic, cultural, or social identity of that natural person.

16.4.1.2 Data Subject

The subject is a natural person, who is identified or identifiable. The identifiable natural person is one who can be identified, directly or indirectly, in particular by reference to his or her personal data.

16.4.1.3 Controller

A natural or legal person, public authority, agency or other body can play this role. This new element under the GDPR is that the controller determines also the conditions of the processing of personal data.

16.4.1.4 Processor

The processor is also an important actor, who is also a natural or legal person, public authority, agency, or other body, which processes personal data on behalf of the controller.

16.4.1.5 Pseudonymization

It is a new term, which means the processing of personal data in such a manner that the personal data can no longer be attributed to a specific data subject without the use of additional information, provided that such additional information is kept separately and is subject to technical and organizational measures to ensure that the personal data are not attributed to an identified or identifiable natural person.

16.4.1.6 Limitation

What has a great importance among the principles relating to personal data processing is the limitation. Purpose of the collection, the quality of the data, and the duration of the storage are all limited based on their necessity. New elements are, in particular, the transparency principle, the clarification of the data minimization principle, and the establishment of a comprehensive responsibility and liability of the controller.

16.4.1.7 Consent

In order for personal data processing to be lawful, it has to be on the basis of the consent of the data subject for one or more specific purposes. The processing should be necessary for the performance of a contract in which the data subject is party or in order to take steps at the request of the data subject prior to entering into a contract. More specifically:

- Processing is necessary for compliance with a legal obligation of the controller.
- Processing is necessary in order to protect the vital interests of the data subject.
- Processing is necessary for the performance of a task carried out in the public interest or in the exercise of official authority vested in the controller.
- Processing is necessary for the purposes of the legitimate interests pursued by a controller, except where such interests are overridden by the interests or fundamental rights and freedoms of the data subject that requires protection of personal data, in particular where the data subject is a child. This shall not apply to processing carried out by public authorities in the performance of their tasks.

Regarding the conditions for consent, the data subject shall have the right to withdraw his or her consent at any time. In this case, the lawfulness of the former processing should not be affected by the withdrawal of consent. Consent shall not provide a legal basis for the processing, where there is a significant imbalance between the position of the data subject and the controller. In order to have one single and consistent definition, the GDPR contains that "consent" means any freely given, specific, informed and unambiguous agreement of the data subject to the processing of personal data relating to him or her. It could be given either by a statement or by a clear affirmative action. So it should be given explicitly by any appropriate method enabling a freely given specific and informed indication of the data subject's wishes. Therefore, silence or inactivity should not create the consent. Consent has to cover all processing activities carried out for the same purpose. The processing of the personal data of a child shall be lawful where the child is at least 16 years old and after his or her consent was given. Where the child is below the age of 16 years, such processing shall be lawful only if the consent was given or authorized by the holder of parental responsibility over the child.

16.4.1.8 Right to Be Forgotten

The GDPR further elaborates and specifies the data subject's right of erasure and provides the conditions of the right to be forgotten, when the data are no longer necessary in relation to the purposes for which they were collected or otherwise processed. Another case is when the data subject withdraws consent on which the processing is based, or when the storage period consented to has expired, and where there is no other legal ground for the processing of the data. This means the obligation of the controller that has made the personal data public to inform third parties to erase any links to, or copy or replication of that personal data. In relation to a third party publication of personal data, the controller should be considered responsible for the publication, where the controller has authorized the publication by the third party. The controller shall carry out the erasure without delay, but there are some exceptions when the retention of the personal data is necessary (e.g. for exercising the right of freedom of expression or for reasons of public interest in the area of public health; for historical, statistical and scientific research purposes, etc). Where the erasure is carried out, the controller shall not otherwise process such personal data.

This right is particularly relevant, when the data subject has given their consent as a child, when not being fully aware of the risks involved by the processing, and later wants to remove such personal data especially from the Internet.

16.4.1.9 Data Portability

The GDPR introduces the data subject's right to data portability (i.e. to transfer data from one electronic processing system to, such as a social network, into another, without being prevented from doing so by the controller). As a precondition and in order to improve access of individuals to their personal data, it provides the right to obtain from the controller those data in a structured and commonly used electronic format. This option could apply where the data subject provided the data to the automated processing system, based on their consent or in the performance of a contract.

16.4.2 Obligations Defined by the GDPR

The data subject has the right to object a measure based on profiling solely on automated processing intended to evaluate certain personal aspects relating to this natural person or to analyze or predict in particular the natural person's performance at work, economic situation, location, health, personal preferences, reliability, or behavior.

16.4.2.1 Obligations of the Controller

The GDPR introduces the obligation on controllers to provide transparent, easily accessible and understandable information, inspired in particular by the Madrid Resolution on international standards on the protection of personal data and privacy (Madrid Resolution, 2009). Another obligation of the controller is to provide procedures and mechanism for exercising the data subject's rights, including means for electronic requests, requiring response to the data subject's request within a defined deadline (at the latest within one month of receipt of the request), and the motivation of refusals.

The controller has an obligation regarding information about the data subject, too. The controller shall provide all the information about:

- The identity and the contact details of the controller and, where applicable, of the controller's representative.
- The contact details of the data protection officer, where applicable.
- The purposes of the processing for which the personal data are intended as well as the legal basis for the processing.
- Where the processing is necessary for the purposes of the legitimate interests, about the legitimate interests themselves pursued by the controller or by a third party.
- The recipients or categories of recipients of the personal data, if any.
- Where applicable, the fact that the controller intends to transfer personal data to a third country or international organization and the existence or absence of an adequacy decision by the commission.

There are some additional pieces of information that shall be given by the controller: the storage period; the right to withdraw the consent any time; access to and rectification or erasure of personal data or restriction of processing concerning the data subject or to object to processing as well as the right to data portability; the right to lodge a complaint with a supervisory authority; and the existence of automated decision-making, including profiling as well as the significance and the envisaged consequences of such processing for the data subject. The data subject could request a confirmation from the controller at any time, whether or not personal data relating to the data subject are being processed.

Data protection by design and by default. To ensure privacy and data security, the GDPR introduces a new term called Data Protection by Design (or Privacy by Design in the draft proposal of the GDPR). It means that the controller shall, both at the time of the determination of the means for processing and at the time of the processing itself, implement appropriate technical and organizational measures and procedures considering the state of the art and the cost of implementation, in such a way that the processing will meet the requirements of the GDPR and ensure the protection of the rights of the data subject. Such measures should include minimizing the processing of personal data and applying pseudonymization on the personal data as soon as possible. The appropriate system should also enable the data subject to monitor the data processing and the controller to create and improve security features. This principle and the named measures are particularly important in designing fog environments. These measures shall be steps to protect personal data against accidental or unlawful destruction or accidental loss and to prevent any unlawful forms of processing in particular any unauthorized disclosure, dissemination or access, or alteration of personal data. We further detail these issues in the next section.

Regarding to the state of the art and the cost of implementation, the controller shall, both at the time of the determination of the means for processing and at the time of the processing itself, implement appropriate technical and organizational measures and procedures in such a way that the processing will meet the requirements of the GDPR and ensure the protection of the rights of the data subject.

Articles 26 and 27 address some of the issues raised by cloud computing, more specifically from cloud federations. While these provisions do not indicate whether outsourcers are joint data controllers, they acknowledge the fact that there may be more than one data controller. The provision of the GDPR clarifies the responsibilities of joint controllers as regards their internal relationship and toward the data subject. Where a controller determines the purposes, conditions, and means of the processing of personal data jointly with others, the joint controllers shall determine their respective responsibilities for compliance with the obligations under the GDPR, by means of an arrangement between them.

Those controllers or processors, who are not established in the European Union, have an obligation to designate a representative in the EU in a written form, where the GDPR applies to their processing activities. The exceptions are when the data processing is occasional and includes not special categories of data or when the controller is a public authority or body. The representative should act on behalf of the controller or processor and may be addressed by any supervisory authority.

The main establishment of a controller in the EU should be determined according to objective criteria and should imply the effective and real exercise of management activities determining the main decisions as to the purposes, conditions and means of processing through stable arrangements. Only the presence and use of technical means and technologies for processing personal data do not constitute such main establishment themselves and are therefore not determining criteria for a main establishment. The main establishment of a controller or a processor should be the place of its central administration in the EU and implies the effective and real exercise of activity through stable arrangements according to the GDPR.

16.4.2.2 Obligations of the Processor

The GDPR also clarifies the position and obligation of processors adding new elements, including that a processor who processes data beyond the controller's instructions is to be considered as a joint controller. The Regulation requires that the controller shall use only processors providing sufficient guarantees to implement appropriate technical and organizational measures in such a manner that processing will meet ensure the protection of the rights of the data subject. The processor shall not apply another processor without prior specific or general written authorization of the controller.

The carrying out of processing by a processor shall be governed by a written contract or other legal act, including in electronic form, binding the processor to the controller and stipulating in particular that the processor shall:

- Act only on instructions from the controller, in particular, where the transfer of the personal data used is prohibited.
- Employ only staff who have committed themselves to confidentiality or are under a statutory obligation of confidentiality.
- Take all required measures.
- Enlist another processor only with the prior permission of the controller.
- Insofar as this is possible given the nature of the processing, create in agreement with the controller the necessary technical and organizational requirements for the fulfilment of the controller's obligation.
- Assist the controller in ensuring compliance with the obligations.
- At the choice of the controller, delete or return all the personal data to the controller after the end of the provision of services relating to processing, and delete existing copies unless EU or member state law requires storage of the personal data.
- Make available to the controller and the supervisory authority all information necessary to control compliance with the obligations laid down in the GDPR.

This contract or legal act should contain, in whole or in part, on standard contractual clauses, including when they are part of a certification granted to the controller or processor in accordance with the provisions of the GDPR regarding the certification. The European Commission could lay down additional standard contractual clauses.

The controller and the processor shall document in writing the controller's instructions and the processor's obligations. The processor shall be considered to be a controller in respect of that processing and shall be subject to the rules on joint controllers, if a processor processes personal data other than as instructed by the controller.

GDPR introduces the obligation for controllers and processors to maintain a record of processing operations under their responsibility in written and in electronic form, instead of a general notification to the supervisory authority required by the former Data Protection Directive of the EU. It shall contain some relevant information such as the purpose of the data processing, the name and contact details of the controller or the processor, and description of the categories of data subjects and of the categories of personal data, etc.

The GDPR introduces an obligation to notify personal data breaches, building on the personal data breach notification in Article 4(3) of the e-privacy Directive 2002/58/EC. Moreover, the former DPD provided for a general obligation to notify processing of personal data to the supervisory authorities, which notification could create administrative and financial burdens. According to the Commission, this general obligation should be replaced by effective procedures. Therefore, the new Regulation introduces a new element, namely the obligation of controllers and processors to carry out a data protection impact assessment prior to risky processing operations, which could present specific and high risks to the rights and freedoms of data subjects by virtue of their nature, their scope, or their purposes. According to the GDPR, the following processing operations in particular present specific risks:

- "A systematic and extensive evaluation of personal aspects relating to natural persons which is based on automated processing, including profiling, and on which decisions are based that produce legal effects concerning the natural person or similarly significantly affect the natural person;
- Processing on a large scale of special categories of data or of personal data relating to criminal convictions and offences;
- A systematic monitoring of a publicly accessible area on a large scale."

About those processing operations that require data protection impact assessment, a public list should be created by the supervisory authority. The impact assessment shall contain at least:

- A detailed description of the envisaged processing operations and the purposes of the processing, including, where applicable, the legitimate interest pursued by the controller.
- An assessment of the necessity and proportionality of the processing operations in relation to the purposes.
- An assessment of the risks to the rights and freedoms of data subjects.
- The measures envisaged to address the risks. These include safeguards, security measures, and mechanisms to ensure the protection of personal data and

to demonstrate compliance with the GDPR, taking into account the rights and legitimate interests of data subjects and other persons concerned.

This provision should, in particular, apply to newly established large-scale filing systems, which aim at processing a considerable amount of personal data at regional, national, or supranational levels and which could affect a large number of data subjects.

The GDPR states that the controller shall consult with the supervisory authority prior to processing where a data protection impact assessment indicates that the processing would result in a high risk in the absence of measures taken by the controller to mitigate the risk, building on the concept of prior checking in Article 20 of DPD.

The new Regulation, based on Article 18(2) of DPD, also introduces the function of a mandatory data protection officer, who should be designated when the processing carried out for the public sector or for large enterprises, or where the core activities of the controller or processor consist of processing operations that require regular and systematic monitoring or consist of processing on a large scale of special categories of data. The data protection officer may be employed by the controller or processor, or fulfill his or her tasks on the basis of a service contract.

Article 40 concerns codes of conduct, building on the concept of Article 27(1) of DPD, clarifying the content of the codes and the procedures. The member states, the Commission, the supervisory authorities, and the Board shall encourage, in particular at European level, the establishment of data protection certification mechanisms and of data protection seals and marks, allowing data subjects to quickly assess the level of data protection provided by controllers and processors. The monitoring of compliance with a code of conduct may be carried out by a body which has an appropriate level of expertise in relation to the subject-matter of the code and is accredited for that purpose by the competent supervisory authority.

16.4.3 Data Transfers Outside the EU

16.4.3.1 Data Transfers to Third Countries

Chapter V of the GDPR contains the rules for transfers of personal data to third countries or international organizations. According to the new provisions, transfer could be carried out only when an adequate level of protection is ensured by the third country, or a territory or a processing sector within that third country, or international organization in question. The new provision now confirms explicitly that the European Commission is in the position to decide whether this adequate level of protection is provided by a territory or a processing sector within a third country.

The criteria that shall be taken into account for the Commission's assessment of an adequate or not adequate level of protection include expressly the rule of law, respect for human rights and fundamental freedoms, relevant legislation, and independent supervision. It is also important that the international commitments the third country or international organization concerned has entered into, or other obligations, arise from legally binding conventions or instruments as well as from its participation in multilateral or regional systems, in particular in relation to the protection of personal data.

Where the Commission decides that an adequate level of protection is ensured, a so-called implementing act shall be created for a mechanism for a periodic review, at least every four years, which shall take into account all relevant developments in the third country or international organization. The Commission has the duty to monitor these developments.

A list of those third countries, territories, and processing sectors within a third country and international organizations, where it has decided that an adequate level of protection is or is not ensured, should be published by the Commission in the Official Journal of the European Union.

When no such adequacy decision has been adopted by the Commission, the GDPR requires for transfers to third countries that appropriate safeguards be provided. In particular:

- A legally binding and enforceable instrument between public authorities or bodies;
- Binding corporate rules;
- Standard data protection clauses adopted by the Commission or by a supervisory authority;
- An approved code of conduct together with binding and enforceable commitments of the controller or processor in the third country to apply the appropriate safeguards, including as regards data subjects' rights; or
- An approved certification mechanism together with binding and enforceable commitments of the controller or processor in the third country to apply the appropriate safeguards, including as regards data subjects' rights.

The GDPR explicitly provides for international cooperation mechanisms such as mutual assistance for the protection of personal data between the Commission and the supervisory authorities of third countries.

The draft version of the GDPR contained provisions such that if the Commission decided the adequate level was not ensured in a third country or a territory such as third country or a territory within that third country, or the international organization, any transfer of personal data to that place in question should be prohibited. In this case, the Commission should enter into consultations with this third country or international organization to remedy the situation resulting from this inadequacy decision. This statement of the Commission was missing from the final version of the GDPR.

In the absence of an adequacy decision or of appropriate safeguards, including binding corporate rules, a transfer or a set of transfers of personal data to a third country or an international organization shall take place under only one of the following conditions:

1. The data subject has explicitly consented to the proposed transfer, after having been informed of the possible risks of such transfers for the data subject due to the absence of an adequacy decision and appropriate safeguards.
2. The transfer is necessary for the performance of a contract between the data subject and the controller or the implementation of pre-contractual measures taken at the data subject's request.
3. The transfer is necessary for the conclusion or performance of a contract concluded in the interest of the data subject between the controller and other natural or legal person.
4. The transfer is necessary for important reasons of public interest.
5. The transfer is necessary for the establishment, exercise, or defense of legal claims.
6. The transfer is necessary in order to protect the vital interests of the data subject or of other persons, where the data subject is physically or legally incapable of giving consent.
7. The transfer is made from a register that, according to union or member state law, is intended to provide information to the public and that is open to consultation either by the public in general or by any person who can demonstrate a legitimate interest, but only to the extent that the conditions laid down by union or member state law for consultation are fulfilled in the particular case.

The controller shall inform the supervisory authority of the transfer. The controller shall also provide the information to the data subject of the transfer regarding the compelling legitimate interests pursued.

16.4.3.2 Remedies, Liabilities, and Sanctions

The Regulation contains provisions for remedies, liabilities, and sanctions. The new Regulation concerns the right to a judicial remedy against a controller or processor, providing a choice to go to court in the member state where the defendant is established or where the data subject has his or her habitual residence, unless the controller or processor is a public authority of a member state acting in the exercise of its public powers.

If material or nonmaterial damage was caused by an infringement of the GDPR, the controller or processor shall provide compensation for the damage suffered. One of the possible penalties could be administrative fines; besides that, other penalties should be laid down by the member states.

16.4.4 Summary

As a summary, due to the legal nature of a regulation under EU law, the GDPR established a single rule that applies directly and uniformly. EU regulations are the most direct form of EU law. A regulation is directly binding on the member states and is directly applicable within the member states. As soon as a regulation is entered into force, it automatically becomes the part of the national legal system of each member state, and it is not allowed to create a new or different legislative text by each member state. Contrarily, EU directives are flexible tools of the EU legislation; they are used to harmonize the different national laws in line with each other. Directives prescribe only an end result that must be achieved by every member state; the form and methods of implementing the principles included in a directive are a matter for each member state to decide for itself. Each member state must implement the directive into its legal system, but can do so in its own words. A directive only takes effect through national legislation that implements the measures.

We revealed in a former work on cloud federations [14] that according to the Article 4 of the former DPD, the location of the data controller's establishment determined the national law applicable, which could be variable, as we have seen in specific cloud use cases. However, the GDPR with its unified rules, must be applied in every member state in the same way, so there will be and can be no discrepancy among them. Moreover, where the national law of a member state applies by virtue of public international law, this Regulation also applies to a controller not established in the EU, such as in a member state's diplomatic mission or consular post (Preamble (22) of GDPR).

In the next section we further detail the data protection by design principle, and discuss its implementation needs and its possible causes.

16.5 Data Protection by Design Principles

The Privacy by Design (PbD) concept was comprehensively explained in the 1990s by Ann Cavoukian, who is the former information and privacy commissioner for the Canadian province of Ontario. Her philosophy received a high level of attention not only from the privacy scholars but by the legislators, too. Such that, Article 25 was placed into the GDPR, which legally binds the data controllers to take several technical and organizational measures to comply with the related law. The GDPR uses the title "Data Protection by Design (DPbD)" as it focuses only on the data protection; however, there is no difference between the two terms, both in the legal and practical meaning. We will also follow the GDPR's notion in this chapter.

Cavoukian [15] uses the Fair Information Practices Principle as a basis for the DPbD principles. These principles could, as found in the GDPR, be counted as follows: data minimization; data retention and data usage limit (purpose specification); individual consent; notice responsibility (transparency); stored data security; right to access to own personal data; and accountability. Her solution for the serious privacy risks in the highly growing technological environment foresees developments of systems that are not interrupted by the privacy rules, but make these rules an integral part of the "organizational priorities, project objectives, design processes, and planning operations." In order to do that, the DPbD philosophy should be adopted from the beginning of the system design [16], and should follow the system's life cycle until it becomes useless. Today, system design does not only mean the technical part of the system creation, such as code developing. Many different technological solutions offered by the IT companies consider organizational aspects to the legal compliances, during the system design. For this reason, it is possible to say that the concept of the DPbD is in relation with both legal and technical, as well as organizational aspects. It is legal, because the legal developments trigger the adoption of the DPbD. It is organizational, because it means self-assessment, self-regulation, and self-reaction to reach the privacy-friendly technologies. It is technical, because as a result of the legal requirements and the organizational planning, tangible steps are required toward privacy-friendly systems. This step generally requires technical solutions involved with the system, which are privacy enhancing technologies (PETs). Through PETs, the DPbD becomes visible by the end user. Altogether, DPbD means to draw the map of the personal data collection, usage, transmission, access, storage, and shortly any processing activity, as well as the business models behind the personal data, taking the necessary technical safeguards to ensure security of the data in a certain system that reduces users' data protection concerns.

16.5.1 Reasons for Adopting Data Protection Principles

Before we go into the details, one may ask why DPbD principles should be adopted. First, if data protection is a fundamental right and if it should be taken into account from the beginning of the system design, then the DPbD concept is the one that can create the "data protection first" [17] culture. This culture leads the company to gain user trust. Whenever the Internet users share personal data online, they trust the promises of the service providers on data collection, usage, storage, and safety. Only with user trust can the Internet economy grow [18] because more DPbD friendly systems will be used by more people [19]. This might be one of the reason why Apple grows, because "at Apple, our trust means everything to them" and "that's why they respect our privacy and protect it with strong encryption…" [20] and other techniques.

Second, the organizations will fully comply with the legal obligations so they will not be faced with huge amount of fines and will not lose money. Similarly, as much as it is possible to foresee the risks, the organizations will spend fewer money to fix them than after launching the product [21]. More sanctions lead to more reputation loss, either [22]. In addition, the organizations will create a data protection culture [23] automatically in the company. Moreover, as the technology changes and develops so quickly, it is not easy to control the privacy concerns during the system usage. It is necessary to foresee such dangers from the beginning of the basic system design and simply do the right thing. Additionally, the whole philosophy can contribute to the global data protection, which is missing because of different data protection understanding and implementations.

Finally, DPbD helps reduce the world data protection asymmetry, power games, and political conflicts, and also promotes free flow of information, national security, and democracy. It is a philosophy to be embraced against surveillance, misuses, and illegal uses. Thanks to the globalization and to the Internet, where there is a product like Facebook, the data protection leading countries' legal pressure benefits everyone – the data protection shields they earn cover every country in the world [16].

16.5.2 Privacy Protection in the GDPR

Now, taking a closer look at the principles of the DPbD could give a more comprehensive understanding of what exactly is indicated by the DPbD as a privacy protection. Regardless of its orders, the first principle appears to be the logic of the whole DPbD understanding, which points out the current problem of being unable to fix the data leakage once it happens. Interconnected online networks do not seem helpful for fixing unwanted data disclosures due to the fact that it is not possible to find out all the possible connections of an online personal data. Once data go online, it is almost impossible to destroy them.

Proactive and preventative approaches to personal data protection lower the risk of such disclosures by adopting even higher standards than the already known ones, creating privacy network between the users and partners, and realizing the privacy-weak points of the systems. From the system point of view, this requires embedding privacy into the system's architecture. One way to find out exactly what kind of privacy tools should be embedded into the systems is to look at the privacy impact assessments. Article 35 of the GDPR brings data protection impact assessment (DPIA) responsibility to the data controllers when data processing is "likely to result in a high risk to the rights and freedoms of natural persons." The fact is that any system dealing with personal data carries some level of risk. In order to define the level of risk and take necessary steps, DPIA is the first attempt on the way to PDbD, and success of the PDbD depends highly on a successful DPIA [23].

DPIA is a systematic way of assessing the risks that will lead the businesses, together with their stakeholders and employees, to know what and how to handle the risks related to data protection in a certain system(s). The DPIA helps the organizations to create a complete picture of the personal data collection, storage, usage, transfer, and management of the risks appearing in these processes. The relationship between the DPbD and the DPIA is twofold; they feed each other, because in the end, data protection measures and the techniques will be proactively built into the systems. The result of the assessment helps decision-makers to have a plan on how to strengthen the data security, which directs them to decide on what PETs to implement.

PETs are perhaps described best in the EU literature especially from the data protection point of view:

> It is a system of Information and Communication Technologies measures protecting informational privacy by eliminating or minimizing personal data thereby preventing unnecessary or unwanted processing of personal data, without the loss of the functionality of the information system [24].

The PETs are not newly referred in the EU data protection literature; however, the GDPR was widened to explain them (Recital 78). They are the technical tools that help organizations to reduce the risks revealed through the DPIA. These tools are, in general, encryption, email privacy tools, anonymization and pseudonymization tools, authentication tools, cookie cutters, and The Platform for Privacy Preferences, etc. The list is not exhaustive, due to the fact that privacy protection and especially data protection technology will continue to grow ever faster now that the GDPR is in full force.

16.5.3 Data Protection by Default

Secured systems combining with the data processing principles consists of privacy by default or, with the GDPR words, data protection by default (DPbD). Basically, DPbD is related to the data minimization principle and orders to the data controller to collect the minimum possible personal data during the services. This should not interrupt the system functionality and does not prevent the data controllers from collecting the necessary data to run the system. There might be functions that could only be available if the user shares some of the personal data. These functions should not be available to the users without obtaining their consent to process the necessary personal data. Indeed, the consent should be given in an informed basis, freely, should be specific to the purpose of the specific function, should be unambiguous or explicit (depending on a type of personal data e.g. whether sensitive data or not), and should be given with an affirmative action (Recital 32 of the GDPR). The latter criterion is an opt-in process, which is more or less the same meaning as privacy by default. Opt-in action ensures that the necessary personal data being collected and further data processing activity are left to be decided by the data subjects, manually. The data subject should have options to choose between giving consent or leaving processing activity out of the functions. This is very significant in view of the example of Facebook in 2008 and 2017. Previously, users creating an individual Facebook profile were expected to share lots of personal information, including sensitive information such as their religion, political views, and nationality. There was no setting available for the users to choose whether they would like to display such information on their profile (personal data management tool). Moreover, the users were not given a choice to restrict whom they would like to display their own profile, which may include their pictures, posts, and videos as well as the other information that they gave away during the profile creation. Since 2014, Facebook has changed its "everything should be public" approach to "everything should be private and manageable" approach. Now, besides the default privacy settings, Facebook users can manage third-party data disclosures, set the public-private post rule at the time of posting, and manage whole privacy settings in an understandable, user-friendly interface. Altogether, Facebook has drawn its own borders of data collection, usage, and disclosure.

Creating successful privacy-friendly systems could be possible with cooperation between the stakeholders, as well as their cooperation with the individuals. Principle of visibility and transparency create personal data protection policies and procedure documents, and to share them with the related entities and individuals. In this case, providing comprehensive, understandable, and clear information to the individuals about their rights (Articles 12–23) and the remedies (Articles 77–80, Article 82 of the GDPR). It is also crucial for the data controllers to inform the Data Protection Authority (DPA) about these policies because in the end, they will be monitored by the DPA whether they are in compliant with the law. While compliance is an important issue, all the steps are taken toward with respect for user privacy. Cavoukian suggests to "keep the design user centric" by providing the necessary tools and information to the users to be able to execute their own data self-management. The GDPR strengthens many of these tools, such as by interpreting the conditions for consent clearly (Article 7), introducing consent mechanisms for children (Article 8), introducing the right to be forgotten (Article 17), and right to data portability (Article 20). The companies offer users creative and user-friendly interfaces to access and manage their data. Google offers data management and privacy check platforms designed with figures and animations, including short and understandable documents, and a control panel to manage all related data and information collected by Google. As much as personal data is being processed, such data control panels should be design so that all users can understand and use them.

Finally, one may wonder, what will happen if DPbD principles are implemented? First of all, the systems as well as the data will be secured in life cycle protection, which stresses the importance of the continuous and standard data security applications and their balance between the functionality of the system and users' rights. As long as new technological developments, such as artificial intelligence and robots, become a part of people's daily life, there is a positive signal for ongoing changes and improvements in the data protection field both from a legal and practical point of view. For this reason, data protection is a dynamic field that requires constant system monitoring to keep the level of protection or implementations, or to create even higher protection tools.

Second, if the DPbD principles are followed, any actors involving the data processing activity will find themselves in a win-win position. In this way, users could use the system without any doubt about how their data are being used, and as a result of the DPbD, the system stakeholders ensure an adequate level of data security within the systems, which they can reflect to the users and data protection authorities, whether or not they are in compliance with the privacy policies, rules, and legislation.

To summarize these thoughts, we argue that all parties of operating and using a fog application related to a member state of the EU should be aware of the GDPR, and PET is an approach that could be applied in IoT/fog/cloud environments. The possible fog use cases we depicted in Figure 16.1 highlight that multitenancy is even more existent in IoT and fog environments than in purely cloud setups, and the number of participating entities is also higher (specifically in multiple fog regions), which means that the correct identification of controller and processor roles is crucial.

16.6 Future Research Directions

The result of our investigation shows that the DPbD principle could reduce possible privacy harms of IoT applications in cloud and fog environments by combining the data protection impact assessment and data protection enhancing technologies. In the future, we plan to further analyze IoT, fog, and cloud use cases and perform legal role mappings to reveal responsibilities and provide hints for designing and operating applications in these fields.

16.7 Conclusions

Following the recent technological trends, IoT environments generate unprecedented amounts of data that should be stored, processed, and analyzed. Cloud and fog technologies can be used to aid these tasks, but their application give birth to complex systems, where data management raises legal issues. The European Commission continues to modernize its legal system for the protection of personal data to respond to the use of these new technologies to strengthen users' influence on their personal data, to reduce administrative formalities, and to improve the clarity and coherence of the EU rules for personal data protection. To achieve these goals, the Commission created the General Data Protection Regulation, which we analyzed in this chapter in detail.

In this chapter we also introduced fog characteristics and security challenges in the light of the new European legislation. We further detailed data protection by design principle, and suggested the use of privacy enhancing technologies to comply with the regulation and to ease the management of fog environments.

Acknowledgment

The research leading to these results was supported by the UNKP-17-4 New National Excellence Program of the Ministry of Human Capacities of Hungary, and by the Hungarian government and the European Regional Development Fund under the grant number GINOP-2.3.2-15-2016-00037 ("Internet of Living Things").

References

1 R. Buyya, C. S. Yeo, S. Venugopal, J. Broberg, and I. Brandic. Cloud computing and emerging IT platforms: vision, hype, and reality for delivering computing as the 5th utility. *Future Generation Computer Systems* 25: 599–616, 2009.
2 H. Sundmaeker, P. Guillemin, P. Friess, and S. Woelfflé. Vision and challenges for realising the Internet of Things. CERP IoT – Cluster of European Research Projects on the Internet of Things, CN: KK-31-10-323-EN-C, March 2010.
3 European Commission. REGULATION (EU) 2016/679 of the European Parliament and of the Council of 27 April 2016 on the protection of natural persons with regard to the processing of personal data and on the free movement of such data, and repealing Directive 9546EC (General Data Protection Regulation). Official Journal of the European Union, Last visited on June 17, 2017.
4 Directive 95/46/EC of the European Parliament and of the Council of 24 October 1995 on the protection of individuals with regard to the processing of personal data and on the free movement of such data. *Official Journal L,* 281,: 31–50, 1995.
5 A. V. Dastjerdi, R. Buyya. Fog computing: Helping the Internet of Things realize its potential. *Computer,* 49: 112–116, August 2016.
6 B. Escribano. Privacy and security in the Internet of Things: Challenge or opportunity. OLSWANG. http://www.olswang.com/media/48315339/privacy_and_security_in_the_iot.pdf. Accessed November 2014.
7 Opinion 8/2014 on the Recent Developments on the Internet of Things. http://ec.europa.eu/justice/data-protection/article-29/documentation/opinion-recommendation/files/2014/wp223_en.pdf. Accessed October 2014.
8 S. Yi, Z. Qin, and Q. Li. Security and privacy issues of fog computing: A survey. In *International Conference on Wireless Algorithms, Systems, and Applications* (pp. 685–695). Springer, Cham, August 2015.

9 K. Lee, D. Kim, D. Ha, and H. Oh. On security and privacy issues of fog computing supported Internet of Things environment. In *IEEE 6th International Conference on the Network of the Future (NOF)*, September 2015: 1–3.

10 M. Mukherjee et al. Security and privacy in fog computing: challenges. *IEEE Access*, 5: 19293–19304, 2017.

11 A. Kertesz. Characterizing cloud federation approaches. In *Cloud Computing: Challenges, Limitations and R&D Solutions. Computer Communications and Networks*. Springer, Cham, 2014, pp. 277–296.

12 R. Want and S. Dustdar. Activating the Internet of Things. *Computer*, 48(9): 16–20, 2015.

13 L. Atzori, A. Iera, and G. Morabito. The Internet of Things: A Survey. *Computer Network*, 54(15): 2787–2805, 2010.

14 A. Kertesz, Sz. Varadi. Legal aspects of data protection in cloud federations. In S. Nepal and M. Pathan (Ed.). *Security, Privacy and Trust in Cloud Systems*. Berlin, Heidelberg. Springer-Verlag, 2014, pp. 433–455.

15 A. Cavoukian. *Privacy by Design: The 7 Foundational Principles Implementation and Mapping of Fair Information Practices*, 2011. http://www.ontla.on.ca/library/repository/mon/24005/301946.pdf.

16 I. Rubinstein. Regulating Privacy by Design. *Berkeley Technology Law Journal* 26 (2011): 1409.

17 E. Everson. Privacy by Design: Taking CTRL of big data. *Cleveland State Law Review*, 65: 27–44, 2016.

18 A. Rachovitsa. Engineering and lawyering Privacy by Design: understanding online privacy both as a technical and an international human rights issue. *International Journal of Law and Information Technology*, 24(4): 374–399, 2016.

19 P. Schaar. Privacy by Design. *Identity in the Information Society*, 3(2): 267–274, 2010.

20 Apple Inc. Apple's commitment to your privacy. Available: https://www.apple.com/privacy/. December 2017.

21 Information Commissioner's Office (ICO). Conducting privacy impact assessments code of practice, 2014. Available: https://ico.org.uk/media/for-organisations/documents/1595/-pia-code-of-practice.pdf.

22 N. Hodge. The EU: Privacy by default analysis. *In-House Perspective* 8: 19–22, 2012.

23 K.A. Bamberger and D.K. Mulligan. PIA requirements and privacy decision-making in us government agencies. In *Privacy Impact Assessment*. D. Wright and P. De Hert, Eds. Dordrecht: Springer Netherlands, 2012, pp. 225–250.

24 Privacy and data protection by design – from policy to engineering. European Union Agency for Network and Information Security (ENISA), 2014.

17

Modeling and Simulation of Fog and Edge Computing Environments Using iFogSim Toolkit

Redowan Mahmud and Rajkumar Buyya

17.1 Introduction

Relying on rapid advancement of hardware and communication technology, Internet of Things (IoT) is consistently promoting every sphere of cyber-physical environments. Consequently, different IoT-enabled systems such as smart healthcare, smart city, smart home, smart factory, smart transport, and smart agriculture are getting significant attention across the world. Cloud computing is considered as the base stone for offering infrastructure, platform, and software services to develop IoT-enabled systems [1]. However, cloud datacenters reside at a multihop distance from the IoT data sources that increase latency in data propagation. This issue also adversely impacts the service delivery time of IoT-enabled systems, and for real-time use cases such as monitoring health of critical patients, emergency fire, and traffic management, this is quite unacceptable.

In addition, IoT devices are geographically distributed and can generate a huge amount of data in per unit time. If every single IoT-data point is sent to the cloud for processing, the global Internet will be overloaded. To overcome these challenges, involvement of edge computational resources to serve IoT-enabled systems can be a potential solution [2].

Fog computing, interchangeably defined as edge computing, is a very recent inclusion in the domain of computing paradigms that targets offering cloud-like services at the edge network to assist large number of IoT devices. In fog computing, heterogeneous devices such as Cisco IOx networking equipment, micro-datacenter, nano-server, smart phone, personal computer and cloudlets, commonly known as fog node, create a wide distribution of services to process IoT-data closer to the source. Hence, fog computing plays a significant role in minimizing the service delivery latency of different IoT-enabled systems and relaxing the network from dealing a huge amount of data-load [3]. Compared to cloud datacenters, fog nodes are not resource

Fog and Edge Computing: Principles and Paradigms, First Edition.
Edited by Rajkumar Buyya and Satish Narayana Srirama.

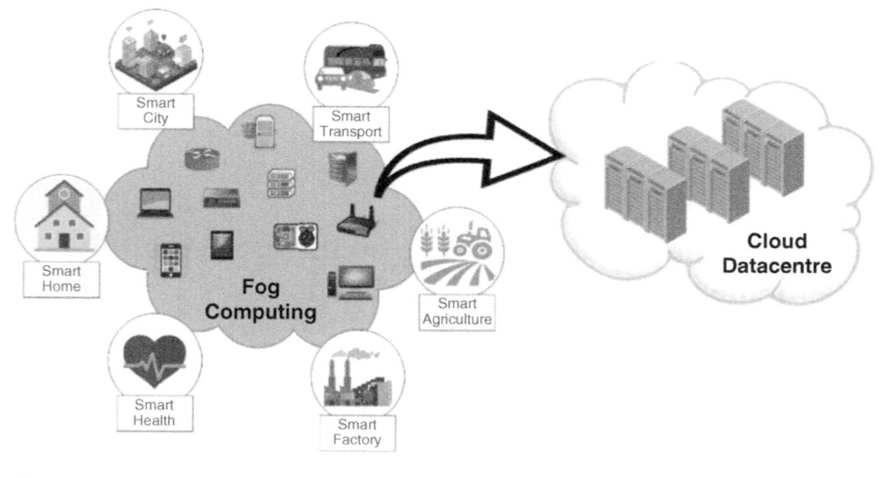

Figure 17.1 Interactions among IoT-enabled systems, fog and cloud computing.

enriched. Therefore, most often, fog and cloud computing paradigm work in integrated manner (Figure 17.1) to tackle both resource and quality of service (QoS) requirements of large -scale IoT-enabled systems [4].

Resource management in fog computing is very complicated as it engages significant number of diverse and resource constraint fog nodes to meet computational demand of IoT-enabled systems in distributed manner. Its integration with cloud triggers further difficulties in combined resource management. Different sensing frequency of IoT devices, distributed application structure and their coordination also influence resource management in fog computing environment [5]. For advancement of fog and its resource management, the necessity of extensive research in beyond question.

In order to develop and evaluate different ideas and resource management policies, empirical analysis on fog environment is the key. Since fog computing environment incorporates IoT devices, fog nodes and cloud datacenters along with huge amount of IoT-data and distributed applications, real-world implementation of fog environment for research will be very costly. Moreover, modification of any entity in real-world fog environment will be tedious. In this circumstance, simulating the fog computing environment can be very helpful. Simulation toolkits not only provide frameworks to design customized experiment environment but also assist in repeatable evaluation. There exists a certain number of simulators such as Edgecloudsim [6], SimpleIoTSimulator [7], and iFogSim [8] for modeling the fog computing environment and running experiments. In this chapter we focus on delivering a tutorial on iFogSim. iFogSim is currently getting remarkable attention from fog computing researchers and we believe this chapter will offer them a simplified way to apply iFogSim in their research works.

In later sections of the chapter, we briefly discuss the iFogSim simulator and its basic components. We revisit the way of installing iFogSim and provide a guideline to model fog environment. Some fog scenarios and their corresponding user extensions are also included in this chapter. Finally, we conclude the chapter with simulation of a simple application placement policy and a case study.

17.2 iFogSim Simulator and Its Components

iFogSim simulation toolkit is developed upon the fundamental framework of CloudSim [9]. CloudSim is one the wildly adopted simulators to model cloud computing environments [10, 11]. Extending the abstraction of basic CloudSim classes, iFogSim offers scopes to simulate customized fog computing environment with large number of fog nodes and IoT devices (e.g. sensors, actuators). However, in iFogSim the classes are annotated in such a way that users, having no prior knowledge of CloudSim, can easily define the infrastructure, service placement, and resource allocation policies for fog computing. iFogSim applies *Sense-Process-Actuate* and distributed dataflow model while simulating any application scenario in fog computing environment. It facilitates evaluation of end-to-end latency, network congestion, power usage, operational expenses, and QoS satisfaction. In a significant amount of research works, iFogSim has already been used for simulating resource [12], mobility [13], latency [14], quality of experience (QoE) [15], energy [16], security [17], and QoS-aware [18] management of fog computing environment. iFogSim is composed of three basic components.

17.2.1 Physical Components

Physical components include fog devices (fog nodes). The fog devices are orchestrated in hierarchical order. The lower-level fog devices are directly connected with associated sensors and actuators. Fog devices act like the datacenters in a cloud computing paradigm by offering memory, network, and computational resources. Each fog device is created with specific instruction processing rate and power consumption attributes (busy and idle power) that reflects its capability and energy efficiency.

The sensors in iFogSim generate tuples that can be referred as tasks in cloud computing. The creation of tuples (tasks) is event driven, and the interval between generating two tuples is set following deterministic distribution while creating the sensors.

17.2.2 Logical Components

Application modules (*AppModules*) and Application edges (*AppEdges*) are the logical components of iFogSim. In iFogSim, applications are considered as a collection of interdependent AppModules that consequently promote the concept of distributed applications. The dependency between two modules is defined by the features of AppEdges. In the domain of cloud computing, the AppModules can be mapped with virtual machines (VMs) and the AppEdges are the logical dataflow between two VMs. In iFogSim, each AppModule (VM) deals with a particular type of tuples (tasks) coming from the predecessor AppModule (VM) of the dataflow. The tuple forwarding between two AppModules can be periodic, and upon reception of a tuple of a particular type, whether a module will trigger another tuple (different type) to the next module is determined by fractional selectivity model.

17.2.3 Management Components

Management component of iFogSim consists of Controller and Module Mapping objects. The Module Mapping object according to the requirements of the AppModules, identifies available resources in the fog devices and place them within it. By default, iFogSim support hierarchical placement of modules. If a fog device is unable to meet the requirements of a module, the module is sent to upper level fog device. The controller object launches the AppModules on their assigned fog devices following the placement information provided by Module Mapping object and periodically manages the resources of fog devices. When the simulation is terminated, the Controller object gather results of cost, network usage and energy consumption during the simulation period from the fog devices. The interaction between iFogSim components are represented in Figure 17.2.

17.3 Installation of iFogSim

iFogSim is an open-source Java-based simulator developed by Cloud Computing and Distributed Systems (CLOUDS) Laboratory at University of Melbourne. The download link for iFogSim source code is given in their website. A very simple way to install iFogSim is described below:

1. Download iFogSim source zip file from https://github.com/Cloudslab/iFogSim or http://cloudbus.org/cloudsim/.
2. Extract the zip file named *iFogSim-master*.
3. Install Java Standard Edition Development Kit (jdk) / Runtime Environment (jre) 1.7 or more and Eclipse Juno / latest releases in personal computer.
4. Define workspace for Eclipse.

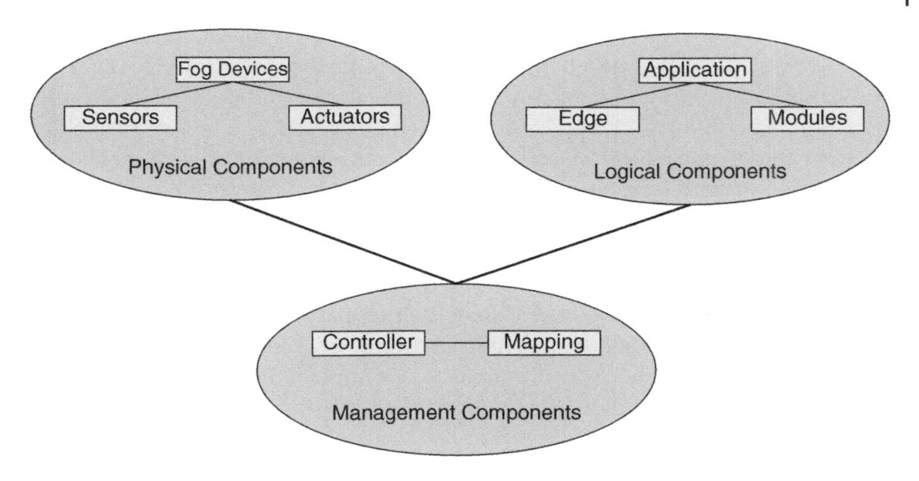

Figure 17.2 High-level view of interactions among iFogSim components.

5. Create a folder in workspace.
6. Copy and Paste all contents from *iFogSim-master* to the newly created folder.
7. Open Eclipse application wizard and create a new Java Project with the same name of newly created folder.
8. From *src* (source) of the project, open org.fog.test.perfeval package and run any of the example simulation codes.

17.4 Building Simulation with iFogSim

In this section, high-level steps to model and simulate fog computing environment in iFogSim are explored.

1. The physical components are created with specific configuration. The configuration parameters include ram, processing capability in million instructions per second (MIPS), cost per million instruction processing, uplink and downlink bandwidth, busy and idle power along with their hierarchical level. While creating lower-level fog devices, the associate IoT devices (sensors and actuators) need to be created. Particular value in the *transmitDistribution* object is set in creating a IoT sensor has to do with its sensing interval. In addition, the creation of sensors and actuators require the reference of application ID and broker ID.
2. Next, the logical components such as AppModule, AppEdge, and AppLoop are required to be created. While creating the AppModules, their configurations are provided and the AppEdge objects include information regarding tuples' type, their direction, CPU, and networking length, along with the

reference of source and destination module. In background, different types of tuples are created based on the given specification on AppEdge objects.

3. Management components (module mapping) are initiated to define different scheduling and AppModule placement policies. Users can consider total energy consumption, service latency, network usage, operational cost, and device heterogeneity while assigning AppModules to fog devices and can extend the abstraction of module mapping class accordingly. Based on the information of AppEdges, the requirements of an AppModule must be aligned with the specification of corresponding tuple type and satisfied by the available fog resources. Once the mapping of AppModules and fog devices are conducted, the information of physical and logical components are forwarded to the controller object. The controller object later submits the whole system to CloudSim engine for simulation.

17.5 Example Scenarios

To start with iFogSim, it is recommended to follow the built-in example codes such as VRGameFog and DCNSFog. Here, we discuss several fog environment scenarios that can be simulated through iFogSim.

17.5.1 Create Fog Nodes with Heterogeneous Configurations

The *FogDevice* class of iFogSim offers users a public constructor to create different types of fog nodes. A sample code snippet to create heterogeneous fog devices (nodes) on a particular hierarchical level is given below:

Code Snippet-1

- To be placed in Main Class

```
static int numOfFogDevices = 10;
static List<FogDevice> fogDevices = new ArrayList<FogDevice>();
static Map<String, Integer> getIdByName = new HashMap <String,
   Integer>();
private static void createFogDevices() {
   FogDevice cloud = createAFogDevice("cloud", 44800, 40000, 100,
      10000, 0, 0.01, 16*103, 16*83.25);
   cloud.setParentId(-1);
   fogDevices.add(cloud);
   getIdByName.put(cloud.getName(), cloud.getId());
   for(int i=0;i<numOfFogDevices;i++){
       FogDevice device = createAFogDevice("FogDevice-"+i,
          getValue(12000, 15000), getValue(4000, 8000),
                    getValue(200, 300), getValue(500, 1000), 1, 0.01,
                                getValue(100,120), getValue(70, 75));
```

```
            device.setParentId(cloud.getId());
            device.setUplinkLatency(10);
            fogDevices.add(device);
            getIdByName.put(device.getName(), device.getId());}
}
private static FogDevice createAFogDevice(String nodeName, long mips,
    int ram, long upBw, long downBw, int level, double ratePerMips,
    double busyPower, double idlePower) {
    List<Pe> peList = new ArrayList<Pe>();
    peList.add(new Pe(0, new PeProvisionerOverbooking(mips)));
    int hostId = FogUtils.generateEntityId();
    long storage = 1000000;
    int bw = 10000;
    PowerHost host = new PowerHost(hostId,
        new RamProvisionerSimple(ram), new
        BwProvisionerOverbooking(bw), storage, peList,
        new StreamOperatorScheduler(peList),
        new FogLinearPowerModel(busyPower, idlePower));
    List<Host> hostList = new ArrayList<Host>();
    hostList.add(host);
    String arch = "x86";
    String os = "Linux";
    String vmm = "Xen";
    double time_zone = 10.0;
    double cost = 3.0;
    double costPerMem = 0.05;
    double costPerStorage = 0.001;
    double costPerBw = 0.0;
    LinkedList<Storage> storageList = new LinkedList<Storage>();
    FogDeviceCharacteristics characteristics = new
        FogDeviceCharacteristics(arch, os, vmm, host, time_zone, cost,
                    costPerMem, costPerStorage, costPerBw);
    FogDevice fogdevice = null;
    try {
        fogdevice = new FogDevice(nodeName, characteristics,
                    new AppModuleAllocationPolicy(hostList),
                    storageList, 10, upBw, downBw, 0, ratePerMips);}
    catch (Exception e) {
          e.printStackTrace();}
    fogdevice.setLevel(level);
    return fogdevice;}
```

Code Snippet-1 creates a certain number of fog nodes having configurations within a fixed range.

17.5.2 Create Different Application Models

Different types of application models can be simulated through iFogSim. In the following subsections, we discuss two types of such an application model.

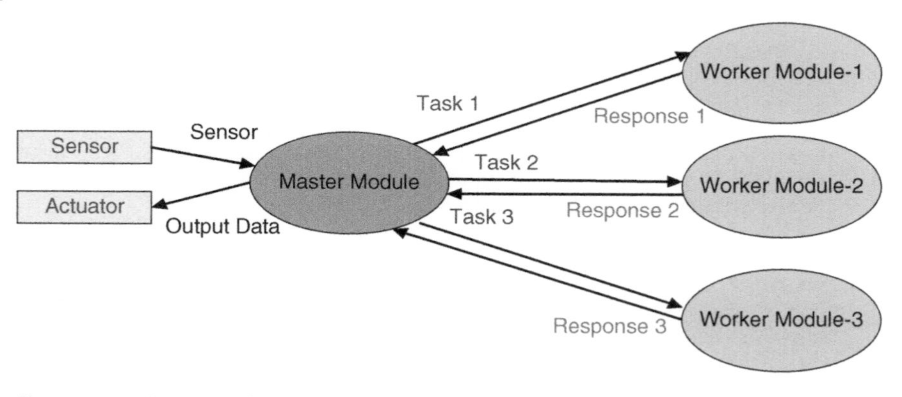

Figure 17.3 Master–worker application model.

17.5.2.1 Master–Worker Application Models

The interaction among application modules on master–worker application model are represented in Figure 17.3.

To model such application in iFogSim, Code Snippet-2 can be used. Note that the name of an IoT sensor and the name of its emitted tuple type should be the same.

Code Snippet-2

- To be placed in Main Class

```
private static Application createApplication(String appId,
   int brokerId){
      Application application = Application.createApplication(appId,
         brokerId);
      application.addAppModule("MasterModule", 10);
      application.addAppModule("WorkerModule-1", 10);
      application.addAppModule("WorkerModule-2", 10);
      application.addAppModule("WorkerModule-3", 10);

      application.addAppEdge("Sensor", "MasterModule", 3000, 500,
        "Sensor", Tuple.UP, AppEdge.SENSOR);
      application.addAppEdge("MasterModule", "WorkerModule-1", 100,
         1000, "Task-1", Tuple.UP, AppEdge.MODULE);
      application.addAppEdge("MasterModule", "WorkerModule-2", 100,
         1000, "Task-2", Tuple.UP, AppEdge.MODULE);
      application.addAppEdge("MasterModule", "WorkerModule-3", 100,
         1000, "Task-3", Tuple.UP, AppEdge.MODULE);
      application.addAppEdge("WorkerModule-1", "MasterModule",20,
         50, "Response-1", Tuple.DOWN, AppEdge.MODULE);
      application.addAppEdge("WorkerModule-2", "MasterModule",20,
         50, "Response-2", Tuple.DOWN, AppEdge.MODULE);
      application.addAppEdge("WorkerModule-3", "MasterModule",20,
         50, "Response-3", Tuple.DOWN, AppEdge.MODULE);
```

```
application.addAppEdge("MasterModule", "Actuators", 100, 50,
   "OutputData", Tuple.DOWN, AppEdge.ACTUATOR);

application.addTupleMapping("MasterModule", " Sensor ",
  "Task-1", new FractionalSelectivity(0.3));
application.addTupleMapping("MasterModule", "Sensor ",
  "Task-2", new FractionalSelectivity(0.3));
application.addTupleMapping("MasterModule", " Sensor ",
  "Task-3", new FractionalSelectivity(0.3));
application.addTupleMapping("WorkerModule-1", "Task-1",
  "Response-1", new FractionalSelectivity(1.0));
application.addTupleMapping("WorkerModule-2", "Task-2",
  "Response-2", new FractionalSelectivity(1.0));
application.addTupleMapping("WorkerModule-3", "Task-3",
  "Response-3", new FractionalSelectivity(1.0));
application.addTupleMapping("MasterModule", "Response-1",
  "OutputData", new FractionalSelectivity(0.3));
application.addTupleMapping("MasterModule", "Response-2",
  "OutputData", new FractionalSelectivity(0.3));
application.addTupleMapping("MasterModule", "Response-3",
  "OutputData", new FractionalSelectivity(0.3));

  final AppLoop loop1 = new AppLoop(new ArrayList<String>(){{
     add("Sensor");add("MasterModule");add("WorkerModule-1");
     add("MasterModule");add("Actuator");}});
  final AppLoop loop2 = new AppLoop(new ArrayList<String>(){{
     add("Sensor");add("MasterModule");add("WorkerModule-2");
     add("MasterModule");add("Actuator");}});
  final AppLoop loop3 = new AppLoop(new ArrayList<String>(){{
     add("Sensor");add("MasterModule");add("WorkerModule-3");
     add("MasterModule");add("Actuator");}});
  List<AppLoop> loops = new ArrayList<AppLoop>(){{add(loop1);
     add(loop2);add(loop3);}};
     application.setLoops(loops);

  return application;}
```

17.5.2.2 Sequential Unidirectional Dataflow Application Model

Figure 17.4 depicts a sample sequential unidirectional application model.
Code Snippet-3 refers the instructions to model such applications in iFogSim.

Code Snippet-3

- To be placed in Main Class

```
private static Application createApplication(String appId,
   int brokerId){
     Application application = Application.createApplication(appId,
        brokerId);
     application.addAppModule("Module1", 10);
```

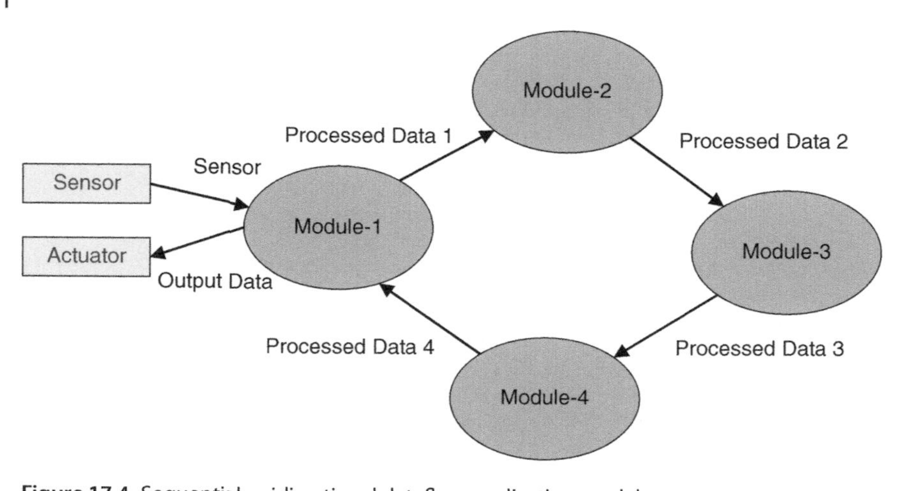

Figure 17.4 Sequential unidirectional dataflow application model.

```
application.addAppModule("Module2", 10);
application.addAppModule("Module3", 10);
application.addAppModule("Module4", 10);

application.addAppEdge("Sensor", "Module1", 3000, 500,
   "Sensor", Tuple.UP, AppEdge.SENSOR);
application.addAppEdge("Module1", "Module2", 100, 1000,
   "ProcessedData-1", Tuple.UP, AppEdge.MODULE);
application.addAppEdge("Module2", "Module3", 100, 1000,
   "ProcessedData-2", Tuple.UP, AppEdge.MODULE);
application.addAppEdge("Module3", "Module4", 100, 1000,
   "ProcessedData-3", Tuple.UP, AppEdge.MODULE);
application.addAppEdge("Module4", "Module1", 100, 1000,
   "ProcessedData-4", Tuple.DOWN, AppEdge.MODULE);
application.addAppEdge("Module1", "Actuators", 100, 50,
   "OutputData", Tuple.DOWN, AppEdge.ACTUATOR);

application.addTupleMapping("Module1", "Sensor",
   "ProcessedData-1", new FractionalSelectivity(1.0));
application.addTupleMapping("Module2", "ProcessedData-1",
   "ProcessedData-2", new FractionalSelectivity(1.0));
application.addTupleMapping("Module3", "ProcessedData-2",
   "ProcessedData-3", new FractionalSelectivity(1.0));
application.addTupleMapping("Module4", "ProcessedData-3",
   "ProcessedData-4", new FractionalSelectivity(1.0));
application.addTupleMapping("Module1", "ProcessedData-4",
   "OutputData", new FractionalSelectivity(1.0));

final AppLoop loop1 = new AppLoop(new ArrayList<String>(){{
   add("Sensor");add("Module1");add("Module2");add("Module3");
   add("Module4");add("Module1");add("Actuator");}});
```

```
List<AppLoop> loops = new ArrayList<AppLoop>(){}add(loop1);}};
application.setLoops(loops);
return application;}
```

17.5.3 Application Modules with Different Configuration

The following Code Snippet-4 creates modules with different configurations.

Code Snippet-4

- To be placed in Main Class

```
private static Application createApplication(String appId,
    int brokerId){
      Application application = Application.createApplication(appId,
         brokerId);
      application.addAppModule("ClientModule", 20,500, 1024, 1500);
      application.addAppModule("MainModule", 100, 1200, 4000, 100);

      application.addAppEdge("Sensor", "ClientModule", 3000, 500,
         "Sensor", Tuple.UP, AppEdge.SENSOR);
      application.addAppEdge("ClientModule", "MainModule", 100,
         1000, "PreProcessedData", Tuple.UP, AppEdge.MODULE);
      application.addAppEdge("MainModule", "ClientModule", 100,
         1000, "ProcessedData", Tuple.DOWN, AppEdge.MODULE);
      application.addAppEdge("ClientModule", "Actuators", 100,
         50, "OutputData", Tuple.DOWN, AppEdge.ACTUATOR);

      application.addTupleMapping("ClientModule", "Sensor",
         "PreProcessedData", new FractionalSelectivity(1.0));
      application.addTupleMapping("MainModule", "PreProcessedData",
         "ProcessedData", new FractionalSelectivity(1.0));
      application.addTupleMapping("ClientModule", "ProcessedData",
         "OutputData", new FractionalSelectivity(1.0));

      final AppLoop loop1 = new AppLoop(new ArrayList<String>(){{
         add("Sensor");add("ClientModule");add("MainModule");
         add("Actuator");}});
      List<AppLoop> loops = new ArrayList<AppLoop>(){{add(loop1);}};
      application.setLoops(loops);
      return application;}
```

- To be placed in Application Class

```
public void addAppModule(String moduleName, int ram, int mips,
   long size, long bw){
      String vmm = "Xen";
      AppModule module = new AppModule(FogUtils.generateEntityId(),
         moduleName, appId, userId, mips, ram, bw, size, vmm,
         new TupleScheduler(mips, 1), new HashMap<Pair<String,
         String>, SelectivityModel>());
      getModules().add(module);
   }
```

17.5.4 Sensors with Different Tuple Emission Rate

To create sensors with different tuple emission rate, Code Snippet-5 can be used.

Code Snippet-5

- To be placed in Main Class

```
private static FogDevice addLowLevelFogDevice(String id,
   int brokerId, String appId, int parentId){
      FogDevice lowLevelFogDevice = createAFogDevice
         ("LowLevelFogDevice-"+id, 1000, 1000, 10000, 270, 2, 0,
         87.53, 82.44);
       lowLevelFogDevice.setParentId(parentId);
       getIdByName.put(lowLevelFogDevice.getName(),
          lowLevelFogDevice.getId());}
       Sensor sensor = new Sensor("s-"+id, "Sensor", brokerId,
          appId, new DeterministicDistribution(getValue(5.00)));
       sensors.add(sensor);
       Actuator actuator = new Actuator("a-"+id, brokerId, appId,
          "OutputData");
       actuators.add(actuator);
       sensor.setGatewayDeviceId(lowLevelFogDevice.getId());
       sensor.setLatency(6.0);
       actuator.setGatewayDeviceId(lowLevelFogDevice.getId());
       actuator.setLatency(1.0);
       return lowLevelFogDevice;}

   private static double getValue(double min) {
      Random rn = new Random();
      return rn.nextDouble()*10 + min;}
```

17.5.5 Send Specific Number of Tuples from a Sensor

Code Snippet-6 enables sensors to create a specific number of tuples.

Code Snippet-6

- To be placed in Sensor Class

```
static int numOfMaxTuples = 100;
static int tuplesCount = 0;
public void transmit(){
   System.out.print(CloudSim.clock()+": ");
   if(tuplesCount<numOfMaxTuples){
      AppEdge _edge = null;
      for(AppEdge edge : getApp().getEdges()){
         if(edge.getSource().equals(getTupleType()))
            _edge = edge;
      }
      long cpuLength = (long) _edge.getTupleCpuLength();
```

```
        long nwLength = (long) _edge.getTupleNwLength();
        Tuple tuple = new Tuple(getAppId(), FogUtils.generateTupleId(),
            Tuple.UP, cpuLength, 1, nwLength, outputSize,
                new UtilizationModelFull(), new UtilizationModelFull(),
                new UtilizationModelFull());
         tuple.setUserId(getUserId());
         tuple.setTupleType(getTupleType());
         tuple.setDestModuleName(_edge.getDestination());
         tuple.setSrcModuleName(getSensorName());
         Logger.debug(getName(), "Sending tuple with tupleId = "
            +tuple.getCloudletId());
         int actualTupleId = updateTimings(getSensorName(),
            tuple.getDestModuleName());
         tuple.setActualTupleId(actualTupleId);
         send(gatewayDeviceId, getLatency(), FogEvents.TUPLE_ARRIVAL,
            tuple);
         tuplesCount++;
        }
}
```

17.5.6 Mobility of a Fog Device

In hierarchical order, each fog device of particular level is connected with upper-level fog nodes. Code Snippet-7 represents how to deal with mobility issues in iFogSim. Here, we have considered mobility of arbitrary lower level fog devices to a certain destination.

Code Snippet-7

- To be placed in Main Class

```
static Map<Integer, Pair<Double, Integer>> mobilityMap = new HashMap
   <Integer, Pair<Double, Integer>>();
   static String mobilityDestination = "FogDevice-0";
   private static FogDevice addLowLevelFogDevice(String id,
      int brokerId, String appId, int parentId){
      FogDevice lowLevelFogDevice = createAFogDevice
         ("LowLevelFogDevice-"+id, 1000, 1000, 10000, 270, 2, 0,
         87.53, 82.44);
      lowLevelFogDevice.setParentId(parentId);
      getIdByName.put(lowLevelFogDevice.getName(),
         lowLevelFogDevice.getId());

      if((int)(Math.random()*100)%2==0){
         Pair<Double, Integer> pair = new Pair<Double,
         Integer>(100.00, getIdByName.get(mobilityDestination));
         mobilityMap.put(lowLevelFogDevice.getId(), pair);}

      Sensor sensor = new Sensor("s-"+id, "Sensor", brokerId, appId,
         new DeterministicDistribution(getValue(5.00)));
      sensors.add(sensor);
```

```
        Actuator actuator = new Actuator("a-"+id, brokerId, appId,
           "OutputData");
        actuators.add(actuator);
        sensor.setGatewayDeviceId(lowLevelFogDevice.getId());
        sensor.setLatency(6.0);
        actuator.setGatewayDeviceId(lowLevelFogDevice.getId());
        actuator.setLatency(1.0);
        return lowLevelFogDevice;}
```

- Inclusion in the Main method

```
Controller controller = new Controller("master-controller",
   fogDevices, sensors, actuators);
controller.setMobilityMap(mobilityMap);
```

- To be placed in Controller Class

```
private static Map<Integer, Pair<Double, Integer>> mobilityMap;
public void setMobilityMap(Map<Integer, Pair<Double,
   Integer>> mobilityMap) {
      this.mobilityMap = mobilityMap;
   }
   private void scheduleMobility(){
      for(int id: mobilityMap.keySet()){
         Pair<Double, Integer> pair = mobilityMap.get(id);
         double mobilityTime = pair.getFirst();
         int mobilityDestinationId = pair.getSecond();
         Pair<Integer, Integer> newConnection = new Pair<Integer,
            Integer>(id, mobilityDestinationId);
         send(getId(), mobilityTime, FogEvents.FutureMobility,
            newConnection);
      }
   }
private void manageMobility(SimEvent ev) {

      Pair<Integer, Integer>pair =
        (Pair<Integer, Integer>)ev.getData();
      int deviceId = pair.getFirst();
      int newParentId = pair.getSecond();
      FogDevice deviceWithMobility = getFogDeviceById(deviceId);
      FogDevice mobilityDest = getFogDeviceById(newParentId);
      deviceWithMobility.setParentId(newParentId);
      System.out.println(CloudSim.clock()+" "+deviceWithMobility
        .getName()+" is now connected to "+mobilityDest.getName());}
```

- Inclusion in Controller startEntity method

```
scheduleMobility();
```

- Inclusion in Controller processEvent method

```
case FogEvents.FutureMobility:
        manageMobility(ev);
        break;
```

- To be placed in FogEvents Class

```
public static final int FutureMobility = BASE+26;
```

In Code Snippet-7, users can add other required instructions on manageMobility method to deal with the mobility driven issues such as AppModule migration and connection with latency.

17.5.7 Connect Lower-Level Fog Devices with Nearby Gateways

Code Snippet-8 refers to a simple way to connect low-level fog devices to nearby gateway fog devices. Here the gateway fog devices are created with corresponding x- and y-coordinate values.

Code Snippet-8

- To be placed in Main Class

```
private static FogDevice addLowLevelFogDevice(String id,
   int brokerId, String appId){
      FogDevice lowLevelFogDevice = createAFogDevice
         ("LowLevelFogDevice-"+id, 1000, 1000, 10000, 270, 2, 0,
         87.53, 82.44);
      lowLevelFogDevice.setParentId(-1);
      lowLevelFogDevice.setxCoordinate(getValue(10.00));
      lowLevelFogDevice.setyCoordinate(getValue(15.00));
      getIdByName.put(lowLevelFogDevice.getName(),
         lowLevelFogDevice.getId());
      Sensor sensor = new Sensor("s-"+id, "Sensor", brokerId,
         appId, new DeterministicDistribution(getValue(5.00)));
      sensors.add(sensor);
      Actuator actuator = new Actuator("a-"+id, brokerId,
         appId, "OutputData");
      actuators.add(actuator);
      sensor.setGatewayDeviceId(lowLevelFogDevice.getId());
      sensor.setLatency(6.0);
      actuator.setGatewayDeviceId(lowLevelFogDevice.getId());
      actuator.setLatency(1.0);
      return lowLevelFogDevice;}

private static double getValue(double min) {
      Random rn = new Random();
      return rn.nextDouble()*10 + min;}
```

- To be placed in Constructor Class

```
private void gatewaySelection() {
      // TODO Auto-generated method stub
      for(int i=0;i<getFogDevices().size();i++){
         FogDevice fogDevice = getFogDevices().get(i);
         int parentID=-1;
         if(fogDevice.getParentId()==-1) {
            double minDistance = Config.MAX_NUMBER;
            for(int j=0;j<getFogDevices().size();j++){
               FogDevice anUpperDevice = getFogDevices().get(j);
               if(fogDevice.getLevel()+1==anUpperDevice.getLevel()){
                  double distance = calculateDistance(fogDevice,
                     anUpperDevice);
                  if(distance<minDistance){
                     minDistance = distance;
                     parentID = anUpperDevice.getId();}
                  }
               }
            }
            fogDevice.setParentId(parentID);
      }
   }
private double calculateDistance(FogDevice fogDevice,
     FogDevice anUpperDevice) {
      // TODO Auto-generated method stub
      return Math.sqrt(Math.pow(fogDevice.getxCoordinate()-
         anUpperDevice.getxCoordinate(), 2.00)+
            Math.pow(fogDevice.getyCoordinate()-anUpperDevice
               .getyCoordinate(), 2.00));}
```

- To be placed in FogDevice Class

```
protected double xCoordinate;
protected double yCoordinate;

public double getxCoordinate() {
   return xCoordinate;}

public void setxCoordinate(double xCoordinate) {
   this.xCoordinate = xCoordinate;}

public double getyCoordinate() {
   return yCoordinate;}

public void setyCoordinate(double yCoordinate) {
   this.yCoordinate = yCoordinate;}
```

- Inclusion in Controller constructor method

```
gatewaySelection();
```

- Inclusion in Config Class

```
public static final double MAX_NUMBER = 9999999.00;
```

17.5.8 Make Cluster of Fog Devices

In Code Snippet-9, we draw a very simple principle for creating clusters of fog devices. Here, two fog devices, residing in the same level and connected with identical upper-level fog nodes, if located at a threshold distance, are considered belonging to the same fog cluster.

Code Snippet-9

- To be placed in Controller Class

```
static Map<Integer, Integer> clusterInfo = new HashMap<Integer,
   Integer>();
static Map<Integer, List<Integer>> clusters = new HashMap<Integer,
   List<Integer>>();
   private void formClusters() {
      for(FogDevice fd: getFogDevices()){
         clusterInfo.put(fd.getId(), -1);
      }

      int clusterId = 0;

      for(int i=0;i<getFogDevices().size();i++){
         FogDevice fd1 = getFogDevices().get(i);
         for(int j=0;j<getFogDevices().size();j++) {
            FogDevice fd2 = getFogDevices().get(j);
            if(fd1.getId()!=fd2.getId()&&
               fd1.getParentId()==fd2.getParentId()
                   &&calculateDistance(fd1,fd2)<Config.CLUSTER_
                     DISTANCE && fd1.getLevel()==fd2.getLevel())
            {
            int fd1ClusteriD = clusterInfo.get(fd1.getId());
            int fd2ClusteriD = clusterInfo.get(fd2.getId());
            if(fd1ClusteriD==-1 && fd2ClusteriD==-1){
               clusterId++;
               clusterInfo.put(fd1.getId(), clusterId);
               clusterInfo.put(fd2.getId(), clusterId);

            }
            else if(fd1ClusteriD==-1)
              clusterInfo.put(fd1.getId(),
              clusterInfo.get(fd2.getId()));
            else if(fd2ClusteriD==-1)
              clusterInfo.put(fd2.getId(),
              clusterInfo.get(fd1.getId()));
            }
         }
      }

      for(int id:clusterInfo.keySet()){
         if(!clusters.containsKey(clusterInfo.get(id))){
            List<Integer>clusterMembers = new ArrayList<Integer>();
```

```
            clusterMembers.add(id);
            clusters.put(clusterInfo.get(id), clusterMembers);
        }
        else
        {
            List<Integer>clusterMembers = clusters.get
              (clusterInfo.get(id));
            clusterMembers.add(id);
             clusters.put(clusterInfo.get(id), clusterMembers);
        }
    }

    for(int id:clusters.keySet())
        System.out.println(id+" "+clusters.get(id));

}
```

- Inclusion in Controller constructor method:

```
formClusters();
```

- Inclusion in Config Class:

```
public static final double CLUSTER_DISTANCE = 2.00;
```

17.6 Simulation of a Placement Policy

In this section, we discuss a simple application placement scenario and implement the placement policy in iFogSim simulated Fog environment.

17.6.1 Structure of Physical Environment

In the fog environment, the devices are orchestrated in three-tier hierarchical order (Figure 17.5). The lower-level end fog devices are connected to the IoT sensors and actuators. The gateway fog devices bridge the cloud datacenter and end fog devices to execute a modular application. For simplicity, the fog devices of same hierarchical levels are considered homogeneous. The sensing frequency is same for all sensors.

17.6.2 Assumptions for Logical Components

The application model is depicted in Figure 17.6. Here we assume that ClientModule is placed in end fog devices and StorageModule is placed in the cloud. The MainModule requires certain amount of computational resources to be initiated. To serve the demand of different end devices within their deadline, additional resources can be requested by end devices to connected gateway fog devices.

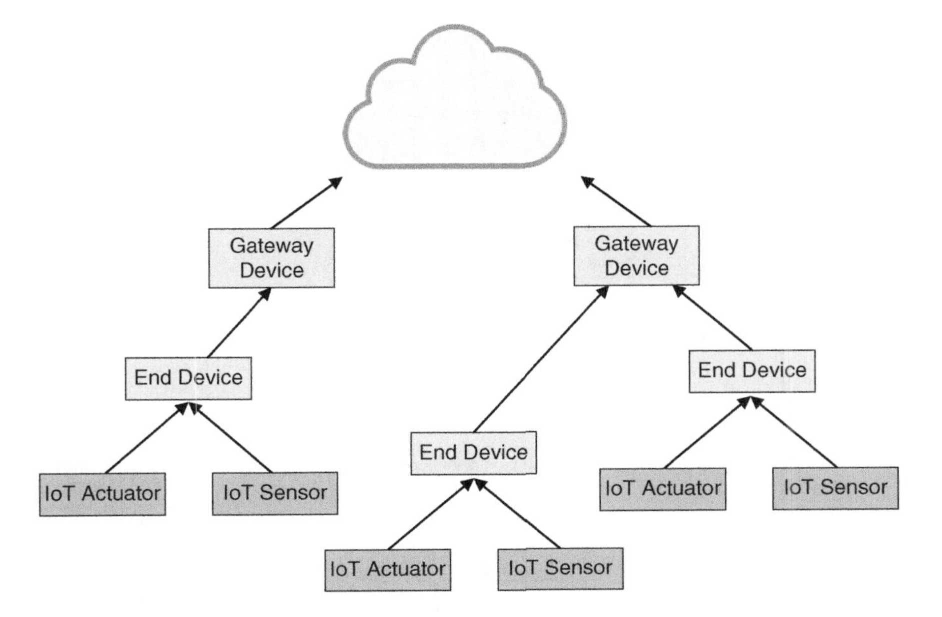

Figure 17.5 Network topology for the placement policy.

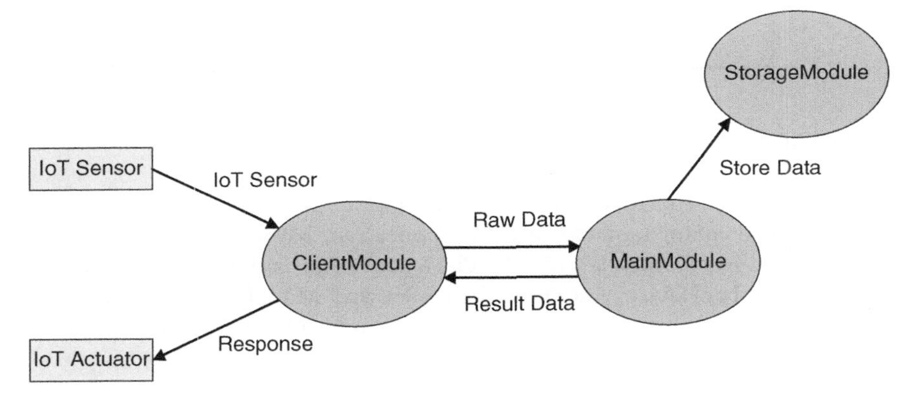

Figure 17.6 Application model for the placement policy.

17.6.3 Management (Application Placement) Policy

We target the MainApplication modules in gateway fog devices for different end devices based on their deadline requirement and resource availability in the host devices. For easier understanding, the flowchart of the application placement policy is represented in Figure 17.7.

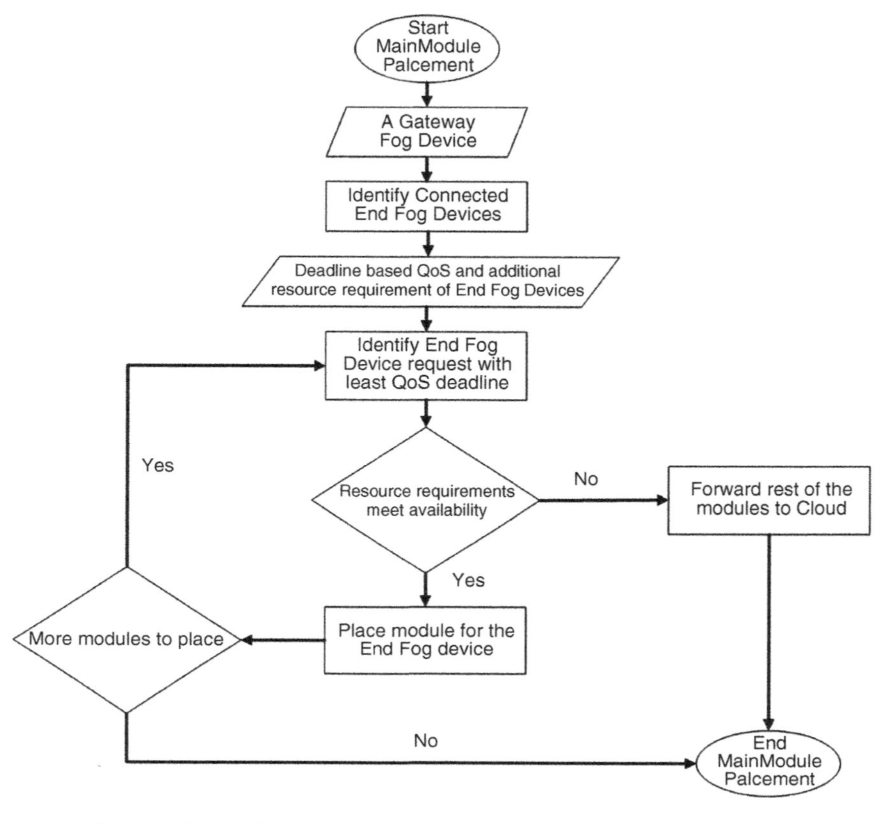

Figure 17.7 Flowchart of the application placement policy.

Code Snippet-10 represents necessary instruction to simulate the case scenario in iFogSim toolkit. Here MyApplication, MySensor, MyFogDevice, MyActuator, My Controller, and MyPlacement class is the same as Application, Sensor, FogDevice, Actuator, Controller, and ModulePlacement class of iFogSim packages, respectively. The inclusions are explicitly mentioned.

Code Snippet-10

- Main Class

```
public class TestApplication {
   static List<MyFogDevice> fogDevices = new
      ArrayList<MyFogDevice>();
   static Map<Integer,MyFogDevice> deviceById =
      new HashMap<Integer,MyFogDevice>();
   static List<MySensor> sensors = new ArrayList<MySensor>();
   static List<MyActuator> actuators = new ArrayList<MyActuator>();
```

```
static List<Integer> idOfEndDevices = new ArrayList<Integer>();
static Map<Integer, Map<String, Double>> deadlineInfo =
   new HashMap<Integer, Map<String, Double>>();
static Map<Integer, Map<String, Integer>> additionalMipsInfo =
   new HashMap<Integer, Map<String, Integer>>();

static boolean CLOUD = false;

static int numOfGateways = 2;
static int numOfEndDevPerGateway = 3;
static double sensingInterval = 5;

public static void main(String[] args) {

   Log.printLine("Starting TestApplication...");

   try{
      Log.disable();
      int num_user = 1;
      Calendar calendar = Calendar.getInstance();
      boolean trace_flag = false;
      CloudSim.init(num_user, calendar, trace_flag);
      String appId = "test_app";
      FogBroker broker = new FogBroker("broker");

      createFogDevices(broker.getId(), appId);

      MyApplication application = createApplication(appId,
         broker.getId());
      application.setUserId(broker.getId());

      ModuleMapping moduleMapping = ModuleMapping
         .createModuleMapping();

      moduleMapping.addModuleToDevice("storageModule", "cloud");
      for(int i=0;i<idOfEndDevices.size();i++)
      {
          MyFogDevice fogDevice = deviceById.get
             (idOfEndDevices.get(i));
          moduleMapping.addModuleToDevice("clientModule",
             fogDevice.getName());
      }


      MyController controller = new MyController
         ("master-controller", fogDevices, sensors, actuators);

      controller.submitApplication(application, 0,
         new MyModulePlacement(fogDevices, sensors,
         actuators, application, moduleMapping,"mainModule"));
```

```
            TimeKeeper.getInstance().setSimulationStartTime
                (Calendar.getInstance().getTimeInMillis());

            CloudSim.startSimulation();

            CloudSim.stopSimulation();

            Log.printLine("TestApplication finished!");
        } catch (Exception e) {
              e.printStackTrace();
              Log.printLine("Unwanted errors happen");
        }
    }

    private static double getvalue(double min, double max)
    {
        Random r = new Random();
        double randomValue = min + (max - min) * r.nextDouble();
        return randomValue;
    }

    private static int getvalue(int min, int max)
    {
        Random r = new Random();
        int randomValue = min + r.nextInt()%(max - min);
        return randomValue;
    }

    private static void createFogDevices(int userId, String appId) {
        MyFogDevice cloud = createFogDevice("cloud", 44800, 40000,
            100, 10000, 0, 0.01, 16*103, 16*83.25);
        cloud.setParentId(-1);
        fogDevices.add(cloud);
        deviceById.put(cloud.getId(), cloud);

        for(int i=0;i<numOfGateways;i++){
            addGw(i+"", userId, appId, cloud.getId());
        }
    }

    private static void addGw(String gwPartialName, int userId,
        String appId, int parentId){
        MyFogDevice gw = createFogDevice("g-"+gwPartialName, 2800,
            4000, 10000, 10000, 1, 0.0, 107.339, 83.4333);
        fogDevices.add(gw);
        deviceById.put(gw.getId(), gw);
        gw.setParentId(parentId);
        gw.setUplinkLatency(4);
        for(int i=0;i<numOfEndDevPerGateway;i++){
            String endPartialName = gwPartialName+"-"+i;
            MyFogDevice end  = addEnd(endPartialName, userId,
```

```
            appId, gw.getId());
        end.setUplinkLatency(2);
        fogDevices.add(end);
        deviceById.put(end.getId(), end);
        }
}

private static MyFogDevice addEnd(String endPartialName,
    int userId, String appId, int parentId){
    MyFogDevice end = createFogDevice("e-"+endPartialName, 3200,
        1000, 10000, 270, 2, 0, 87.53, 82.44);
    end.setParentId(parentId);
    idOfEndDevices.add(end.getId());
    MySensor sensor = new MySensor("s-"+endPartialName,
      "IoTSensor", userId, appId, new DeterministicDistribution
        (sensingInterval));
      // inter-transmission time of EEG sensor follows a
         deterministic distribution sensors.add(sensor);
    MyActuator actuator = new MyActuator("a-"+endPartialName,
        userId, appId, "IoTActuator");
    actuators.add(actuator);
    sensor.setGatewayDeviceId(end.getId());
    sensor.setLatency(6.0);  // latency of connection between
      EEG sensors and the parent Smartphone is 6 ms
    actuator.setGatewayDeviceId(end.getId());
    actuator.setLatency(1.0);  // latency of connection between
        Display actuator and the parent Smartphone is 1 ms
    return end;
}

private static MyFogDevice createFogDevice(String nodeName,
    long mips, int ram, long upBw, long downBw, int level,
    double ratePerMips, double busyPower, double idlePower) {
    List<Pe> peList = new ArrayList<Pe>();
    peList.add(new Pe(0, new PeProvisionerOverbooking(mips)));
    int hostId = FogUtils.generateEntityId();
    long storage = 1000000;
    int bw = 10000;

    PowerHost host = new PowerHost(
            hostId,
            new RamProvisionerSimple(ram),
            new BwProvisionerOverbooking(bw),
            storage,
            peList,
            new StreamOperatorScheduler(peList),
            new FogLinearPowerModel(busyPower, idlePower)
        );
    List<Host> hostList = new ArrayList<Host>();
    hostList.add(host);
    String arch = "x86";
```

```
      String os = "Linux";
      String vmm = "Xen";
      double time_zone = 10.0;
      double cost = 3.0;
      double costPerMem = 0.05;
      double costPerStorage = 0.001;
      double costPerBw = 0.0;
      LinkedList<Storage> storageList = new LinkedList<Storage>();
      FogDeviceCharacteristics characteristics =
         new FogDeviceCharacteristics(
         arch, os, vmm, host, time_zone, cost, costPerMem,
         costPerStorage, costPerBw);

      MyFogDevice fogdevice = null;
      try {
         fogdevice = new MyFogDevice(nodeName, characteristics,
            new AppModuleAllocationPolicy(hostList),
            storageList, 10, upBw, downBw, 0, ratePerMips);
      } catch (Exception e) {
         e.printStackTrace();}
      fogdevice.setLevel(level);
      fogdevice.setMips((int) mips);
      return fogdevice;}

   @SuppressWarnings({"serial" })
   private static MyApplication createApplication(String appId,
      int userId){

      MyApplication application = MyApplication.createApplication
         (appId, userId);
      application.addAppModule("clientModule",10, 1000, 1000, 100);
      application.addAppModule("mainModule", 50, 1500, 4000, 800);
      application.addAppModule("storageModule", 10, 50, 12000, 100);

      application.addAppEdge("IoTSensor", "clientModule", 100, 200,
         "IoTSensor", Tuple.UP, AppEdge.SENSOR);
      application.addAppEdge("clientModule", "mainModule", 6000,
         600  , "RawData", Tuple.UP, AppEdge.MODULE);
      application.addAppEdge("mainModule", "storageModule", 1000,
         300, "StoreData", Tuple.UP, AppEdge.MODULE);
      application.addAppEdge("mainModule", "clientModule", 100, 50,
         "ResultData", Tuple.DOWN, AppEdge.MODULE);
      application.addAppEdge("clientModule", "IoTActuator", 100, 50,
         "Response", Tuple.DOWN, AppEdge.ACTUATOR);

      application.addTupleMapping("clientModule", "IoTSensor",
         "RawData", new FractionalSelectivity(1.0));
      application.addTupleMapping("mainModule", "RawData",
         "ResultData", new FractionalSelectivity(1.0));
      application.addTupleMapping("mainModule", "RawData",
         "StoreData", new FractionalSelectivity(1.0));
```

```
        application.addTupleMapping("clientModule", "ResultData",
            "Response", new FractionalSelectivity(1.0));

        for(int id:idOfEndDevices)
        {
            Map<String,Double>moduleDeadline = new HashMap
              <String,Double>();
            moduleDeadline.put("mainModule", getvalue(3.00, 5.00));
            Map<String,Integer>moduleAddMips = new HashMap<String,
                Integer>();
            moduleAddMips.put("mainModule", getvalue(0, 500));
            deadlineInfo.put(id, moduleDeadline);
            additionalMipsInfo.put(id,moduleAddMips);}

        final AppLoop loop1 = new AppLoop(new ArrayList<String>(){{
            add("IoTSensor");add("clientModule");add("mainModule");
            add("clientModule");add("IoTActuator");}});
        List<AppLoop> loops = new ArrayList<AppLoop>(){{add(loop1);}};
        application.setLoops(loops);
        application.setDeadlineInfo(deadlineInfo);
        application.setAdditionalMipsInfo(additionalMipsInfo);
        return application;}
}
```

- Inclusion in MyApplication Class

```
private Map<Integer, Map<String, Double>> deadlineInfo;
private Map<Integer, Map<String, Integer>> additionalMipsInfo;

public Map<Integer, Map<String, Integer>> getAdditionalMipsInfo() {
        return additionalMipsInfo;
    }
public void setAdditionalMipsInfo(
        Map<Integer, Map<String, Integer>> additionalMipsInfo) {
        this.additionalMipsInfo = additionalMipsInfo;
    }
public void setDeadlineInfo(Map<Integer, Map<String, Double>>
      deadlineInfo) {
       this.deadlineInfo = deadlineInfo;
    }

public Map<Integer, Map<String, Double>> getDeadlineInfo() {
        return deadlineInfo;
    }
public void addAppModule(String moduleName,int ram, int mips,
    long size, long bw){
        String vmm = "Xen";
        AppModule module = new AppModule(FogUtils.generateEntityId(),
           moduleName, appId, userId, mips, ram, bw, size, vmm,
           new TupleScheduler(mips, 1), new HashMap<Pair<String,
           String>, SelectivityModel>());

          getModules().add(module);    }
```

- Inclusion in MyFogDevice Class

```
private int mips;

   public int getMips() {
      return mips;
   }

   public void setMips(int mips) {
      this.mips = mips;
   }
```

- MyModulePlacement Class

```
public class MyModulePlacement extends MyPlacement{

protected ModuleMapping moduleMapping;
protected List<MySensor> sensors;
protected List<MyActuator> actuators;
protected String moduleToPlace;
protected Map<Integer, Integer> deviceMipsInfo;

public MyModulePlacement(List<MyFogDevice> fogDevices,
    List<MySensor> sensors, List<MyActuator> actuators,
         MyApplication application, ModuleMapping
            moduleMapping, String moduleToPlace){
      this.setMyFogDevices(fogDevices);
      this.setMyApplication(application);
      this.setModuleMapping(moduleMapping);
      this.setModuleToDeviceMap(new HashMap<String,
         List<Integer>>());
      this.setDeviceToModuleMap(new HashMap<Integer,
         List<AppModule>>());
      setMySensors(sensors);
      setMyActuators(actuators);
      this.moduleToPlace = moduleToPlace;
      this.deviceMipsInfo = new HashMap<Integer, Integer>();
      mapModules();
   }

   @Override
   protected void mapModules() {

         for(String deviceName : getModuleMapping().
            getModuleMapping().keySet()){
            for(String moduleName : getModuleMapping().
               getModuleMapping().get(deviceName)){
               int deviceId = CloudSim.getEntityId(deviceName);
               AppModule appModule = getMyApplication().
                  getModuleByName(moduleName);
               if(!getDeviceToModuleMap().containsKey(deviceId))
               {
```

```
                List<AppModule>placedModules = new ArrayList
                    <AppModule>();
                placedModules.add(appModule);
                getDeviceToModuleMap().put(deviceId,
                    placedModules);
            }
            else
            {
                List<AppModule>placedModules =
                    getDeviceToModuleMap().get(deviceId);
                placedModules.add(appModule);
                getDeviceToModuleMap().put(deviceId,
                     placedModules);
            }
        }
}
for(MyFogDevice device:getMyFogDevices())
{
    int deviceParent = -1;
    List<Integer>children = new ArrayList<Integer>();

    if(device.getLevel()==1)
    {
        if(!deviceMipsInfo.containsKey(device.getId()))
            deviceMipsInfo.put(device.getId(), 0);
        deviceParent = device.getParentId();
        for(MyFogDevice deviceChild:getMyFogDevices())
        {
            if(deviceChild.getParentId()==device.getId()){
                children.add(deviceChild.getId());}
        }
        Map<Integer, Double>childDeadline = new HashMap<Integer,
            Double>();
        for(int childId:children)
            childDeadline.put(childId,getMyApplication().
            getDeadlineInfo().get(childId).get(moduleToPlace));

        List<Integer> keys = new ArrayList <Integer>
             (childDeadline.keySet());

        for(int i = 0; i<keys.size()-1; i++)
        {
            for(int j=0;j<keys.size()-i-1;j++)
            {
                 if(childDeadline.get(keys.get(j))>childDeadline
                     .get(keys.get(j+1))){
                   int tempJ = keys.get(j);
                   int tempJn = keys.get(j+1);
                   keys.set(j, tempJn);
                   keys.set(j+1, tempJ);
            }
```

```
            }
        }
        int baseMipsOfPlacingModule = (int)getMyApplication().
          getModuleByName(moduleToPlace).getMips();
        for(int key:keys)
        {
            int currentMips = deviceMipsInfo.get(device.getId());
            AppModule appModule = getMyApplication()
                .getModuleByName(moduleToPlace);
            int additionalMips = getMyApplication().
               getAdditionalMipsInfo().get(key).get(moduleToPlace);
            if(currentMips+baseMipsOfPlacingModule+additionalMips
                                      <device.getMips())
            {
              currentMips = currentMips+baseMipsOfPlacingModule+
                 additionalMips;
              deviceMipsInfo.put(device.getId(), currentMips);
              if(!getDeviceToModuleMap().containsKey
                   (device.getId()))
              {
                 List<AppModule>placedModules = new
                    ArrayList<AppModule>();
                 placedModules.add(appModule);
                 getDeviceToModuleMap().put(device.getId(),
                    placedModules);

              }
              else
              {
              List<AppModule>placedModules =
                 getDeviceToModuleMap().get(device.getId());
              placedModules.add(appModule);
              getDeviceToModuleMap().put(device.getId(),
                 placedModules);
              }

          }
          else
          {
               List<AppModule>placedModules =
                  getDeviceToModuleMap().get(deviceParent);
               placedModules.add(appModule);
               getDeviceToModuleMap().put(deviceParent,
               placedModules);
               }
          }
        }
    }
}

public ModuleMapping getModuleMapping() {
```

```
        return moduleMapping;
    }

    public void setModuleMapping(ModuleMapping moduleMapping) {
        this.moduleMapping = moduleMapping;
    }


    public List<MySensor> getMySensors() {
        return sensors;
    }

      public void setMySensors(List<MySensor> sensors) {
          this.sensors = sensors;
    }

    public List<MyActuator> getMyActuators() {
        return actuators;
    }

    public void setMyActuators(List<MyActuator> actuators) {
        this.actuators = actuators;
    }
}
```

17.7 A Case Study in Smart Healthcare

IoT's role in healthcare solutions currently lies in handheld or body-connected IoT devices such as pulse oximeter, ECG monitor, smart watches, etc., which perceive health context of the users through a client application module. The IoT devices are usually connected with smart phones. The smart phones act as the application gateway node for the corresponding application. These nodes pre-process the IoT-device-sensed data. If resource availability in the application gateway node meets the requirements, the data analysis and event management operation of the application is conducted there. Otherwise, the operations are executed in upper-level fog computational nodes. For the second case, application gateway nodes select suitable computational nodes to deploy other application modules and initiate actuators based on the result coming from those modules. Extending such cases of IoT-enabled healthcare solution [5], we discuss the ways to simulate the corresponding fog environment in iFogSim. The system architecture and application model for the IoT-enabled healthcare solutions is represented in Figure 17.8 and 17.9, respectively. Features of the system and the application, along with required guidelines to model them in iFogSim, are listed below:

- It is an *n*-tier hierarchical fog environment. As the rank of fog levels goes higher, the number of fog devices residing at that level gets lower. Fog

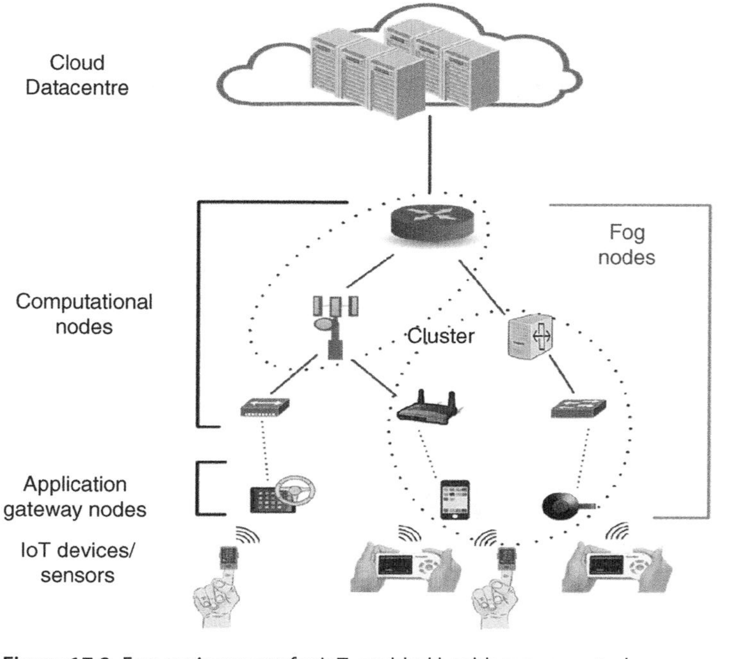

Figure 17.8 Fog environment for IoT-enabled healthcare case study.

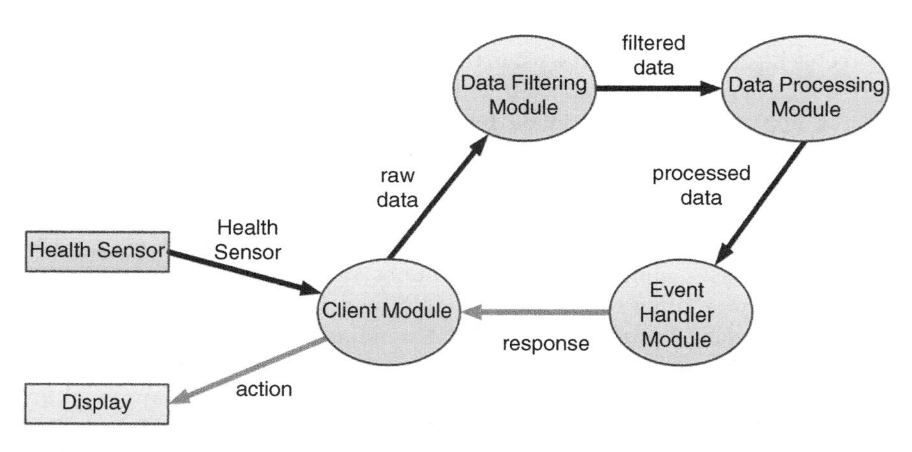

Figure 17.9 Application model for IoT-enabled healthcare case study.

devices form clusters among themselves and can be mobile. IoT devices (pulse oximeter, ECG monitor, etc.) are connected to lower-level fog devices. The sensing frequency of the IoT devices are different. There are three steps to model these physical entities:

1. Create FogDevice object and define n-tier hierarchical fog environment by following Code Snippet-1 and -10.
2. Create Sensor object with different sensing intervals and transmission of a particular number of tuples using Code Snippet-5 and -6.
3. Model mobility of and form cluster of the fog devices by modifying Code Snippet-7 and -9, respectively.

- The application model consists of four modules with a sequential unidirectional data flow. The requirements of the application modules are different, and each application modules can request for additional resources from the host fog devices to process a data within QoS-defined deadline. There are also three steps to model these logical entities:
  1. Define application object for the discussing IoT-enabled healthcare application through Code Snippet-2 and -3.
  2. Create ApplicationModule object with different requirements using Code Snippet-4.
  3. Deal with additional requirements and deadline expectations of the ApplicationModule objects following Code Snippet-10.
- The application module placement in this case study should be done is such a way that takes least possible amount of time for the application to generate response for an event. In this case latency-aware placement of the modules on constrained Fog devices can be very effective [14]. Steps to model these management issues are:
  1. Connect Application gateway nodes with low latency fog computational nodes modifying Code Snippet-8.
  2. Implement user-defined latency-aware application module placement policy following Code Snippet-10.

17.8 Conclusions

In this chapter, we highlighted key features of iFogSim along with providing instructions to install it and simulate a fog environment. We discussed some example scenarios and corresponding code snippets. Finally, we demonstrated how to implement custom application placement in a iFogSim simulated fog environment and provided an IoT-enabled smart healthcare case study.

The simulation source codes of example scenarios and placement policy discussed in this chapter are available from CLOUDS Laboratory GitHub webpage, https://github.com/Cloudslab/iFogSimTutorials.

References

1 J. Gubbi, R. Buyya, S. Marusic, and M. Palaniswami. Internet of Things (IoT): A vision, architectural elements, and future directions. *Future Generation Computer Systems,* 29(7): 1645–1660, 2013.

2 R. Mahmud, K. Ramamohanarao, and R. Buyya. Fog computing: A taxonomy, survey and future directions. *Internet of Everything: Algorithms, Methodologies, Technologies and Perspectives.* Di Martino Beniamino, Yang Laurence, Kuan-Ching Li, et al. (eds.), ISBN 978-981-10-5861-5, Springer, Singapore, Oct. 2017.

3 F. Bonomi, R. Milito, J. Zhu, and S. Addepalli. Fog computing and its role in the Internet of things. In *Proceedings of the first edition of the MCC workshop on Mobile Cloud computing (MCC '12),* pp. 13–16, Helsinki, Finland, Aug. 17–17, 2012.

4 A. V. Dastjerdi and R. Buyya. Fog computing: Helping the Internet of Things realize its potential. *IEEE Computer,* 49(8):112–116, 2016.

5 R. Mahmud, F. L. Koch, and R. Buyya. Cloud-fog interoperability in IoT-enabled healthcare solutions. In *Proceedings of the 19th International Conference on Distributed Computing and Networking (ICDCN '18),* pp. 1–10, Varanasi, India, Jan. 4–7, 2018.

6 C. Sonmez, A. Ozgovde, and C. Ersoy. Edgecloudsim. An environment for performance evaluation of edge computing systems. In *Proceedings of the Second International Conference on Fog and Mobile Edge Computing (FMEC'17),* pp. 39–44, Valencia, Spain, May 8–11, 2017.

7 Online: https://www.smplsft.com/SimpleIoTSimulator.html, Accessed April 17, 2018.

8 H. Gupta, A. Dastjerdi, S. Ghosh, and R. Buyya. iFogSim: A toolkit for modeling and simulation of resource management techniques in internet of things, edge and fog computing environments. *Software: Practice and Experience (SPE),* 47(9): 1275–1296, 2017.

9 R.N. Calheiros, R. Ranjan, A. Beloglazov, C.A.F. De Rose, and R. Buyya. CloudSim: A toolkit for modeling and simulation of cloud computing environments and evaluation of resource provisioning algorithms. *Software: Practice and Experience,* 41(1): 23–50, 2011.

10 R. Benali, H. Teyeb, A. Balma, S. Tata, and N. Hadj-Alouane. Evaluation of traffic-aware VM placement policies in distributed cloud using CloudSim. In *Proceedings of the 25th International Conference on Enabling Technologies: Infrastructure for Collaborative Enterprises (WETICE'16),* pp. 95–100, Paris, France, June 13–15, 2016.

11 R. Mahmud, M. Afrin, M.A. Razzaque, M.M. Hassan, A. Alelaiwi and M.A. AlRubaian. Maximizing quality of experience through context-aware mobile application scheduling in cloudlet infrastructure. *Software: Practice and Experience,* 46(11):1525–1545, 2016.

12 M. Taneja and A. Davy. Resource aware placement of IoT application modules in Fog-Cloud Computing Paradigm. In *Proceedings of the IFIP/IEEE Symposium on Integrated Network and Service Management (IM'17),* pp. 1222–1228, Lisbon, Portugal, May 8–12, 2017

13 L.F. Bittencourt, J. Diaz-Montes, R. Buyya, O.F. Rana, and M. Parashar. Mobility-aware application scheduling in fog computing. *IEEE Cloud Computing,* 4(2): 26–35, 2017.

14 R. Mahmud, K. Ramamohanarao, and R. Buyya. Latency-aware application module management for fog computing environments. *ACM Transactions on Internet Technology (TOIT),* DOI: 10.1145/3186592, 2018.

15 R. Mahmud, S. N. Srirama, K. Ramamohanarao, and R. Buyya. Quality of experience (QoE)-aware placement of applications in fog computing environments. *Journal of Parallel and Distributed Computing.* DOI: 10.1016/j.jpdc.2018.03.004, 2018.

16 M. Mahmoud, J. Rodrigues, K. Saleem, J. Al-Muhtadi, N. Kumar, and V. Korotaev. Towards energy-aware fog-enabled cloud of things for healthcare. *Computers & Electrical Engineering,* 67: 58–69, 2018).

17 A. Chai, M. Bazm, S. Camarasu-Pop, T. Glatard, H. Benoit-Cattin and F. Suter. Modeling distributed platforms from application traces for realistic file transfer simulation. In *Proceedings of the 17th IEEE/ACM International Symposium on Cluster, Cloud and Grid Computing (CCGRID'17),* pp. 54–63, Madrid, Spain, May 14–15, 2017.

18 O. Skarlat, M. Nardelli, S. Schulte, and S. Dustdar. Towards QoS-aware fog service placement. In *Proceedings of the 1st IEEE International Conference on Fog and Edge Computing (ICFEC'17),* pp. 89–96, Madrid, Spain, May 14–15, 2017.

Index

a

Advanced Message Queuing Protocol (AMQP) 11
AlexNet 325
Ambient assisted living 3, 17
Anomaly detection 238, 244, 412
Apache Edgent 19
Application module 436, 440, 443, 451
Application placement 96, 211, 435, 451
Artificial neural networks 237, 244
Autonomous vehicle testing 403

b

Bandwidth 3, 59, 71, 90, 109, 181, 192, 199, 277, 359, 437
Bayesian models 57
Benchmarks 154
Big-Data driven analytics 364, 368
Blockchain 146, 160, 163, 343, 400
BLURS 3
  bandwidth 1
  latency 2
  resource-constraint 2
  security 2
  uninterrupted 2

c

Cisco IOX 18, 433
Cisco research 5
Cloud-centric Internet of Things 3
Cloud computing 82
  IaaS 21, 204
  PaaS 10, 145
  SaaS 10
  XaaS 16
Cloud federation 413, 419, 425
Cloudlet 7, 131
Cloud radio access network 82
Cloud-to-thing 51, 355
Clustering 139, 235, 286, 331, 368
Cluster storage 152
Collision avoidance testing 403
Computational complexity 70, 104, 106
Conformance testing 383
Consent 412, 416
Constraint Application Protocol (CoAP) 11
Container orchestration 146, 152, 162
Context as a service 10
Context awareness 82, 137, 193, 268, 352
Co-optimization 118
Cost model 201, 211, 271
Customer-premises equipment 16

d

Data acquisition 126, 175, 284
Data aggregation 173, 368
Data Analytics Engine 274

Fog and Edge Computing: Principles and Paradigms, First Edition.
Edited by Rajkumar Buyya and Satish Narayana Srirama.

Data compression 176, 368
Data controller 415
Data integration 172, 270
Data life cycle 175, 369
Data management 264
Data model 266, 306
Data processing 10, 58, 132, 145, 171, 184, 233, 266, 292, 320, 368, 413
Data processor 415
Data protection by design 412, 425
Data Protection Directive 411
Data protection impact assessment 412, 421, 427
Data provenance 149, 160
DBSCAN 236, 252
Decision-making 5, 78, 367
Decision trees 240, 243
Denial of service 225, 230
Deployment 12, 26, 199, 210, 365
Design issues 123, 320
Device-to-device 131
Discovery 27, 35, 127
Distributed data-flow 435
Distributed denial of service (DDoS) 230
Docker 16, 154
Docker Swarm 156

e

Edge computing 8
Edge networking 30, 79
Edge offloading 67
Edge processing 126, 138
Edge simulation 44
EdgeX Foundry 21
E-health data 292
Energy efficiency 12, 67, 287, 380, 393, 403
Energy efficiency testing 380
Ensemble machine learning 246
Enterprise workflow testing 383
European Commission 411, 422
Extensible Messaging and Presence Protocol (XMPP) 11
Extreme-edge 14

f

Far-edge 14
Fault detection 96, 249, 303, 366
Fault fuzzy-ontology 56, 65
Federated edge 25, 28
Fog analytics 268
Fog application deployment 191, 210
Fog computing 5, 8, 82
Fog data management 174
Fog-engine 269
Fog layers 114, 178, 185, 251, 359
Fog radio access network 83
Fog resource consumption 193, 201, 206
Fon 16
Functionality segmentation 192
Functionality testing 401

g

5G 42, 80, 401
Gaussian distribution function 333
General Data Protection Regulation 414, 418
Geo-distributed data center 5
GoogleNet 325
Gradient computing 334

h

Haar cascaded-feature extraction 321
Healthcare 238, 461
Health monitoring 67, 284, 291
Heart rate variability 293, 309
Heuristics 117, 216, 329, 366
Human fall detection 301, 308
Human object detection 320, 337

i

Identity 12, 133, 149, 160, 230, 283, 415, 418

iFogSim 212, 433, 435
ImageNet 325
IndieFog 16, 285
In-memory analytics 265
Inner-edge 13
Integer linear programming 63, 91
Intelligent traffic management system 369
Intelligent transportation systems 348, 390
Internet of Things 3
Interoperability 42, 299
Interoperability testing 377
Intrusion detection 241, 244, 285, 402
IoT architecture 62, 161, 252
IoT attacks 246
IoT Sensors 127, 279, 391, 450

j

Joint controller 419

k

K-means 235
K-Nearest Neighbors 235

l

Legal system 411, 414, 425
Lightweight 145, 149, 175, 335
Linear regression 235, 333
Load balancing 11, 32, 36, 239

m

Machine-to-Machine (M2M) 15
Markov chains 55, 89
Message Queue Telemetry Transport (MQTT) 11, 133, 276
Middle-edge 14
Middleware architecture 139
Migration 26, 37
Mist computing 7
Mobile ad hoc network 4
Mobile app testing 400
Mobile crowdsensing as a service 57, 67
Mobile edge computing 4, 82, 115, 125
Mobile Web services 7
Mobility 29, 44, 135, 137, 445
Monte Carlo simulation 204
MQL5 Cloud Network distributed computing 16
Multi-component application 192, 197
MxNet 326

n

Naïve Bayes 235, 244
Network-as-a-service 80
Network function-as-a-service 81, 91
Network function virtualization 7, 80
Network slice 42, 79
n-tier 249, 461, 463

o

Object tracking 327, 339
  Kalman filters 330
  Kernel-based tracking 331
  Kernelized Correlation Filters 332
  Multiple hypothesis tracking 331
  Point-based tracking 329
  Silhouette-Based Tracking 332
OpenFog consortium 6, 92
Optimization 61, 103, 107, 113
Optimization algorithm 61, 116, 118, 364
Optimization problem 61, 106, 333
  constraint 106
  domain 106
  objective function 106
  variable 106
Outer-edge 14
Out-of-box experience 17

p

Pareto-optimal solution 107
Peer-to-peer 131
Performance metrics 37
Performance testing 383
Personal data 112, 415, 226, 411, 415
Petri Nets 55, 61
Platform testing 383
Pollution control testing 403
Predictive analysis 191, 197
Privacy 133, 179, 223, 363, 427
Privacy enhancing technologies 426, 430
Privacy testing 383
Privacy threats 229
Probability distribution 193, 204
Problem formalization 106
Profiling 229, 412, 418

q

Quality of experience 10, 112, 352, 435
Quality of service 71, 79, 128, 191, 292, 413, 434

r

Random forest 243
Raspberry Pi cluster 146, 151
Real time applications 4, 12, 36, 109, 134, 148, 171, 212, 240, 319
Reinforcement learning 234, 314, 350
Reliability 29, 51, 60, 104, 133, 360, 375
Reliability testing 393
Remote health monitoring 295
ResNet 326
Resource allocation 26, 67, 91, 285, 347
Resource management 38, 145
Right to be forgotten 417
Roadside units 351, 357

s

SCALE 8
  agility 9
  cognition 8
  efficiency 9
  latency 9
  security 8
SCANC 9
  acceleration 10
  compute 10
  control 12
  networking 11
  storage 9
Security 72, 132, 187, 223, 249, 300, 363
Security testing 383
Sensing as a service 67, 193, 202
Service level agreement 81, 138
Single Shot Multi-box Detector 326
Situational awareness 319, 355
Smart contracts 149, 160
Smart home 241, 277, 376
Smart living 57, 67
Smart nutrition monitoring systems 279
Smart traffic 67, 128, 247, 351
Software testing 386
Software-defined cloud 80, 87, 93
Software-defined network (SDN) 11, 26, 80, 138, 147, 358
Supervised learning 234
Support vector machine 243, 322
Systematic review 58

t

TOSCA 150, 210
Trusted orchestration 162

u

Unsupervised learning 234
Usability testing 383

V
Vehicles mobility 359
Virtualization 33, 85

W
W3C PROV 163
What-if analysis 207
WSO2–IoT server 18

Z
ZigBee 12, 184, 313
Z-Wave 12

Wiley Series on Parallel and Distributed Computing

Series Editor: Albert Y. Zomaya

Parallel and Distributed Simulation Systems
Richard Fujimoto

Mobile Processing in Distributed and Open Environments
Peter Sapaty

Introduction to Parallel Algorithms
C. Xavier and S. S. Iyengar

Solutions to Parallel and Distributed Computing Problems: Lessons from Biological Sciences
Albert Y. Zomaya, Fikret Ercal, and Stephan Olariu (Editors)

Parallel and Distributed Computing: A Survey of Models, Paradigms, and Approaches
Claudia Leopold

Fundamentals of Distributed Object Systems: A CORBA Perspective
Zahir Tari and Omran Bukhres

Pipelined Processor Farms: Structured Design for Embedded Parallel Systems
Martin Fleury and Andrew Downton

Handbook of Wireless Networks and Mobile Computing
Ivan Stojmenović (Editor)

Internet-Based Workflow Management: Toward a Semantic Web
Dan C. Marinescu

Parallel Computing on Heterogeneous Networks
Alexey L. Lastovetsky

Performance Evaluation and Characterization of Parallel and Distributed Computing Tools
Salim Hariri and Manish Parashar

Distributed Computing: Fundamentals, Simulations, and Advanced Topics, *Second Edition*
Hagit Attiya and Jennifer Welch

Smart Environments: Technology, Protocols, and Applications
Diane Cook and Sajal Das

Fundamentals of Computer Organization and Architecture
Mostafa Abd-El-Barr and Hesham El-Rewini

Advanced Computer Architecture and Parallel Processing
Hesham El-Rewini and Mostafa Abd-El-Barr

UPC: Distributed Shared Memory Programming
Tarek El-Ghazawi, William Carlson, Thomas Sterling, and Katherine Yelick

Handbook of Sensor Networks: Algorithms and Architectures
Ivan Stojmenović (Editor)

Parallel Metaheuristics: A New Class of Algorithms
Enrique Alba (Editor)

Design and Analysis of Distributed Algorithms
Nicola Santoro

Task Scheduling for Parallel Systems
Oliver Sinnen

Computing for Numerical Methods Using Visual C++
Shaharuddin Salleh, Albert Y. Zomaya, and Sakhinah A. Bakar

Architecture-Independent Programming for Wireless Sensor Networks
Amol B. Bakshi and Viktor K. Prasanna

High-Performance Parallel Database Processing and Grid Databases
David Taniar, Clement Leung, Wenny Rahayu, and Sushant Goel

Algorithms and Protocols for Wireless and Mobile Ad Hoc Networks
Azzedine Boukerche (Editor)

Algorithms and Protocols for Wireless Sensor Networks
Azzedine Boukerche (Editor)

Optimization Techniques for Solving Complex Problems
Enrique Alba, Christian Blum, Pedro Isasi, Coromoto León, and Juan Antonio Gómez (Editors)

Emerging Wireless LANs, Wireless PANs, and Wireless MANs: IEEE 802.11, IEEE 802.15, IEEE 802.16 Wireless Standard Family
Yang Xiao and Yi Pan (Editors)

High-Performance Heterogeneous Computing
Alexey L. Lastovetsky and Jack Dongarra

Mobile Intelligence
Laurence T. Yang, Augustinus Borgy Waluyo, Jianhua Ma, Ling Tan, and Bala Srinivasan (Editors)

Research in Mobile Intelligence
Laurence T. Yang (Editor)

Advanced Computational Infrastructures for Parallel and Distributed Adaptive Applicatons
Manish Parashar and Xiaolin Li (Editors)

Market-Oriented Grid and Utility Computing
Rajkumar Buyya and Kris Bubendorfer (Editors)

Cloud Computing Principles and Paradigms
Rajkumar Buyya, James Broberg, and Andrzej Goscinski (Editors)

Algorithms and Parallel Computing
Fayez Gebali

Energy-Efficient Distributed Computing Systems
Albert Y. Zomaya and Young Choon Lee (Editors)

Scalable Computing and Communications: Theory and Practice
Samee U. Khan, Lizhe Wang, and Albert Y. Zomaya (Editors)

The DATA Bonanza: Improving Knowledge Discovery in Science, Engineering, and Business
Malcolm Atkinson, Rob Baxter, Michelle Galea, Mark Parsons, Peter Brezany, Oscar Corcho, Jano van Hemert, and David Snelling (Editors)

Large Scale Network-Centric Distributed Systems
Hamid Sarbazi-Azad and Albert Y. Zomaya (Editors)

Verification of Communication Protocols in Web Services: Model-Checking Service Compositions
Zahir Tari, Peter Bertok, and Anshuman Mukherjee

High-Performance Computing on Complex Environments
Emmanuel Jeannot and Julius Žilinskas (Editors)

Advanced Content Delivery, Streaming, and Cloud Services
Mukaddim Pathan, Ramesh K. Sitaraman, and Dom Robinson (Editors)

Large-Scale Distributed Systems and Energy Efficiency
Jean-Marc Pierson (Editor)

Activity Learning: Discovering, Recognizing, and Predicting Human Behavior from Sensor Data
Diane J. Cook and Narayanan C. Krishnan

Large-scale Distributed Systems and Energy Efficiency: A Holistic View
Jean-Marc Pierson

Programming Multicore and Many-core Computing Systems
Sabri Pllana and Fatos Xhafa (Editors)

Fog and Edge Computing: Principles and Paradigms
Rajkumar Buyya and Satish Narayana Srirama (Editors)

Printed and bound by CPI Group (UK) Ltd, Croydon, CR0 4YY
18/08/2022
03142627-0002